pH Sensors, Biosensors and Systems

pH Sensors, Biosensors and Systems

Collection Editors

Pietro Salvo
Lorena Tedeschi

MDPI

Basel • Beijing • Wuhan • Barcelona • Belgrade • Novi Sad • Cluj • Manchester

Collection Editors
Pietro Salvo
Institute of Clinical Physiology
Italian National Research Council
Pisa
Italy

Lorena Tedeschi
Institute of Clinical Physiology
Italian National Research Council
Pisa
Italy

Editorial Office
MDPI AG
Grosspeteranlage 5
4052 Basel, Switzerland

This is a reprint of the Topical Collection, published open access by the journal *Chemosensors* (ISSN 2227-9040), freely accessible at: https://www.mdpi.com/journal/chemosensors/topical_collections/ph_sens_bio_sys.

For citation purposes, cite each article independently as indicated on the article page online and as indicated below:

Lastname, A.A.; Lastname, B.B. Article Title. *Journal Name* **Year**, *Volume Number*, Page Range.

ISBN 978-3-7258-3767-0 (Hbk)
ISBN 978-3-7258-3768-7 (PDF)
https://doi.org/10.3390/books978-3-7258-3768-7

Contents

About the Editors

Pietro Salvo

Pietro Salvo received an M.Sc. degree in Electronics Engineering and a Ph.D. degree in Automation, Robotics, and Bioengineering from the University of Pisa, Italy, in 2004 and 2009, respectively. At present, he is a senior researcher at the Institute of Clinical Physiology, National Council of Research, Pisa, Italy, where he is the head of the lab for Biofluids Analysis. His scientific areas of interest include lab-on-chip technology, sensors and biosensors, and nanomaterials for the non-invasive or minimally invasive study of pathological biomarkers in bodily fluids.

Lorena Tedeschi

Lorena Tedeschi received a Laurea degree in Pharmaceutical Chemistry and Technology and a Ph.D. degree in Molecular Biotechnologies from the University of Pisa, Pisa, Italy, in 1999 and 2003, respectively. She has been a researcher at the CNR Institute of Clinical Physiology, Pisa, since 2007, where she is involved in the design, synthesis, and purification of bioreceptors and their labeling and immobilization on various kinds of transducers using DNA-based and protein-based biosensors.

chemosensors

Editorial

pH Sensors, Biosensors and Systems

Pietro Salvo * and Lorena Tedeschi

Institute of Clinical Physiology, Italian National Research Council, Via Moruzzi 1, 56124 Pisa, Italy; lorena.tedeschi@cnr.it
* Correspondence: pietro.salvo@cnr.it

The study of biological systems for the protection of the environment, food, and health is among the most important research fields in the scientific literature. Many biological systems are sensitive even to small changes in chemo-physical parameters such as pH, temperature and humidity [1,2].

The concept of using the pH value to express the activity of hydrogen ions on a logarithmic scale dates back to 1909, and since then, many measurement methods have been developed [3]. Paper test strips and glass electrodes are the most used types of pH sensors, but the demand for less fragile, highly accurate, miniaturized devices that can be flexible and wearable has prompted research on alternatives [4]. Most of these pH sensors can be grouped into one of three macro areas, i.e., optical, electrochemical and field effect sensors, depending on the measurement method. Vivaldi et al. reviewed these areas in detail [5]. Optical pH sensors have good photo- and chemostability and are often used for non-invasive measurements, especially for biological tissues or cell cultures, e.g., for detecting tumor regions using pH-triggered contrast agents or the metabolic profiles of cancer lines [6,7]. Optical pH sensors are also popular for colorimetric or photometric analyses for food safety and control, e.g., using chromophores, which are suitable for continuous monitoring [8].

Electrochemical pH sensors measure the electrical potential in two- or three-electrode cells. These pH sensors can be micrometric, flexible, biocompatible, or even knitted into garments [9–11]. Many materials have been exploited to fabricate electrochemical pH sensors with different properties. Reduced graphene oxide has been investigated because it contains carboxylic and hydroxyl groups that can react with hydrogen ions [12–14]. Metal oxides have high sensitivity and fast responses but can also exhibit considerable drift and hysteresis [15,16]. Among metal oxides, iridium oxide-based microwires stand out, since they are fabricated with a near-Nernstian response and low drift. These microwires can be knitted, integrated into bandages or used for cell analysis [17,18]. Some polymers have ionizable acidic or basic groups and have thus been used as pH sensors [19,20]. Examples of acidic pH-responsive acidic polymers are polycarboxylic, polysulphonic and polyphosphoric acids, whereas for basic pH vinyl, (meth)acrylamide and (meth)acrylate polymers can be used. Polymers can suffer from reduced stability and degradation over time; however, the most recent formulations might have overcome these limitations. For example, Korostynska et al. reported a polypyrrole-based pH sensor with sensitivity of about 60 mV/pH and a 0.25 mV drift per day [21].

Although they belong to the category of electrochemical pH sensors, field effect transistors (FETs) have been so widely investigated that they can form another group. FETs can be mass-produced, have fast responses and can be miniaturized by means of micro-photolithography. The concentration of hydrogen ions is transduced during a change in the electrical current and depends on their interaction with the transistor gate [22,23].

Received: 25 February 2025
Accepted: 3 March 2025
Published: 5 March 2025

Citation: Salvo, P.; Tedeschi, L. pH Sensors, Biosensors and Systems. *Chemosensors* **2025**, *13*, 90. https://doi.org/10.3390/chemosensors13030090

However, this direct interaction can degrade the gate, since in the liquid solution, there are usually other species, such as ions or biological molecules. To overcome this drawback, new technical approaches have been employed. In particular, the dual gate approach consists of fabricating an additional gate that is the only surface exposed to the solution. Cho et al. and Hyun et al. proposed a coplanar-gate AlGaN/GaN FET and a dual-gate a-IGZO FET with an SnO_2 sensing membrane, respectively [24,25]. Both of these FETs achieved remarkable superNernstian sensitivities. Further improvements were achieved by Kim et al., who adopted a strategy for integrating p- and n-type FETs [26]. The p-type FET has lower noise than the n-type, whereas the n-type FET has higher mobility and lower drift than the p-type. An ambipolar FET showed a superNernstian sensitivity of about 170 mV/pH and a drift of 14 mV/h in the n region.

Although the glass electrode is the gold standard for pH measurement, the devices described in this Collection and in the scientific literature show that there is a constant interest in developing new approaches. This interest depends on the ubiquitous presence of pH as one of the main parameters that can control chemophysical reactions for the sensing and fabrication of sensitive surfaces [27–30]. Therefore, we would like to thank all of the authors who contributed to this Collection for submitting papers demonstrating the broad spectrum of pH sensors and providing new approaches that will pave the way towards further advances in this field.

Acknowledgments: We thank all of the authors who contributed with their work to this Collection.

Conflicts of Interest: The authors declare no conflicts of interest.

References

1. Sensors and Biosensors for C-Reactive Protein, Temperature and pH, and Their Applications for Monitoring Wound Healing: A Review. Available online: https://www.mdpi.com/1424-8220/17/12/2952 (accessed on 21 February 2025).
2. Salvo, P.; Calisi, N.; Melai, B.; Cortigiani, B.; Mannini, M.; Caneschi, A.; Lorenzetti, G.; Paoletti, C.; Lomonaco, T.; Paolicchi, A.; et al. Temperature and pH Sensors Based on Graphenic Materials. *Biosens. Bioelectron.* **2017**, *91*, 870–877. [CrossRef]
3. Myers, R.J. One-Hundred Years of pH. *J. Chem. Educ.* **2010**, *87*, 30–32. [CrossRef]
4. Salvo, P.; Melai, B.; Calisi, N.; Paoletti, C.; Bellagambi, F.; Kirchhain, A.; Trivella, M.G.; Fuoco, R.; Di Francesco, F. Graphene-Based Devices for Measuring pH. *Sens. Actuators B Chem.* **2018**, *256*, 976–991. [CrossRef]
5. Vivaldi, F.; Salvo, P.; Poma, N.; Bonini, A.; Biagini, D.; Del Noce, L.; Melai, B.; Lisi, F.; Francesco, F.D. Recent Advances in Optical, Electrochemical and Field Effect pH Sensors. *Chemosensors* **2021**, *9*, 33. [CrossRef]
6. Kye, H.; Jo, D.; Jeong, S.; Kim, C.; Kim, J. Photoacoustic Imaging of pH-Sensitive Optical Sensors in Biological Tissues. *Chemosensors* **2024**, *12*, 257. [CrossRef]
7. Blaszczak, W.; Tan, Z.; Swietach, P. Cost-Effective Real-Time Metabolic Profiling of Cancer Cell Lines for Plate-Based Assays. *Chemosensors* **2021**, *9*, 139. [CrossRef]
8. Voskoboynikova, O.; Sukhanov, A.; Duerkop, A. Optical pH Sensing in Milk: A Small Puzzle of Indicator Concentrations and the Best Detection Method. *Chemosensors* **2021**, *9*, 177. [CrossRef]
9. Zhou, J.; Zhang, L.; Tian, Y. Micro Electrochemical pH Sensor Applicable for Real-Time Ratiometric Monitoring of pH Values in Rat Brains. *Anal. Chem.* **2016**, *88*, 2113–2118. [CrossRef]
10. Manjakkal, L.; Dang, W.; Yogeswaran, N.; Dahiya, R. Textile-Based Potentiometric Electrochemical pH Sensor for Wearable Applications. *Biosensors* **2019**, *9*, 14. [CrossRef]
11. Manjakkal, L.; Sakthivel, B.; Gopalakrishnan, N.; Dahiya, R. Printed Flexible Electrochemical pH Sensors Based on CuO Nanorods. *Sens. Actuators B Chem.* **2018**, *263*, 50–58. [CrossRef]
12. Poma, N.; Vivaldi, F.; Bonini, A.; Carbonaro, N.; Di Rienzo, F.; Melai, B.; Kirchhain, A.; Salvo, P.; Tognetti, A.; Di Francesco, F. Remote Monitoring of Seawater Temperature and pH by Low Cost Sensors. *Microchem. J.* **2019**, *148*, 248–252. [CrossRef]
13. Salvo, P.; Calisi, N.; Melai, B.; Dini, V.; Paoletti, C.; Lomonaco, T.; Pucci, A.; Di Francesco, F.; Piaggesi, A.; Romanelli, M. Temperature- and pH-Sensitive Wearable Materials for Monitoring Foot Ulcers. *Int. J. Nanomed.* **2017**, *12*, 949–954. [CrossRef]
14. Vivaldi, F.; Santalucia, D.; Poma, N.; Bonini, A.; Salvo, P.; Del Noce, L.; Melai, B.; Kirchhain, A.; Kolivoška, V.; Sokolová, R.; et al. A Voltammetric pH Sensor for Food and Biological Matrices. *Sens. Actuators B Chem.* **2020**, *322*, 128650. [CrossRef]

15. Manjakkal, L.; Szwagierczak, D.; Dahiya, R. Metal Oxides Based Electrochemical pH Sensors: Current Progress and Future Perspectives. *Prog. Mater. Sci.* **2020**, *109*, 100635. [CrossRef]
16. Ghoneim, M.T.; Nguyen, A.; Dereje, N.; Huang, J.; Moore, G.C.; Murzynowski, P.J.; Dagdeviren, C. Recent Progress in Electrochemical pH-Sensing Materials and Configurations for Biomedical Applications. *Chem. Rev.* **2019**, *119*, 5248–5297. [CrossRef]
17. Chawang, K.; Bing, S.; Kwon, K.Y.; Chiao, J.-C. Miniaturized Iridium Oxide Microwire pH Sensor for Biofluid Sensing. *Chemosensors* **2024**, *12*, 168. [CrossRef]
18. Jírů, J.; Hybášek, V.; Michalcová, A.; Korbelová, K.; Koláčný, L.; Fojt, J. Preparation, Characterization and Electrochemical Response of Nanostructured TiAlV with Potentiostatically Deposited IrOx as a pH Sensor for Rapid Detection of Inflammation. *Chemosensors* **2024**, *12*, 109. [CrossRef]
19. Kocak, G.; Tuncer, C.; Bütün, V. pH-Responsive Polymers. *Polym. Chem.* **2017**, *8*, 144–176. [CrossRef]
20. Alam, A.U.; Qin, Y.; Nambiar, S.; Yeow, J.T.W.; Howlader, M.M.R.; Hu, N.-X.; Deen, M.J. Polymers and Organic Materials-Based pH Sensors for Healthcare Applications. *Prog. Mater. Sci.* **2018**, *96*, 174–216. [CrossRef]
21. Korostynska, O.; Arshak, K.; Gill, E.; Arshak, A. Review on State-of-the-Art in Polymer Based pH Sensors. *Sensors* **2007**, *7*, 3027–3042. [CrossRef]
22. Ren, H.; Liang, K.; Li, D.; Chen, Y.; Tang, Y.; Wang, Y.; Li, F.; Liu, G.; Zhu, B. Field-Effect Transistor-Based Biosensor for pH Sensing and Mapping. *Adv. Sens. Res.* **2023**, *2*, 2200098. [CrossRef]
23. Lee, C.-S.; Kim, S.K.; Kim, M. Ion-Sensitive Field-Effect Transistor for Biological Sensing. *Sensors* **2009**, *9*, 7111–7131. [CrossRef]
24. Hyun, T.-H.; Cho, W.-J. Fully Transparent and Highly Sensitive pH Sensor Based on an A-IGZO Thin-Film Transistor with Coplanar Dual-Gate on Flexible Polyimide Substrates. *Chemosensors* **2023**, *11*, 46. [CrossRef]
25. Cho, S.-K.; Cho, W.-J. High-Sensitivity pH Sensor Based on Coplanar Gate AlGaN/GaN Metal-Oxide-Semiconductor High Electron Mobility Transistor. *Chemosensors* **2021**, *9*, 42. [CrossRef]
26. Kim, Y.-U.; Cho, W.-J. High-Performance Bidirectional Chemical Sensor Platform Using Double-Gate Ion-Sensitive Field-Effect Transistor with Microwave-Assisted Ni-Silicide Schottky-Barrier Source/Drain. *Chemosensors* **2022**, *10*, 122. [CrossRef]
27. Amaral, D.S.d.; Tholozan, L.V.; Bonemann, D.H.; Jansen-Alves, C.; Boschetti, W.; Novo, D.L.R.; Carreno, N.L.V.; Pereira, C.M.P. de Algal Biosensors for Detection of Potentially Toxic Pollutants and Validation by Advanced Methods: A Brief Review. *Chemosensors* **2024**, *12*, 235. [CrossRef]
28. Hashtroudi, H.; Yu, A.; Juodkazis, S.; Shafiei, M. Two-Dimensional Dy2O3-Pd-PDA/rGO Heterojunction Nanocomposite: Synergistic Effects of Hybridisation, UV Illumination and Relative Humidity on Hydrogen Gas Sensing. *Chemosensors* **2022**, *10*, 78. [CrossRef]
29. Kuswandi, B.; Asih, N.P.N.; Pratoko, D.K.; Kristiningrum, N.; Moradi, M. Edible pH Sensor Based on Immobilized Red Cabbage Anthocyanins into Bacterial Cellulose Membrane for Intelligent Food Packaging. *Packag. Technol. Sci.* **2020**, *33*, 321–332. [CrossRef]
30. Damaghi, M.; Wojtkowiak, J.W.; Gillies, R.J. pH Sensing and Regulation in Cancer. *Front. Physiol.* **2013**, *4*, 370. [CrossRef]

chemosensors

Review

Recent Advances in Optical, Electrochemical, and Field Effect pH Sensors

Federico Vivaldi [1,2], Pietro Salvo [2,*], Noemi Poma [1], Andrea Bonini [1], Denise Biagini [1], Lorenzo Del Noce [1], Bernardo Melai [1], Fabio Lisi [3] and Fabio Di Francesco [1,4]

[1] Department of Chemistry and Industrial Chemistry, University of Pisa, Via Giuseppe Moruzzi 13, 56124 Pisa, Italy; federicomaria.vivaldi@phd.unipi.it (F.V.); n.pomasajama@studenti.unipi.it (N.P.); andrea.bonini@phd.unipi.it (A.B.); denisebiagini@virgilio.it (D.B.); Lorenzo.delnoce95@gmail.com (L.D.N.); melai@dcci.unipi.it (B.M.); fabio.difrancesco@unipi.it (F.D.F.)
[2] Institute of Clinical Physiology (IFC) of National Research Council (CNR), Via Moruzzi 1, 56124 Pisa, Italy
[3] School of Chemical Engineering and Australian Centre for NanoMedicine, University of New South Wales, Sydney 2052, Australia; fabio.lisi.lj@gmail.com
[4] National Interuniversity Consortium for Materials Science and Technology (INSTM), Via G. Giusti, 9, 50121 Firenze, Italy
* Correspondence: pietro.salvo@ifc.cnr.it; Tel.: +39-050-3152703

Abstract: Although its first definition dates back to more than a century ago, pH and its measurement are still studied for improving the performance of current sensors in everyday analysis. The gold standard is the glass electrode, but its intrinsic fragility and need of frequent calibration are pushing the research field towards alternative sensitive devices and materials. In this review, we describe the most recent optical, electrochemical, and transistor-based sensors to provide an overview on the status of the scientific efforts towards pH sensing.

Keywords: pH; pH sensors; optical pH sensors; voltammetric pH sensors; potentiometric pH sensors

Citation: Vivaldi, F.; Salvo, P.; Poma, N.; Bonini, A.; Biagini, D.; Del Noce, L.; Melai, B.; Lisi, F.; Francesco, F.D. Recent Advances in Optical, Electrochemical, and Field Effect pH Sensors. *Chemosensors* **2021**, *9*, 33. https://doi.org/10.3390/chemosensors9020033

Academic Editor: Núria Serrano
Received: 7 January 2021
Accepted: 4 February 2021
Published: 5 February 2021

1. Introduction

The activity of the hydrogen cation plays an important role in a variety of large- and small-scale processes, ranging from industry to microbial life [1,2]. The measurement of the hydrogen cation concentration is a routine procedure in several fields, such as production control to guarantee the quality and durability of many industrial products, and clinical analysis during sample management and testing. For example, in the pharmaceutical industry, pH control makes the synthesis processes and pharmacokinetics analyses accurate and safe [3], whereas in the clinical field, it can be applied for the monitoring of chronic wounds [4,5], discrimination between healthy and tumor cells [6], and monitoring of exposure to pollution of street traffic-controllers and office-workers [7]. This wide application of pH measurement motivates the scientific research to find new methods and devices that can adapt to different scenarios.

The glass electrode, one of the very first systems to measure pH, is nowadays the gold standard. The first model was proposed in 1909 by F. Haber and Z. Klemensiewicz following the results obtained by M. Cremer, who, in 1906, noticed a potential difference across a thin glass membrane that separated two solutions with different pH values [8]. The most common model of the glass electrode consists of an inner Ag/AgCl reference electrode separated from another Ag/AgCl electrode by a glass membrane. One electrode is in contact with a known pH solution while the other with an unknown pH sample. An overall electric potential difference (E_m) across the membrane appears because the silicate network has an affinity for specific cations, which are adsorbed within the structure. E_m is the sum of contributions from the junctions between the various zones of the glass electrode, including terms, such as the interfacial potentials arising from adsorbed cations, diffusion potentials within the glass membrane, and the asymmetry potential (generated by the natural inhomogeneity of the membrane due to its construction or wear during use). Glass membranes are mostly selective for H_3O^+, but they also respond to other ions, such as Na^+, Li^+, K^+, Ag^+, and NH_4^+.

Figure 1A shows a typical glass electrode. Despite its proven reliability, a glass electrode suffers from several drawbacks, such as frequent calibrations to compensate for the drifts caused by the asymmetry potential, and the intrinsic fragility of the glass membrane, which can degrade because of continuous measurements or harsh environments (e.g., reactors, biosystems, or in vivo applications).

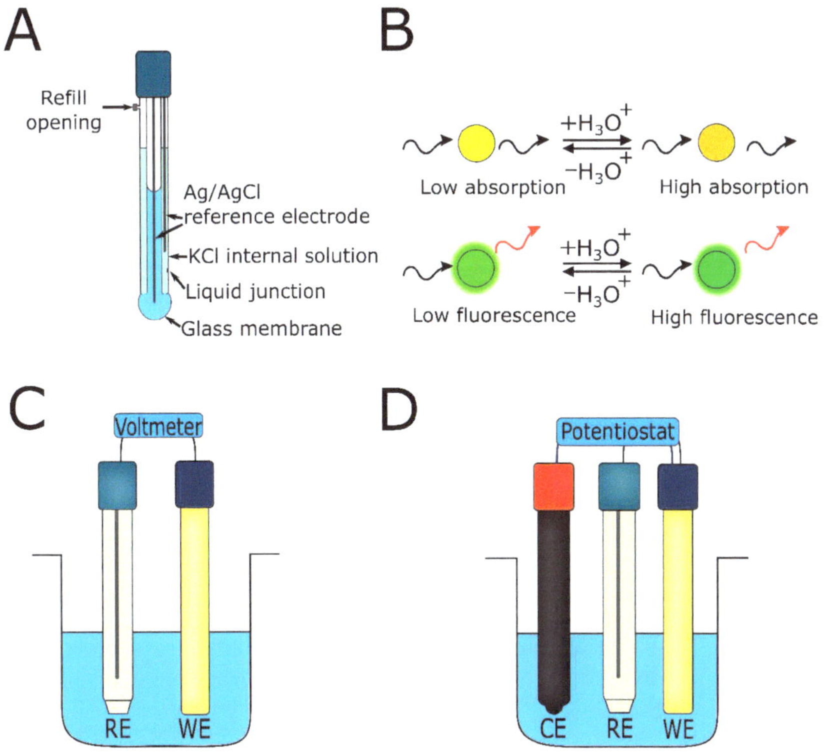

Figure 1. (**A**) Schematic representation of a cell with a glass electrode. Two Ag/AgCl electrodes are immersed in two solutions (with known and unknown pH values, respectively) and separated by a glass membrane. The electrical potential difference between the two electrodes is correlated to the pH of the unknown solution. (**B**) Schematic representation of the principle of optical transduction in an optical pH sensor based on absorption and fluorescence. (**C**) Schematization of a 2-electrode system inside a solution of a given analyte. (**D**) Schematization of a 3-electrode system for a voltammetric measurement.

In this review, we discuss the most promising pH sensors reported in the literature: optical and electrochemical sensors, and field effect transistors (FETs). Optical pH sensors typically use chemical species that modify their optical properties as a function of pH (e.g., index of refraction, polarization, absorbance, etc.) [9–12]. For example, Figure 1B shows a pH-sensitive species that changes its UV-VIS absorbance or its fluorescence emission depending on the protonation equilibrium. Electrochemical pH sensors are mainly potentiometric and voltammetric. Potentiometric sensors measure the potential generated across two electrodes, i.e., the working electrode (WE) and the reference electrode (RE) (Figure 1C). Voltammetric sensors measure the current generated when a potential is imposed in an electrochemical cell through a three-electrode system with WE, RE, and a counter electrode (CE) that supplies current to the cell (Figure 1D). ISFETs (ion-sensitive field effect transistors) detect the change of the electrical field caused by hydrogen ions that modulate the current flowing in the transistor conduction channel. ISFETs' popularity depends on their microscopic size and potential high sensitivity towards pH [12,13].

2. Optical pH Sensors

Optical pH sensors are mostly based on sensitive acidic/basic materials with a specific pKa, a high molar extinction coefficient, absorbance and emission spectra in the visible range, and good stability against light and a chemical environment (e.g., photostability and chemostability) [14]. A common approach is to immobilize a molecular probe with a chromophore on two types of supports, i.e., optical fibers and planar sensors. Optical fibers are flexible, can achieve a microscopic spatial discrimination, and can be quite easily modified. However, they are limited by ionic strength interferences and the photodegradation and leakage of the chromophore [15–18]. Planar sensors are easily fabricated and usually have a simple structure with a large pH-sensitive area [19–22].

In 2000, Jin et al. reported a sensor based on a polyaniline (PANI) film prepared by chemical oxidation at room temperature [23]. The different protonation of the imine nitrogen ion of PANI led to a pH-dependent behavior in the UV-VIS-NIR spectrum. The PANI film was deposited onto a planar plexiglass surface and showed an absorbance shift at 575 and 750 nm, which allowed pH to be measured in the range 2–12. Although this study reported promising data on the temporal stability of PANI, it did not show its application in real matrices. In addition, the presence of a hysteresis effect made this sensor incompatible for a continuous monitoring in wide ranges of pH.

In another study, Kermis et al. proposed a rapid method for the preparation of a sensor based on a fluorescent dye in a matrix obtained by copolymerization of 6-methacryloyl-8-acid hydroxy-1,3-pyren sulfonic with polyethylene-glycol acrylate [24]. The choice of these two polymers stems from to their excellent mechanical properties and the possibility to directly immobilize a dye inside the matrix during the polymerization phase. Although the planar sensor showed a small operating range between pH 6 and 9, the reproducibility was high (<0.10 pH units) and the dependence on ionic strength had a negligible effect (residual standard deviation, RSD < 3.6%). Stahl et al. followed a similar approach [21] and embedded a fluorescent pH-sensitive molecule, 2′,7′-dihexyl-5(6)-*N*-octadecyl-carboxamidofluorescein ethyl ester, in a hydrogel matrix (polyurethane hydrogel "Hydromed D4") with a fluorescent standard (ruthenium(II)-Tris-4,7-diphenyl-1,10-phenanthroline). This setup was used to measure pH in 2-D marine sediments in the pH range 7.3–9.3 (Figure 2A). The concept of spatial measurements for marine samples has also been studied in the pH range 6–8 by Jiang et al., who used a fluorescent pH indicator (5-Hexadecanoylamino-fluorescein) and Hydromed D4 [22]. Recently, Gotor et al. designed several fluorescent dyes (boron−dipyrromethene derivatives) embedded in Hydromed D4 to build an optical array capable of extending the pH measurements to a wide pH range (0–14) [25].

In 2012, Raoufi et al. reported a polyallylamine hydrochloride as a linker for a dye, "Brilliant Yellow", covering the tip of an optical fiber [26]. The polymer was deposited using the layer-by-layer technique, which allows for fine control of the polymer deposition on a solid surface by applying layers with opposite charge. The best sensitivity of the prototype was obtained in a range between pH 6.8 and 9 with an accuracy of 0.2 pH units.

In 2016, Abu-Thabit et al. described a similar system, which measured the pH response of polysulfone membranes coated with PANI nanofibers [27]. The nanoscale PANI structure improved the diffusion of molecules into the nanofiber and provided a response time of 4 s over a pH range from 4 to 12 with a standard deviation ranging from 0.002 to 0.05 pH units. The sensor was also tested in a detergent (Dettol brand), obtaining a relative error of about 1%. Although this sensor was tested for repeatability and was described to work flawlessly from week to week, no numerical data was reported.

Safavi et al. immobilized two dyes, "Victoria Blue" and dipicryamine (pKa 1.82 and 11.2, respectively), onto optically transparent triacetyl-cellulose membranes [28]. The authors used a neural network to extend the measurement range and obtained a sensor that performed better than a glass electrode in extreme pH conditions (RSD < 2% at pH 0–1 and 13–14). In 2006, Hashemi et al. immobilized a "Congo Red" dye on agarose membranes [29]. Although it had high precision (RSD < 0.30%), the sensor had a response time of about 3 min and could only be used in the acidic range (pH 0.5–5). Taweetanavanich et al. described a rhodamine-based colorimetric and fluorimetric sensor that worked between pH 1 and 8 [30]. The interference of several metal cations (Na^+, K^+, Ag^+, Mg^{2+}, Ca^{2+}, Pb^{2+}, Ni^{2+}, Cu^{2+}, Zn^{2+}, Cd^{2+}, Hg^{2+}, Al^{3+}, Cr^{3+}, Fe^{3+}, Au^{3+}, Pt^{2+}, Ru^{2+}) was negligible, thus confirming the selectivity of the sensor towards the hydronium ion. However, the article did not mention any data regarding repeatability, reproducibility, and sensitivity.

In 2013, Li et al. removed the core from a section of an optical fiber and dip-coated the corresponding cladding with a pH-sensitive layer made of tetraethylorthosilicate (TEOS) and a mixture of pH-sensitive dies (cresol red, dichlorophenol red, bromophenol blue) (Figure 2B) [31]. The core removal increased the power of the evanescent wave [32]. This sensor showed a linear response in the pH range 1–13 with a sensitivity of 0.6 dBm/pH, a response time of 40 s, and high repeatability (RSD < 1%). In a study from 2014 by Schyrr et al., the core of an optical fiber was coated with organic modified silicates (ORMOSILs) doped with bromophenol blue [33]. This sensor showed a sigmoidal response between pH 3 and 9 with a linear range between 5 and 7. The response time was 20 min for a pH variation of 1 unit. This sensor was tested in human serum samples and had an accuracy of 0.2 pH when compared with a glass pH meter. Another example of sol-gel technology was presented by Sørensen et al., who entrapped two dyes within a sol-gel matrix, which allows the rapid diffusion of protons [34]. Such a sensor worked in a range between 4.7 and 7.7 pH units with an accuracy <0.1 pH units. Similarly, Jeon et al. used neutral red as a pH-sensitive dye, obtaining a device with a working pH range between 6 and 9 [35]. Vafi et al. proposed an optical absorption sensor based on a sol-gel silica matrix doped with thionines [36]. The sensor had a response time of about 1 min between 11 and 13 pH units with a lifetime of 6 months. Wencel et al. used a sol-gel-based optical pH sensor for real-time monitoring in human tissues [37]. The fluorescent dye 8-hydroxypyrene-1,3,6-trisulfonic acid was used to monitor pH between 6 and 8.5 and had a response time of less than 2 min.

Recently, Gong et al. proposed a fluorescent hydrogel-based optical fiber for lung tumors [38]. A fluorescent dye (5,10,15,20-Tetrakis(4-hydroxyphenyl)-21H,23H-porphine) was embedded in a polymeric matrix (4-hydroxybutylmethacrylate and dimethylaminoethylacrylate) and used to measure the pH (5.5–8) of ovine lung tissue within 30 s. Figure 2C shows another optical fiber sensor with a pH-sensitive hydrogel (acrylamide, *N,N′*-methylene diacrylamide, *N,N,N,N*-tetramethylethylenediamine, and methacrylic acid) [39]. A change in the volume and refractive index of the hydrogel caused by pH led to a shift in the surface plasmon resonance (SPR) wavelength. The highest sensitivity (13 nm/pH) was obtained in the range pH 8–10.

Moradi et al. injected a fluorescent pH-sensitive indicator (8-hydroxypyrene and 1,3,6-trisulfonic acid sodium salt) in a microfluidic serpentine (Figure 2D) [40]. The fluorescent signal was recorded in the pH range 2.5–9; however, the sensitivity was not constant but increased from 6 mV/pH at basic pH up to 42 mV/pH at acidic pH. The sensor had a response time of about 10 s with an optimal indicator concentration of 500 mg/L.

A different approach includes pH-sensitive inorganic materials, such as carbon nanostructures [41], metals (Ms), and metal oxides (MOs) [42–46]. Metallic materials can absorb hydronium ions or can form compounds with the general chemical formula $M_xO_yH^+$ [47]. Because of the hydroxyl and carboxylic groups, reduced graphene oxide (rGO) is an excellent candidate for pH measurement [5,13]. An example was the deposition onto the unclad core of an optical fiber of a silver thin film coated with rGO and PANI [48]. The change in the refractive index increased with OH^- concentration; thus, the sensor had a higher sensitivity at basic pH values (75.09 nm/pH at pH 11.35). In 2017, a metallic nanostructured complex ($ZnLi_2$) was used in combination with sodium tetraphenylborate, dibutylphthalate, and polyvinyl chloride [49]. This absorbance sensor showed two linear ranges between pH 4 and 8 and between 5 and 8 depending on the wavelength (393 and 570 nm, respectively). The average response time was 4 min, with 1.14% measurement repeatability and 4.06% reproducibility. The sensor lifetime was more than 2 months when stored in water.

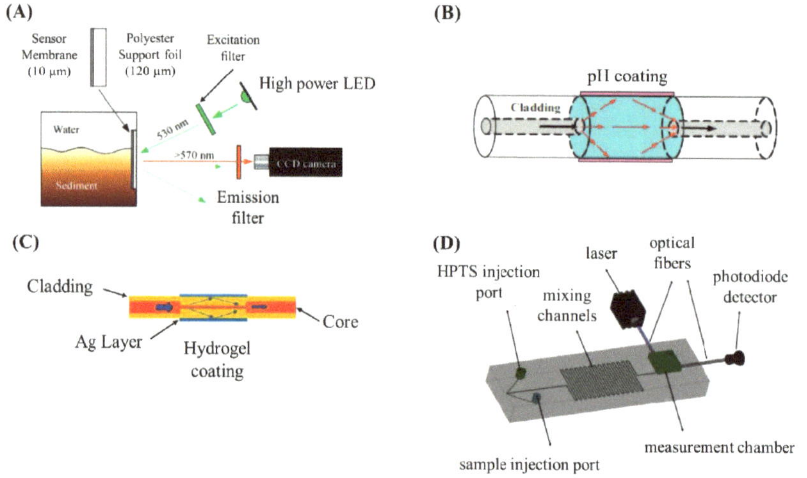

Figure 2. (**A**) Setup for spatial pH measurement of marine sediment (reprinted from [21] with the permission of Wiley). (**B**) Optical fiber without a core used for pH sensing (reprinted from [31] with the permission of Elsevier). (**C**) Hydrogel base fiber optic structure (reprinted from [39] with the permission of Elsevier). (**D**) Microfluidic device for pH measurements reported by Vahid et al. (reprinted from [40] with the permission of Elsevier).

Although there are some examples with inorganic materials, the most common optical pH sensors mainly consist of a dye embedded in a polymer. These sensors achieved a wide measurement range and, apart from some specific applications, such as the monitoring of human tissue, optical fibers are quite popular because they are relatively inexpensive and do not usually need a complex electronic readout system. However, there is the need for materials capable of reducing the leakage of the pH-sensitive optical species into the medium. Such research would allow for long-lasting sensors for pH monitoring.

3. Potentiometric pH Sensors

Potentiometric sensors measure the electrical potential between the WE and RE in a solution. The distribution of electric charge is a time-dependent phenomenon that is a function of several properties of the investigated system, e.g., the bulk composition, the composition of the sensitive layer, and the thermodynamic and kinetic properties [50].

Polymers are often used because they can be functionalized or embedded with pH-sensitive molecules. Lakard et al. studied the responses to pH of five different polymers (monomers: 1,3-diaminopropane, diethylene triamine, pyrrole, p-phenylene diamine, and aniline) electroplated on a platinum wire [51]. All sensors a showed linear response in the range pH 2–10, with the highest sensitivity of 52 mV/pH obtained for PANI films, close to that of the glass pH meter. The same authors used [52] photolithography to fabricate micro-supports for the electrodeposition of a polypyrrole film. This sensor proved that pH sensitivity depended on the polypyrrole thickness. A larger range was obtained from the fabrication of a pH-sensitive membrane by electropolymerization of poly-bis phenol on an indium tin oxide electrode [53]. This method allowed robust systems to be fabricated with a sensitivity of around 57 mV/pH, a working range of 1–14 pH units, and a low interference from other common cations and anions (Na^+, K^+, Cl^-, $SO_4{}^{2-}$). These membranes were also tested in real matrices, such as milk and fruit juice. Guinovart et al. modified a commercial adhesive bandage for monitoring wound status. The sensor consisted of screen-printed Ag/AgCl tracks and an electropolymerized PANI WE. A polyvinyl butyral (PVB) membrane coated the RE to protect it against wound exudate (Figure 3A) [54]. The sensor worked in a physiological pH range (5.5–8 pH), and proved to have extremely efficient mechanical resistance, repeatability, and reproducibility of the measurements. The authors analyzed the sensitivity of the device after sterilization with an autoclave and observed a slight variation that depended on the glass transition of the membrane used as RE. A PANI nanofiber tested in buffer solution led to a superNernstian response (63 mV/pH) [55], whereas a slightly smaller sensitivity was obtained after coating gold interdigitated electrodes with PANI (58.57 mV/pH in the range pH 5.45–8.62, Figure 3B) [56].

One of the first examples of metallic-based potentiometric pH sensors used two forms of lead oxide to fabricate a pH-sensitive membrane on an aluminum substrate [57]. In the pH range 1–12, this sensor had a sensitivity of about 58 mV/pH with good repeatability (RSD < 2.7%) and was not affected from interferents, such as Li^+, Na^+, Mg^{2+}, Ca^{2+}, NH^{4+}, and HCO_3^-. Manjakkal et al. proposed a potentiometric pH sensor based on ruthenium oxides and tantalum oxide as a pH-sensitive layer screen-printed on alumina substrates [58]. The sensors showed short response times (<15 s), long lifetime, sensitivities close to the Nernstian limit, and low interference towards metal cations, such as Li^+, Na^+, and K^+. The sensor was tested in real matrices like lemon juice and river water, obtaining values close to commercial devices. Another superNernstian response in a potentiometric sensor was obtained by Khalil et al., who electroplated iridium oxide nanoparticles on a gold substrate [59]. This method led to a sensitivity of 73 mV/pH in a wide pH range (pH 1.68–12.36). Tanumihardja et al. used ruthenium oxide nanorods to fabricate an integrated sensor in an organ on chip system [60]. A near Nernstian sensitivity was achieved (58 mV/pH), with a short response time (2 s) and a drift of 0.013 pH/h. Choi et al. coupled tungsten oxide nanofibers with a dual-channel differential amplifier, which improved the signal-to-noise ratio and allowed the sensitivity to be increased to 377.5 mV/pH in the pH range 6.9–8.94 [61]. Crespo et al. used multi-walled carbon nanotubes (MWCNTs) in combination with an ion-selective membrane (tridodecylamine and potassium tetrakis [3,5-bis(trifluoromethyl)phenyl] borate in a methyl methacrylate and an n-butyl acrylate matrix) and achieved a good sensitivity of about 58 mV/pH, although the presence of Li^+, Na^+, K^+, NH_4^+, Mg^{2+}, and Ca^{2+} was noted [41]. Smith et al. proposed a combination of poly(3,4-ethylene-dioxythiophene)-poly(styrene sulfonate) (PEDOT:PSS), MWCNTs, and PANI on a flexible cotton yarn [62]. The yarn worked in a wide pH range (2–12) and in artificial sweat with an almost ideal sensitivity (61 ± 2 mV/pH).

Graphene and its derivatives have also been studied for potentiometric pH sensing [13,63,64]. Graphene oxide (GO) has several carboxylic, epoxy, and alcoholic functional groups that make it sensitive to pH changes. Melai et al. used GO for chronic wounds monitoring using a flexible support with screen-printed silver tracks and carbon WE. The sensor had a sensitivity of 31.8 mV/pH [65]. For clinical wound monitoring, Rahimi et al. proposed a pH sensor consisting of PANI deposited onto a combination of a laser-scribed polyimide sheet and a highly deformable ecoflex support [66].

Poma and coworkers decorated rGO with different organic molecules (4-aminobenzoic acid and 4-amino-fenilacetic acid) and improved the sensitivity up to 45 mv/pH in sea water, blood serum, and exudate [67–70]. A similar rGO-based sensor functionalized with 3-(4-aminophenil)propionic acid was efficiently transferred on paper and tested against the same matrices (Figure 3C) [71]. Manjakkal et al. proposed a wearable pH sensor on a cellulose-polyester cloth coated with a polyurethane-graphite WE and an Ag/AgCl RE [72]. This sensor worked in the pH range 6–9 but had low sensitivity (4 mV/pH).

In 2009, Marxer et al. presented a sensor based on an Ag/AgCl electrode coated with a hybrid xerogel (aminosilanes/alkylsilane) of aminosilanes [73]. The working range was in the physiological pH range, whereas its sensitivity was amino silane dependent and ranged from 44 to 55 mV/pH.

The coupling of carbon-nanostructured material with a polymeric film was performed by Zuaznabar-Gardona et al., who electropolymerized a polydopamine film on a multi-layer carbon nano-onion substrate deposited on an glassy carbon electrode [74]. This composite demonstrated low interference from monovalent cations and a sensitivity close to that of the glass electrode. The device was capable of measuring pH in different matrices (milk, sea water, pineapple juice, and vinegar).

Sulka et al. proposed hydroquinone monosulfonate-doped polypyrrole nanowires and obtained a sensitivity of 46–49 mV/pH, which was not far from that of the glass electrode [75].

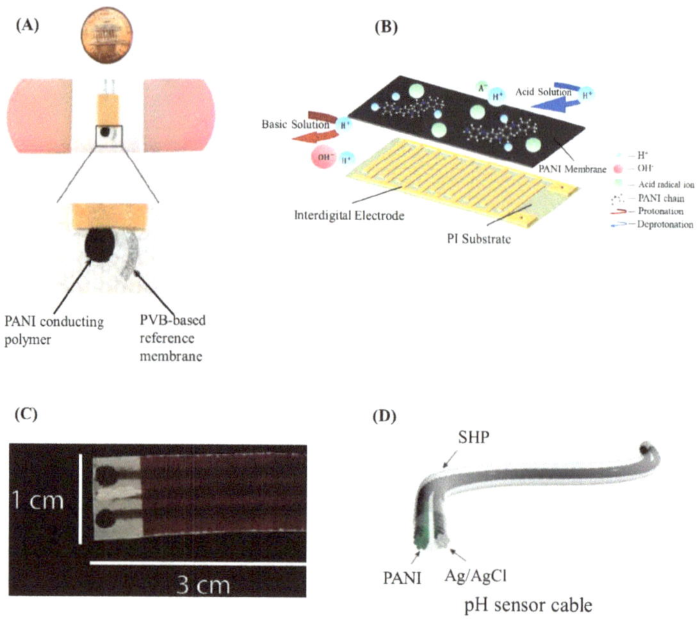

Figure 3. (**A**) PANI-based pH sensor for wound monitoring (reprinted from [54] with the permission of Wiley). (**B**) PANI-based interdigitate gold electrodes onto a polyimide (PI) substrate for pH sensing (reprinted from [56], published by The Royal Society of Chemistry). (**C**) Paper-based pH sensor using rGO functionalized with 3-(4-aminophenil)propionic acid (reprinted from [71] with the permission of IEEE). (**D**) Self-healing PANI-based (SHP) pH sensor (reprinted from [76] with the permission of Elsevier).

Yoon et al. developed a remarkable pH sensor that consisted of two carbon fiber threads coated with a self-healing polymer (poly(1,4-cyclohexanedimethanol succinate-co-citrate)) (Figure 3D) [76]. One thread was coated with a PANI layer, whereas the other with Ag/AgCl as RE. The device showed an almost ideal sensitivity (58.1 mV/pH) with a 5-s response time in the range of pH 4–10. Furthermore, measurements were performed in real matrices (urine, saliva, sweat, and human tears) and results were comparable with those of a commercial pH meter.

4. Voltammetric pH Sensors

Voltammetric pH systems are less common than potentiometric and optical chemical pH sensors because of the few suitable compounds for this type of transduction. These measurements usually depend on sensitive materials capable of involving hydrogen ions in the electrochemical reaction. Therefore, by measuring the current generated during the redox process, it is possible to establish a correlation with pH. In voltammetry, the sensitivity can be expressed as mV/pH since the current/potential profile, called a voltammogram, is often chosen to monitor pH.

Stred'Ansky et al. studied the redox quinone-hydroquinone pair, which was known to be pH dependent [77]. In 2002, Wildgoose et al. described a more accurate work, where carbon particles were covalently modified with anthraquinone and immobilized onto pyrolytic graphite electrodes [78]. This sensor had a sensitivity of about 58 mV/pH over a range of about 1 to 9 pH units. The article analyzed the effect of temperature on the sensor sensitivity in the range 20–70 °C, recording a variation consistent with the Nernst law. Makos et al. developed a carbon fiber microelectrode using the chemistry of para-quinone [79]. Although having the advantage of a miniaturized system, the sensor showed a low sensitivity, about 38 mV/pH, and a small working range (6.5 at 8 pH). Later, Amiri et al. electrodeposited polydopamine in aqueous solution onto glassy carbon electrodes [80]. As for quinone, polydopamine is involved in a two electron–two proton exchange process, which allows for the transduction of the H^+ concentration into current. The sensor showed a sensitivity of approximately 58 mv/pH in a range from 1 to 12 pH units with a reproducibility of 0.83%. Chaisiwamongkhol et al. reported a sensor based on porous graphitic carbon fibers functionalized with quinone groups [81]. This sensor had a superNernstian sensitivity of 65 mV/pH in the pH range 2 to 8 by performing scans on a potential range from −0.2 to 0.8 V. When tested in a real saliva sample, the performances were comparable with those of a glass electrode.

Recently, Vivaldi et al. fabricated a voltammetric pH sensor using an indoaniline derivative as a sensitive layer (Figure 4A) [82]. Exploiting the two electron–two proton reaction mechanism typical of quinones, this molecule allowed for pH measurement in low potential windows (−0.4–0.2 V) with a sensitivity of 56 mV/pH. This sensor experienced no leakage of the sensitive molecule in the medium thanks to the binding properties of the indoaniline derivative. Furthermore, the device was successfully used in biological samples (urine, saliva, and blood) and in beverages (orange juice, milk, and tea) with an accuracy better than 0.1 pH unit. Genotoxicity tests were carried out on the synthetized indoaniline derivative, proving the absence of genotoxic effects on living cells.

Chaisiwamongkhol et al. used a metal oxide (iridium oxide) for pH monitoring in blood samples [83]. This sensor had a sensitivity of about 63 mV/pH and high reproducibility (RSD < 2%) in the working potential range −0.2–0.8 V. In 2019, Tham et al. used riboflavin as a pH-dependent molecule and a vitamin E derivative as a pH-independent reference to fabricate a biocompatible pH sensor that proposed an alternative to the typical Ag/AgCl RE. Thanks to the reversible $2e^-/2H^+$ process in riboflavin, this sensor had a sensitivity of about 50 mV/pH in buffered media, whereas the response in unbuffered media was negligible because of the low concentration of localized H^+ (Figure 4B) [84].

pH-sensitive carbon materials were reported by Zhu et al., who fabricated a voltammetric pH sensor using an electrochemically modified nanocrystalline graphite-like amorphous carbon [85]. The application of positive voltage (2 V vs. an Ag/AgCl RE) in sulfuric acid allowed for the formation of pH-dependent quinonic groups on the carbon surface. This sensor had a superNernstian response of 63.3 mV and negligible effect from K^+ and Na^+. However, the surface required electrochemical regeneration after 20 measurements. Hu et al. described a quinone-functionalized tryptophane for the development of a pH sensor on a graphite electrode [86]. Interestingly, this sensor could be used both in voltammetric and potentiometric modes, obtaining a sensitivity of 52 mV/pH in a wide pH range (1–12). A comparison with a glass electrode led to accurate measurements in complex matrices like milk and cola.

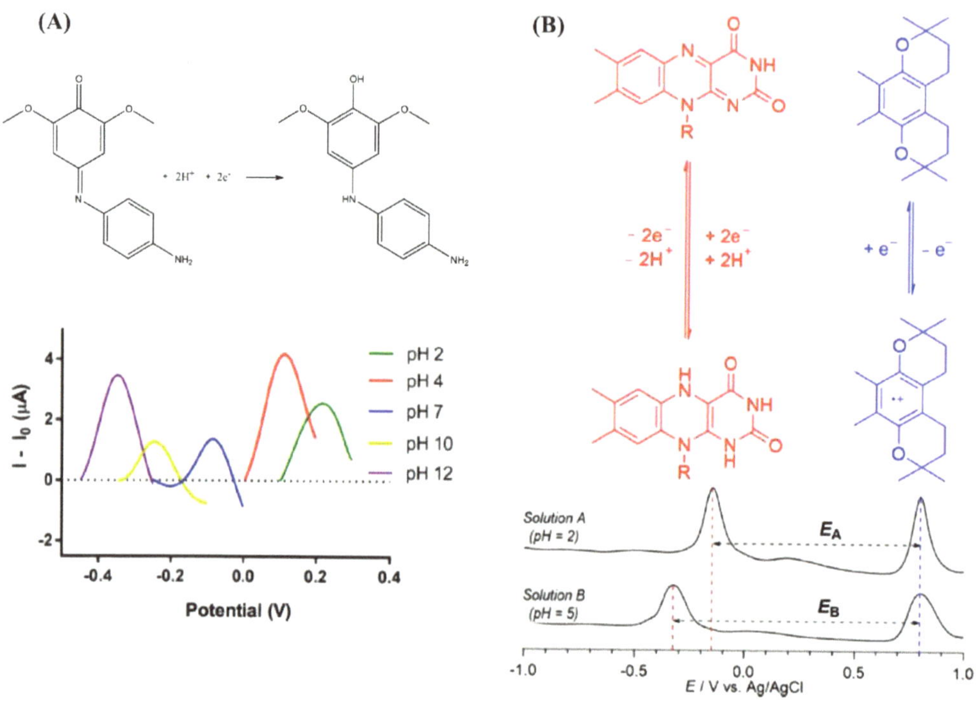

Figure 4. (**A**) Response of an indoaniline-based voltammetric pH sensor (reprinted from [82] with the permission of Elsevier). (**B**) Square wave voltammetry of the vitamin-based pH sensor (reprinted from [84] with the permission of Elsevier).

Instead of the potential, Gao et al. measured the current flowing in a PANI film electropolymerized onto a graphite electrode [87]. This sensor worked between 2 and 10 pH units; however, sensitivity was low. In another study, Sha et al. assessed by cyclic voltammetry the pH dependence of an electrodeposited PANI layer onto a graphene substrate [88]. The device showed two linear response ranges, one from pH 1 to 5, and a second from 7 to 11 with a sensitivity of about −50 and 139 $\mu A/(pH \cdot cm^2)$, respectively, with potential scans ranging from −0.25 to 1 V. Further work is needed to assess whether this sensor can be used as a linear pH sensor in the full range.

5. pH Sensors Based on Field Effect Transistors

An ISFET is capable of transducing the H^+ concentration into a current via the field effect [13], which depends on the interaction between an H^+-sensitive layer deposited onto the gate. The gate is often removed and the solution under measurement becomes the gate. These devices are easily miniaturized and mass-produced, and have proven to be effective in the quantitative measurement of analytes as early as 1970 [89].

At the beginning of the 2000s, a mix of tin oxide and metallic aluminum was used as the gate [90]. This ISFET showed a sensitivity of about 58 mV/pH working in a pH range from 2 to 10. However, the article did not analyze any other analytical parameter since it focused on the study of the sensor's electro-technical apparatus.

In 2002, Shitashima et al. fabricated an FET using SiO_2/Si_3N_4 as a pH-sensitive gate insulator, which was coupled with an ion-sensitive electrode for the chloride ion to monitor marine pH [91]. The article only reported the calibration method and the results in real matrices. In another paper, a microfluidic system was used to supply standard pH solution for the calibration of an ISFET that measured marine pH [92]. Rani et al. developed an ISFET by changing the type of gate from single to "multi finger" to improve the transistor performances [93]. However, the sensitivity was rather low (about 45 mV/pH) and the response time was long (600 s).

Parizi et al. obtained a superNernstian response using an ISFET based on an aluminum oxide layer deposited on an aluminum extended gate [94]. A remarkable sensitivity of 130 mV/pH was achieved in the range pH 4–10. In 2014, Das et al. proposed a superNernstian pH (62 mV/pH) sensor using an EGFET (extended gate field effect transistor) with palladium oxide (Figure 5A) [95]. The EGFET worked in the pH range 2–12, with a hysteresis of about 7.4 mV. The drift was 2.4 mV/h.

Cho et al. fabricated an extended gate of indium tin oxide further covered with a layer of tin oxide on paper [96]. The device showed a sensitivity of 57 mV/pH in the range pH 4–10. This study demonstrated the feasibility of the integration into systems that aim to be more environmentally friendly than classic silicon-based systems.

An article by Kang et al. reduced the RE noise coupling an ISFET with an REFET (reference field effect transistor), which has low pH sensitivity [97]. Integrating a platinum RE, the fluctuation of the solution potential was removed by a differential measurement between the ISFET and the REFET.

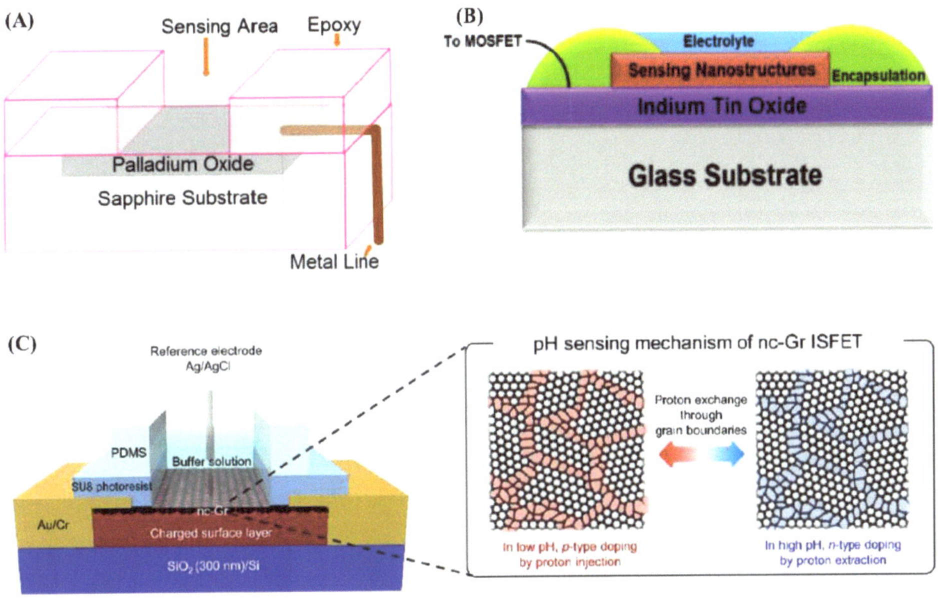

Figure 5. (**A**) Palladium oxide extended gate FET (reprinted from [95] with the permission of Elsevier). (**B**). Na_3BiO_4 and Bi_2O_3 ISFET schematics (reprinted from [98] with the permission of Elsevier) (**C**) Nanocrystalline graphene-based pH sensor (polydimethylsiloxane, PDMS) (reprinted from [99] with the permission of ACS).

Sharma et al. reported a mixed layer of Na_3BiO_4 and Bi_2O_3 nanostructures on an indium tin oxide electrode (Figure 5B) [98]. This oxide layer allowed for a sensitivity of 49.63 mV/pH in the pH range 7–12. Jung et al. coated an SiO_2 layer with nanocrystalline graphene that, when exposed to H^+, changed its conductivity and allowed for a sensitivity of 140 mV/pH in the pH range 6–7.6 (Figure 5C) [99].

Table 1 summarizes the analytical parameters of the sensors reported in this review.

Table 1. Analytical parameters of the cited works. Reproducibility is defined as the relative standard deviation over different sensors. Precision is defined as the RSD on the same sensor.

Transduction	Reproducibility	Precision	Sensitivity	Hysteresis	Interferents	Working Range	Matrix	Response Time	Life Time	Ref
Optical	-	-	-	-	-	2–8	-	1 s	1 month	[23]
Optical	<0.1 pH units	-	-	-	-	6–9	-	180 s	-	[24]
Optical	-	-	-	-	-	7.3–9.3	Marine sediments	200 s	>3 days	[21]
Optical	-	-	-	-	-	6–8	Marine sediments	16 s	1 week	[22]
Optical	<0.20 pH units	0.16 pH units	-	-	-	0–14	-	-	-	[25]
Optical	-	-	-	-	-	6.80–9.00	-	-	-	[26]
Optical	-	-	-	-	-	4–12	Detergent	<5 s	>1 month	[27]
Optical	-	<2%	-	-	-	1–14 with Neural network	-	54 s	-	[28]
Optical	-	<0.3%	-	-	-	0.5–5	-	180 s	>3 months	[29]
Optical	-	-	-	-	Na^+, K^+, Ag^+, Mg^{2+}, Ca^{2+}, Pb^{2+}, Co^{2+}, Ni^{2+}, Cu^{2+}, Zn^{2+}, Cd^{2+}, Hg^{2+}, Al^{3+}, Cr^{3+}, Fe^{3+}, Au^{3+}, Pt^{2+}, Ru^{2+}	1–8	-	-	-	[30]
Optical	-	<1%	6.6 dBm/pH	-	-	1–13	-	40 s	-	[31]
Optical	-	±0.20 pH units	-	-	Ionic Strength	5–8	Human serum	1200 s	>2 months	[33]
Optical	-	<0.1 pH units	-	-	-	4.7–7.7	-	-	-	[34]
Optical	-	-	-	-	-	6–9	-	20 s	-	[35]
Optical	-	-	-	-	-	11–13	-	50 s	>7 months	[36]
Optical	-	-	-	-	-	6–8.5	Subcutaneous tissue	<120 s	28 days	[37]
Optical	-	0.1 pH units	-	-	-	5.5–8	Ovine lung tissue	30 s	-	[38]
Optical			13 nm/pH	0.5%	-	1–12	-	20 s	10 days	[39]
Optical	-	-	6–42 mV/pH	-	-	2.5–9	-	10 s	-	[40]
Optical	-	-	24.93–11.35 nm/pH	-	-	2.4–11.35	-	-	-	[48]
Optical	4.06%	1.14%	-	-	Ionic Strength	4–8 5–8	-	240 s	>2 months	[49]
Potentiometric	-	-	34–52 mV/pH	-	-	2–11	-	-	-	[51]
Potentiometric	-	-	40–50 mV/pH	-	-	2–11	-	-	>1 month	[52]
Potentiometric	2.4–2.9%	-	56.7–58.6 mV/pH	-	Na^+, K^+, Cl^-, SO_4^{2-}	1–15	Milk, orange juice	<20 s	-	[53]
Potentiometric	0.66%	1.9%	54–56 mV/pH	-	Na^+, K^+, Cl^-, SO_4^{2-}	5.5–8	-	<20 s	-	[54]
Potentiometric	-	-	62.4 mV/pH	5.6 mV	Li^+, Na^+, K^+, Mg^{2+}, Ca^{2+}, NH_4^+	4–10	milk	12.8 s	-	[55]
Potentiometric	2.39%	8%	58.57 mV/pH	12%	-	5.45–8.62	-	45 s	-	[56]
Potentiometric	2.7%	-	58 mV/pH	-	Li^+, Na^+, Mg^{2+}, Ca^{2+}, NH^{4+}, HCO^{3-}	1–12	Cola	-	>1 months	[57]

Table 1. *Cont.*

Transduction	Reproducibility	Precision	Sensitivity	Hysteresis	Interferents	Working Range	Matrix	Response Time	Life Time	Ref
Potentiometric	±1 mV/pH	-	56 mV/pH	±3 mV acid region, ±8 mV basic region	Li^+, Na^+, K^+	2–10	River water, lemon juice	<15 s	-	[58]
Potentiometric	-	-	73 mV/pH	-	-	2–12	-	<10 s	-	[59]
Potentiometric	-	-	58 mV/pH	-	Li^+, Ca^{2+}, Cl^-, SO_4^{2-}	2–10	-	<2 s	-	[60]
Potentiometric	-	-	377.5 mV/pH	-	-	6.90–8.94	Artificial sea water	-	-	[61]
Potentiometric	-	±0.4 mV/pH	58 mV/pH	-	Li^+, Na^+, K^+, Mg^{2+}, Ca^{2+}, NH^{4+}	2.89–9.90	-	10 s	-	[41]
Potentiometric	±2 mV/pH	-	61 mV/pH	-	Na^+, K^+, Mg^{2+}, Ca^{2+}, NH^{4+}	2–12	Artificial sweat	60 ± 20 s	-	[62]
Potentiometric	-	-	31.8 mV/pH	-	-	4–10	Wound exudate	-	>4 days	[65]
Potentiometric	-	-	53 mV/pH	-	-	4–10	-	-	-	[66]
Potentiometric	5%	5%	45 mV/pH	-	-	4–10	sea water	-	>7 days	[67]
Potentiometric	-	-	4 mV/pH	0.5 mV	Glucose, Urea	6–9	-	5 s	-	[72]
Potentiometric	-	-	44–55 mV/pH	-	Na^+, K^+	3–8	-	<3 s	>2 weeks	[73]
Potentiometric	-	-	58.3–60.1 mV/pH	-	Li^+, Na^+, K^+	2–10	Milk, sea water, pineapple juice, vinegar	-	-	[74]
Potentiometric	-	-	46–49 mV/pH	-	-	2–12	-	-	>2 months	[75]
Potentiometric	-	-	58.7 mV/pH	5.6	K^+, Na^+, Ca^+, NH_4^+	4–10	Urine, Saliva, Sweat, Tears	5 s		[76]
Voltammetric	-	-	58 mV/pH	-	-	1–9	-	-	-	[78]
Voltammetric	-	-	38 mV/pH	-	Mg^+, Ca^+, K^+	6.5–8	Bacteria broth	1.6 s	-	[79]
Voltammetric	<0.8%	±0.09 mV/pH	58 mV/pH	-	-	1–12	-	-	-	[80]
Voltammetric	-	-	65 mV/pH	-	-	2–8	Synthetic saliva and saliva	-	-	[81]
Voltammetric	5%	7%	56 mV/pH	-	Li^+, Na^+	2–12	Biological and food matrix	-	>2 months	[82]
Voltammetric	<2%	-	62.7 mV/pH	-	-	5–9	Animal blood	-	-	[83]
Voltammetric	-	-	50 mV/pH	-	-	1–12	-	-	-	[84]
Voltammetric	-	-	63.3 mV/pH	-	Na^+, K^+, Dissolved oxygen	0–11	Apple cider vinegar	-	-	[85]
Voltammetric and potentio-metric	-	-	52 mV/pH	-	-	1–12	Cola, Milk	-	16 days	[86]
Voltammetric	-	0.5%	32,4 mA/pH 15.9 mA/pH	-	K^+, Na^+, Li^+	2–5.5 5.5–10	-	<8 s	-	[87]

Table 1. *Cont.*

Transduction	Reproducibility	Precision	Sensitivity	Hysteresis	Interferents	Working Range	Matrix	Response Time	Life Time	Ref
Voltammetric	<3.4%	-	−50.14 µA/pH cm^2 −139.2 µA/pH cm^2	-	-	1–5, 7–11	-	-	-	[88]
Transistor	-	-	58 mV/pH	-	-	2–10	-	-	-	[90]
Transistor	-	-	45.1 mV/pH	24 mV 12 mV	-	-	-	600 s	-	[92]
Transistor	-	-	130 mV/pH	-	-	4–10	-	-	-	[93]
Transistor	-	-	62 mV/pH	7.4 mV	-	2–12	-	-	-	[94]
Transistor	-	-	57 mV/pH	25 mV	-	4–10	-	-	-	[95]
Transitor	-	-	49.63 mV/pH	-	-	7–12	-	-	-	[98]
Transistor	-	-	140 mV/pH	-	-	6–7.6	-	-	-	[99]

6. Conclusions and Outlook

The optical pH sensors reported in the literature have achieved high precision, but the working range is often rather narrow, or the linear response is poor. To prevent these drawbacks, the combination of several chromophores can improve performances but at the cost of a more complex fabrication. A major objective emerging from the analysis of the literature is the need for matrices capable of reducing the leakage of the pH-sensitive optical species into the medium. Such research is presently focused on new embedding matrices, such as hydrogels, which would allow for long-lasting sensors for pH monitoring. The choice between optical fibers and planar structures is application oriented. Optical fibers seem more suitable for dynamic sensors with reduced dimensions for in situ measurements. Optical planar structures are being successfully employed mainly for static analysis, where the use of specific equipment (e.g., Charge-Coupled Device cameras) allows for spatial pH measurement.

Potentiometric sensors share the same transduction mechanism with the glass electrode, and they can hardly overcome the limits dictated by Nernst's law. Furthermore, several examples have been reported where the impact of interferents (e.g., cations of alkaline and earth alkaline metals) is hardly removed or not studied at all. So far, PANI is the most used pH-sensitive material because the fabrication of a pH sensor is simplified by the electropolymerization of the monomer onto an electrode. The structure of graphene, on the other hand, allows for easy surface functionalization, improving the sensitivity towards pH.

Voltammetric sensors are an evolution of potentiometric sensors. The literature on this topic is limited and the available studies have shown the importance of designing better electroactive molecules to obtain competitive devices. ISFETs are valid alternative systems to the classical pH measurement methods, capable of overcoming the Nernstian limitation and achieving small dimensional scales thanks to microfabrication procedures. However, the available analytical data are not detailed enough to evaluate their performance, especially in real matrices.

Although glass electrodes are still far from being outdated by new pH sensors, the research for new materials could boost the development of the devices. In particular, a full integration of a reference electrode into an ISFET could pave the way towards a wider use of this technology for pH measurement.

Funding: This research was funded by the Italian National Interuniversity *Consortium* of Materials Science and Technology (INSTM), grant number INSTMPI004.

Conflicts of Interest: The authors declare no conflict of interest.

References

1. Sarkawi, S. An Overview on pH Measurement Technique and Application in Biomedical and Industrial Process. In Proceedings of the 2015 2nd International Conference on Biomedical Engineering (ICoBE), Penang, Malaysia, 30–31 March 2015.
2. Jin, Q.; Kirk, M.F. pH as a Primary Control in Environmental Microbiology: 1. Thermodynamic Perspective. *Front. Environ. Sci.* **2018**, *6*, 21. [CrossRef]
3. Allen, L.; Ansel, H.C. *Ansel's Pharmaceutical Dosage Forms and Drug Delivery Systems*; Lippincott Williams & Wilkins: Philadelphia, PA, USA, 2013; ISBN 1451188765.
4. Salvo, P.; Dini, V.; Di Francesco, F.; Romanelli, M. The role of biomedical sensors in wound healing. *Wound Med.* **2015**, *8*, 15–18. [CrossRef]
5. Salvo, P.; Calisi, N.; Melai, B.; Dini, V.; Paoletti, C.; Lomonaco, T.; Pucci, A.; Di Francesco, F.; Piaggesi, A.; Romanelli, M. Temperature-and pH-sensitive wearable materials for monitoring foot ulcers. *Int. J. Nanomed.* **2017**, *12*, 949. [CrossRef] [PubMed]
6. Zhang, X.; Lin, Y.; Gillies, R.J. Tumor pH and its measurement. *J. Nucl. Med.* **2010**, *51*, 1167–1170. [CrossRef] [PubMed]
7. de Lima, T.M.; Kazama, C.M.; Koczulla, A.R.; Hiemstra, P.S.; Macchione, M.; Godoy Fernandes, A.L.; de Paula Santos, U.; Bueno-Garcia, M.L.; Zanetta, D.M.; Saldiva de André, C.D.; et al. PH in exhaled breath condensate and nasal lavage as a biomarker of air pollution-related inflammation in street traffic-controllers and office-workers. *Clinics* **2013**, *68*, 1488–1494. [CrossRef]
8. Cremer, M. Über die Ursache der elektromotorischen Eigenschaften der Gewebe, zugleich ein Beitrag zur Lehre von polyphasischen Elektrolytketten. *Z. Biol.* **1906**, *47*, 562–608.

9. Leiner, M.J.P.; Hartmann, P. Theory and practice in optical pH sensing. *Sens. Actuators B. Chem.* **1993**. [CrossRef]
10. Richter, A.; Paschew, G.; Klatt, S.; Lienig, J.; Arndt, K.F.; Adler, H.J.P. Review on hydrogel-based pH sensors and microsensors. *Sensors* **2008**, *8*, 561–581. [CrossRef]
11. Weston, M.; Kuchel, R.P.; Ciftci, M.; Boyer, C.; Chandrawati, R. A polydiacetylene-based colorimetric sensor as an active use-by date indicator for milk. *J. Colloid Interface Sci.* **2020**, *572*, 31–38. [CrossRef]
12. Yuqing, M.; Jianrong, C.; Keming, F. New technology for the detection of pH. *J. Biochem. Biophys. Methods* **2005**, *63*, 1–9. [CrossRef]
13. Salvo, P.; Melai, B.; Calisi, N.; Paoletti, C.; Bellagambi, F.; Kirchhain, A.; Trivella, M.G.; Fuoco, R.; Di Francesco, F. Graphene-based devices for measuring pH. *Sens. Actuators B Chem.* **2018**, *256*, 976–991. [CrossRef]
14. Wencel, D.; Abel, T.; McDonagh, C. Optical chemical pH sensors. *Anal. Chem.* **2014**, *86*, 15–29. [CrossRef]
15. Baldini, F. Critical review of pH sensing with optical fibres. In *Chemical, Biochemical, and Environmental Fiber Sensors X*; International Society for Optics and Photonics: Bellingham, WA, USA, 1999; Volume 3540, pp. 2–9.
16. Butler, T.M.; MacCraith, B.D.; McDonagh, C. Leaching in sol-gel-derived silica films for optical pH sensing. *J. Non. Cryst. Solids* **1998**, *224*, 249–258. [CrossRef]
17. Lev, O.; Tsionsky, M.; Rabinovich, L.; Glezer, V.; Sampath, S.; Pankratov, I.; Gun, J. Organically Modified Sol-Gel Sensors. *Anal. Chem.* **1995**, *67*, 22A–30A. [CrossRef]
18. Plaschke, M.; Czolk, R.; Ache, H.J. Fluorimetric determination of mercury with a water-soluble porphyrin and porphyrin-doped sol-gel films. *Anal. Chim. Acta* **1995**, *304*, 107–113. [CrossRef]
19. Zhu, Q.; Aller, R.C. Planar fluorescence sensors for two-dimensional measurements of H2S distributions and dynamics in sedimentary deposits. *Mar. Chem.* **2013**. [CrossRef]
20. Zhu, Q.; Aller, R.C.; Fan, Y. High-performance planar pH fluorosensor for two-dimensional pH measurements in marine sediment and water. *Environ. Sci. Technol.* **2005**, *39*, 8906–8911. [CrossRef] [PubMed]
21. Stahl, H.; Glud, A.; Schröder, C.R.; Klimant, I.; Tengberg, A.; Glud, R.N. Time-resolved pH imaging in marine sediments with a luminescent planar optode. *Limnol. Oceanogr. Methods* **2006**, *4*, 336–345. [CrossRef]
22. Jiang, Z.; Yu, X.; Hao, Y. Design and fabrication of a ratiometric planar optode for simultaneous imaging of pH and oxygen. *Sensors* **2017**, *17*, 1316. [CrossRef]
23. Jin, Z.; Su, Y.; Duan, Y. An improved optical pH sensor based on polyaniline. *Sens. Actuators B Chem.* **2000**, *71*, 118–122. [CrossRef]
24. Kermis, H.R.; Kostov, Y.; Rao, G. Rapid method for the preparation of a robust optical pH sensor. *Analyst* **2003**, *128*, 1181–1186. [CrossRef] [PubMed]
25. Gotor, R.; Ashokkumar, P.; Hecht, M.; Keil, K.; Rurack, K. Optical pH Sensor Covering the Range from pH 0-14 Compatible with Mobile-Device Readout and Based on a Set of Rationally Designed Indicator Dyes. *Anal. Chem.* **2017**, *89*, 8437–8444. [CrossRef] [PubMed]
26. Raoufi, N.; Surre, F.; Sun, T.; Rajarajan, M.; Grattan, K.T.V. Wavelength dependent pH optical sensor using the layer-by-layer technique. *Sens. Actuators, B Chem.* **2012**, *169*, 374–381. [CrossRef]
27. Abu-Thabit, N.; Umar, Y.; Ratemi, E.; Ahmad, A.; Abuilaiwi, F.A. A flexible optical pH sensor based on polysulfone membranes coated with pH-responsive polyaniline nanofibers. *Sensors* **2016**, *16*, 986. [CrossRef] [PubMed]
28. Safavi, A.; Bagheri, M. Novel optical pH sensor for high and low pH values. *Sens. Actuators B Chem.* **2003**, *90*, 143–150. [CrossRef]
29. Hashemi, P.; Abolghasemi, M.M. Preparation of a novel optical sensor for low pH values using agarose membranes as support. *Sens. Actuators B Chem.* **2006**, *115*, 49–53. [CrossRef]
30. Taweetanavanich, T.; Wanno, B.; Tuntulani, T.; Pulpoka, B.; Kaewtong, C. A pH optical and fluorescent sensor based on rhodamine modified on activated cellulose paper. *J. Chinese Chem. Soc.* **2019**, *66*, 493–499. [CrossRef]
31. Li, W.; Cheng, H.; Xia, M.; Yang, K. An experimental study of pH optical sensor using a section of no-core fiber. *Sens. Actuators A Phys.* **2013**, *199*, 260–264. [CrossRef]
32. Jung, Y.; Kim, S.; Lee, D.; Oh, K. Compact three segmented multimode fibre modal interferometer for high sensitivity refractive-index measurement. *Meas. Sci. Technol.* **2006**, *17*, 1129. [CrossRef]
33. Schyrr, B.; Pasche, S.; Scolan, E.; Ischer, R.; Ferrario, D.; Porchet, J.A.; Voirin, G. Development of a polymer optical fiber pH sensor for on-body monitoring application. *Sens. Actuators B Chem.* **2014**, *194*, 238–248. [CrossRef]
34. Sørensen, T.J.; Rosenberg, M.; Frankær, C.G.; Laursen, B.W. An Optical pH Sensor Based on Diazaoxatriangulenium and Isopropyl-Bridged Diazatriangulenium Covalently Bound in a Composite Sol–Gel. *Adv. Mater. Technol.* **2019**. [CrossRef]
35. Jeon, D.; Yoo, W.J.; Seo, J.K.; Shin, S.H.; Han, K.T.; Kim, S.G.; Park, J.Y.; Lee, B. Fiber-optic pH sensor based on sol-gel film immobilized with neutral red. *Opt. Rev.* **2013**, *20*, 209–213. [CrossRef]
36. Mahboubeh, V.; Rounaghi, G.H.; Es'haghi, Z.; Moradi, Z. Design and Application of an Optical pH Sensor Based on Thionine Doped Modified Sol–Gel Film. *Russ. J. Phys. Chem. A* **2019**, *93*, 1389–1393. [CrossRef]
37. Wencel, D.; Kaworek, A.; Abel, T.; Efremov, V.; Bradford, A.; Carthy, D.; Coady, G.; McMorrow, R.C.N.; McDonagh, C. Optical Sensor for Real-Time pH Monitoring in Human Tissue. *Small* **2018**, *14*, 1–8. [CrossRef]
38. Gong, J.; Tanner, M.G.; Venkateswaran, S.; Stone, J.M.; Zhang, Y.; Bradley, M. A hydrogel-based optical fibre fluorescent pH sensor for observing lung tumor tissue acidity. *Anal. Chim. Acta* **2020**, *1134*, 136–143. [CrossRef]
39. Zhao, Y.; Lei, M.; Liu, S.X.; Zhao, Q. Smart hydrogel-based optical fiber SPR sensor for pH measurements. *Sens. Actuators, B Chem.* **2018**, *261*, 226–232. [CrossRef]

40. Moradi, V.; Akbari, M.; Wild, P. A fluorescence-based pH sensor with microfluidic mixing and fiber optic detection for wide range pH measurements. *Sens. Actuators A Phys.* **2019**, *297*, 111507. [CrossRef]
41. Crespo, G.A.; Gugsa, D.; MacHo, S.; Rius, F.X. Solid-contact pH-selective electrode using multi-walled carbon nanotubes. *Anal. Bioanal. Chem.* **2009**, *395*, 2371–2376. [CrossRef]
42. Gill, E.; Arshak, K.; Arshak, A.; Korostynska, O. Mixed metal oxide films as pH sensing materials. *Microsyst. Technol.* **2008**, *14*, 499–507. [CrossRef]
43. Fog, A.; Buck, R.P. Electronic semiconducting oxides as pH sensors. *Sens. Actuators* **1984**, *5*, 137–146. [CrossRef]
44. Manjakkal, L.; Szwagierczak, D.; Dahiya, R. Metal oxides based electrochemical pH sensors: Current progress and future perspectives. *Prog. Mater. Sci.* **2020**, *109*, 100635. [CrossRef]
45. Wei, A.; Pan, L.; Huang, W. Recent progress in the ZnO nanostructure-based sensors. *Mater. Sci. Eng. B Solid-State Mater. Adv. Technol.* **2011**, *176*, 1409–1421. [CrossRef]
46. Huang, X.J.; Choi, Y.K. Chemical sensors based on nanostructured materials. *Sens. Actuators B Chem.* **2007**, *122*, 659–671. [CrossRef]
47. Głáb, S.; Hulanicki, A.; Edwall, G.; Folke, F.; Ingman, I.; Koch, W.F. Metal-Metal Oxide and Metal Oxide Electrodes as pH Sensors. *Crit. Rev. Anal. Chem.* **1989**, *21*, 29–47. [CrossRef]
48. Semwal, V.; Gupta, B.D. Highly sensitive surface plasmon resonance based fiber optic pH sensor utilizing rGO-Pani nanocomposite prepared by in situ method. *Sens. Actuators B Chem.* **2019**, *283*, 632–642. [CrossRef]
49. Shahamirifard, S.A.; Ghaedi, M.; Montazerozohori, M. Application of nanostructure ZnLI2 complex in construction of optical pH sensor. *Appl. Organomet. Chem.* **2018**, *32*, 1–11. [CrossRef]
50. Bobacka, J.; Ivaska, A.; Lewenstam, A. Potentiometric ion sensors. *Chem. Rev.* **2008**, *108*, 329–351. [CrossRef] [PubMed]
51. Lakard, B.; Herlem, G.; Lakard, S.; Guyetant, R.; Fahys, B. Potentiometric pH sensors based on electrodeposited polymers. *Polymer* **2005**, *46*, 12233–12239. [CrossRef]
52. Lakard, B.; Segut, O.; Lakard, S.; Herlem, G.; Gharbi, T. Potentiometric miniaturized pH sensors based on polypyrrole films. *Sens. Actuators, B Chem.* **2007**, *122*, 101–108. [CrossRef]
53. Li, Q.; Li, H.; Zhang, J.; Xu, Z. A novel pH potentiometric sensor based on electrochemically synthesized polybisphenol A films at an ITO electrode. *Sens. Actuators B Chem.* **2011**, *155*, 730–736. [CrossRef]
54. Guinovart, T.; Valdés-Ramírez, G.; Windmiller, J.R.; Andrade, F.J.; Wang, J. Bandage-Based Wearable Potentiometric Sensor for Monitoring Wound pH. *Electroanalysis* **2014**, *26*, 1345–1353. [CrossRef]
55. Park, H.J.; Yoon, J.H.; Lee, K.G.; Choi, B.G. Potentiometric performance of flexible pH sensor based on polyaniline nanofiber arrays. *Nano Converg.* **2019**, *6*. [CrossRef]
56. Li, Y.; Mao, Y.; Xiao, C.; Xu, X.; Li, X. Flexible pH sensor based on a conductive PANI membrane for pH monitoring. *RSC Adv.* **2019**, *10*, 21–28. [CrossRef]
57. Eftekhari, A. pH sensor based on deposited film of lead oxide on aluminum substrate electrode. *Sens. Actuators B Chem.* **2003**, *88*, 234–238. [CrossRef]
58. Manjakkal, L.; Zaraska, K.; Cvejin, K.; Kulawik, J.; Szwagierczak, D. Potentiometric RuO2-Ta2O5 pH sensors fabricated using thick film and LTCC technologies. *Talanta* **2016**, *147*, 233–240. [CrossRef]
59. Khalil, M.; Liu, N.; Lee, R.L. Super-Nernstian potentiometric pH sensor based on the electrodeposition of iridium oxide nanoparticles. *Int. J. Technol.* **2018**, *57*, 446–454. [CrossRef]
60. Tanumihardja, E.; Olthuis, W.; van den Berg, A. Ruthenium oxide nanorods as potentiometric ph sensor for organs-on-chip purposes. *Sensors* **2018**, *18*, 2901. [CrossRef] [PubMed]
61. Choi, S.J.; Savagatrup, S.; Kim, Y.; Lang, J.H.; Swager, T.M. Precision pH Sensor Based on WO3 Nanofiber-Polymer Composites and Differential Amplification. *ACS Sens.* **2019**, *4*, 2593–2598. [CrossRef] [PubMed]
62. Smith, R.E.; Totti, S.; Velliou, E.; Campagnolo, P.; Hingley-Wilson, S.M.; Ward, N.I.; Varcoe, J.R.; Crean, C. Development of a novel highly conductive and flexible cotton yarn for wearable pH sensor technology. *Sens. Actuators B Chem.* **2019**, *287*, 338–345. [CrossRef]
63. Salvo, P.; Calisi, N.; Melai, B.; Cortigiani, B.; Mannini, M.; Caneschi, A.; Lorenzetti, G.; Paoletti, C.; Lomonaco, T.; Paolicchi, A.; et al. Temperature and pH sensors based on graphenic materials. *Biosens. Bioelectron.* **2017**, *91*, 870–877. [CrossRef]
64. Salvo, P.; Dini, V.; Kirchhain, A.; Janowska, A.; Oranges, T.; Chiricozzi, A.; Lomonaco, T.; Di Francesco, F.; Romanelli, M. Sensors and biosensors for C-reactive protein, temperature and pH, and their applications for monitoring wound healing: A review. *Sensors* **2017**, *17*, 2952. [CrossRef]
65. Melai, B.; Salvo, P.; Calisi, N.; Moni, L.; Bonini, A.; Paoletti, C.; Lomonaco, T.; Mollica, V.; Fuoco, R.; Di Francesco, F. A graphene oxide pH sensor for wound monitoring. In Proceedings of the 2016 38th Annual International Conference of the IEEE Engineering in Medicine and Biology Society (EMBC), Orlando, FL, USA, 16–20 August 2016; pp. 1898–1901.
66. Rahimi, R.; Ochoa, M.; Tamayol, A.; Khalili, S.; Khademhosseini, A.; Ziaie, B. Highly Stretchable Potentiometric pH Sensor Fabricated via Laser Carbonization and Machining of Carbon−Polyaniline Composite. *ACS Appl. Mater. Interfaces* **2017**, *9*, 9015–9023. [CrossRef]
67. Poma, N.; Vivaldi, F.; Bonini, A.; Carbonaro, N.; Di Rienzo, F.; Melai, B.; Kirchhain, A.; Salvo, P.; Tognetti, A.; Di Francesco, F. Remote monitoring of seawater temperature and pH by low cost sensors. *Microchem. J.* **2019**, *148*, 248–252. [CrossRef]

68. Bonini, A.; Di Francesco, F.; Salvo, P.; Vivaldi, F.; Herrera, E.; Melai, B.; Kirchhain, A.; Poma, N.; Mattonai, M.; Caprioli, R.; et al. A Graphenic Biosensor for Real-Time Monitoring of Urea during Dialysis. *IEEE Sens. J.* **2020**. [CrossRef]
69. Herrera, E.G.; Bonini, A.; Vivaldi, F.; Melai, B.; Salvo, P.; Poma, N.; Santalucia, D.; Kirchhain, A.; Di Francesco, F. A Biosensor for the Detection of Acetylcholine and Diazinon. *Proc. Annu. Int. Conf. IEEE Eng. Med. Biol. Soc. EMBS* **2019**, 1159–1162. [CrossRef]
70. Poma, N.; Vivaldi, F.; Bonini, A.; Salvo, P.; Kirchhain, A.; Melai, B.; Bottai, D.; Tavanti, A.; Di Francesco, F. A graphenic and potentiometric sensor for monitoring the growth of bacterial biofilms. *Sens. Actuators B Chem.* **2020**. [CrossRef]
71. Vivaldi, F.; Bonini, A.; Melai, B.; Poma, N.; Kirchhain, A.; Santalucia, D.; Salvo, P.; Di Francesco, F. A graphene-based pH sensor on paper for human plasma and seawater. In Proceedings of the 2019 41st Annual International Conference of the IEEE Engineering in Medicine and Biology Society (EMBC), Berlin, Germany, 23–27 July 2019; pp. 1563–1566. [CrossRef]
72. Manjakkal, L.; Dang, W.; Yogeswaran, N.; Dahiya, R. Textile-based potentiometric electrochemical PH sensor for wearable applications. *Biosensors* **2019**, *9*, 14. [CrossRef]
73. Marxer, S.M.; Schoenfisch, M.H. Sol-gel derived potentiometric pH sensors. *Anal. Chem.* **2005**, *77*, 848–853. [CrossRef] [PubMed]
74. Zuaznabar-Gardona, J.C.; Fragoso, A. A wide-range solid state potentiometric pH sensor based on poly-dopamine coated carbon nano-onion electrodes. *Sens. Actuators B Chem.* **2018**, *273*, 664–671. [CrossRef]
75. Sulka, G.D.; Hnida, K.; Agnieszka Brzózka, A. PH sensors based on polypyrrole nanowire arrays. *Electrochim. Acta* **2013**, *104*, 536–541. [CrossRef]
76. Yoon, J.H.; Kim, S.M.; Park, H.J.; Kim, Y.K.; Oh, D.X.; Cho, H.W.; Lee, K.G.; Hwang, S.Y.; Park, J.; Choi, B.G. Highly self-healable and flexible cable-type pH sensors for real-time monitoring of human fluids. *Biosens. Bioelectron.* **2020**, *150*, 111946. [CrossRef] [PubMed]
77. Stred'Anský, M.; Pizzariello, A.; Stred'Anská, S.; Miertuš, S. Amperometric pH-sensing biosensors for urea, penicillin, and oxalacetate. *Anal. Chim. Acta* **2000**, *415*, 151–157. [CrossRef]
78. Wildgoose, G.G.; Pandurangappa, M.; Lawrence, N.S.; Jiang, L.; Jones, T.G.J.; Compton, R.G. Anthraquinone-derivatised carbon powder: Reagentless voltammetric pH electrodes. *Talanta* **2003**, *60*, 887–893. [CrossRef]
79. Makos, M.A.; Omiatek, D.M.; Ewing, A.G.; Heien, M.L. Development and characterization of a voltammetric carbon-fiber microelectrode pH sensor. *Langmuir* **2010**, *26*, 10386–10391. [CrossRef] [PubMed]
80. Amiri, M.; Amali, E.; Nematollahzadeh, A.; Salehniya, H. Poly-dopamine films: Voltammetric sensor for pH monitoring. *Sens. Actuators B Chem.* **2016**, *228*, 53–58. [CrossRef]
81. Chaisiwamongkhol, K.; Batchelor-Mcauley, C.; Compton, R.G. Amperometric micro pH measurements in oxygenated saliva. *Analyst* **2017**, *142*, 2828–2835. [CrossRef]
82. Vivaldi, F.; Santalucia, D.; Poma, N.; Bonini, A.; Salvo, P.; Del Noce, L.; Melai, B.; Kirchhain, A.; Kolivoška, V.; Sokolová, R.; et al. A voltammetric pH sensor for food and biological matrices. *Sens. Actuators B Chem.* **2020**. [CrossRef]
83. Chaisiwamongkhol, K.; Batchelor-Mcauley, C.; Compton, R.G. Optimising amperometric pH sensing in blood samples: An iridium oxide electrode for blood pH sensing. *Analyst* **2019**, *144*, 1386–1393. [CrossRef]
84. Tham, G.X.; Fisher, A.C.; Webster, R.D. A vitamin-based voltammetric pH sensor that functions in buffered and unbuffered media. *Sens. Actuators B Chem.* **2019**, *283*, 495–503. [CrossRef]
85. Zhu, H.; Hassan, T.; Kabir, H.; May, J.; Hamal, K.; Lopez, R.; Smith, H.J.; Nicholas, N.W.; Sankaran, P.; McIlroy, D.N.; et al. Voltammetric pH sensor based on electrochemically modified pseudo-graphite. *Analyst* **2020**, *145*, 7252–7259. [CrossRef] [PubMed]
86. Hu, G.; Li, N.; Zhang, Y.; Li, H. A novel pH sensor with application to milk based on electrochemical oxidative quinone-functionalization of tryptophan residues. *J. Electroanal. Chem.* **2020**, *859*, 113871. [CrossRef]
87. Gao, W.; Song, J. Polyaniline film based amperometric pH sensor using a novel electrochemical measurement system. *Electroanalysis* **2009**, *21*, 973–978. [CrossRef]
88. Sha, R.; Komori, K.; Badhulika, S. Amperometric pH Sensor Based on Graphene-Polyaniline Composite. *IEEE Sens. J.* **2017**, *17*, 5038–5043. [CrossRef]
89. Bergveld, P. Development of an Ion-Sensitive Solid-State Device for Neurophysiological Measurements. *IEEE Trans. Biomed. Eng.* **1970**, *BME-17*, 70–71. [CrossRef]
90. Chin, Y.L.; Chou, J.C.; Sun, T.P.; Liao, H.K.; Chung, W.Y.; Hsiung, S.K. A novel SnO2/Al discrete gate ISFET pH sensor with CMOS standard process. *Sens. Actuators B Chem.* **2001**, *75*, 36–42. [CrossRef]
91. Shitashima, K.; Kyo, M.; Koike, Y.; Henmi, H. Development of in situ pH sensor using ISFET Environmental Science Department. In Proceedings of the 2002 Interntional Symposium on Underwater Technology (Cat. No.02EX556), Tokyo, Japan, 19 April 2002; pp. 106–108. [CrossRef]
92. Fukuba, T.; Tamai, Y.; Kyo, M.; Shitashima, K.; Koike, Y.; Fujii, T. Development and field evaluation of ISFET pH sensor integrated with self-calibration device for deep-sea oceanography applications. In Proceedings of the 12th International Conference on Miniaturized Systems for Life Sciences, San Diego, CA, USA, 12–16 October 2008; pp. 1983–1985.
93. Rani, R.A.; Syono, M.I.; Ramli, A.S. Multi-finger gate ISFET (Mf-ISFET) for pH sensor application. In Proceedings of the 2008 IEEE International Conference on Semiconductor Electronics, Johor Bahru, Malaysia, 25–27 November 2008; pp. 350–353. [CrossRef]
94. Parizi, K.B.; Yeh, A.J.; Poon, A.S.Y.; Wong, H.S.P. Exceeding Nernst limit (59mV/pH): CMOS-based pH sensor for autonomous applications. In Proceedings of the 2012 International Electron Devices Meeting, San Francisco, CA, USA, 10–13 December 2012. [CrossRef]

95. Das, A.; Ko, D.H.; Chen, C.H.; Chang, L.B.; Lai, C.S.; Chu, F.C.; Chow, L.; Lin, R.M. Highly sensitive palladium oxide thin film extended gate FETs as pH sensor. *Sens. Actuators B Chem.* **2014**, *205*, 199–205. [CrossRef]
96. Cho, W.J.; Lim, C.M. Sensing properties of separative paper-based extended-gate ion-sensitive field-effect transistor for cost effective pH sensor applications. *Solid. State. Electron.* **2018**, *140*, 96–99. [CrossRef]
97. Kang, T.; Lee, I.; Oh, S.; Jang, T.; Kim, Y.; Ahn, H.; Kim, G.; Shin, S.U.; Jeong, S.; Sylvester, D.; et al. A 1.74.12 mm^3 Fully Integrated pH Sensor for Implantable Applications using Differential Sensing and Drift-Compensation. In Proceedings of the 2019 Symposium on VLSI Circuits, Kyoto, Japan, 9–14 June 2019. [CrossRef]
98. Sharma, P.; Gupta, S.; Singh, R.; Ray, K.; Kothari, S.L.; Sinha, S.; Sharma, R.; Mukhiya, R.; Awasthi, K.; Kumar, M. Hydrogen ion sensing characteristics of Na_3BiO_4–Bi_2O_3 mixed oxide nanostructures based EGFET pH sensor. *Int. J. Hydrogen Energy* **2020**, *45*, 18743–18751. [CrossRef]
99. Jung, S.H.; Seo, Y.M.; Gu, T.; Jang, W.; Kang, S.G.; Hyeon, Y.; Hyun, S.H.; Lee, J.H.; Whang, D. Super-Nernstian pH Sensor Based on Anomalous Charge Transfer Doping of Defect-Engineered Graphene. *Nano Lett.* **2020**, *21*, 34–42. [CrossRef] [PubMed]

chemosensors

MDPI

Review

Photoacoustic Imaging of pH-Sensitive Optical Sensors in Biological Tissues

Hyunjun Kye [1,†], Dongyoung Jo [1,†], Sanghwa Jeong [2], Chulhong Kim [3] and Jeesu Kim [1,*]

[1] Departments of Cogno-Mechatronics Engineering and Optics & Mechatronics Engineering, Pusan National University, Busan 46241, Republic of Korea; hyunjuijui@pusan.ac.kr (H.K.); jdy9321@pusan.ac.kr (D.J.)
[2] School of Biomedical Convergence Engineering, Pusan National University, Yangsan 50612, Republic of Korea; sanghwa.jeong@pusan.ac.kr
[3] Departments of Electrical Engineering, Convergence IT Engineering, Medical Science and Engineering, Mechanical Engineering, and Medical Device Innovation Center, Pohang University of Science and Technology (POSTECH), Pohang 37673, Republic of Korea; chulhong@postech.edu
* Correspondence: jeesukim@pusan.ac.kr
† These authors contributed equally to this work.

Abstract: Photoacoustic imaging is an emerging biomedical imaging technique that enables non-invasive visualization of the optical absorption properties of biological tissues in vivo. Although numerous studies have used contrast agents to achieve high-contrast imaging in deep tissues, targeting specific areas remains a challenge when using agents that are continuously activated. Recent research has focused on developing triggered contrast agents that are selectively activated in target areas. This review delves into the use of pH-triggered contrast agents in photoacoustic imaging, which take advantage of the lower pH of the tumor microenvironment compared to normal tissues. The paper discusses the mechanisms of pH-triggered contrast agents that contribute to improving depth and contrast in photoacoustic tumor imaging. In addition, the integration of functionalities, such as photothermal therapy and drug delivery monitoring, into these agents demonstrates significant potential for biomedical applications.

Keywords: photoacoustic imaging; pH-sensitive sensor; nanomaterials; tumor detection

Citation: Kye, H.; Jo, D.; Jeong, S.; Kim, C.; Kim, J. Photoacoustic Imaging of pH-Sensitive Optical Sensors in Biological Tissues. *Chemosensors* **2024**, *12*, 257. https://doi.org/10.3390/chemosensors12120257

Received: 28 October 2024
Revised: 4 December 2024
Accepted: 7 December 2024
Published: 9 December 2024

1. Introduction

Photoacoustic imaging (PAI) is an advanced, non-invasive, and non-ionizing technique that uses the photoacoustic (PA) effect, wherein ultrasound (US) waves are generated through light absorption and subsequent heat release [1]. One of the most significant advantages of PAI is its molecular functional imaging capability, which is achieved by analyzing multispectral responses of biological tissues in vivo [2]. Recently, PAI has been used extensively in biomedical imaging, particularly focusing on cancer diagnosis [3–6], theragnosis [7–10], and drug delivery tracking [11].

PAI is increasingly being applied in biomedical studies owing to its compact and cost-effective implementation compared to conventional biomedical imaging modalities such as X-ray computed tomography and magnetic resonance imaging. Its non-ionizing characteristics ensure a safer and more straightforward experimental setup. Compared with other non-ionizing imaging techniques, PAI has several distinct advantages. While US imaging is limited to visualizing structural anatomy using acoustic impedance differences, PAI provides functional information by detecting variations in optical absorption within tissues. Similarly, although optical imaging techniques such as fluorescence imaging and two-photon microscopy offer molecular functional analysis, they are often constrained by shallow imaging depths owing to significant optical scattering in tissues. In contrast, PAI achieves signals through acoustic wave propagation, which has much less scattering, thus providing a much deeper imaging depth than pure optical imaging methods.

PAI can also obtain label-free images from endogenous chromophores by selecting the appropriate laser wavelengths. For example, oxyhemoglobin and deoxyhemoglobin serve as primary contrast agents in vascular imaging because of their strong absorption in the visible to near-infrared (NIR) range. Multi-wavelength analysis of these hemoglobins enables the assessment of oxygen saturation (sO_2) levels in biological tissues, which is important for numerous applications [12–15]. Melanin, another intrinsic contrast agent, is commonly used to image melanomas because of its broad absorption spectrum, ranging from ultraviolet to NIR [16–19]. Lipids can also be used as endogenous markers because of their remarkable absorption capacity in the NIR region [20–22].

Despite the ability of PAI to perform label-free imaging, the use of exogenous contrast agents is widespread, primarily to enhance signal-to-noise ratios in biological tissues [23,24]. Contrast-enhanced PAI enables the visualization of biological structures with low optical absorption, such as lymphatic networks [25–29], the liver [30], and the bladder [31], which are difficult to image with label-free PAI. In addition, various contrast agents have been used for deep-tissue imaging, particularly for longer wavelengths in the NIR region [32–34].

Contrast-enhanced PAI is critical for the diagnosis and monitoring of diseases with targeted agents [35–37]. For this purpose, contrast agents activated under specific conditions have been developed to improve target detection efficiency [38,39]. For instance, Chen et al. demonstrated that heat-sensitive capsules released internal agents upon laser-induced heating, thereby generating PA signals [40]. Lin et al. synthesized a highly volatile agent that produced amplified PA signals when vaporized under laser irradiation [41].

In the context of tumor detection, contrast agents that selectively alter their optical absorption characteristics in response to the unique properties of the tumor microenvironment have attracted significant interest. The tumor microenvironment, which is characterized by increased cellular proliferation and metabolic activity, has a significant impact on tumor initiation and progression. This review focuses on pH levels within the tumor microenvironment, which are typically lower (pH 6.5–6.8) than those in normal tissues (approximately pH 7.4) [42,43]. By using this acidic tumor environment to induce changes in their absorption spectra, pH-sensitive contrast agents offer a non-invasive and highly selective approach to tumor detection without additional external stimulation.

Herein, we provide an overview of contrast agents designed to achieve high contrast in PAI, particularly for tumor identification. We focus on the nanoparticles that remain "inactivated" in normal tissue but become "activated" in the acidic environment by altering their optical absorption characteristics in response to pH variations. Mechanisms such as nanoparticle aggregation or morphological changes in acidic conditions are discussed, as well as the release of light-absorbing chromophores caused by the destruction of pH-sensitive capsules. This review highlights the potential of PAI for detecting tumor regions and monitoring treatment progress using pH-triggered contrast agents, thereby expanding its application in biomedical research.

2. Principles of Photoacoustic Imaging

PAI operates based on the fundamental principle of energy transduction from light to acoustic waves through the PA effect (Figure 1) [1]. When short pulses of light illuminate a target tissue, chromophores within it absorb energy. This absorption excites the electrons in the chromophores from their ground state; as they return to their original state, a portion of the absorbed energy is released as heat. This rapid heating induces thermoelastic expansion in the tissue, owing to the extremely short duration of the light pulses (typically less than 10 ns). After a brief period of illumination, the expanded tissue rapidly contracts back to its original volume, generating wideband acoustic waves that propagate through the tissue. The initial pressure (P) of the generated acoustic wave, known as the PA wave, is directly proportional to the following four parameters:

$$P \propto \Gamma(T) \times \sigma \times \mu_a \times F, \quad (1)$$

where Γ denotes the Grüneisen parameter that is related to the local temperature (T), σ denotes the thermal conversion efficiency, μ_a denotes the optical absorption coefficient, and F denotes the fluence of light reaching the local region. The pH-sensitive agents discussed in this paper modify the thermal conversion efficiency or optical absorption in tumor areas, where the pH levels are lower than those in normal tissues. Because these parameters are directly proportional to the initial pressure of the PA waves, contrast-enhanced PA images can be obtained when these agents interact with the tumor microenvironment.

In addition, pH-sensitive agents alter their optical absorption peaks, and their distribution can be effectively detected using multi-wavelength PA responses. An efficient technique for this purpose is ratiometric PA analysis, which calculates the ratio of PA signals at different wavelengths [44]. This method provides a robust and calibration-free measurement, delivering quantitative information about the agents. Although ratiometric PA analysis does not quantify the molecular concentration of agents, as can be achieved through spectral unmixing techniques [45], it provides the straightforward and effective detection of target agents with minimal computational complexity.

For deep tissue imaging, PA agents are often designed to absorb light in the second NIR (NIR-II: 1000–1700 nm) region, where photon scattering is significantly reduced. In addition, the increased fluence of optical illumination, which can generate stronger PA signals, is available due to a higher maximum permissible exposure of light compared to the other wavelength regions [46]. Consequently, optical probes absorbing in the NIR-II region have been demonstrated to enhance imaging depth [47].

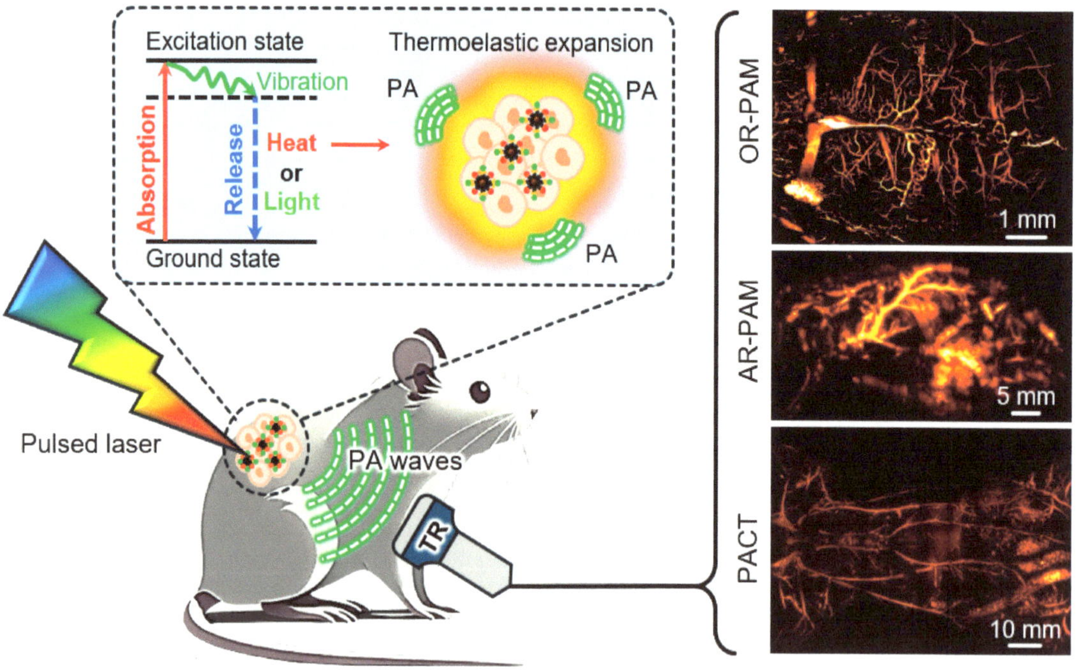

Figure 1. Schematic illustration for the principles of photoacoustic imaging. PA, photoacoustic; TR, ultrasound transducer; OR-PAM, optical resolution photoacoustic microscopy; AR-PAM, acoustic resolution photoacoustic microscopy; PACT, photoacoustic computed tomography. The images are reproduced with permission from [48–50].

One unique characteristic of PAI compared with other biomedical imaging techniques is its ability to provide multiscale resolution and imaging depth. The generated PA waves have a wide frequency range of several to tens of MHz, allowing for an adjustable resolution and imaging depth based on the selected frequency range of the transducer [51]. In general, high-frequency transducers are used for high-resolution imaging, while low-frequency transducers are used for deeper imaging.

Beyond the transducer selection, the resolution of the PAI system can be further tuned through the configuration of the optical and acoustic foci. PAI systems are classified into three types: optical-resolution photoacoustic microscopy (OR-PAM), acoustic-resolution photoacoustic microscopy (AR-PAM), and photoacoustic computed tomography (PACT). OR-PAM achieves high resolution by focusing light on a spot smaller than the acoustic focus of the transducer [52]. This method enables a very high lateral resolution, particularly when using an objective lens with a large numerical aperture, although it is limited by a shallow imaging depth of approximately 1 mm from the surface. AR-PAM operates by illuminating a broad area with light and detecting PA signals within the focal zone of the transducer. Although AR-PAM offers a lower lateral resolution than OR-PAM, it allows greater imaging depth owing to its extended depth of focus. In contrast to AR-PAM and OR-PAM, PACT uses multi-element array transducers, enabling the acquisition of multiple data points using a single laser pulse [53]. Although this method requires complex mathematical calculations for image reconstruction, it provides a significantly greater imaging depth and can be performed in real time. These configurations of PAI systems have been used effectively in various biomedical applications to visualize pH-sensitive agents.

3. Mechanisms of pH-Sensitive Sensors for Photoacoustic Imaging

We categorized and summarized the pH-sensitive sensors based on their mechanisms and responsiveness to pH changes, particularly in mildly acidic tumor microenvironments (Figure 2). These sensors typically undergo dynamic protonation, which is driven by their chemical structures. Consequently, they may experience structural changes such as aggregation, hydrolysis, or the formation of zwitterionic surfaces [54,55]. These changes often cause shifts in the absorption spectrum or the release of active agents, altering the PA amplitude at specific excitation wavelengths. This enables selective imaging of tumors with enhanced contrast. This section focuses on the mechanisms that trigger PA signal generation in response to pH changes, which can be categorized into three phenomena: (1) aggregation driven by interparticle attraction, (2) separation involving the release of encapsulated agents, and (3) protonation-induced changes within the agents.

Aggregation occurs when attractive interparticle forces are activated, resulting in the formation of larger clusters [56]. Under acidic conditions, pH-sensitive sensors often exhibit either positive or negative surface charges owing to the hydrolysis of pH-sensitive side chains or interactions between mixed side chains. This creates an electrostatic attraction between oppositely charged sensors, causing them to aggregate and increase in size. Such aggregation shifts the absorption spectrum to longer wavelengths, thereby enhancing the PA signals at specific excitation wavelengths. Nanoparticle-based pH-sensitive sensors are particularly effective at accumulating in tumor tissues, facilitating both imaging and phototherapy [57].

The separation mechanism involves the release of substances that generate or enhance PA signals [58]. Under acidic conditions, structural changes in pH-sensitive materials result in the release of ions and molecules. These substances can trigger reactions such as gas formation that amplify the PA signal. In some cases, an altered pH environment lowers the stability or melting point of micelle-like structures, releasing encapsulated agents with strong optical absorption and further enhancing the PA signals. This release process can also be induced by external stimulation, enabling the controlled delivery of signal-enhancing compounds [59].

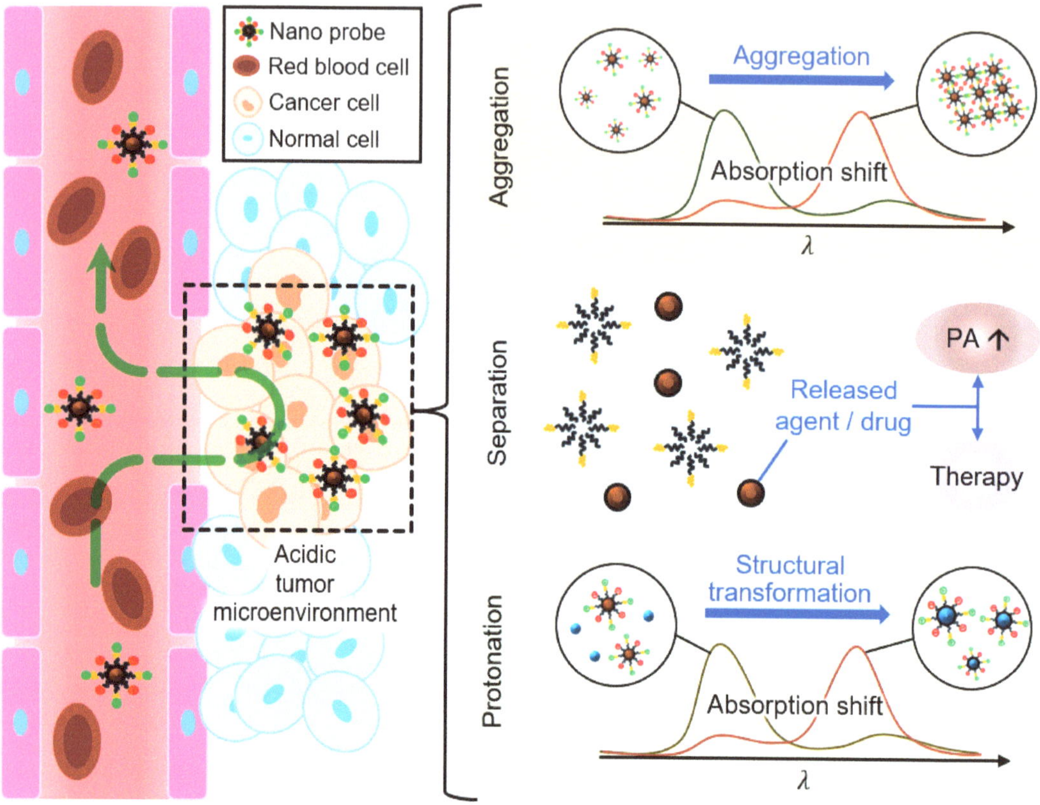

Figure 2. Schematic illustration for the PA signal-enhancing mechanisms of the pH-sensitive sensors in the tumor microenvironment. PA, photoacoustic.

Protonation refers to the intrinsic changes in absorption that occur because of structural transformations [60]. In acidic environments, some pH-sensitive sensors undergo protonation and electron transfer within their heterocyclic structures, resulting in a spectral shift in the absorption peak. For example, in polymers with functional groups such as carboxyl and imine, protonation can convert the base form to the salt form [61]. This conversion contributes to a redshift in the absorption spectrum, further enhancing the PA response.

4. Contrast-Enhanced Photoacoustic Imaging of pH-Sensitive Sensors

4.1. Aggregation

Gold nanoparticles (AuNPs) have been extensively studied for biological and medical applications because of their well-established synthesis protocols. The size- and shape-dependent optical properties of AuNPs allow for the alteration of PA signals upon aggregation. Song et al. synthesized pH-sensitive AuNPs by conjugating ligands composed of dithiol and citraconic amide [62]. In mildly acidic environments, citraconic amide is hydrolyzed, converting the carboxylic acid into a primary amine, which promotes AuNP aggregation owing to electrostatic attraction (Figure 3a). While the AuNPs exhibited an absorption peak at 520 nm at pH 7.4, a red-shifted optical absorption was observed at pH 5.5 (Figure 3b). The potential for tumor-specific detection using these conjugated AuNPs was validated using the PAI of the phantoms (Figure 3c). HeLa breast cancer cells and NIH 3T3 fibroblasts were co-incubated with AuNP conjugates and fixed in agar solutions for

PAI. The results showed that the PA signal of cancerous HeLa cells was 1.7 times stronger than that of normal NIH 3T3 cells, indicating the aggregation of AuNPs in the acidic tumor microenvironment.

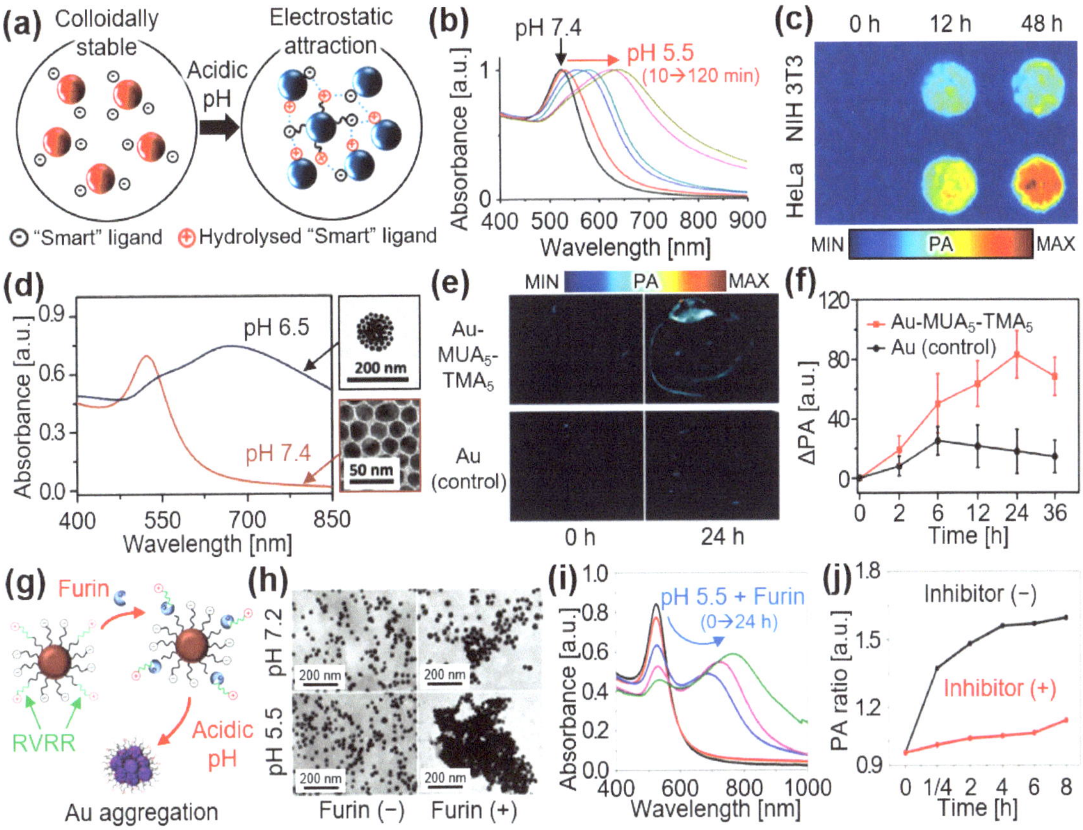

Figure 3. Representative pH-sensitive sensors that aggregate in an acidic tumor microenvironment. (**a**) Schematic illustration of a pH-sensitive AuNP that aggregates in acidic pH by electrostatic attraction. (**b**) Optical absorption spectra of AuNP conjugates along exposure time at pH 5.5. (**c**) PA images of AuNP conjugates, which were co-incubated with HeLa breast cancer cells and NIH 3T3 fibroblast cells. (**d**) Optical absorption of AuNP-based MUA-TMA conjugates at pH 6.5 (black) and 7.4 (red). Insets are TEM images of AuNP conjugates. (**e**) PA images of AuNP conjugates after injection into tumors. (**f**) PA signal enhancement in tumors with AuNP conjugates and control AuNPs. (**g**) Schematic illustration of AuNPs that synergistically aggregate by the furin enzyme and the acidic conditions. (**h**) TEM images of AuNP conjugates with and without the furin enzyme. (**i**) Optical absorption spectra of AuNP conjugates after triggering furin and acidic conditions. (**j**) PA ratio at tumor after injection of AuNP conjugates. PA, photoacoustic; TEM, transmission electron microscopy; AuNP, gold nanoparticle; MUA, mercaptoundecanoic acid; TMA, (10-mercaptodecyl)trimethylammonium bromide; Au-MUA_5-TMA_5, Au conjugates with 5:5 ratio of MUA to TMA; RVRR, Arg-Val-Arg-Arg peptide. The images are reproduced with permission from [62–64].

Zhang et al. [63] developed pH-sensitive agents by conjugating AuNPs with mercaptoundecanoic acid (MUA) and (10-mercaptodecyl)trimethylammonium bromide (TMA). The conjugates showed pH-dependent aggregation, with sensitivities varying based on the ratio of MUA to TMA. At a 5:5 ratio, sharp changes in the absorption spectrum were observed as the pH decreased from 6.7 to 5.0, with a noticeable shift in the absorption peak from 524 nm to the NIR region (approximately 650 to 700 nm) (Figure 3d). At pH 7.4, the conjugates remained approximately 15 nm in size but aggregated to approximately 180 nm at pH 6.5. At an excitation wavelength of 808 nm, contrast-enhanced PA images were obtained from subcutaneously transplanted U87MG glioblastoma tumors in mice following tail vein injection of conjugates (Figure 3e). The results showed an approximately 4-fold greater signal enhancement in tumors treated with the conjugates than in the control group (Figure 3f). In addition, after 10 min of 808 nm laser illumination at a fluence of 1.0 W/cm^2, the tumor was successfully ablated, demonstrating the potential of the developed conjugates for potential photothermal therapy (PTT).

Cheng et al. developed Arg-Val-Arg-Arg (RVRR) peptide-conjugated AuNPs that are selectively aggregated under the synergistic influence of the furin enzyme and acidic conditions of the tumor microenvironment [64]. The furin enzyme efficiently cleaved the RVRR peptides from the conjugates, promoting AuNP aggregation in the mildly acidic tumor environment (Figure 3g). In the presence of furin, the conjugates maintained their original size of approximately 63 nm at pH 7.2; however, they aggregated to approximately 327 nm after 24 h at pH 5.5 (Figure 3h). This aggregation caused a shift in the absorption peak from 520 nm to approximately 800 nm, enabling detection by multispectral PAI (Figure 3i). The potential of these conjugates was further validated by measuring the temperature increase in HCT-116 colon tumor-bearing mice in vivo. HCT-116 tumors were subcutaneously xenografted onto the dorsal region of mice, and pH-sensitive conjugates were administered via tail vein injection. After 8 h, the PA signal showed a 1.6-fold increase compared to that of the control group (Figure 3j), demonstrating the signal-switching ability of the conjugates through aggregation triggered by the combined effects of the furin enzyme and the acidic tumor microenvironment. In addition, the local tumor temperature increased by approximately 26.9 °C after 6 min of 808 nm laser illumination with a fluence of 0.75 W/cm^2, confirming the potential for PTT.

Li et al. designed surface-modified AuNPs to improve tumor detection accuracy [65]. In the tumor microenvironment, the carboxylate anions of the AuNPs are protonated into amines and subsequently hydrolyzed, leading to nanoparticle aggregation (Figure 4a). This aggregation intensified at lower pH levels, resulting in a redshift in the absorption wavelengths (Figure 4b). At pH 7.4, the AuNPs maintained a size of approximately 16 nm with an absorption peak between 550 and 600 nm. However, at pH 5.8, the particle size increased to over 500 nm, and the absorption shifted above 600 nm, resulting in a 3.6-fold increase in PA amplitude at a wavelength of 680 nm (Figure 4c). In a subsequent study, they demonstrated a pH-sensitive nanoparticle that showed a clear absorption shift in a mildly acidic tumor microenvironment [66]. They synthesized poly(ethylene glycol)-poly(lactic acid/glycolic acid) copolymeric nanoparticles encapsulating croconazole dyes. As the pH decreased, the absorption of nanoparticles at 815 nm increased. The PA signals also increased, showing a 4-fold increase in the signal at pH 5.8 compared to that at pH 7.4. After tail vein injection of the nanoparticles into MDA-MB-231 breast cancer xenografted mice, in vivo PA images showed a 2.8-fold signal increase compared with the control group. Moreover, when the accumulated nanoparticles in the tumor were irradiated with an 808 nm laser at a fluence of 1.5 W/cm^2, the local tumor temperature rose to approximately 49 °C, leading to significant tumor volume reduction, indicating the potential of this nanoparticle for PTT.

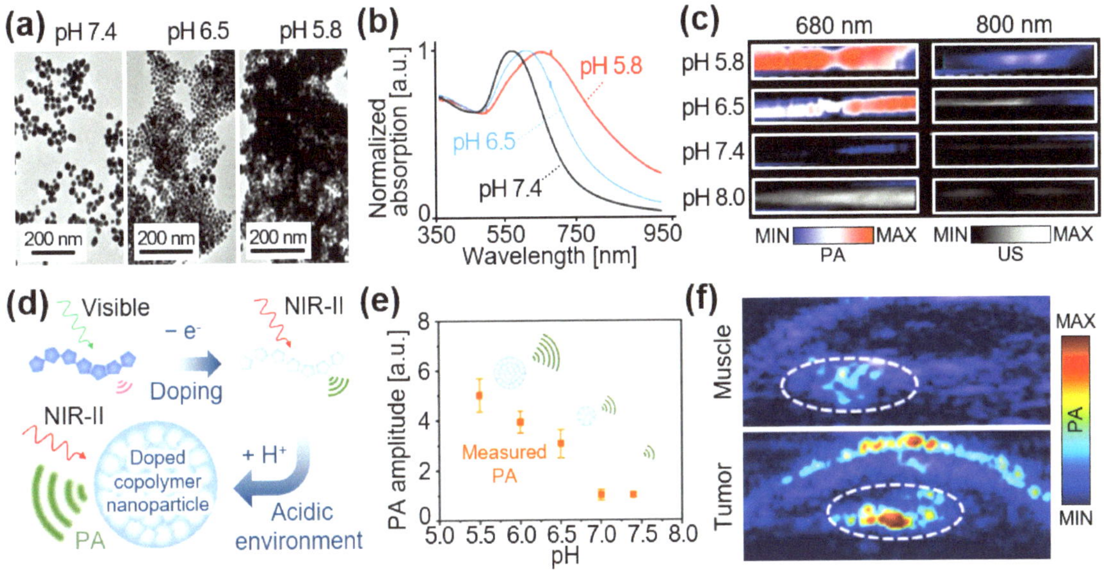

Figure 4. PA signal-enhancing nanoparticles that aggregate in response to low pH levels. (**a**) TEM images of the surface-modified gold nanoparticles at different pH levels of phosphate buffer solutions. (**b**) Normalized optical absorption spectra of the nanoparticles after 1 h incubation. (**c**) PA images of the nanoparticles contained in polyethylene tubes. (**d**) Schematic illustration for acid-induced aggregation mechanism of copolymer nanoparticle that absorbs light in the NIR-II region. (**e**) PA amplitudes of the doped copolymer nanoparticle at different pH levels. (**f**) Contrast-enhanced PA images of mice after local injection of the doped copolymer nanoparticle into muscle and tumor area. Dashed circles indicate enhanced PA signals from the doped copolymer nanoparticle. PA, photoacoustic; US, ultrasound; TEM, transmission electron microscopy; NIR-II, second near-infrared. The images are reproduced with permission from [65,67].

Wu et al. demonstrated another copolymer nanoparticle that absorbs light in the second NIR (NIR-II: 1000–1700 nm) region, which is particularly promising for deep-tissue PAI owing to reduced photon scattering in this range [67]. The nanoparticles aggregated upon oxygen doping, causing a redshift in the absorption peak from the visible (~600 nm) to the NIR-II (~1000 nm) region (Figure 4d). At an excitation wavelength of 1064 nm, PA signals were significantly enhanced, showing 3.1-fold and 5.0-fold increases at pH 6.5 and 5.5, respectively, compared to pH 7.4 (Figure 4e). To confirm nanoparticle aggregation in the tumor microenvironment in vivo, the nanoparticles were intratumorally injected into subcutaneously implanted PC-3 prostate tumors in mice. The PA signals from the tumor were compared with those from muscles in which the nanoparticles were intramuscularly injected (Figure 4f). The results showed a 3.4-fold higher PA signal in the tumor, indicating successful nanoparticle aggregation in an acidic tumor environment.

In this section, the aggregation mechanism of the pH-sensitive nanoparticles is explored (Table 1). These nanoparticles exhibited red shifts in their absorption spectra when aggregated in mildly acidic tumor microenvironments, resulting in enhanced PA signals in the NIR region. Various strategies, such as ligand conjugation, surface modification, and copolymer encapsulation, have been employed to design nanoparticles that aggregate under acidic conditions, leading to improved tumor detection and potential for PTT. Further advancements in nanoparticle design could focus on enhancing both the sensitivity and specificity of aggregation mechanisms, enabling more precise and effective tumor imaging and therapy while ensuring biocompatibility and minimizing off-target effects in clinical applications.

Table 1. Summary of pH-sensitive PA agents that exhibit absorption shift by aggregation in tumor acidic microenvironment. PA, photoacoustic; λ_a, peak absorption wavelength; λ_{ex}, excitation wavelength for PA imaging; AR-PAM, acoustic resolution PA microscopy; PACT, PA computed tomography; MUA, mercaptoundecanoic acid; TMA, trimethylammonium bromide; RVRR, Arg-Val-Arg-Arg; PPE, polyphenylene Ether.

Base	pH-Sensitive Material	Size [nm] (pH)		λ_a [nm] (pH)		PA Imaging			Ref.
		Normal	Tumor	Normal	Tumor	Configuration	λ_{ex} [nm]	Application	
Au	Citraconic amide	10 (7.4)	-	520 (7.4)	-	AR-PAM	680	HeLa breast tumor	[62]
Au	MUA-TMA	15 (7.4)	180 (6.5)	524 (7.4)	650–700 (6.5)	PACT	808	U87MG glioblastoma	[63]
Au	RVRR	63 (7.2)	327 (5.5)	520 (7.2)	700–900 (5.5)	PACT	750	HCT-116 colon tumor	[64]
Au	Citraconic amide	16.4 (7.4)	≥500 (5.8)	550–600 (7.4)	650 (5.8)	PACT	680, 800	U87MG glioblastoma	[65]
Croconaine	Croconaine	185 (7.4)	190 (6.5)	630 (7.4)	815 (6.5)	PACT	770	MDA-MB-231 breast tumor	[66]
PPE	Doped PPE	-	4000 (6)	1150 (7.4)	1100 (6)	PACT	1064	PC-3 prostate tumor	[67]

4.2. Separation

In acidic tumor environments, the protonation of pH-sensitive sensors causes the disassembly of micelle structures or surface-bound functional groups, resulting in the release of encapsulated agents. This process induces or enhances PA signaling in the tumor region. In addition, this mechanism facilitates drug delivery to the tumor site. Zhong et al. demonstrated pH-responsive nanoparticles designed to release encapsulated inhibitors of heat shock proteins at acidic pH levels, thereby reducing tumor thermoresistance and enhancing PTT efficiency [68]. They synthesized self-assembled silver sulfide (Ag_2S) nanoparticles loaded with the inhibitor quercetin (QE) and a pH-responsive polymer (Poly(ethylene glycol)$_{5k}$-poly(β-aminoesters)$_{10k}$), denoted as QE-PEG-Ag_2S (Figure 5a). The release of QE exhibited significantly higher efficiency at lower pH levels, with release rates of 50% and 72% at pH 6.5 and 5.5, respectively (Figure 5b). To validate the PA signal enhancement in tumors, QE-PEG-Ag_2S nanoparticles were intravenously injected into HepG2 liver tumor-bearing mice through the tail vein. The tumor accumulation of nanoparticles was monitored by acquiring cross-sectional PA and US images at various time points after injection (Figure 5c). The PA amplitude in the tumor region reached a maximum value 24 h after injection, showing a 3.8-fold increase compared to the pre-injection values. Furthermore, the enhanced PTT efficiency was evaluated by monitoring the tumor size after the injection of different nanoparticles. These results confirm that the inhibitor-loaded nanoparticles efficiently ablated tumors without recurrence, demonstrating their potential for cancer treatment (Figure 5d).

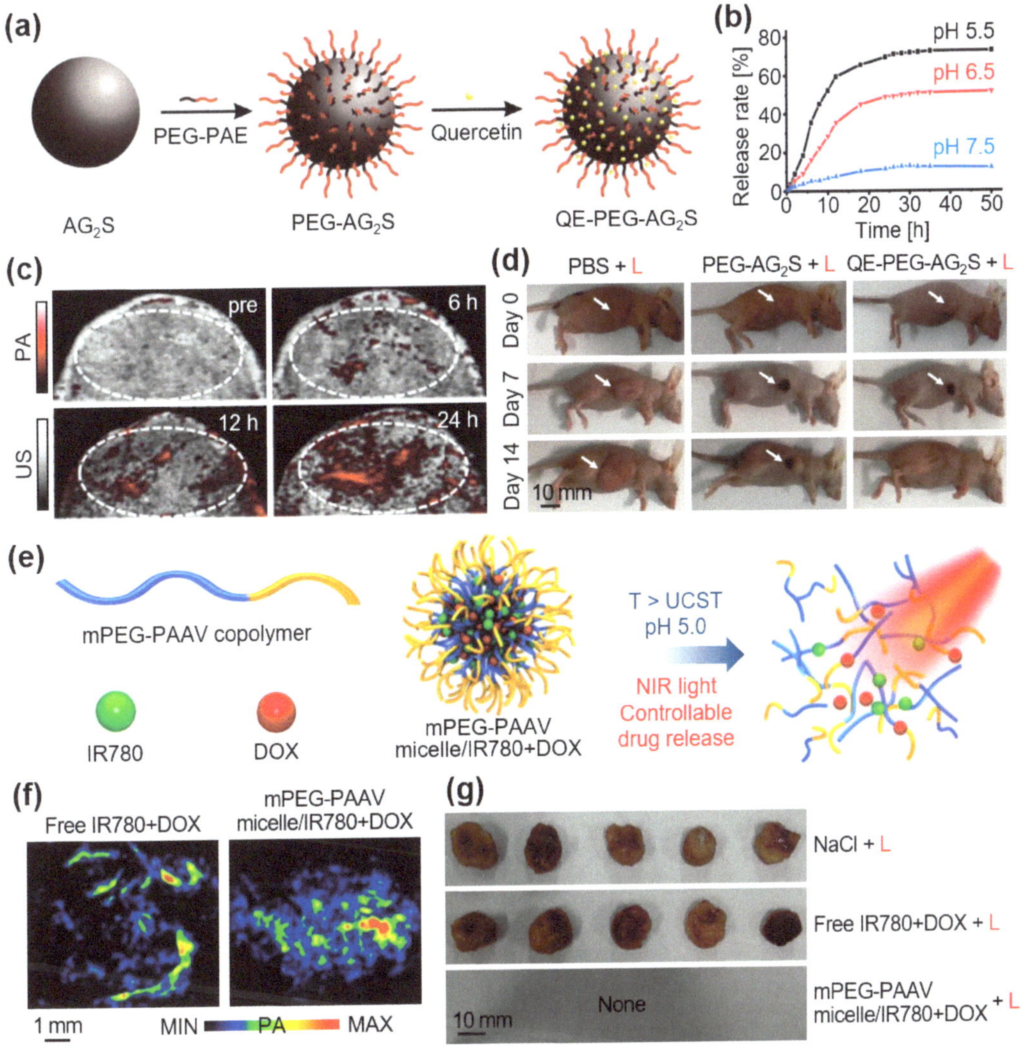

Figure 5. Drug-release nanoparticles triggered by an acidic tumor microenvironment. (**a**) Schematic illustration for the synthesis of pH-responsive nanoparticles designed to release encapsulated inhibitors of heat shock proteins. (**b**) Quercetin release ratio of QE-PEG-Ag_2S nanoparticles at different pH levels. (**c**) Overlaid PA and US images of the tumor region before and after injection of QE-PEG-Ag_2S nanoparticles. White dashed circles indicate the tumor region. (**d**) Photographs of HepG2 liver tumor-bearing mice after different treatment conditions. An 808 nm laser with a fluence of 1.5 W/cm^2 was irradiated for 10 min. (**e**) Schematic illustration for the synthesis of the pH-responsive micellar system that responds to both pH levels and external heat generated by NIR laser irradiation. (**f**) PA images of tumor 24 h after injection of different drug formulations. (**g**) Photographs of excised 4T1 breast tumors at the end point of treatment. Phototherapy was performed with an 808 nm laser at 1.0 W/cm^2 for 5 min. PA, photoacoustic; US, ultrasound; AG_2S, silver sulfide; NaCl, sodium chloride; PEG-PAE, poly(ethylene glycol)$_{5k}$-poly(β-aminoesters)$_{10k}$; mPEG-PAAV, amphiphilic poly(ethylene glycol)-*b*-poly(acrylamide-*co*-acrylonitrile-*co*-vinylimidazole); DOX, doxorubicin; T, temperature; L, laser illumination; NIR, near-infrared; UCST, upper critical solution temperature. The images are reproduced with permission from [68,69].

To improve the controllability of drug release, researchers have explored nanoparticles that can be activated by both external triggers and the internal pH environment. Yang et al. reported a micellar system that responded to both pH levels and external heat generated by NIR laser irradiation [69]. They synthesized an amphiphilic poly(ethylene glycol)-*b*-poly(acrylamide-*co*-acrylonitrile-*co*-vinylimidazole) copolymer (mPEG-PAAV) with an upper critical solution temperature (UCST) that decreased under acidic conditions. The mPEG-PAAV copolymer was used to form a micellar system, encapsulating both doxorubicin (DOX) and IR780 dye (Figure 5e). Upon tail vein injection into mice bearing 4T1 breast tumors, the micellar system demonstrated a 1.5-fold enhancement in the PA signal compared to free IR780 dye, indicating efficient accumulation at the tumor site (Figure 5f). In addition to enhancing the PA contrast, the IR780 dye converts the absorbed light energy into heat. Once the temperature exceeds the UCST, the micelles disassemble, resulting in the rapid release of DOX. After 5 min illumination of an 808 nm laser at 1.0 W/cm^2, the tumors were successfully eradicated within 9 days (Figure 5g). The results presented an effective combined treatment for tumors.

Recently, Xu et al. demonstrated an innovative tumor treatment strategy that combined synergistic PTT and US therapy [70]. They synthesized two types of nanoparticles (Figure 6a): gold nanorods modified with polyethylene glycol (PEG) and polyethyleneimine (PEI), denoted as mPEG-PEI-AuNRs, and carbonate nanoparticles conjugated with PEG and PEI via electrostatic interactions (mPEG-PEI/CaNPs). The latter released carbon dioxide (CO_2) bubbles in acidic environments. The phantom experiment showed that CO_2 bubbles significantly enhanced the PA signal of the mPEG-PEI-AuNRs (Figure 6b). For in vivo validation, both types of nanoparticles were intravenously injected into MCF-7 breast tumor-xenografted mice, and cross-sectional PA images were acquired at various time intervals (Figure 6c). The results showed 3-fold enhanced PA signals in the tumor 24 h after injection. Under the irradiation of an 808 nm laser with a fluence of 1.0 W/cm^2, the local temperature of the tumor rapidly increased to 51 °C within 2 min, confirming the potential of the nanoparticles for effective PTT. Moreover, the CO_2 bubbles released from mPEF-PEI/CaNPs were used for US therapy, causing tumor cell necrosis through a bubble explosion triggered by US irradiation. When therapeutic US waves with a frequency of 1 Hz and fluence of 1.5 W/cm^2 were applied to the PTT-treated tumor, the combined approach achieved successful tumor eradication, highlighting the potential for synergistic PTT and US treatment strategies.

In addition, to enhance therapeutic efficacy, efforts have been directed toward achieving deeper drug penetration through extracellular release mechanisms. Lee et al. synthesized a charge-convertible nanomedicine for penetrating solid tumors through transcytosis [71]. This albumin-based calcium phosphate nanomedicine, denoted as mAlb-820@CaP, was loaded with IR820 dye and used as both a contrast agent for PAI and an optical absorber for PTT (Figure 6d). Following extravasation through the enhanced permeability and retention effect of the tumor vasculature, the nanoparticles decomposed into cationic albumins and released calcium ions (Ca^{2+}). This decomposition was triggered by low pH levels that transformed the modified albumins into cationic albumins. The released Ca^{2+} ions in the lysosomes induced lysosomal exocytosis, further enabling the delivery of cationic albumins to the surrounding cells. To verify the deep penetration of the nanoparticles, PA images of 4T1 breast tumor-xenografted mice were acquired after tail vein injection of the nanoparticles (Figure 6e). The PA signal in the tumor region gradually increased, reaching a 2.3-fold enhancement 24 h after the mAlb-820@CaP nanoparticle injection (Figure 6f). In contrast, the PA signals from the free IR820 dye increased 1.6-fold at 4 h post-injection; however, they declined after 24 h, indicating less efficient accumulation. These results confirmed the enhanced accumulation of nanoparticles in the tumor tissue. Subsequent PTT with 5 min of 808 nm laser illumination at 1.0 W/cm^2 demonstrated tumor regression, highlighting the potential of mAlb-820@CaP nanoparticles for PA imaging-guided tumor therapy.

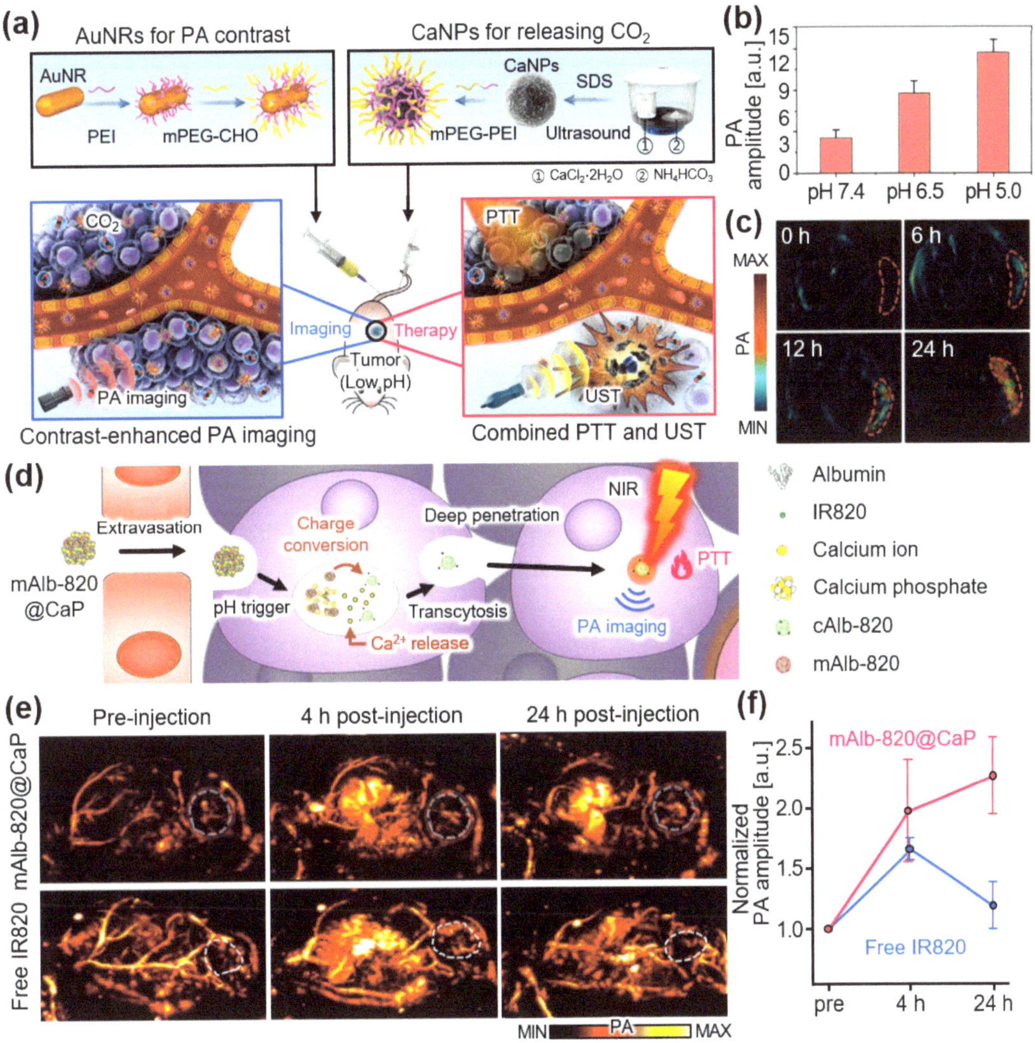

Figure 6. Drug-release nanoparticles for effective phototherapy. (**a**) Schematic illustration for tumor treatment strategy combining synergistic PTT and UST using two nanoparticles: AuNRs for PA contrast and CaNPs for releasing CO_2 bubbles. (**b**) PA amplitude of the mixture of the nanoparticles at different pH conditions. (**c**) Contrast-enhanced PA images of tumor region after injection of the two nanoparticles. (**d**) Schematic illustration for the mechanism of the deep penetrating nanoparticles for effective tumor treatment. (**e**) PA images of the whole body of mice after injection of the mAlb-820@CaP nanoparticles and free IR820 dyes. White dashed areas indicate tumor regions delineated by corresponding ultrasound images. (**f**) Normalized PA amplitudes of tumor region, measured before, 4 h after, and 24 h after injection. PA, photoacoustic; PTT, photothermal therapy; UST, ultrasound therapy; AuNR, gold nanorod; CaNP, carbonate nanoparticle; CO_2, carbon dioxide; PEI, polyethyleneimine; mPEG-CHO, aldehyde-modified polyethylene glycol; mPEG-PEI, polyethylene glycol- and polyethyleneimine modification; SDS, sodium dodecyl sulfate; $CaCl_2 \cdot 2H_2O$, calcium chloride dihydrate; NH_4HCO_3, ammonium bicarbonate; mAlb-820, modified albumin-IR820 complex; cAlb-820, cationic albumin-IR820 complex; CaP, calcium phosphate; Ca^{2+}, calcium ion. The images are reproduced with permission from [70,71].

This section explores pH-sensitive nanoparticles that release drugs and PA contrast agents through separation mechanisms, particularly in acidic tumor environments (Table 2). These nanoparticles disassemble their micelle structures or surface-bound functional groups, releasing encapsulated agents and enhancing PA signals in the tumor region. The strategies described in this section collectively highlight the potential of pH-sensitive nanoparticles for tumor-specific drug release, contrast-enhanced PAI, and effective PTT. Optimizing the balance among nanoparticle stability, drug release control, and therapeutic efficiency is crucial for advancing the application of these systems for precise cancer treatment.

Table 2. Summary of pH-sensitive nanoparticles that release PA contrast and drugs in tumor acidic microenvironment. PA, photoacoustic; λ_{th}, wavelength of therapeutic laser illumination; λ_{ex}, excitation wavelength for PA imaging; AR-PAM, acoustic-resolution PA microscopy; PACT, PA computed tomography; Ag_2S, silver sulfide; DOX, doxorubicin; Au, gold; mAlb, modified albumin; PEG, polyethylene glycol; mPEG-PAVV, amphiphilic poly(ethylene glycol)-*b*-poly(acrylamide-*co*-acrylonitirle-*co*-vinylimidazole); DMMA, dimethyl maleic amide.

pH-Sensitive Nanoparticles		Therapy			PA Imaging			Ref.
Base	pH-Sensitive Material	λ_{th} [nm]	Power [W/cm^2]	Time [min]	Configuration	λ_{ex} [nm]	Application	
Ag_2S	PEG	808	1.5	10	PACT	680	HepG2 liver tumor	[68]
DOX + IR780	mPEG-PAAV	808	1.0	5	PACT	720	4T1 breast tumor	[69]
Au	PEG	808	1.0	6	PACT	830	MCF-7 breast tumor	[70]
mAlb	DMMA	808	1.0	5	AR-PAM	830	4T1 breast tumor	[71]

4.3. Protonation

Protonation and deprotonation are chemical reactions in which hydrogen ions are exchanged in response to changes in pH. In the previous sections, the aggregation and separation mechanisms involved protonation and deprotonation. The previous sections focused on how these processes lead to the generation of PA signals. In this section, we discuss the protonation mechanisms that change the energy absorption spectrum by altering the electron density within a molecule.

Tian et al. demonstrated pH-responsive polyaniline assemblies for efficient PTT and tumor accumulation monitoring using contrast-enhanced PAI (Figure 7a) [72]. They conjugated bovine serum albumin with polyaniline (BSA-PANI), which exhibited a shift in absorption spectrum over a range of pH 3.0–8.0. Under acidic conditions, the emeraldine base in polyaniline was converted to an emeraldine salt state. Although this conversion typically occurred under strongly acidic conditions (i.e., pH < 4.0), the developed BSA-PANI assemblies achieved this transformation at pH < 7.0, which represents a mildly acidic tumor microenvironment. This conversion resulted in a red shift in the absorption and an enhanced PA signal in the NIR region (Figure 7b). To evaluate contrast enhancement, PA images of xenografted 4T1 breast tumors in mice were acquired after intravenous injection of BSA-PANI nanoparticles (Figure 7c). The PA amplitude reached a maximum of 12 h after injection, with approximately a 3.6-fold increase compared to the pre-injection level. In subsequent PTT experiments, the local temperature of the tumor increased to approximately 56 °C under a 5 min irradiation of an 808 nm laser at 1.0 W/cm^2. The treatment resulted in complete tumor eradication within 16 days (Figure 7d), demonstrating the promising potential of BSA-PANI assemblies for PA-guided PTT.

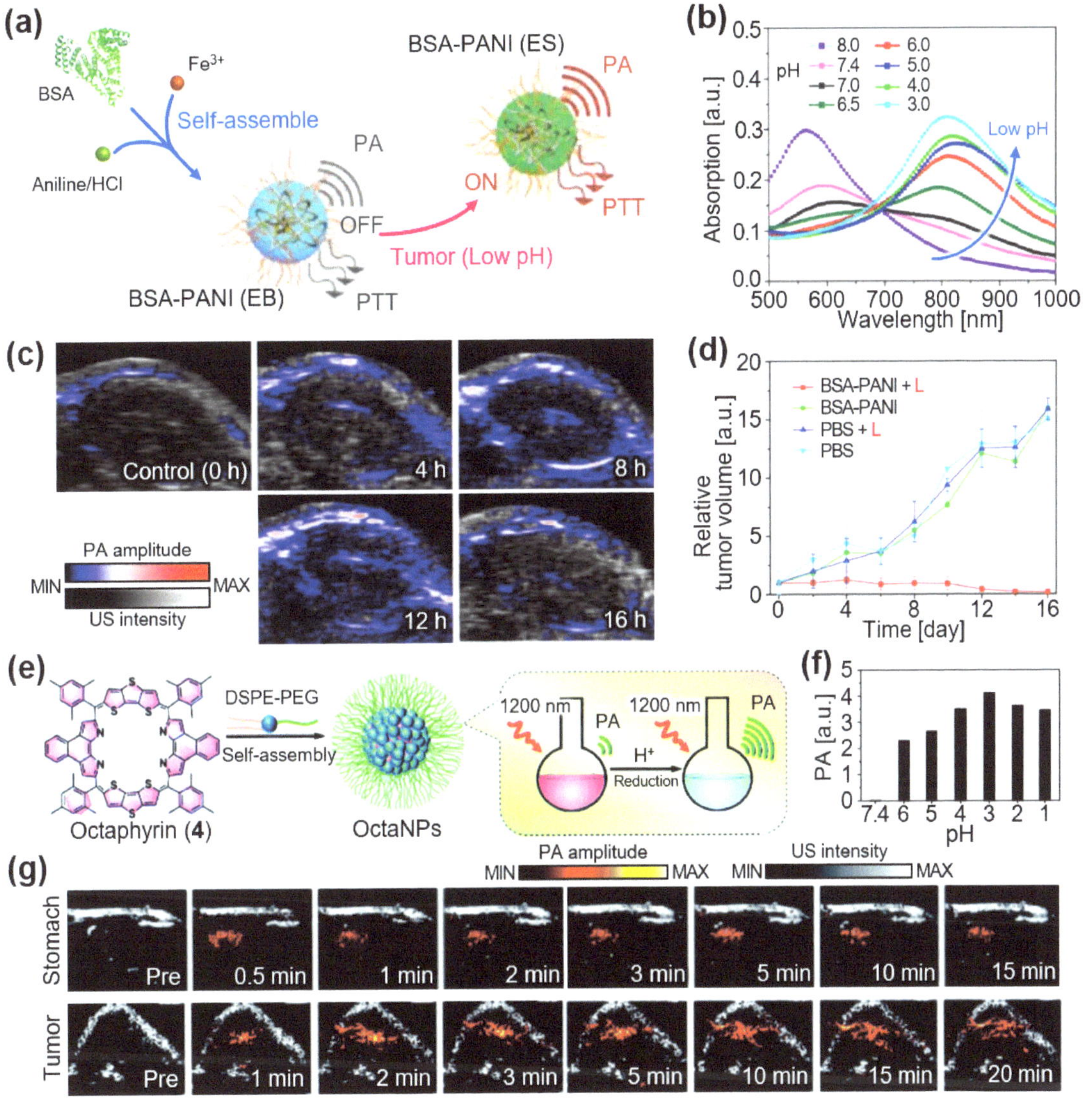

Figure 7. pH-responsive nanoparticles that exhibit redshift of absorption spectrum under acidic conditions. (**a**) Schematic illustration of pH-responsive BSA-PANI assemblies, which show enhanced PA contrast and PTT efficiency. (**b**) Absorption spectra of BSA-PANI assemblies dispersed in buffer solutions with different pH values. (**c**) PA images achieved before and after injection of BSA-PANI assemblies into tumor-bearing mice. (**d**) Relative tumor volume of different treatment groups. (**e**) Schematic illustration of the preparation of OctaNPs, which generate strong PA signals at 1200 nm. (**f**) Quantified PA amplitudes at varying pH values in tetrahydrofuran. (**g**) PA images of stomach and tumor region before and after injection of OctaNPs. PA, photoacoustic; US, ultrasound; PTT, photothermal therapy; BSA, bovine serum albumin; PANI, polyaniline; PBS, phosphate-buffered saline; L, laser illumination; OctaNP, nanoparticles synthesized from octaphyrin 4; DSPE-PEG, 1,2-distearoyl-*sn*-glycero-3-phosphoethanolamine-poly(ethylene glycol); H^+, hydrogen ion. The images are reproduced with permission from [72,73].

To enhance the imaging depth, ongoing studies have focused on developing PA agents that absorb light in the NIR-II region, which is advantageous for deep tissue imaging because of its reduced optical scattering compared to the visible or NIR-I regions. Chen et al. developed NIR-II-absorbing nanoparticles that exhibited enhanced PA signals under acidic conditions (Figure 7e) [73]. They synthesized nanoparticles, denoted as OctaNPs, that encapsulated octaphyrin (4) within 1,2-distearoyl-*sn*-glycero-3-phosphoethanolamine-poly(ethylene glycol) (DSPE-PEG). DSPE-PEG exhibits pH-responsive properties with proton-coupled electron transfer, converting octaphyrin (4) into its form (6) in acidic environments, thereby generating strong PA signals at an excitation wavelength of 1200 nm (Figure 7f). PA images acquired after the intragastric injection of OctaNPs confirmed enhanced contrast in the low-pH environment of the stomach (Figure 7g). To target the mildly acidic tumor microenvironment, OctaNPs were modified with high concentrations of glutathione to enhance optical absorption and then delivered to subcutaneously implanted HepG2 liver tumors. The resulting PA images demonstrated contrast enhancement in the tumor region, indicating the potential of OctaNPs for tumor imaging in vivo.

While single-wavelength PAI has shown promising feasibility for monitoring drug delivery in tumors, ratiometric PA analysis using multi-wavelength PA responses can more efficiently delineate drug distribution. When the absorption peak of nanoparticles shifts under the acidic conditions of the tumor, the ratio of PA signals varies significantly compared to that in normal tissue. Chen et al. demonstrated this approach by synthesizing self-assembled albumin-based nanoparticles that responded to two wavelengths of light [74]. They conjugated human serum albumin (HSA) with two NIR-absorbing dyes: pH-active benzo[*a*]phenoxazine (BPOx), which exhibits a red shift in its absorption spectrum owing to protonation under acidic conditions, and pH-inert IR825, which absorbs 825 nm light and remains unaffected by pH changes (Figure 8a). Glutaraldehyde was used to induce covalent cross-linking of HSA, thereby enhancing the stability of the nanoparticles. In acidic environments, the self-assembled C-HSA-BPOx-IR825 nanoparticles showed increased absorption in the 600–700 nm range, whereas no significant change was observed at 825 nm (Figure 8b). For ratiometric PA analysis, PA signals were measured from subcutaneously implanted 4T1 breast tumors in mice after the intravenous injection of C-HSA-BPOx-IR825 nanoparticles (Figure 8c). The PA signal at 680 nm increased significantly, whereas only marginal changes were observed at 825 nm (Figure 8d), resulting in a significant increase in the PA ratio (Figure 8e). These results highlight the potential of ratiometric PA analysis for high-sensitivity tumor detection in vivo.

Yang et al. also performed ratiometric PA analysis using a perylene diimide (PDI)-based theranostic platform (Figure 8f) [75]. They synthesized THPDINs by encapsulating IR825 dye and doxorubicin (DOX) using pH-sensitive PDIs. In this system, the pH-inert IR825 served as a reference, producing stable PA amplitudes in the NIR region regardless of pH changes, whereas the pH-sensitive PDI exhibited a blue shift to below 600 nm. Consequently, the absorption of the THPDIN nanoparticles decreased significantly at 680 nm while showing minimal changes at 825 nm (Figure 8g). After the intravenous injection of THPDINs into U87MG glioblastoma tumor-bearing mice, PA images were acquired at wavelengths of 680 and 825 nm (Figure 8h). The PA amplitudes in the tumor region confirmed the pH-responsive characteristics of the THPDINs, with the PA signal ratio peaking at 24 h post-injection and reaching approximately 1.74-fold higher than that of the control group (Figure 8i). At 24 h post-injection, the ratio of PA signals reached a maximum, which was approximately 1.74-fold higher than that in the control group (Figure 8j). In addition, THPDINs suppressed tumor growth, with 100% survival observed 45 days after treatment, highlighting the therapeutic potential of this platform.

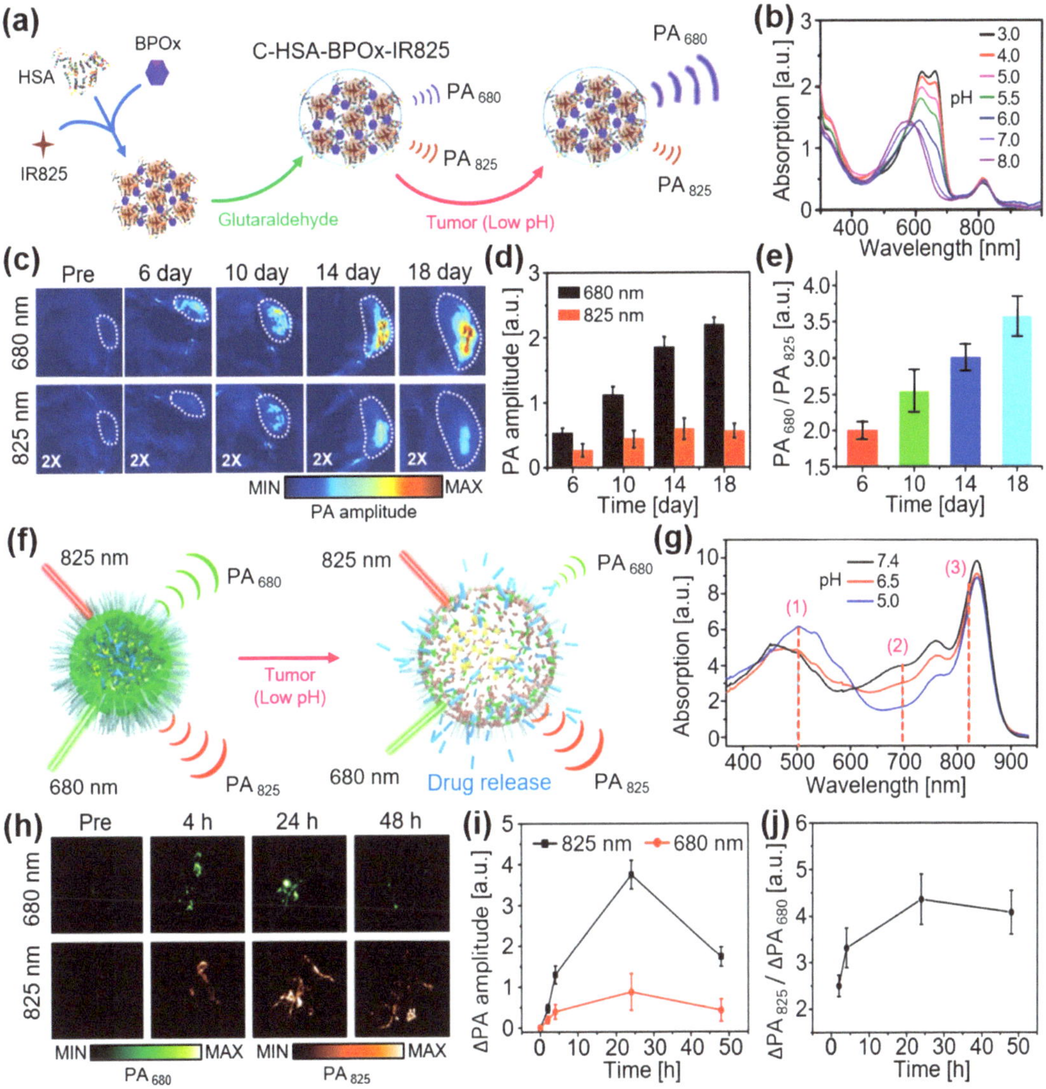

Figure 8. Representative ratiometric PA analysis of pH-sensitive nanoparticles in tumor acidic condition. (**a**) Schematic illustration for the synthesis of C-HSA-BPOx-IR825 nanoparticles and its response in a tumor acidic environment. (**b**) Optical absorption spectra of C-HSA-BPOx-IR825 nanoparticles measured in buffers with different pH values. (**c**) PA images, (**d**) PA amplitudes, and (**e**) PA ratio of 4T1 breast tumors in mice after intravenous injection of C-HSA-BPOx-IR825 nanoparticles. (**f**) Schematic illustration for PA response of THPDINs in a tumor acidic environment. (**g**) Optical absorption spectra of the THPDINs with different pH values. Red dashed lines indicate representative wavelengths: (1) 500, (2) 680, and (3) 825 nm. (**h**) PA images, (**i**) PA amplitude increase, and (**j**) PA ratio of U87MG glioblastoma in a tumor acidic environment after injection of THPDINs. PA, photoacoustic; PA_{680}, PA signal at 680 nm; PA_{825} PA signal at 825 nm; HSA, human serum albumin; BPOx, benzo[*a*]phenoxazine; C-HSA-BPOx-IR825, covalent cross-link induced albumin-based nanoparticles; THPDIN, perylene diimide-based theranostic platform. The images are reproduced with permission from [74,75].

Recently, Liu et al. synthesized nanoparticles based on a heptamethine cyanine dye and indoline for ratiometric PA analysis [76]. Nanoparticles use the pH sensitivity of nitrogen atoms within the dye, which undergo protonation and deprotonation. These nanoparticles displayed stable PA signals at 680 nm across varying pH levels, but a notable increase was observed in the PA signal at 760 nm as the pH decreased, enabling ratiometric PA analysis. Following intravenous injection, the PA signals of the nanoparticles were monitored in xenografted MCF-7 breast tumors in mice. The highest PA signal at 760 nm was observed at 18 h post-injection, demonstrating a clear distinction from the control group. In addition to PA imaging, the fluorescence signals of nanoparticles exhibited similar trends, confirming the feasibility of multimodal imaging. Furthermore, 10 min of 808 nm laser illumination at a fluence of 0.8 W/cm^2 elevated the local temperature to 62 °C, effectively eradicating the tumor. These results highlight the potential of the nanoparticles for image-guided PTT.

This section explores pH-sensitive agents that use protonation mechanisms to alter their energy absorption spectra, enhancing both PA contrast and therapeutic efficacy (Table 3). This section also highlights ratiometric PAI, in which multi-wavelength responses in acidic conditions enable precise tumor detection by analyzing PA signal ratios. These approaches show great potential for improving tumor imaging and treatment. Future efforts should focus on optimizing the biocompatibility and stability of these agents and refining ratiometric PAI for highly sensitive diagnostics and effective cancer therapies.

Table 3. Summary of pH-sensitive PA agents that exhibit absorption shift by protonation in tumor acidic microenvironment. PA, photoacoustic; λ_a, peak absorption wavelength; λ_{ex}, exciation wavelength for PA imaging; PANI, polyaniline; PACT, PA computed tomography; BPOx, benzo[*a*]phenoxazine; PDI, perylene diimide.

Base	λ_a [nm] (pH)		PA Imaging			Ref.
	Normal	Tumor	Configuration	λ_{ex}	Application	
PANI	570 (8)	800 (3)	PACT	-	4T1 breast tumor	[72]
Octaphyrin	1200 (7.4)	1200 (5)	PACT	1200	HepG2 liver tumor	[73]
BPOx, IR825	580 (7)	620 (5.5)	PACT	680, 825	4T1 breast tumor	[74]
PDI	450 (7.4)	500 (5)	PACT	680, 825	U87MG glioblastoma	[75]
Heptamethine cyanine dye	680 (10)	760 (4)	PACT	680, 760	MCF-7 breast tumor	[76]

5. Conclusions

This review explores the recent advancements in the design and application of pH-sensitive PA sensors, highlighting their potential to improve the sensitivity, specificity, and depth of tumor imaging and therapy. Various pH-responsive agents have been synthesized to enhance the precision of PAI-guided diagnoses and therapies. Using protonation and deprotonation mechanisms, these agents provide measurable changes in optical absorption and, consequently, stronger PA signals. This enables targeted imaging, particularly in acidic tumor microenvironments. The integration of multi-wavelength lasers further enhances their versatility, allowing for more precise ratiometric analyses and multimodal detection.

Several key strategies for designing pH-sensitive agents have emerged and are highlighted in this review. For instance, polyaniline-based assemblies and BSA-conjugated nanoparticles efficiently shift absorption spectra in response to the pH levels of the tumor microenvironment, enhancing PA signals in the NIR region for deeper imaging. Similarly, the use of NIR-II-absorbing agents such as OctaNPs demonstrated the potential for deeper tissue penetration and improved imaging contrast. Notably, ratiometric PAI analysis using multi-wavelength responses offered a more straightforward understanding of drug distribution and tumor targeting. These multi-wavelength approaches, particularly in the NIR region, provided a more accurate delineation of tumor boundaries, allowing for enhanced specificity in both diagnosis and therapy monitoring.

Compared to other activation methods, pH-triggering offers exceptional selectivity for tumor detection. While lasers can precisely activate agents at specific locations and time points, their effectiveness is limited by penetration depth, restricting their applicability for deeply posed tumors. Similarly, enzyme-triggered agents respond to specific enzymes present in tumors, but their contrast enhancement may be limited in tumors with low expression levels of these enzymes, reducing their diagnostic efficacy.

Despite the promising results obtained with pH-sensitive agents, several challenges remain to be addressed for their successful translation from preclinical models to clinical practice. Key concerns include ensuring biocompatibility, minimizing toxicity, and achieving reproducibility in nanoparticle synthesis. In addition, the optimization of the excitation wavelengths to maximize both the imaging depth and signal-to-noise ratio should be further refined for clinical applications.

Currently, the clinical trials of PAI are mainly based on label-free approaches, such as sO_2 level analysis in cancerous regions [77–79]. While these studies have shown potential for enhancing diagnostic accuracy, the use of tumor-responsive contrast agents could significantly improve the specificity and sensitivity in identifying cancerous tissues, once approved for human use. Therefore, with continued research on nanoparticle engineering, pH-sensitive photoacoustic agents can be used in cancer diagnostics and image-guided therapies in clinical applications.

From a systems perspective, although multi-element array transducers in PAI have enabled the capture of depth images in real-time, 3D volumetric imaging still relies heavily on mechanical scanning operations, which presents challenges for clinical applications. Human trials, in particular, face limitations owing to the high costs and operational complexity associated with mechanical scanning. Recent efforts have been aimed at overcoming these obstacles through the development of more efficient and portable handheld PAI systems, which could significantly enhance accessibility and reduce scanning time [80,81]. The use of volumetric array transducers also enables real-time, high-quality PAI to monitor the distribution of agents in vivo [82].

To generate strong PA signals capable of achieving deep imaging depth, delivering maximum laser power is crucial. However, the laser power must remain within the maximum permissible exposure limits guided by the safety standard [46] to prevent tissue damage. Notably, the current laser power used in most systems is much below the safety thresholds, indicating a margin for potential optimization to enhance imaging performance while ensuring safety.

From the perspective of image generation, tissue motion during acquisition can degrade PA image quality. This challenge can be addressed by incorporating US imaging. Structural information from US images can be used to correct motion-induced distortions in corresponding PA images [49]. Additionally, US images can be also used for laser power compensation, enabling more accurate spectral analysis and enhancing the reliability of quantitative imaging [83].

The accuracy of PAI reconstruction is based on the speed of sound across different biological tissues, and current systems often use predefined constants. However, this can result in image distortion, particularly in heterogeneous soft tissues. To address this issue, advanced computational techniques, such as GPU-accelerated processing and artificial intelligence-driven algorithms, are being explored to correct these distortions [84]. These innovations have the potential to enhance image accuracy and reliability, thereby enabling more precise diagnostic capabilities for PAI.

Author Contributions: Conceptualization, J.K., C.K. and S.J.; investigation, H.K. and D.J.; resources, H.K., D.J. and S.J.; writing—original draft preparation, H.K. and D.J.; writing—review and editing, J.K., C.K. and S.J. All authors have read and agreed to the published version of the manuscript.

Funding: This work was supported by the National Research Foundation (NRF) grants (2021R1A5A1032937; 2023R1A2C3004880; 2021M3C1C3097624; 2020R1A6A1A03047902) and the Korea Health Technology R&D Project (HR20C0026) funded by the Korean government (Ministry of Science and ICT; Ministry of Education; Ministry of Health & Welfare). This work was also supported by BK21 FOUR program and Glocal 30 University Projects (Pohang University of Science and Technology).

Institutional Review Board Statement: Not applicable.

Informed Consent Statement: Not applicable.

Data Availability Statement: Not applicable.

Conflicts of Interest: C.K. has financial interests in OPTICHO, which did not support this work.

References

1. Bell, A.G. The Photophone. *Science* **1880**, *1*, 130–134. [CrossRef] [PubMed]
2. Liu, C.; Wang, L. Functional Photoacoustic Microscopy of Hemodynamics: A Review. *Biomed. Eng. Lett.* **2022**, *12*, 97–124. [CrossRef] [PubMed]
3. Mallidi, S.; Luke, G.P.; Emelianov, S. Photoacoustic Imaging in Cancer Detection, Diagnosis, and Treatment Guidance. *Trends Biotechnol.* **2011**, *29*, 213–221. [CrossRef]
4. Zhao, Z.; Swartchick, C.B.; Chan, J. Targeted Contrast Agents and Activatable Probes for Photoacoustic Imaging of Cancer. *Chem. Soc. Rev.* **2022**, *51*, 829–868. [CrossRef]
5. Jokerst, J.V.; Cole, A.J.; Van de Sompel, D.; Gambhir, S.S. Gold Nanorods for Ovarian Cancer Detection with Photoacoustic Imaging and Resection Guidance via Raman Imaging in Living Mice. *ACS Nano* **2012**, *6*, 10366–10377. [CrossRef]
6. Agarwal, A.; Huang, S.; O'Donnell, M.; Day, K.; Day, M.; Kotov, N.; Ashkenazi, S. Targeted gold nanorod contrast agent for prostate cancer detection by photoacoustic imaging. *J. Appl. Phys.* **2007**, *102*, 064701. [CrossRef]
7. Liu, S.; Pan, X.; Liu, H. Two-Dimensional Nanomaterials for Photothermal Therapy. *Angew. Chem.* **2020**, *132*, 5943–5953. [CrossRef]
8. Guo, T.; Tang, Q.; Guo, Y.; Qiu, H.; Dai, J.; Xing, C.; Zhuang, S.; Huang, G. Boron Quantum Dots for Photoacoustic Imaging-Guided Photothermal Therapy. *ACS Appl. Mater. Interfaces* **2020**, *13*, 306–311. [CrossRef]
9. Xu, M.; Wang, L.V. Photoacoustic Imaging in Biomedicine. *Rev. Sci. Instrum.* **2006**, *77*, 041101. [CrossRef]
10. Apriyanto, D.K.; Setiawan, A.; Widyaningrum, R. Therapeutic and Contrast Agents for Photoacoustic Imaging-Guided Photothermal Therapy: A Narrative Review. *Nanotheranostics* **2024**, *8*, 506. [CrossRef]
11. Das, S.S.; Bharadwaj, P.; Bilal, M.; Barani, M.; Rahdar, A.; Taboada, P.; Bungau, S.; Kyzas, G.Z. Stimuli-Responsive Polymeric Nanocarriers for Drug Delivery, Imaging, and Theragnosis. *Polymers* **2020**, *12*, 1397. [CrossRef] [PubMed]
12. Xavierselvan, M.; Cook, J.; Duong, J.; Diaz, N.; Homan, K.; Mallidi, S. Photoacoustic Nanodroplets for Oxygen Enhanced Photodynamic Therapy of Cancer. *Photoacoustics* **2022**, *25*, 100306. [CrossRef] [PubMed]
13. Li, M.; Tang, Y.; Yao, J. Photoacoustic Tomography of Blood Oxygenation: A Mini Review. *Photoacoustics* **2018**, *10*, 65–73. [CrossRef] [PubMed]
14. Ning, B.; Kennedy, M.J.; Dixon, A.J.; Sun, N.; Cao, R.; Soetikno, B.T.; Chen, R.; Zhou, Q.; Shung, K.K.; Hossack, J.A. Simultaneous Photoacoustic Microscopy of Microvascular Anatomy, Oxygen Saturation, and Blood Flow. *Opt. Lett.* **2015**, *40*, 910–913. [CrossRef] [PubMed]
15. Zhang, H.F.; Maslov, K.; Sivaramakrishnan, M.; Stoica, G.; Wang, L.V. Imaging of Hemoglobin Oxygen Saturation Variations in Single Vessels In Vivo using Photoacoustic Microscopy. *Appl. Phys. Lett.* **2007**, *90*, 053901. [CrossRef]
16. Kim, J.; Kim, Y.H.; Park, B.; Seo, H.M.; Bang, C.H.; Park, G.S.; Park, Y.M.; Rhie, J.W.; Lee, J.H.; Kim, C. Multispectral Ex Vivo Photoacoustic Imaging of Cutaneous Melanoma for Better Selection of the Excision Margin. *Br. J. Dermatol.* **2018**, *179*, 780–782. [CrossRef]
17. Attia, A.B.E.; Chuah, S.Y.; Razansky, D.; Ho, C.J.H.; Malempati, P.; Dinish, U.; Bi, R.; Fu, C.Y.; Ford, S.J.; Lee, J.S.-S. Noninvasive Real-Time Characterization of Non-Melanoma Skin Cancers with Handheld Optoacoustic Probes. *Photoacoustics* **2017**, *7*, 20–26. [CrossRef]
18. Wang, Y.; Xu, D.; Yang, S.; Xing, D. Toward In Vivo Biopsy of Melanoma Based on Photoacoustic and Ultrasound Dual Imaging with an Integrated Detector. *Biomed. Opt. Express* **2016**, *7*, 279–286. [CrossRef]
19. Langhout, G.C.; Grootendorst, D.J.; Nieweg, O.E.; Wouters, M.W.J.M.; van der Hage, J.A.; Jose, J.; van Boven, H.; Steenbergen, W.; Manohar, S.; Ruers, T.J.M. Detection of Melanoma Metastases in Resected Human Lymph Nodes by Noninvasive Multispectral Photoacoustic Imaging. *Int. J. Biomed. Imaging* **2014**, *2014*, 163652. [CrossRef]

20. Karlas, A.; Kallmayer, M.; Bariotakis, M.; Fasoula, N.-A.; Liapis, E.; Hyafil, F.; Pelisek, J.; Wildgruber, M.; Eckstein, H.-H.; Ntziachristos, V. Multispectral Optoacoustic Tomography of Lipid and Hemoglobin Contrast in Human Carotid Atherosclerosis. *Photoacoustics* **2021**, *23*, 100283. [CrossRef]
21. Cao, Y.; Kole, A.; Hui, J.; Zhang, Y.; Mai, J.; Alloosh, M.; Sturek, M.; Cheng, J.-X. Fast Assessment of Lipid Content in Arteries In Vivo by Intravascular Photoacoustic Tomography. *Sci. Rep.* **2018**, *8*, 2400. [CrossRef] [PubMed]
22. Buma, T.; Conley, N.C.; Choi, S.W. Multispectral Photoacoustic Microscopy of Lipids using a Pulsed Supercontinuum Laser. *Biomed. Opt. Express* **2018**, *9*, 276–288. [CrossRef]
23. Weber, J.; Beard, P.C.; Bohndiek, S.E. Contrast Agents for Molecular Photoacoustic Imaging. *Nat. Methods* **2016**, *13*, 639–650. [CrossRef]
24. Kilian, H.I.; Ma, C.; Zhang, H.; Chen, M.; Nilam, A.; Quinn, B.; Tang, Y.; Xia, J.; Yao, J.; Lovell, J.F. Intraperitoneal Administration for Sustained Photoacoustic Contrast Agent Imaging. *Photoacoustics* **2022**, *28*, 100406. [CrossRef] [PubMed]
25. Lee, C.; Kim, J.; Zhang, Y.; Jeon, M.; Liu, C.; Song, L.; Lovell, J.F.; Kim, C. Dual-Color Photoacoustic Lymph Node Imaging using Nanoformulated Naphthalocyanines. *Biomaterials* **2015**, *73*, 142–148. [CrossRef] [PubMed]
26. Vonk, J.; Kukačka, J.; Steinkamp, P.; de Wit, J.; Voskuil, F.; Hooghiemstra, W.; Bader, M.; Jüstel, D.; Ntziachristos, V.; van Dam, G. Multispectral Optoacoustic Tomography for In Vivo Detection of Lymph Node Metastases in Oral Cancer Patients using an EGFR-Targeted Contrast Agent and Intrinsic Tissue Contrast: A Proof-of-Concept Study. *Photoacoustics* **2022**, *26*, 100362. [CrossRef]
27. Liu, C.; Chen, J.; Zhang, Y.; Zhu, J.; Wang, L. Five-Wavelength Optical-Resolution Photoacoustic Microscopy of Blood and Lymphatic Vessels. *Adv. Photonics* **2021**, *3*, 016002. [CrossRef]
28. Forbrich, A.; Heinmiller, A.; Zemp, R.J. Photoacoustic Imaging of Lymphatic Pumping. *J. Biomed. Opt.* **2017**, *22*, 106003. [CrossRef]
29. Nagaoka, R.; Tabata, T.; Yoshizawa, S.; Umemura, S.-i.; Saijo, Y. Visualization of Murine Lymph Vessels using Photoacoustic Imaging with Contrast Agents. *Photoacoustics* **2018**, *9*, 39–48. [CrossRef]
30. Lee, D.; Beack, S.; Yoo, J.; Kim, S.K.; Lee, C.; Kwon, W.; Hahn, S.K.; Kim, C. In Vivo Photoacoustic Imaging of Livers Using Biodegradable Hyaluronic Acid-Conjugated Silica Nanoparticles. *Adv. Funct. Mater.* **2018**, *28*, 1800941. [CrossRef]
31. Wang, Z.; Lee, C.S.; Waltzer, W.C.; Liu, J.; Xie, H.; Yuan, Z.; Pan, Y. In vivo Bladder Imaging with Microelectromechanical-Systems-Based Endoscopic Spectral Domain Optical Coherence Tomography. *J. Biomed. Opt.* **2007**, *12*, 034009. [CrossRef] [PubMed]
32. Ding, Y.; Park, B.; Ye, J.; Wang, X.; Liu, G.; Yang, X.; Jiang, Z.; Han, M.; Fan, Y.; Song, J. Surfactant-Stripped Semiconducting Polymer Micelles for Tumor Theranostics and Deep Tissue Imaging in the NIR-II Window. *Small* **2022**, *18*, 2104132. [CrossRef] [PubMed]
33. Wang, S.; Lin, J.; Wang, T.; Chen, X.; Huang, P. Recent Advances in Photoacoustic Imaging for Deep-Tissue Biomedical Applications. *Theranostics* **2016**, *6*, 2394. [CrossRef] [PubMed]
34. Chitgupi, U.; Nyayapathi, N.; Kim, J.; Wang, D.; Sun, B.; Li, C.; Carter, K.; Huang, W.C.; Kim, C.; Xia, J. Surfactant-Stripped Micelles for NIR-II Photoacoustic Imaging through 12 cm of Breast Tissue and Whole Human Breasts. *Adv. Mater.* **2019**, *31*, 1902279. [CrossRef] [PubMed]
35. Singh, S.; Giammanco, G.; Hu, C.-H.; Bush, J.; Cordova, L.S.; Lawrence, D.J.; Moran, J.L.; Chitnis, P.V.; Veneziano, R. Size-Tunable ICG-based Contrast Agent Platform for Targeted Near-Infrared Photoacoustic Imaging. *Photoacoustics* **2022**, *29*, 100437. [CrossRef]
36. Wei, W.; Zhang, X.; Zhang, S.; Wei, G.; Su, Z. Biomedical and Bioactive Engineered Nanomaterials for Targeted Tumor Photothermal Therapy: A Review. *Mater. Sci. Eng. C* **2019**, *104*, 109891. [CrossRef]
37. Beztsinna, N.; Tsvetkova, Y.; Jose, J.; Rhourri-Frih, B. Photoacoustic Imaging of Tumor Targeting with Riboflavin-Functionalized Theranostic Nanocarriers. *Int. J. Nanomed.* **2017**, *12*, 3813. [CrossRef]
38. He, L.; Nie, T.; Xia, X.; Liu, T.; Huang, Y.; Wang, X.; Chen, T. Designing Bioinspired 2D $MoSe_2$ Nanosheet for Efficient Photothermal-Triggered Cancer Immunotherapy with Reprogramming Tumor-Associated Macrophages. *Adv. Funct. Mater.* **2019**, *29*, 1901240. [CrossRef]
39. Lin, L.-S.; Yang, X.; Niu, G.; Song, J.; Yang, H.-H.; Chen, X. Dual-Enhanced Photothermal Conversion Properties of Reduced Graphene Oxide-Coated Gold Superparticles for Light-Triggered Acoustic and Thermal Theranostics. *Nanoscale* **2016**, *8*, 2116–2122. [CrossRef]
40. Chen, Z.; Tu, Y.; Zhang, D.; Liu, C.; Zhou, Y.; Li, X.; Wu, X.; Liu, R. A Thermosensitive Nanoplatform for Photoacoustic Imaging and NIR Light Triggered Chemo-Photothermal Therapy. *Biomater. Sci.* **2020**, *8*, 4299–4307. [CrossRef]
41. Lin, S.; Shah, A.; Hernández-Gil, J.; Stanziola, A.; Harriss, B.I.; Matsunaga, T.O.; Long, N.; Bamber, J.; Tang, M.-X. Optically and Acoustically Triggerable Sub-Micron Phase-Change Contrast Agents for Enhanced Photoacoustic and Ultrasound Imaging. *Photoacoustics* **2017**, *6*, 26–36. [CrossRef] [PubMed]
42. Arneth, B. Tumor Microenvironment. *Medicina* **2019**, *56*, 15. [CrossRef] [PubMed]
43. Webb, B.A.; Chimenti, M.; Jacobson, M.P.; Barber, D.L. Dysregulated pH: A Perfect Storm for Cancer Progression. *Nat. Rev. Cancer* **2011**, *11*, 671–677. [CrossRef] [PubMed]
44. Yin, L.; Sun, H.; Zhang, H.; He, L.; Qiu, L.; Lin, J.; Xia, H.; Zhang, Y.; Ji, S.; Shi, H. Quantitatively Visualizing Tumor-Related Protease Activity In Vivo using a Ratiometric Photoacoustic Probe. *J. Am. Chem. Soc.* **2019**, *141*, 3265–3273. [CrossRef]
45. Olefir, I.; Tzoumas, S.; Restivo, C.; Mohajerani, P.; Xing, L.; Ntziachristos, V. Deep Learning-Based Spectral Unmixing for Optoacoustic Imaging of Tissue Oxygen Saturation. *IEEE Trans. Med. Imaging* **2020**, *39*, 3643–3654. [CrossRef]

46. *ANSI Z136.1*; American National Standard for the Safe Use of Lasers. American National Standards Institute: Washington, DC, USA, 2022.
47. Zhang, W.; Sun, X.; Huang, T.; Pan, X.; Sun, P.; Li, J.; Zhang, H.; Lu, X.; Fan, Q.; Huang, W. 1300 nm Absorption Two-Acceptor Semiconducting Polymer Nanoparticles for NIR-II Photoacoustic Imaging System Guided NIR-II Photothermal Therapy. *Chem. Commun.* **2019**, *55*, 9487–9490. [CrossRef]
48. Lee, H.; Park, S.M.; Park, J.; Cho, S.-W.; Han, S.; Ahn, J.; Cho, S.; Kim, C.; Kim, C.-S.; Kim, J. Transportable Multispectral Optical-Resolution Photoacoustic Microscopy using Stimulated Raman Scattering Spectrum. *IEEE Trans. Instrum. Meas.* **2024**, *73*, 4502309. [CrossRef]
49. Lee, H.; Han, S.; Park, S.; Cho, S.; Yoo, J.; Kim, C.; Kim, J. Ultrasound-Guided Breath-Compensation in Single-Element Photoacoustic Imaging for Three-Dimensional Whole-Body Images of Mice. *Front. Phys.* **2022**, *10*, 457. [CrossRef]
50. Choi, S.; Yang, J.; Lee, S.Y.; Kim, J.; Lee, J.; Kim, W.J.; Lee, S.; Kim, C. Deep Learning Enhances Multiparametric Dynamic Volumetric Photoacoustic Computed Tomography In Vivo (DL-PACT). *Adv. Sci.* **2023**, *10*, 2202089. [CrossRef]
51. Cho, S.; Kim, M.; Ahn, J.; Kim, Y.; Lim, J.; Park, J.; Kim, H.H.; Kim, W.J.; Kim, C. An Ultrasensitive and Broadband Transparent Ultrasound Transducer for Ultrasound and Photoacoustic Imaging In-Vivo. *Nat. Commun.* **2024**, *15*, 1444. [CrossRef]
52. Yao, J.; Wang, L.V. Photoacoustic Microscopy. *Laser Photonics Rev.* **2013**, *7*, 758–778. [CrossRef] [PubMed]
53. Park, J.; Choi, S.; Knieling, F.; Clingman, B.; Bohndiek, S.; Wang, L.V.; Kim, C. Clinical Translation of Photoacoustic Imaging. *Nat. Rev. Bioeng.* **2024**. [CrossRef]
54. Kim, J.; Anthony, M.Y.; Kubelick, K.P.; Emelianov, S.Y. Gold Nanoparticles Conjugated with DNA Aptamer for Photoacoustic Detection of Human Matrix Metalloproteinase-9. *Photoacoustics* **2022**, *25*, 100307. [CrossRef]
55. Yuan, Y.Y.; Mao, C.Q.; Du, X.J.; Du, J.Z.; Wang, F.; Wang, J. Surface Charge Switchable Nanoparticles based on Zwitterionic Polymer for Enhanced Drug Delivery to Tumor. *Adv. Mater.* **2012**, *24*, 5476–5480. [CrossRef] [PubMed]
56. Nam, J.; Won, N.; Jin, H.; Chung, H.; Kim, S. pH-Induced Aggregation of Gold Nanoparticles for Photothermal Cancer Therapy. *J. Am. Chem. Soc.* **2009**, *131*, 13639–13645. [CrossRef]
57. Chen, Y.; Xu, C.; Cheng, Y.; Cheng, Q. Photostability Enhancement of Silica-Coated Gold Nanostars for Photoacoustic Imaging Guided Photothermal Therapy. *Photoacoustics* **2021**, *23*, 100284. [CrossRef]
58. Torchilin, V.P. Multifunctional, Stimuli-Sensitive Nanoparticulate Systems for Drug Delivery. *Nat. Rev. Drug Discov.* **2014**, *13*, 813–827. [CrossRef]
59. Cao, Y.; Dumani, D.S.; Hallam, K.A.; Emelianov, S.Y.; Ran, H. Real-Time Monitoring of NIR-Triggered Drug Release from Phase-Changeable Nanodroplets by Photoacoustic/Ultrasound Imaging. *Photoacoustics* **2023**, *30*, 100474. [CrossRef]
60. Yang, Y.; Fan, Z.; Zheng, K.; Shi, D.; Su, G.; Ge, D.; Zhao, Q.; Fu, X.; Hou, Z. A Novel Self-Targeting Theranostic Nanoplatform for Photoacoustic Imaging-Monitored and Enhanced Chemo-Sonodynamic Therapy. *J. Mater. Chem. B* **2021**, *9*, 5547–5559. [CrossRef]
61. Sarma, T.; Kim, G.; Sen, S.; Cha, W.-Y.; Duan, Z.; Moore, M.D.; Lynch, V.M.; Zhang, Z.; Kim, D.; Sessler, J.L. Proton-Coupled Redox Switching in an Annulated π-Extended Core-Modified Octaphyrin. *J. Am. Chem. Soc.* **2018**, *140*, 12111–12119. [CrossRef]
62. Song, J.; Kim, J.; Hwang, S.; Jeon, M.; Jeong, S.; Kim, C.; Kim, S. "Smart" Gold Nanoparticles for Photoacoustic Imaging: An Imaging Contrast Agent Responsive to the Cancer Microenvironment and Signal Amplification via pH-Induced Aggregation. *Chem. Commun.* **2016**, *52*, 8287–8290. [CrossRef] [PubMed]
63. Zhang, R.; Wang, L.; Wang, X.; Jia, Q.; Chen, Z.; Yang, Z.; Ji, R.; Tian, J.; Wang, Z. Acid-Induced In Vivo Assembly of Gold Nanoparticles for Enhanced Photoacoustic Imaging-Guided Photothermal Therapy of Tumors. *Adv. Healthc. Mater.* **2020**, *9*, 2000394. [CrossRef]
64. Cheng, X.; Zhou, X.; Xu, J.; Sun, R.; Xia, H.; Ding, J.; Chin, Y.E.; Chai, Z.; Shi, H.; Gao, M. Furin Enzyme and pH Synergistically Triggered Aggregation of Gold Nanoparticles for Activated Photoacoustic Imaging and Photothermal Therapy of Tumors. *Anal. Chem.* **2021**, *93*, 9277–9285. [CrossRef]
65. Li, S.; Lui, K.-H.; Tsoi, T.-H.; Lo, W.-S.; Li, X.; Hu, X.; Tai, W.C.-S.; Hung, C.H.-L.; Gu, Y.-J.; Wong, W.-T. pH-Responsive Targeted Gold Nanoparticles for In Vivo Photoacoustic Imaging of Tumor Microenvironments. *Nanoscale Adv.* **2019**, *1*, 554–564. [CrossRef] [PubMed]
66. Li, S.; Lui, K.-H.; Li, X.; Fang, X.; Lo, W.-S.; Gu, Y.-J.; Wong, W.-T. pH-Triggered poly (ethylene glycol)–poly (lactic acid/glycolic acid)/croconaine Nanoparticles-Assisted Multiplexed Photoacoustic Imaging and Enhanced Photothermal Cancer Therapy. *ACS Appl. Bio Mater.* **2021**, *4*, 4152–4164. [CrossRef]
67. Wu, J.; You, L.; Chaudhry, S.T.; He, J.; Cheng, J.-X.; Mei, J. Ambient Oxygen-Doped Conjugated Polymer for pH-Activatable Aggregation-Enhanced Photoacoustic Imaging in the Second Near-Infrared Window. *Anal. Chem.* **2021**, *93*, 3189–3195. [CrossRef] [PubMed]
68. Zhong, Y.; Zou, Y.; Liu, L.; Li, R.; Xue, F.; Yi, T. pH-Responsive Ag_2S Nanodots Loaded with Heat Shock Protein 70 Inhibitor for Photoacoustic Imaging-Guided Photothermal Cancer Therapy. *Acta Biomater.* **2020**, *115*, 358–370. [CrossRef]
69. Yang, Z.; Cheng, R.; Zhao, C.; Sun, N.; Luo, H.; Chen, Y.; Liu, Z.; Li, X.; Liu, J.; Tian, Z. Thermo-and pH-Dual Responsive Polymeric Micelles with Upper Critical Solution Temperature Behavior for Photoacoustic Imaging-Guided Synergistic Chemo-Photothermal Therapy Against Subcutaneous and Metastatic Breast Tumors. *Theranostics* **2018**, *8*, 4097. [CrossRef]
70. Xu, C.; Wang, Y.; Wang, E.; Yan, N.; Sheng, S.; Chen, J.; Lin, L.; Guo, Z.; Tian, H.; Chen, X. Effective Eradication of Tumors by Enhancing Photoacoustic-Imaging-Guided Combined Photothermal Therapy and Ultrasonic Therapy. *Adv. Funct. Mater.* **2021**, *31*, 2009314. [CrossRef]

71. Lee, H.; Park, B.; Lee, J.; Kang, Y.; Han, M.; Lee, J.; Kim, C.; Kim, W.J. Transcytosis-Inducing Multifunctional Albumin Nanomedicines with Deep Penetration Ability for Image-Guided Solid Tumor Treatment. *Small* **2023**, *19*, 2303668. [CrossRef]
72. Tian, Q.; Li, Y.; Jiang, S.; An, L.; Lin, J.; Wu, H.; Huang, P.; Yang, S. Tumor pH-Responsive Albumin/Polyaniline Assemblies for Amplified Photoacoustic Imaging and Augmented Photothermal Therapy. *Small* **2019**, *15*, 1902926. [CrossRef] [PubMed]
73. Chen, J.; Sedgwick, A.C.; Sen, S.; Ren, Y.; Sun, Q.; Chau, C.; Arambula, J.F.; Sarma, T.; Song, L.; Sessler, J.L. Expanded Porphyrins: Functional Photoacoustic Imaging Agents that Operate in the NIR-II Region. *Chem. Sci.* **2021**, *12*, 9916–9921. [CrossRef] [PubMed]
74. Chen, Q.; Liu, X.; Chen, J.; Zeng, J.; Cheng, Z.; Liu, Z. A Self-Assembled Albumin-Based Nanoprobe for In Vivo Ratiometric Photoacoustic pH Imaging. *Adv. Mater.* **2015**, *27*, 6820–6827. [CrossRef]
75. Yang, Z.; Song, J.; Tang, W.; Fan, W.; Dai, Y.; Shen, Z.; Lin, L.; Cheng, S.; Liu, Y.; Niu, G. Stimuli-Responsive Nanotheranostics for Real-Time Monitoring Drug Release by Photoacoustic Imaging. *Theranostics* **2019**, *9*, 526. [CrossRef]
76. Liu, X.; Xiang, C.; Lv, Y.; Xiang, J.; Ma, G.; Li, C.; Hu, Y.; Guo, C.; Sun, H.; Cai, L. Preparation of Near-Infrared Photoacoustic Imaging and Photothermal Treatment Agent for Cancer using a Modifiable Acid-Triggered Molecular Platform. *Analyst* **2024**, *149*, 3064–3072. [CrossRef]
77. Kim, J.; Park, B.; Ha, J.; Steinberg, I.; Hooper, S.M.; Jeong, C.; Park, E.-Y.; Choi, W.; Liang, T.; Bae, J.-S.; et al. Multiparametric Photoacoustic Analysis of Human Thyroid Cancers In Vivo. *Cancer Res.* **2021**, *81*, 4849–4860. [CrossRef]
78. Noltes, M.E.; Bader, M.; Metman, M.J.; Vonk, J.; Steinkamp, P.J.; Kukačka, J.; Westerlaan, H.E.; Dierckx, R.A.; van Hemel, B.M.; Brouwers, A.H. Towards In Vivo Characterization of Thyroid Nodules Suspicious for Malignancy using Multispectral Optoacoustic Tomography. *Eur. J. Nucl. Med. Mol. Imaging* **2023**, *50*, 2736–2750. [CrossRef] [PubMed]
79. Horiguchi, A.; Shinchi, M.; Nakamura, A.; Wada, T.; Ito, K.; Asano, T.; Shinmoto, H.; Tsuda, H.; Ishihara, M. Pilot Study of Prostate Cancer Angiogenesis Imaging using a Photoacoustic Imaging System. *Urology* **2017**, *108*, 212–219. [CrossRef]
80. Kukačka, J.; Metz, S.; Dehner, C.; Muckenhuber, A.; Paul-Yuan, K.; Karlas, A.; Fallenberg, E.M.; Rummeny, E.; Jüstel, D.; Ntziachristos, V. Image Processing Improvements Afford Second-Generation Handheld Optoacoustic Imaging of Breast Cancer Patients. *Photoacoustics* **2022**, *26*, 100343. [CrossRef]
81. Ivankovic, I.; Merčep, E.; Schmedt, C.-G.; Deán-Ben, X.L.; Razansky, D. Real-Time Volumetric Assessment of the Human Carotid Artery: Handheld Multispectral Optoacoustic Tomography. *Radiology* **2019**, *291*, 45–50. [CrossRef]
82. Lin, L.; Hu, P.; Shi, J.; Appleton, C.M.; Maslov, K.; Li, L.; Zhang, R.; Wang, L.V. Single-Breath-Hold Photoacoustic Computed Tomography of the Breast. *Nat. Commun.* **2018**, *9*, 2352. [CrossRef] [PubMed]
83. Jeng, G.-S.; Li, M.-L.; Kim, M.; Yoon, S.J.; Pitre Jr, J.J.; Li, D.S.; Pelivanov, I.; O'Donnell, M. Real-Time Interleaved Spectroscopic Photoacoustic and Ultrasound (PAUS) Scanning with Simultaneous Fluence Compensation and Motion Correction. *Nat. Commun.* **2021**, *12*, 716. [CrossRef] [PubMed]
84. Shi, M.; Vercauteren, T.; Xia, W. Learning-based Sound Speed Estimation and Aberration Correction for Linear-Array Photoacoustic Imaging. *Photoacoustics* **2024**, *38*, 100621. [CrossRef]

chemosensors

MDPI

Article

Optical pH Sensing in Milk: A Small Puzzle of Indicator Concentrations and the Best Detection Method

Olga Voskoboynikova [1], Aleksey Sukhanov [1,*] and Axel Duerkop [2]

[1] School of Computer Science & Robotics of Tomsk Polytechnic University, Lenin Avenue 30, 634050 Tomsk, Russia; vostok@tpu.ru

[2] Institute of Analytical Chemistry, Chemo and Biosensors, University of Regensburg, Universitätsstraße 31, 93053 Regensburg, Germany; axel.duerkop@chemie.uni-regensburg.de

* Correspondence: suhanov@tpu.ru

Abstract: Optical chemical sensors can yield distinctively different responses that are dependent on the method applied for readout and evaluation. We therefore present a comprehensive study on the pH determined non-continuously with optical sensors in real milk samples by either photometry or colorimetry (via the RGB-readout of digital images) compared to the pH values obtained electrochemically by potentiometry. Additionally, the photometric determination of pH was conducted with single-wavelength and a dual wavelength ratiometric evaluation of the absorbance. It was found that both the precision and accuracy of the pH determined by photometry benefit from lower concentrations of bromocresol purple, which served as the pH indicator inside the sensor membrane. A further improvement is obtained by the ratiometric evaluation of the photometric sensor response. The pH values obtained from the colorimetric evaluation, however, gain in precision and accuracy if a higher concentration of the indicator is immobilized inside the sensor membrane. This has a major impact on the future fabrication of optical pH sensor membranes because they can be better tuned to match to the most precise and accurate range of the planned detection method.

Keywords: pH sensor; optical; photometry; colorimetry; potentiometry; digital image

Citation: Voskoboynikova, O.; Sukhanov, A.; Duerkop, A. Optical pH Sensing in Milk: A Small Puzzle of Indicator Concentrations and the Best Detection Method. *Chemosensors* **2021**, *9*, 177. https://doi.org/10.3390/chemosensors9070177

Academic Editor: Pietro Salvo

Received: 8 June 2021
Accepted: 8 July 2021
Published: 12 July 2021

1. Introduction

Optical pH sensors have gained a key role in pH measurements. The basic principles of the creation of optical pH sensors and recent trends of readouts have been presented in comprehensive reviews [1,2]. Colorimetric pH sensors have become especially popular among optical pH sensors due to their low equipment demands. The signal of colorimetric sensors relies on the variation of color that is dependent on the pH of the analyzed sample. This enables simple readout schemes with printed tables for color comparison that can easily be operated also by non-laboratory users.

One of the trend directions for applying colorimetric pH sensors is in the freshness control of perishable foods. In food packaging, a freshness control of perishable foods is frequently not performed in a lab, instead an expiration date is usually stamped onto the packaging with additional recommendations for storage. This does not consider the effect of various conditions upon later transport and storage, which can significantly alter the shelf-life of perishable foods. The spoilage of perishable foods is accompanied by the formation of various chemical or biological species (e.g., biogenic amines or bacteria) or the change of certain characteristics of the foods (e.g., the color, odor and/or homogeneity), which can serve as freshness indicators that can be determined using the headspace above, in contact with, or inside perishable foods. Moreover, the intrinsic acid-base properties of the new compounds formed upon ageing will change the overall pH of the food sample, and their presence can hence be determined by means of a colorimetric pH sensor. Examples of applying colorimetric pH sensors for freshness control include seafood [3–8], meat products [9–11], dairy [12–17], and other perishable foods [18].

Milk acidity can serve as marker of its age, as can be seen from a sensor comprising a polypyrrole/silver nanocomposite chemically deposited on a polyester fiber [17]. The sensor had a reasonable response and reproducibility towards milk pH. Unfortunately, such fiber-optic sensors are not applicable for the evaluation of milk spoilage by untrained consumers outside the lab. Extracts from grape skin incorporated into a tara gum/cellulose nanocrystal matrix was recently used for visible pH-sensing and to monitor the milk spoiling process [13]. The proposed sensor films showed easily detectable color changes from red to green to monitor the freshness of packed food at acidic or alkaline pH, respectively. A new colorimetric method for the detection of milk spoilage using cysteine biofunctionalized silver nanoparticles was developed [15] where the aggregation of biofunctionalized nanoparticles increased with the increase of the lactic acid content of the sample, which leads to a color change from yellow to orange to red to purple. Further, colorimetric sensors based on an anthocyanin-agarose film were proposed as freshness indicators of full cream milk by means of monitoring the pH [16]. A new colorimetric method called red chromatic shift was applied to evaluate the color response of the proposed sensors towards pH. The activity of lactic acid bacteria leads to an increasing production of lactic acid and thus a decrease in the pH level. At normal levels of pH, the main protein in milk, casein, remains evenly dispersed. At lower pH levels below 4.6, the protein coagulates due to the acid generated from fermentation [19]. Thus, milk acidity can be used to monitor milk ageing.

We have previously demonstrated the maintenance of the acid-base properties of bromocresol purple (BCP) in a polymethacrylate matrix(PMM) and calculated the pK_a value of the BCP in the PMM [20]. The response time of the PMM with immobilized BCP towards the changing pH of an aqueous solution was also determined. A polymethacrylate matrix was taken to tightly immobilize BCP and allow easy permeation of the analyte. There are good examples of applying PMM for the solid-phase quantification of various species [21–25]. The initial 6.5 pH value of spoiled milk [26] coincides with BCP's pK_a value 6.5 ± 0.3 in the PMM. Therefore, PMM with immobilized BCP is suitable to monitor even small changes in the pH of milk. It is known that the visibility of a color transition depends on the concentration of an indicator [27]. Obviously, a specific amount of a sensitive reagent in a sensor membrane can also have an effect on its analytical performance.

In this article, various pH-sensitive sensor membranes based on bromocresol purple immobilized into a polymethacrylate matrix were prepared. Spectrophotometric one-wavelength and ratiometric dual-wavelength evaluations of the sensor response in milk were performed, and the resulting pH values were validated against those obtained electroanalytically using a pH electrode. Additionally, the colorimetric evaluation of digital images was compared to the potentiometric pHs. This yields new insights for choosing an appropriate amount of the pH indicator to achieve optimum accuracy and precision depending on the chosen detection method.

2. Materials and Methods

2.1. Apparatus and Chemicals

The absorption spectra and absorbances of sensor membranes and indicator stock solutions were acquired with a Shimadzu UV1800 spectrophotometer. Blanks of the related solvent or the untreated polymer membrane were subtracted. We used a I-160 ionometer (Izmeritelnaya Tekhnika NPO, Moscow, Russia) with a glass pH electrode to determine the reference pH values.

Lactic acid 85% (CAS No. 50-21-5) and bromocresol purple (CAS No. 115-40-2) were purchased from Chimmed, Russia. Those and all other chemicals were used as received and without further purification. The BCP stock solutions were created by dissolving accurately weighed portions of BCP (as to Table 1) in distilled water.

The polymethacrylate membrane was kindly provided by the group of Prof. Gavrilenko from Tomsk Polytechnic University and was synthesized according to a recent protocol [28]. Transparent 10 cm × 10 cm sheets with a thickness of 0.60 ± 0.04 mm were cut into working membrane pieces of 6.0 mm × 8.0 mm size and a mass of about 0.05 g.

Table 1. Parameters of sensors membranes dependent on immobilization conditions.

Designation	(a^1 ± SD), mg/g	Immobilization Conditions				Sensor Membrane			
		C_0 [2], g/L	t, min	V, mL	m, g (Number of PMM Sheets)	A_{415} [3]	RSD [4], %	A_{600} [3]	RSD [4], %
SM1	1.21 ± 0.07	1.0	1.5	25	1.1316 (25 pcs)	1.31	10	1.29	20
SM2	0.82 ± 0.05	1.0	0.75	25	1.1287 (25 pcs)	0.95	7	0.94	10
SM3	0.268 ± 0.006	0.1	2.5	25	1.1356 (25 pcs)	0.30	9	0.26	16

[1] amount of BCP immobilized under the appropriate conditions; [2] initial concentration of BCP in a solution for immobilization; [3] average absorption of sensor membranes at the 415 nm and 600 nm, respectively; [4] relative standard deviation in 25 membranes.

2.2. Preparation of the Sensor Membrane

The preparation of the sensor membrane followed a recent protocol [20]. First, pre-weighed PMM work sheets (25 pcs) were immersed into 25 mL of a stock solution of BCP with a definite concentration and shaken for a set time. Immobilization conditions are presented in Table 1. The amount of BCP immobilized into PMM was calculated according to

$$a = \frac{(C_0 - C) \times V}{m} \tag{1}$$

where C_0 and C are concentrations of BCP in the solution before and after contact with PMM, respectively; V is the volume of the solution used for immobilization, m is the mass of the sheets (made of PMM) for immobilization. Prepared membranes were stored in a desiccator under dark conditions after the removal of excess BCP solution by dabbing the sensor membrane with filter paper.

2.3. Preparation of Milk Samples

Packed pasteurized cow milk (Derevenskoye molochko, Tomsk, Russia) was bought from a local supermarket. The method of the milk pasteurization was heating it to 90 °C for 20 s, which is termed as very high temperature (VHT) pasteurization according to the literature [29]. Milk samples with different pH were prepared with the addition of lactic acid dropwise to imitate a spoilage process [16]. All experiments using milk were done on the date of its manufacture as indicted on the package, at room temperature (24 ± 2 °C).

2.4. Procedure for Determination of pH in Milk Samples

Sensor membranes were immersed in 15 mL of milk samples with adjusted pH value and shaken for 5 min (according to preliminary investigations [20], the response time of similar membranes does not exceed 4 min). The related sensor membranes were then photographed, and the absorption spectra were recorded. Images were captured using a smartphone (Xiaomi Redmi 9) using the standard camera mode. The images were captured at a distance of 10 cm under constant vertical illumination by the built-in flash lamp. The pH of the milk samples was then immediately measured by means of a pH meter.

The algorithm for processing the images of the sensor membranes to extract the colorimetric parameters was created using the software "Vision builder for Automated Inspection (version 3.6.1, 2015)" by National Instruments. The total color difference (TCD) of the sensor membranes was calculated in the RGB color space according to the typical equation

$$\text{TCD} = ((R_0 - R_i)^2 + (G_0 - G_i)^2 + (B_0 - B_i)^2)^{1/2} \tag{2}$$

where the index "0" denotes the color parameter set of the sensor membrane after contact with the milk sample when the milk is fresh; index "i" denotes the color parameter set for the sensor membrane after contact with the milk sample with the adjusted pH. As shown in earlier work [20], such a sensor membrane is stable against repetitive cycling of pH between 1.5 and 11.3 in a flow-through cell for over 30 min. The sensors are not intended for prolonged and continuous contact with food.

3. Results

3.1. Preparation of Sensor Membranes

The amount of BCP in the PMM membrane depends on the immobilization conditions. A higher concentration of BCP in the solution for membrane doping leads to a higher concentration of BCP inside the sensor membrane (see Table 1). More BCP is also adsorbed by the sensor membrane when the immobilization time is increased. Compared with the stirring times usually required for the preparation of sensor cocktails for knife coating or electrospinning (several hours [30]), the current immobilization time is very short, being only a few minutes. This is very suitable for a potential mass production of sensors. According to their BCP content, the obtained sensor membranes were divided into three types. The characterization of the sensor membranes (as to photometric measurements) prepared under different immobilization conditions are presented in Table 1.

3.2. Spectrophotometric Response of Sensor Membranes towards Milk of Various pH

The response of a sensor membrane towards milk samples with various pH values was investigated by photometry. For this purpose, a sensor membrane was immersed into milk samples of adjusted pH and gently shaken. After 5 min, the sensor membrane was dabbed with filter paper, and the absorption spectrum was acquired in transmission mode. The related pH-dependent absorption spectra of the sensor membranes of the SM2 type after contact with the milk samples are presented in Figure 1. It is obvious that a decrease of pH (as it occurs upon the spoiling of the milk) is accompanied by an increase in the absorbance at 415 nm whereas a decrease in the absorbance is found at 600 nm. This is accompanied by a visual color transition from green to yellow. An isosbestic point is found at 485 nm. While the absorption spectra of BCP in a solution show maxima at 430 nm and 590 nm, those shift to 415 nm and 600 nm, respectively, after BCP immobilization in PMM. Please note that the additional reflected light (observed color) is determined by the molar absorbances at both maxima, which also experience ongoing change from the solution into the polymer [20]. This translates into the formation of a green mixing color on the sensor membranes at around pH 7. The absorption spectra of the other type of sensor membranes (SM1 and SM3) are similar with respect to the positions of the absorption maxima and the pH-dependent response. This suggests that the absorbance at these wavelengths or their ratio (A_{415}/A_{600}) may serve as analytical parameters for the solid-phase spectrophotometric determination for the pH of pasteurized milk. The color transition can also be used to visually follow the spoiling of milk.

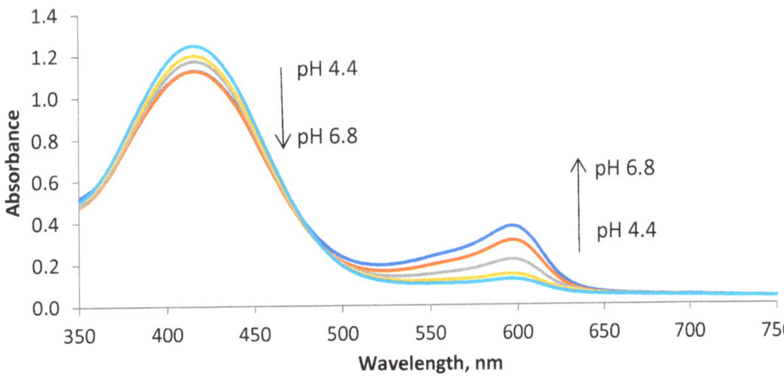

Figure 1. Absorption spectra of SM2-type sensors membranes after contact with milk samples at different pH.

3.3. Effect of the Content of BCP on Photometric Sensor Response

The changes of the photometric response of the sensor membranes with the changing pH of the milk samples should obey the Lambert–Beer law, both at 415 nm and at 600 nm. Therefore, the responses of the three different membrane types towards the milk pH were plotted at both wavelengths, and linear calibration plots were constructed (see Figure S1). A dependence of the analytical signal can be characterized by the linear equation

$$A = k \times pH + C \quad (3)$$

where A is the analytical signal (i.e., the absorbance at the respective wavelength), pH is the pH value of the milk sample, and C is a constant. The related values of k and C for each analytical signal and the corresponding correlation coefficient (r) for each type of sensor membrane are presented in Table 2.

Table 2. Linear response of the analytical signal of the sensor membranes towards milk pH at various detection wavelengths and related correlation coefficients of the linear fit.

Type of Sensor Membrane	k	C	r
Analytical signal: absorbance at 415 nm			
SM1	−0.1225	2.3498	0.914
SM2	−0.039	1.3751	0.927
SM3	−0.0109	0.4889	0.858
Analytical signal: absorbance at 600 nm			
SM1	0.2099	−0.7004	0.983
SM2	0.0982	−0.3503	0.976
SM3	0.0303	−0.0757	0.992
Analytical signal: A_{415}/A_{600}			
SM1	−1.7544	13.671	0.959
SM2	−2.9821	23.48	0.994
SM3	−1.3798	12.546	0.998

As the shortwave absorbance of BCP decreases with pH, a negative slope is found. The use of the absorbance at 415 nm as the analytical signal for the determination of milk pH is undesirable, as low correlation coefficients are found (see Table 2). Moreover, the overall change in the signal is less pronounced at this wavelength (see Figure 1). The absorbance at 600 nm and the ratio of the absorbance at 415 nm to the absorbance at 600 nm (A_{415}/A_{600}), however, can be used as an analytical signal for the solid phase spectrophotometric determination of milk pH. As for the correlation coefficients, A_{600} shows much better results for all three compositions of the sensor membrane. The best correlation is found for the ratiometric response of A_{415}/A_{600} that is dependent on milk pH. Here, SM2 and SM3 are the preferred membrane compositions.

3.4. Validation of Photometric Sensor Response

The pH values obtained from the photometric response of the sensor membranes was then validated against the pH acquired with a commercial pH-meter. As both the absorbance at 600 nm and A_{415}/A_{600} had proved to be suitable for the determination of milk pH, both were used to judge the accuracy of the pH values determined with the sensor membranes. The obtained results are presented in Table 3. The maximum value of the relative error for every type of sensor membrane is indicated in bold. It is clearly obvious that membrane SM1 performs less well than membranes SM2 and SM3, if the related relative errors and the deviations between the average pH-vales as to potentiometry and photometry, respectively, are compared. Here, deviations up to 0.34 pH units are found. These deviations are much smaller for membrane SM2, where the deviation does not exceed 0.23 pH units (if potentiometry and A_{600} are compared) or 0.17 pH units, respectively (A_{415}/A_{600} vs. potentiometry). Membrane SM3 offers the best accuracy, where deviations of less than 0.15 and 0.07 pH units (A_{415}/A_{600} vs. potentiometry) are found. The latter deviation is already within the error range of the electrochemical pH determination, which is why the pH values determined with both methods can be regarded as equal. The average standard deviation of the photometric determination of pH over the whole range of pH 4.5–7.0 can be as little as 0.2 pH units (for membrane SM2), which is remarkably low for a colorimetric sensor. As a result, both accuracy and reproducibility of the sensor membrane SM2 and SM3 can be regarded very good.

Table 3. Validation of the pH obtained in milk by spectrophotometry with the pH from potentiometry (n = 3).

Type of SM	pH Determined by Potentiometry	pH Determined by Spectrophotometry			
		A_{600}	δ^{600}	A_{415}/A_{600}	$\delta^{415/600}$
SM1	6.75 ± 0.03	6.73 ± 0.73	0.3	6.57 ± 0.14	2.6
	6.28 ± 0.01	6.17 ± 0.07	1.7	6.24 ± 0.12	0.6
	5.83 ± 0.05	6.07 ± 0.31	**4.1**	6.18 ± 0.29	**6**
	5.35 ± 0.06	5.27 ± 0.37	1.5	5.31 ± 0.30	0.8
	4.89 ± 0.03	4.86 ± 0.31	0.6	4.65 ± 0.54	4.9
SM2	6.91 ± 0.12	7.16 ± 0.34	3.5	6.82 ± 0.11	1.3
	6.43 ± 0.20	6.26 ± 0.39	2.7	6.44 ± 0.22	0.1
	5.71 ± 0.07	5.59 ± 0.15	2.2	5.88 ± 0.17	**2.9**
	5.16 ± 0.17	5.00 ± 0.06	3.1	5.08 ± 0.22	1.6
	4.55 ± 0.16	4.78 ± 0.10	**4.9**	4.52 ± 0.08	0.7
SM3	7.01 ± 0.08	7.04 ± 0.23	0.4	6.96 ± 0.14	0.8
	6.69 ± 0.03	6.74 ± 0.14	0.8	6.72 ± 0.41	0.5
	6.07 ± 0.07	5.99 ± 0.59	1.4	6.14 ± 0.24	**1.1**
	5.37 ± 0.03	5.22 ± 0.16	**2.8**	5.31 ± 0.33	1
	4.78 ± 0.06	4.91 ± 0.23	2.7	4.79 ± 0.16	0.2

3.5. Colorimetric Analysis of RGB Readout

The absorbance changes at both 415 nm and 600 nm of the sensor membranes after contact with the milk samples visually translate into an observable color transition from green to yellow (see Figure 2b). This color transition can be analyzed by taking images with a digital camera and the subsequent evaluation of the data contained in the three color channels. There are several recent examples using the RGB readout of digital camera images for the quantification of color transitions [30,31]. In our evaluation scheme, the total color difference (TCD) derived from the individual color differences in all three channels was used as the analytical signal (see Section 2.4) in the colorimetric analysis of pH (see Figure 2a). This was done for the three membrane types, which is why the effect of the BCP content on the quality of the RGB readout of sensor membranes could be derived. Equations describing the TCD response from the colorimetric analysis that is dependent on the pH of milk and the corresponding correlation coefficients are presented in Table 4. For the SM3 sensor membrane, a meaningful determination of pH using a linear equation is not possible although a difference between pH 7.0 and pH values less than 6.5 can visually be observed (Figure 2b). The fact that SM3 delivers a reliable result using spectrophotometry (as to Table 3) in contrast to the photographic readout is due to the differences in the light detection of both methods. Spectrophotometry detects the negative logarithm of the ratio of the intensity of the transmitted light emanating from the sample with respect to the intensity of the light irradiated into the sample. Both intensities are detected at a wavelength where the maximum change of light intensity is expected. This makes photometry amenable to detect even small changes of the absorbance of a sample i.e., of the pH. The photographic evaluation, however, uses the differences of the intensities of the reflected light of three wavelength ranges (i.e., the RGB colors), each of them integrated over a broad wavelength range. Further, the photometric signal is acquired by the absorption of light within the entire volume of the sensor membrane. The colorimetric analysis of the image only uses the light reflected from the surface of the sensor, i.e., from a much lower volume of the sensor membrane. Finally, the sensitivity of the photomultiplier in the photometer is higher than the one of a CCD chip in a commercial digital camera. All of this contributes to a lower overall working range of SM3 with photographic readout.

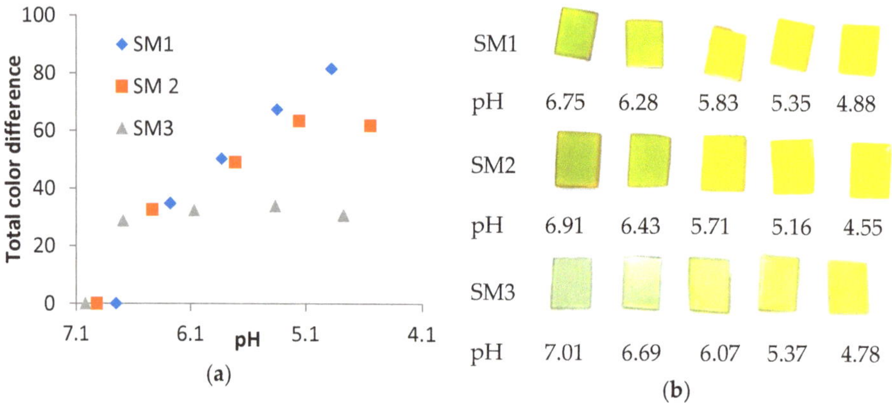

Figure 2. Colorimetric response of sensor membranes: (**a**) relationship of the total color difference (TCD) versus pH of milk samples; (**b**) corresponding images of the sensor membranes.

Table 4. Equations describing the dependence of the TCD on pH.

Type of SM	Range of pH	Equation	*r*
SM1	6.75–4.88	TCD = − 42.059 pH + 291.6 TCD = − 13.275 pH^2 + 112.38 pH − 151.85	0.981 0.996
SM2	6.91–5.16 6.91–5.16	TCD = − 34.077 pH + 242.62 TCD = − 15.667 pH^2 + 155.03 pH − 321.02	0.967 0.990
SM3	Not available	Linear equation not applicable	N/A

Higher contents of BCP in a sensor membrane are accompanied by a wider pH range that correlates linearly to the TCD data. This is also obvious from the best correlation coefficient found for the linear evaluation of the TCD for membrane SM1. Hence the widest measurement range for the colorimetric evaluation via TCD is found for membrane SM1. The pH-range of SM2 is smaller. Using a 2nd order polynomial fit further improves the results of the colorimetric pH evaluation via TCD for SM1 and SM2 to yield very good correlation coefficients that are better than 0.990.

3.6. Validation of Colorimetricmetric Sensor Response

We then compared the pH values determined by colorimetric analysis via the TCD with the potentiometric data (see Table 5). A very good correlation was found specifically for membrane SM1 at all pH values except for the highest one (pH 6.75 vs. pH 6.94). Upon comparing the average deviation of the pH of membrane SM1 determined with either potentiometry or one of the two optical evaluation methods, it becomes clear that the determined colorimetrically determined pHs match the potentiometry data better than the photometry determined pH values (compare data of SM1 in Tables 3 and 5). The standard deviations of the pH determined via colorimetry and photometry, however, are very similar and are typically in a range of 0.1–0.3 pH units. The colorimetrically determined pH for membrane SM2 matches less well with the potentiometry data than that of membrane SM1.

Moreover, the pH-range for the colorimetric evaluation of SM2 is smaller (pH 5.2–7.0) than that of the same membrane evaluated with photometry (pH 4.5–7.0, see Table 3).

Table 5. Validation of the pH obtained in milk by colorimetry with the pH from potentiometry (n = 3).

Type of SM	Potentiometric	Colorimetric	δ,%
SM1	6.75 ± 0.03	6.94 ± 0.13	2.9
	6.28 ± 0.01	6.12 ± 0.25	2.6
	5.83 ± 0.05	5.70 ± 0.44	2.3
	5.35 ± 0.06	5.35 ± 0.21	0.06
	4.89 ± 0.03	4.99 ± 0.13	2.2
SM2	6.91 ± 0.12	7.13 ± 0.30	3.1
	6.43 ± 0.20	6.16 ± 0.22	4.2
	5.71 ± 0.07	5.68 ± 0.10	0.7
	5.16 ± 0.17	5.25 ± 0.11	1.8

4. Conclusions

We demonstrated that the precision and accuracy of the pH determined with an optical sensor based on PMM with immobilized BCP in real milk samples is dependent on the readout method and the concentration of the pH indicator. Two quantification methods of pH were used: (i) spectrophotometry (single wavelength and double wavelength ratiometric) with a standard device and (ii) colorimetric readout based on the evaluation of the total color difference of the sensor images obtained by means of a smartphone camera. It was found that sensor membranes with high BCP content are more preferable for colorimetric pH determination, while for spectrophotometry, pH values gain in precision and accuracy when using lower concentrations of BCP inside the sensor membrane. The validation of the pH obtained by the optical readout of the sensor membrane against the electrochemically determined pH values with a glass electrode showed very similar values. In the case of the ratiometric evaluation of the photometric sensor response, the differences between electrochemical and optical readout were as low as 0.07 pH units, which is equal to the average standard deviation of both methods. Additionally, these sensor membranes can be used for the visual distinction between unspoiled and spoiled milk (pH of less than 6.0). This shows that the pH readings from optical pH sensors in real samples can reach the accuracy and precision of potentiometry as a reference method provided that the concentration of the indicator is tuned and adapted with respect to the optical readout method. A future application of these sensors could be monitoring water deacidification during the preparation of drinking water.

Supplementary Materials: The following are available online at https://www.mdpi.com/article/10.3390/chemosensors9070177/s1. Figure S1. Dependence of an analytical signal on the pH of the milk samples: (a) absorbance at 415 nm; (b) absorbance at 600 nm; (c) ratio A_{415}/A_{600}, n = 3, error bars represent standard deviation.

Author Contributions: Conceptualization, O.V. and A.S.; investigation, O.V. and A.S.; methodology, A.S. and A.D.; validation, O.V. and A.D.; writing—original draft, A.S. and A.D.; writing—review and editing, A.S. and A.D. All authors will be informed about each step of manuscript processing, including submission, revision, revision reminders, etc., via emails from our system or assigned Assistant Editor. All authors have read and agreed to the published version of the manuscript.

Funding: This research was supported by TPU development program.

Conflicts of Interest: The authors declare that they have no known competing financial interests or personal relationships that could appear to influence the work reported in this paper.

References

1. Wencel, D.; Abel, T.; McDonagh, C. Optical Chemical pH Sensors (Review). *Anal. Chem.* **2014**, *86*, 15–29. [CrossRef] [PubMed]
2. Steinegger, A.; Wolfbeis, O.S.; Borisov, S.M. Optical Sensing and Imaging of pH Values: Spectroscopies, Materials, and Applications. *Chem. Rev.* **2020**, *120*, 12357–12489. [CrossRef] [PubMed]
3. Zaragozá, P.; Fuentes, A.; Fernández-Segovia, I.; Vivancos, J.-L.; Rizo, A.; Ros-Lis, J.V.; Barat, J.M.; Martínez-Máñez, R. Evaluation of sea bream (Sparus aurata) shelf life using an optoelectronic nose. *Food Chem.* **2013**, *138*, 1374–1380. [CrossRef]
4. Morsy, M.K.; Zór, K.; Kostesha, N.; Alstrøm, T.S.; Heiskanen, A.; El-Tanahi, H.; Sharoba, A.; Papkovsky, D.; Larsen, J.; Khalaf, H.; et al. Development and validation of a colorimetric sensor array for fish spoilage monitoring. *Food Control* **2016**, *60*, 346–352. [CrossRef]
5. Zhang, X.; Lu, S.; Chen, X. A visual pH sensing film using natural dyes from Bauhinia blakeana Dunn. *Sens. Actuators B Chem.* **2014**, *198*, 268–273. [CrossRef]
6. Byrne, L.; Lau, K.T.; Diamond, D. Monitoring of headspace total volatile basic nitrogen from selected fish species using reflectance spectroscopic measurements of pH sensitive films. *Analyst* **2002**, *127*, 1338–1341. [CrossRef]
7. Kuswandi, B.; Jayus, J.; Restyana, A.; Abdullah, A.; Heng, L.Y.; Ahmad, M. A novel colorimetric food package label for fish spoilage based on polyaniline film. *Food Control* **2012**, *25*, 184–189. [CrossRef]
8. Huang, X.; Xin, J.; Zhao, J. A novel technique for rapid evaluation of fish freshness using colorimetric sensor array. *J. Food Eng.* **2011**, *105*, 632–637. [CrossRef]
9. Lee, K.; Park, H.; Baek, S.; Han, S.; Kim, D.; Chung, S.; Yoon, J.-Y.; Seo, J. Colorimetric array freshness indicator and digital color processing for monitoring the freshness of packaged chicken breast. *Food Packag. Shelf Life* **2019**, *22*, 100408. [CrossRef]
10. Rukchon, C.; Nopwinyuwong, A.; Trevanich, S.; Jinkarn, T.; Suppakul, P. Development of a food spoilage indicator for monitoring freshness of skinless chicken breast. *Talanta* **2014**, *130*, 547–554. [CrossRef]
11. Salinas, Y.; Ros-Lis, J.V.; Vivancos, J.-L.; Martínez-Máñez, R.; Marcos, M.D.; Aucejo, S.; Herranz, N.; Lorente, I.; Garcia, E. A novel colorimetric sensor array for monitoring fresh pork sausages spoilage. *Food Control* **2014**, *35*, 166–176. [CrossRef]
12. Genovese, M.E.; Abraham, S.; Caputo, G.; Nanni, G.; Kumaran, S.K.; Montemagno, C.D.; Athanassiou, A.; Fragouli, D. Photochromic Paper Indicators for Acidic Food Spoilage Detection. *ACS Omega* **2018**, *3*, 13484–13493. [CrossRef] [PubMed]
13. Ma, Q.; Wang, L. Preparation of a visual pH-sensing film based on tara gum incorporating cellulose and extracts from grape skins. *Sens. Actuators B Chem.* **2016**, *235*, 401–407. [CrossRef]
14. Goodarzi, M.M.; Moradi, M.; Tajik, H.; Forough, M.; Ezati, P.; Kuswandi, B. Development of an easy-to-use colorimetric pH label with starch and carrot anthocyanins for milk shelf life assessment. *Int. J. Biol. Macromol.* **2020**, *153*, 240–247. [CrossRef]
15. Lakade, A.J.; Sundar, K.; Shetty, P.H. Nanomaterial-based sensor for the detection of milk spoilage. *LWT Food Sci. Technol.* **2017**, *75*, 702–709. [CrossRef]
16. Weston, M.; Phan, M.A.T.; Arcot, J.; Chandrawati, R. Anthocyanin-based sensors derived from food waste as an active use-by date indicator for milk. *Food Chem.* **2020**, *326*, 127017. [CrossRef]
17. Pirsa, S.; Zandi, M.; Almasi, H. Determination of Quality and Spoilage of Milk by Synthesized Polypyrrole-Ag Nanocomposite Fiber at Room Temperature. *J. Food Process. Eng.* **2015**, *39*, 266–272. [CrossRef]
18. Nopwinyuwong, A.; Trevanich, S.; Suppakul, P. Development of a novel colorimetric indicator label for monitoring freshness of intermediate-moisture dessert spoilage. *Talanta* **2010**, *81*, 1126–1132. [CrossRef] [PubMed]
19. Lu, M.; Shiau, Y.; Wong, J.; Lin, R.; Kravis, H.; Blackmon, T.; Pakzad, T.; Jen, T.; Cheng, A.; Chang, J.; et al. Milk Spoilage: Methods and Practices of Detecting Milk Quality. *Food Nutr. Sci.* **2013**, *4*, 113–123. [CrossRef]
20. Gavrilenko, N.A.; Saranchina, N.V.; Sukhanov, A.; Fedan, D.A. Reversible pH-sensitive element based on bromocresol purple immobilized into the polymethacrylate matrix. *Mendeleev Commun.* **2018**, *28*, 450–452. [CrossRef]
21. Saranchina, N.V.; Damzina, A.A.; Ermolaev, Y.E.; Urazov, E.V.; Gavrilenko, N.A.; Gavrilenko, M.A. Determination of antioxidant capacity of medicinal tinctures using cuprac method involving Cu(II) neocuproine immobilized into polymethacrylate matrix. *Spectrochim. Acta Part A Mol. Biomol. Spectrosc.* **2020**, *240*, 118581. [CrossRef]
22. Gavrilenko, N.A.; Fedan, D.A.; Saranchina, N.V.; Gavrilenko, M.A. Solid phase colorimetric determination of iodine in food grade salt using polymethacrylate matrix. *Food Chem.* **2019**, *280*, 15–19. [CrossRef]
23. Gavrilenko, N.A.; Volgina, T.N.; Pugachev, E.V.; Gavrilenko, M.A. Visual determination of malachite green in sea fish samples. *Food Chem.* **2019**, *274*, 242–245. [CrossRef]
24. Dudkina, A.A.; Volgina, T.N.; Saranchina, N.V.; Gavrilenko, N.A.; Gavrilenko, M.A. Colorimetric determination of food colourants using solid phase extraction into polymethacrylate matrix. *Talanta* **2019**, *202*, 186–189. [CrossRef]
25. Saranchina, N.V.; Slizhov, Y.G.; Vodova, Y.M.; Murzakasymova, N.S.; Ilyina, A.M.; Gavrilenko, N.A.; Gavrilenko, M.A. Smartphone-based colorimetric determination of fluoride anions using polymethacrylate optode. *Talanta* **2021**, *226*, 122103. [CrossRef] [PubMed]
26. Ziyaina, M.; Govindan, B.N.; Rasco, B.; Coffey, T.; Sablani, S.S. Monitoring Shelf Life of Pasteurized Whole Milk Under Refrigerated Storage Conditions: Predictive Models for Quality Loss. *J. Food Sci.* **2018**, *83*, 409–418. [CrossRef] [PubMed]
27. Bishop, E. *Indicators*; Pergamon Press: New York, NY, USA, 1972.
28. Gavrilenko, M.A.; Gavrilenko, N.A. Method of Producing Polymethyl Methacrylate for Solid-Phase Extraction. Worldwide Patent Application No RU2638929C1, 19 December 2017.

29. Brodziak, A.; Król, J.; Litwińczuk, Z.; Zaborska, A.; Czernecki, T. Effect of storage time under home refrigeration conditions on the quality of opened drinking milk. *Mljekarstvo* **2017**, *67*, 283–296. [CrossRef]
30. Mobarez, S.N.; Wongkaew, N.; Simsek, M.; Baeumner, A.J.; Duerkop, A. Dipsticks with Reflectometric Readout of an NIR Dye for Determination of Biogenic Amines. *Chemosensors* **2020**, *8*, 99. [CrossRef]
31. Schanzenbach, C.; Ilver, D.; Blomgren, J.; Jonasson, C.; Johansson, C.; Krozer, A.; Ye, L.; Rustas, B.-O. Preparation and characterisation of a sensing system for wireless pH measurements in vivo, in a rumen of a cow. *Sens. Actuators B Chem.* **2017**, *242*, 637–644. [CrossRef]

chemosensors

MDPI

Article

Preparation, Characterization and Electrochemical Response of Nanostructured TiAlV with Potentiostatically Deposited IrOx as a pH Sensor for Rapid Detection of Inflammation

Jitřenka Jírů [1], Vojtěch Hybášek [1], Alena Michalcová [1], Klára Korbelová [1], Lukáš Koláčný [2] and Jaroslav Fojt [1,*]

[1] Department of Metals and Corrosion Engineering, University of Chemistry and Technology Prague, Technická 5, 166 28 Prague, Czech Republic; jiruj@vscht.cz (J.J.); hybasekv@vscht.cz (V.H.); michalca@vscht.cz (A.M.); korbelok@vscht.cz (K.K.)

[2] Department of Organic Technology, University of Chemistry and Technology Prague, Technická 5, 166 28 Prague, Czech Republic; kolacnyl@vscht.cz

* Correspondence: fojtj@vscht.cz

Abstract: Electrochemical pH sensors have a wide range of industrial applications such as in medicine due to their fast response and high sensitivity to pH changes. This work focuses on the preparation of samples based on the nanostructure of TiO_2 with potentiostatically deposited particles of iridium and its oxides (IrO_2), using a Ti-6Al-4V alloy as the base material, and subsequent surface characterization. Transmission electron microscopy and secondary ion mass spectroscopy showed Ir particles distributed in the nanotubes. Using a potentiostatic method, a stable pH sensor was prepared. By monitoring the open circuit potential, it was shown that this sensor is usable even without being kept in a storage medium and does not react to changes in the redox potential of the solution.

Keywords: pH sensor; Ir oxide; implant; electrochemistry; titanium

Citation: Jírů, J.; Hybášek, V.; Michalcová, A.; Korbelová, K.; Koláčný, L.; Fojt, J. Preparation, Characterization and Electrochemical Response of Nanostructured TiAlV with Potentiostatically Deposited IrOx as a pH Sensor for Rapid Detection of Inflammation. *Chemosensors* **2024**, *12*, 109. https://doi.org/10.3390/chemosensors12060109

Received: 17 May 2024
Revised: 7 June 2024
Accepted: 10 June 2024
Published: 11 June 2024

1. Introduction

Infection is one of the possible causes of implant failure in the human body. As with all invasive surgical procedures, arthroplasty is not completely risk-free. Medical device-associated infection is a worldwide health problem because it is very difficult to diagnose, very difficult to treat and very expensive to treat. The complexity of diagnosing these infections adds to the challenge, often relying on clinical symptoms reported by patients, such as pain, swelling, or compromised joint mobility. However, these symptoms can be non-specific and may overlap with other post-operative complications, necessitating a comprehensive evaluation for accurate diagnosis. Moreover, infection detection often relies on non-specific markers such as elevated leukocyte count and heightened C-reactive protein (CRP) levels, further complicating the diagnostic landscape [1–3]. The efficacy of antibiotic therapy is also a growing concern, given the emergence of antibiotic-resistant strains, posing significant challenges in infection management. Despite the administration of comprehensive treatment, infections often result in surgical removal of the implant. It is also associated with a high mortality of patients [4–6].

The spread of infection occurs due to the preference of bacteria to exist on the implant surface in established communities known as biofilms. Biofilm formation occurs by the attachment and aggregation of free-floating bacteria on the implant surface. The bacteria that exist in the biofilm are part of a complex multicellular community enclosed in extracellular polymeric substances (EPSs). The EPSs produced by the bacteria form a 'slime layer' around the cells, consisting mainly of water and a range of polysaccharides, nucleic acids, proteins and lipids. Bacterial biofilms can resist antibiotics, disinfectants, phagocytosis and other components of the host's innate and adaptive immune system. The main microorganisms that cause infection and subsequent inflammation include *Staphylococcus aureus*,

S. epidermidis and *S. hominis*, among others. Infecting organisms can be implanted with biomaterials into the body [1,7,8]. These bacteria are known to produce acid metabolites, including lactic acid, which exert a significant impact on the pH levels of the surrounding tissue [9,10].

The pH value stands as a pivotal parameter across numerous domains including industry, agriculture, biology, environment and medicine. Traditionally, pH determination has relied on the utilization of glass electrodes. However, despite its widespread use, the glass electrode method presents several drawbacks. High impedance, significant cost, bulky dimensions, restricted shape versatility and susceptibility to mechanical damage are among the primary limitations associated with this conventional approach. These constraints not only impede the widespread deployment of pH measurement systems but also hinder their adaptability to diverse environmental and operational conditions. Consequently, the exploration of alternative pH sensing technologies has emerged as a critical endeavor, aiming to overcome these limitations and deliver more robust, cost-effective and versatile solutions for pH monitoring across various applications [9,11–14].

Over the past two decades, a primary objective has been the exploration of materials with the potential to serve as effective pH sensors. Notably, metal oxide (MOx) electrodes have emerged as promising candidates due to their ability to respond to changes in pH [15]. Unlike traditional glass electrodes, MOx electrodes offer the advantage of adjustability in dimensions, addressing concerns related to size and shape limitations. Consequently, researchers have extensively investigated various MOx electrodes, including titanium dioxide (TiO_2), aluminium oxide (Al_2O_3), platinum dioxide (PtO_2), iridium dioxide (IrO_2) and tungsten trioxide (WO_3), among others, for their suitability as pH sensors. These investigations have revealed promising properties of MOx electrodes, including enhanced sensitivity, improved stability and greater versatility, positioning them as viable alternatives to conventional pH measurement technologies [16–22].

TiO_2 thin film is an n-type semiconductor renowned for its exceptional chemical stability, rendering it a promising candidate for pH sensing applications, particularly in highly acidic or alkaline solutions. Its inherent qualities include excellent chemical stability, making it a sought-after material in various industrial and biomedical contexts. TiO_2 boasts a reputation for being non-toxic, environmentally friendly and corrosion-resistant, further enhancing its appeal for diverse applications. Of particular note is its remarkable biocompatibility, making it invaluable for medical applications, where materials must interact safely with biological systems [23,24]. Moreover, TiO_2 exhibits unique ionic and electrical properties, distinguishing it from other oxides and amplifying its utility in sensor technologies. Multiple fabrication techniques exist for producing TiO_2 nanostructures, among which sol–gel and electrochemical methods are prominent [22,23,25].

IrO_2, among the array of metal oxides considered for pH sensing, has emerged as a standout candidate due to its remarkable stability across a broad pH spectrum, even under high-pressure conditions and in aggressive environments. Notably, IrO_2 exhibits rapid response characteristics, even in non-aqueous solutions, making it exceptionally versatile for various pH measurement applications. Crystallizing in a rutile structure akin to that of ruthenium dioxide (RuO_2), IrO_2 demonstrates a lower catalytic activity compared to RuO_2. Despite this distinction, IrO_2-based electrodes have garnered considerable attention, particularly in biomedical contexts, owing to their impressive stability and performance metrics [11,19–21].

In numerous biomedical applications, iridium oxide-based electrodes have proven invaluable. Compared to other materials such as antimony, titanium, ruthenium or palladium-based electrodes, iridium oxide layers offer distinct advantages. These include faster response times across a wide pH range and superior biocompatibility, underscoring their suitability for interfacing with biological systems. The utilization of IrO_2 electrodes not only enhances the precision and reliability of pH measurements but also aligns with the stringent requirements of biomedical applications, where accuracy and compatibility are paramount considerations [6,14,16,22,25].

In this work, the main objective was to reproducibly create a pH sensor based on TiO_2 nanotubes with deposited iridium. This prepared sensor could one day be part of an implant for the rapid detection of inflammation, and subsequently, its surface could be characterized and its electrochemical response to pH change could be investigated. The use of this sensor without the need for a storage medium was also tested. The use of a metal sensor without a storage medium would reduce two disadvantages of conventional glass pH electrodes.

2. Materials and Methods

As a base material, the Ti-6Al-4V ELI alloy (extra-low interstitials, Ti grade 23) was selected. The diameter of the samples was 16 mm and the height was 3 mm. The surface was ground with a series of abrasive papers up to FEPA 2500 grit. The samples were then rinsed with distilled water, ethanol and acetone. By anodic oxidation, the surface of the prepared samples was nanostructured in a solution of 1 mol/L $(NH_4)_2SO_4$ (Lach:ner, Neratovice, Czech Republic) and 0.2 wt.% NH_4F (Lach:ner) in a PTFE cell. A standard three-electrode setup consists of a silver/silver chloride electrode (3 mol/L KCl, further labeled as SSCE), a glassy carbon counter electrode, and the sample as the working electrode. The nanostructure preparation process consisted of two parts: in the first part, the potential was increased to 20 V/SSCE at a rate of 100 mV/s, and the second part was potentiostatic for 2200 s. The nanostructured alloy was subsequently cleaned in distilled water and ethanol in an ultrasonic cleaner and dried with hot air.

For the deposition of iridium oxide on the nanostructured Ti-6Al-4V surface, a solution of 0.2 mmol/L $IrCl_3 \cdot xH_2O$ (Sigma-Aldrich, St. Louis, MO, USA), 1 mmol/L $C_2H_2O_4 \cdot 2H_2O$ (oxalic acid, Lach:ner) and 5 mmol/L K_2CO_3 (Lach:ner) was used. The solution was aged for 4 days at 37 °C and stored at 5 °C prior to use. The deposition was carried out using potentiostatic mode (−0.4 V/SSCE, 7200 s; further labeled as Ir PS) with an exposure area of 1 cm^2, and the three-electrode setup was used. After deposition, the samples were ultrasonically cleaned in distilled water.

Nanostructured Ti-6Al-4V and Ir PS underwent an oxidation by cyclic polarization (E_{OCP} = 3400 s, −0.05 V/E_{OCP} to 0.1 V/E_{OCP} and back to 0.01 V/E_{OCP}, 20 cycles) in a physiological solution (9 g/L NaCl, Lach:ner). These samples will be further labeled as TiAlV and Ir PS+CP.

Scanning electron microscopy (SEM, TESCAN MIRA 3 LMU with OXFORD INCA 350 EDS analyzer, Brno, Czech Republic) was used for the morphological analysis of the prepared samples. Further analysis was performed using a transmission electron microscope (TEM, EFTEM Jeol 2200 FS, Tokyo, Japan). Samples for TEM (lamellas) were created using SEM (TESCAN LYRA 3, Brno, Czech Republic) and a focused ion beam (FIB) using gallium ions. Secondary ion mass spectroscopy (SIMS) was measured on the same SEM. Image analysis was performed manually using Image J 2.15.0 software.

Electrochemical measurements were realized with potentiostat Gamry Instrument Reference 600 (Warminster, USA) at 37 °C, with the same setup as mentioned earlier, with an exposed area of 1 cm^2. For pH changes, a physiological solution (9 g/L NaCl) with biological buffer TES (N tris(hydroxymethyl)methyl-2-aminoethanesulfonic acid, 5.9 g/L, Sigma-Aldrich, St. Louis, MO, USA) was used. For pH changes, a scale of 7.6, 7.4, 7.0 and 6.5 was chosen and the pH was adjusted with diluted NaOH. The electrochemical response of the prepared samples to the pH change was detected using the monitoring of open circuit potential (E_{OCP}). In order to investigate the effect of the redox potential of the system on the detection of pH change, measurements were performed with a platinum electrode as the working electrode and a silver chloride electrode as the reference electrode. At the same time, a sample was connected in the same cell as the working electrode and a silver/silver chloride electrode as a reference. During the measurement, the solution at pH 7.6 and 6.5 was bubbled first with N_2 and then with air. The potential recording was carried out simultaneously on two potentiostats.

3. Results and Discussion

3.1. Base Material

A nanotubular TiO_2 structure was formed on the Ti-6Al-4V surface by anodic oxidation in a fluoride solution. This procedure was consistent with published procedures [26,27]. The surface morphology is shown in Figure 1. It can be seen that the nanostructure is not homogenous as dissolved regions of the β-phase are present, which is due to the fact that this phase is richer in vanadium than the α-phase, which is richer in aluminum [28]. In a fluoride environment, vanadium dissolves faster than aluminum, this leads to a faster dissolution of the β-phase and the formation of nanotubes is not uniform.

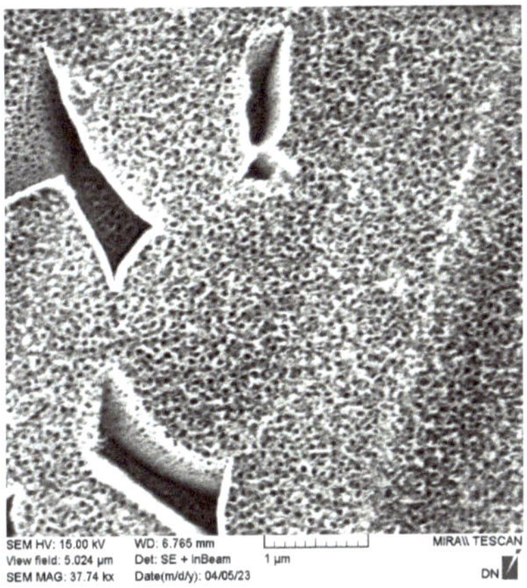

Figure 1. Native 20V TiO_2 nanotubes on the Ti-6Al-4V alloy.

After surface preparation, the sample was further oxidized by cyclic polarization, the course of cyclic polarization is shown in Figure 2a. The TiAlV sample was immersed in the physiological solution at pH 7.6 and the potential course was monitored (Figure 2b). During this time, the nanotubes were flooded and hydrated compounds formed on their surface. This process took 6 days, after which the sample was put into solutions of different pH and the electrochemical response was monitored. The result is shown in Figure 3. TiAlV exhibited a sub-Nernstian response of 31 ± 1 mV/pH.

(**a**) (**b**)

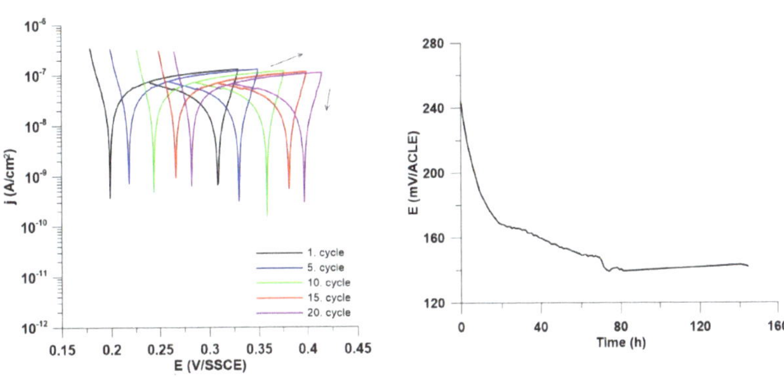

Figure 2. (**a**) The current and potential dependencies of oxidation of TiAlV, arrows indicating polarization direction; (**b**) the process of the potential stabilization of the TiAlV in physiological solution at pH 7.6 at 37 °C.

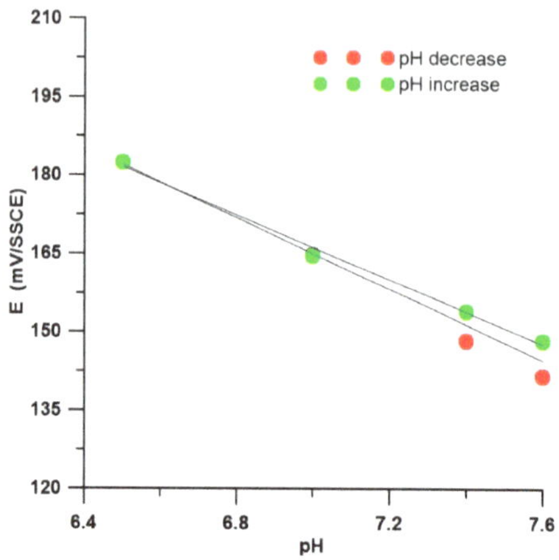

Figure 3. Results of potential change in pH change for TiAlV in physiological solution at 37 °C.

3.2. Sample Preparation and Characterization

The fresh solution of Ir was yellow, but after 4 days of aging at 37 °C, the color changed to blue, still having pH 10. This corresponded to the formation of iridium complexes in the solution, which are in line with the outcomes of Cruz et al. [21]. In their work, they reported that deposited iridium most likely consists of mixed oxides and hydroxides, such as $IrO_{1.1}(OH)_{2.7}{\cdot}0.4H_2O$ and IrO_2. Taking into account the possibility of the presence of IrO_2, there are two possible mechanisms of redox reaction when the pH changes.

$$IrO_2 + 4\,H^+ + 4\,e^- \leftrightarrow Ir + 2\,H_2O$$

$$2\,IrO_2 + 2\,H^+ + 2\,e^- \leftrightarrow Ir_2O_3 + H_2O$$

$$Ir_2O_3 + 6\,H^+ + 6\,e^- \leftrightarrow 2\,Ir + 3\,H_2O$$

This leads to the redox potential according to Equation (1) and subsequently results in a Nernstian response of 59 mV/pH:

$$E = E^0 - 2.303\left(\frac{RT}{nF}\right)pH = E^0 - 0.059\ pH \quad (1)$$

where E is the redox potential (in mV), E^0 is the standard electrode potential (in mV), R is the universal gas constant (8.314 $JK^{-1}mol^{-1}$), T is the absolute temperature (in K), F is the Faraday constant (96,485 $Cmol^{-1}$) and n is the number of electrons transferred in the reaction [16,29].

In the presence of oxohydroxides, the following reactions are possible:

$$Ir(OH)_3 \rightarrow Ir(OH)_2O^- + H^+$$

$$Ir_2O(OH)_6 \rightarrow Ir_2O(OH)_3O_3^{3-} + 3\,H^+$$

$$2\,Ir(OH)_3 + H_2O \leftrightarrow Ir_2O(OH)_6 + 2\,H^+ + 2\,e^-$$

$$2\,Ir(OH)_2O^- + H_2O \leftrightarrow Ir_2O(OH)_3O_3^{3-} + 3\,H^+ + 2\,e^-$$

In this case, the response is equal to 88.5 mV/pH [30].

The current and potential dependency of the potentiostatic deposition is shown in Figure 4a. A blank measurement was carried out using a solution with the same pH as the Ir solution, and equal amounts of $C_2H_2O_4 \cdot 2H_2O$, K_2CO_3 and Cl^- ions using NaCl. From the blank measurements, the total charge of the blank experiment was calculated. That charge is equivalent to the charge of the electrolyte reaction. The total amount of deposited Ir particles was calculated by Faraday's law from the total charge minus the charge from the blank measurement. It was found that the total amount of Ir present on the surface of the samples averaged 0.015 ± 0.007 mg. Figure 2b shows the oxidation process in the physiological solution, which was carried out to oxidize Ir particles on the TiAlV surface.

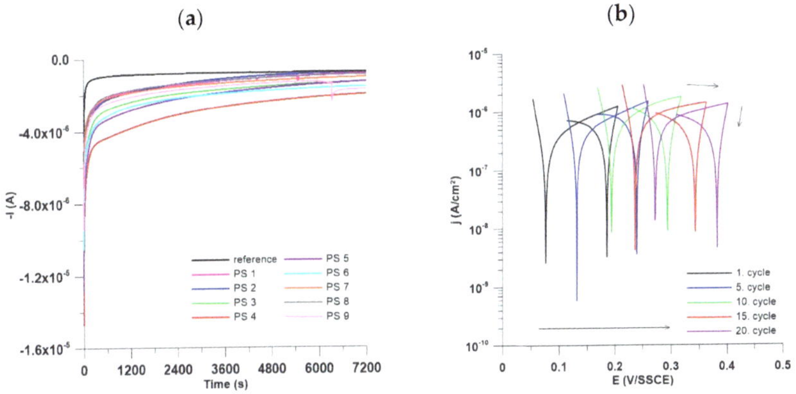

Figure 4. The current and potential dependencies of (**a**) potentiostatic deposition and (**b**) oxidation.

Figure 5a shows Ir PS and Figure 5b displays Ir PS+CP samples. IrO_2 particles were deposited on the samples at the edges of the β-phase cavities and nanotube edges. The diameter of the nanotubes and the width of their walls are summarized in Table 1. The length of the nanotubes was then determined from the TEM image (Figure 5c). These results correspond with those of previously published works [27,31].

Table 1. Diameter, length and wall thickness of nanotube samples Ir PS and Ir PS+CP.

	Nanotube Diameter (nm)	Wall Thickness (nm)	Length (nm)
Ir PS	97.1 ± 7.1	20.9 ± 2.4	429 ± 19
Ir PS+CP	91.0 ± 5.7	19.5 ± 2.5	363.3 ± 8.8

Figure 6 shows the TEM-EDS map of the iridium distribution in the lamella obtained from the (a) Ir PS and (b) Ir PS+CP samples. According to this map, deposited iridium was mainly concentrated on top of the nanotubes in both cases. Only small amounts were detected on the bottom. The main difference between these samples was the amount of Ir detected. Further analyses were subsequently performed to better determine the distribution and the amount of Ir particles.

Figures 7 and 8 show TEM-EDS images with the analyzed areas marked. To confirm the assumption that the deposited Ir occurred predominantly at the tops of the nanotubes, analyses were performed here and at the base of the nanotubes. The results of the analyses are summarized in Tables 2 and 3. The EDS results confirmed that the highest concentration of Ir particles is at the surface of the nanotubes on both samples, while at the same time, the distribution occurs at the bottom of the nanotubes. Platinum was found on the samples from the TEM lamella preparation process, where the surface of the sample is covered with a thin layer of Pt before the lamella is formed and removed, thus protecting the lamella and surrounding area from Ga ions. After the cyclic polarization step, the iridium concentration was reduced as some of it dissolved back into the solution.

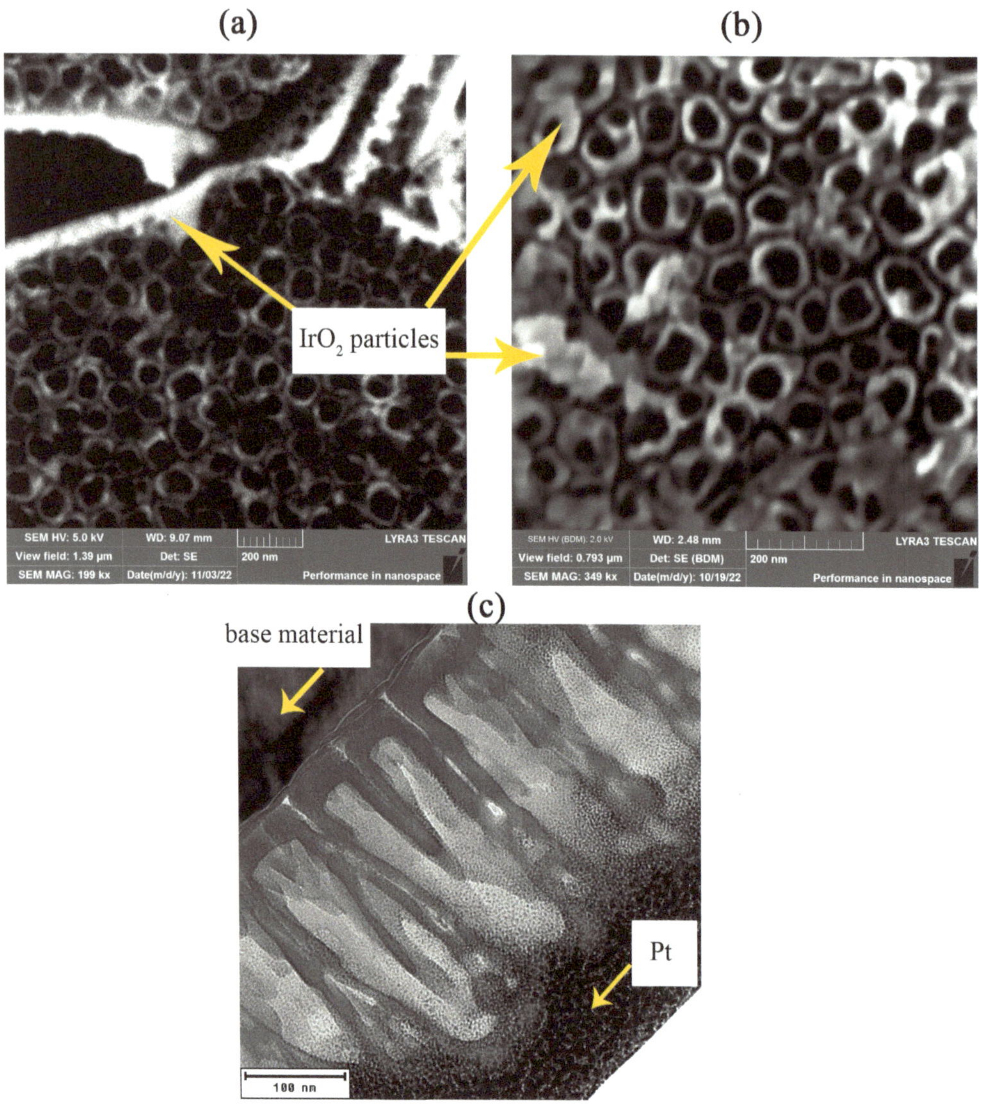

Figure 5. Iridium particles deposited on nanotubes: (**a**) Ir PS (SEM), (**b**) Ir PS+CP (SEM) and (**c**) Ir PS+CP side views of nanotubes (TEM).

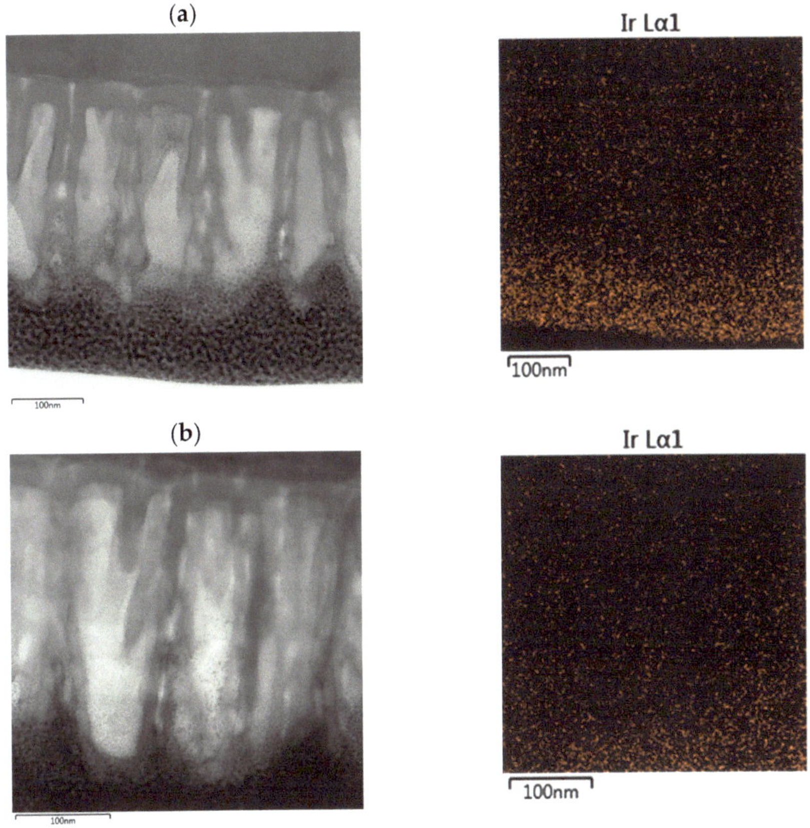

Figure 6. TEM-EDS map of Ir distribution in the lamella from (**a**) Ir PS and (**b**) Ir PS+CP samples.

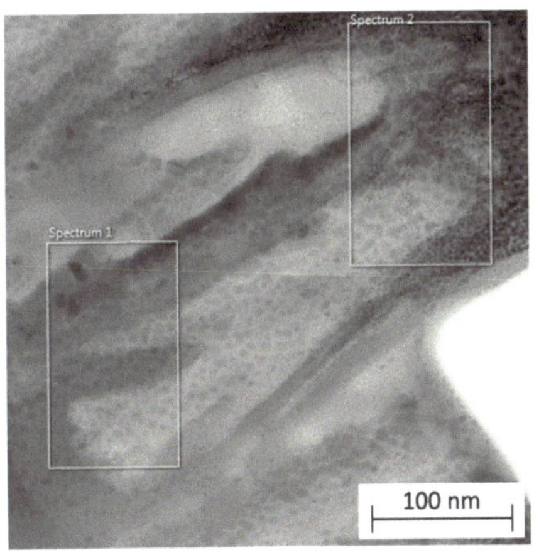

Figure 7. Areas of TEM-EDS analysis of Ir PS for the results shown in Table 2.

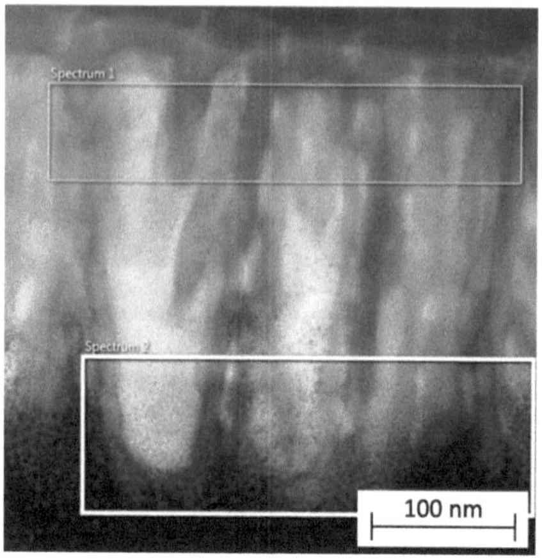

Figure 8. Areas of TEM-EDS analysis of Ir PS+CP for the results shown in Table 3.

Table 2. Results (wt.%) of TEM-EDS analysis of Ir PS.

	Base of Nanotubes (Spectrum 1)	**Top of Nanotubes (Spectrum 2)**
Ti	84.6 ± 1.2	39.9 ± 0.8
Al	7.0 ± 0.4	3.2 ± 0.2
V	3.4 ± 0.5	1.7 ± 0.3
Ir	4.8 ± 0.9	8.7 ± 0.9
Pt	0.9 ± 0.8	46.6 ± 1.0

Table 3. Results (wt.%) of TEM-EDS analysis of Ir PS+CP.

	Base of Nanotubes (Spectrum 1)	**Top of Nanotubes (Spectrum 2)**
Ti	85.6 ± 1.3	47.3 ± 1.6
Al	7.9 ± 0.5	4.6 ± 0.5
V	4.4 ± 0.6	2.1 ± 0.5
Ir	2.1 ± 1.3	3.6 ± 1.9
Pt	-	42.3 ± 1.7

Figure 9 displays the iridium distribution profiles obtained by SIMS for both Ir PS and Ir PS+CP samples. The data unequivocally verify that the highest concentration of iridium is localized on the surface of the nanotubes. Remarkably, the iridium content within the remainder of the nanotube structure is nearly identical for both samples. This intentional distribution of iridium throughout the nanotube architecture is crucial for achieving a pH-sensitive surface, as it ensures a uniform exposure of the sensing material to the surrounding environment. This strategic design not only enhances the sensitivity of the sensor but also promotes consistent and reliable pH measurements across the sensor's surface. Compared to previously published works [11,17–19,25,29,30], our presented IrO_2 deposition procedure achieves a reproducible and uniform distribution of particles on the surface of the titanium substrate. By not sealing the nanotubes with Ir particles, the purpose of nanotube formation is preserved, namely to increase the real area compared to the geometric one and maintain the ability to direct biofilm formation detection [27].

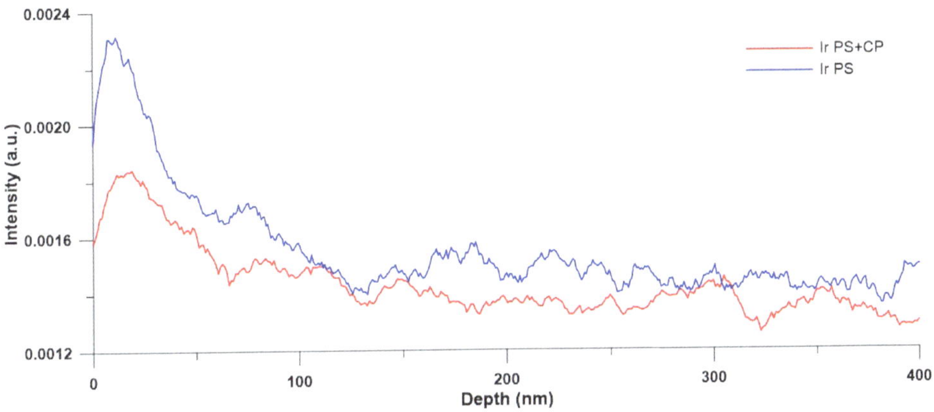

Figure 9. Depth distribution of Ir in Ir PS and Ir PS+CP samples obtained by the SIMS method.

3.3. *Electrochemical Response to pH Change*

To study the electrochemical response of the sample to the pH change, Ir PS+CP was chosen. The pH range from 6.5 to 7.6 was chosen to reflect the physiological environment with and without inflammation [7]. The first measurements were carried out with the sample, which, after preparation and cleaning, was only kept in a sterile container. The results presented in Figure 10 show that the Ir PS+CP surface responded to the change in pH with a sensitivity of 37 ± 3 mV/pH when the first recorded potential data were taken immediately as the pH decreased. When the data were taken after 5 min, the sensitivity was already 43 ± 3 mV/pH. In order to verify the reliability and stability of the measurements, the experiment was continued with a backward pH increase with the same sample. During the experiment, the hydration of the surface was achieved and the resulting sensitivity was 55 ± 2 mV/pH when the pH was increased.

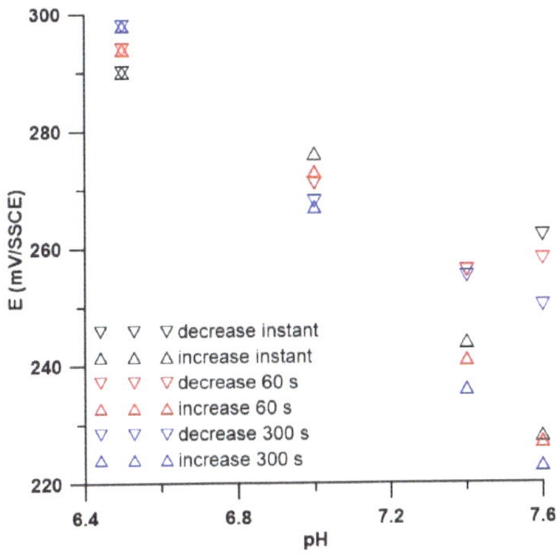

Figure 10. The electrochemical response of Ir PS+CP without initial stabilization.

As the storage medium, the physiological solution with pH 7.6 was chosen. The sample Ir PS+CP was placed in this solution and the potential course was monitored for 40 h (Figure 11). After this time, the surface response to pH change was observed again. Leaving the sample in the storage medium resulted in the hydration of the surface and a more stable response, which was around 45 ± 2 mV/pH for both decreasing and increasing pH values.

The change in pH was recorded by the change in potential immediately after immersion in the solution. However, a better and more stable recording of the electrochemical response was made after 90 s. These data are summarized in Figure 12.

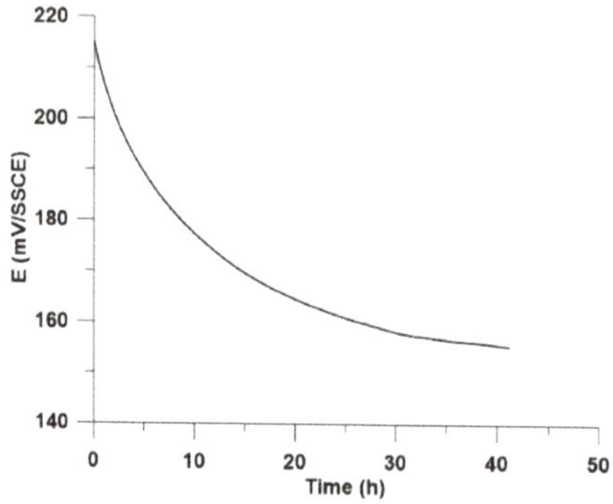

Figure 11. Stabilization of Ir PS+CP sample in a physiological solution at pH 7.6 and 37 °C.

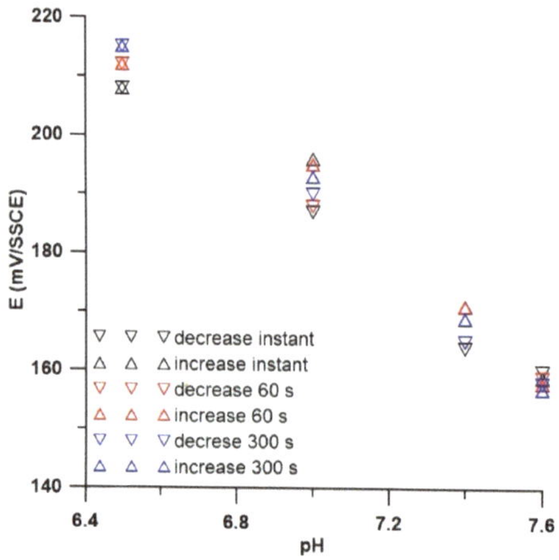

Figure 12. The electrochemical response of Ir PS+CP after 40 h of stabilization.

Another sample was left in the storage medium for 5 days. During that time, the potential course was monitored as an indicator of surface stability (Figure 13). After this time, the electrochemical response was again measured over the selected pH range, as shown in Figure 14. From the first recorded potential value, it was already possible to recognize a change in the pH of the solution. In this case, the surface showed a response of 50 $\pm$ 2 mV/pH. If the potential value after 5 min was taken into account, the recording was already stable and the response was 55 $\pm$ 1 mV/pH. The electrochemical response analysis of both TiAlV and Ir PS+CP samples reveals intriguing insights into the Nernstian behavior observed in the prepared system. Beyond the contribution of iridium oxides, it becomes apparent that titanium oxides also play a significant role in facilitating the Nernstian response exhibited by the Ir PS+CP sample.

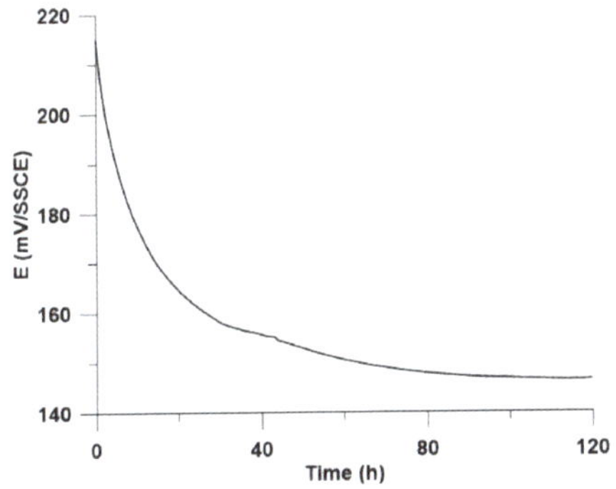

Figure 13. Stabilization of another Ir PS+CP sample in a physiological solution at pH 7.6 and 37 °C.

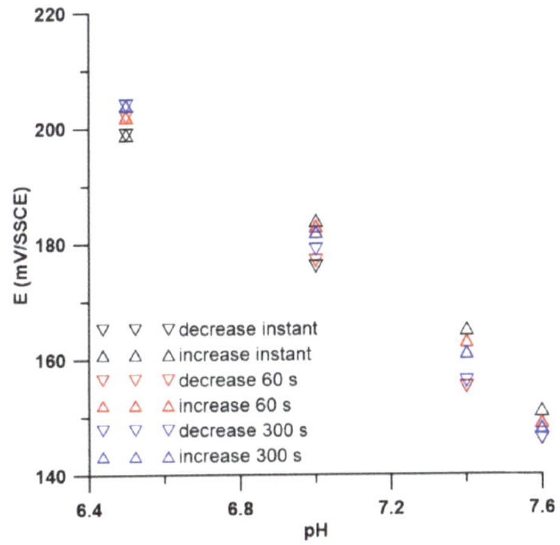

Figure 14. The electrochemical response of Ir PS+CP after 5 days of stabilization.

All previous tests were performed in a 200 mL solution. In the work of Prats-Alfonso et al. [20], the influence of the size of the pH electrode with IrO_2 and their sensitivity was tested. Their work concluded that its size has no effect. On the other hand, the effect of the volume of the tested solution on the sensor response was not tested. In the case of synovial fluid collection in patients with suspected infectious inflammation, specific treatment should be administered. However, in this case, a culture is also needed to determine which bacteria are causing the inflammation around the implant. In this regard, an attempt was made to determine the electrochemical response of the TiO_2-IrO_2 system to a change in pH in a volume of 100 μL. The sample had been in storage solution for a week prior to this experiment. For the detection of pH change, pH values of 7.6 and 6.5 were chosen. Between each measurement, the sample surface was meticulously rinsed with distilled water to mitigate potential contamination and ensure the integrity of subsequent readings, and an SSCE microreference electrode was used as a reference electrode. Remarkably, even within this microvolume setting, the TiO_2-IrO_2 system exhibited remarkable sensitivity to pH changes, with a recorded sensitivity of 55 ± 1 mV/pH. The resulting potential response is shown in Table 4. These results are very positive in view of the intended use in medicine. As discussed in the work of Taheri et al. [16], for pH sensors used in biological environments, rapid response and high sensitivity are very important. An acceptable response to pH change was achieved even in a small volume and short time. Compared to published procedures, potentiostatic deposition appears to be more controllable and economically less demanding [32,33].

Table 4. Results of the pH change monitoring in the microvolume on the surface of Ir PS+CP.

pH	E (mV/SSCE)
7.6	144
6.5	205
7.6	145
6.5	208

Prior to conducting the experiment on the effect of redox potential on the response, the Ir PS+CP sample underwent a preconditioning phase where it was immersed in a storage solution for a week to ensure its stability and consistency. The actual measurement started in a solution of pH 7.6. At the moment when the bubbling of nitrogen started, a potential change was immediately detected at the platinum electrode. In contrast, the prepared Ir PS+CP surface showed no noticeable response to this environmental change. Subsequently, after an hour, when air was reintroduced into the solution, the electrochemical response of Ir PS+CP remained unaffected, contrasting with the detectable changes observed at the platinum electrode. Following this phase, the solution's pH was adjusted to 6.5, prompting an immediate response from the Ir PS+CP sample in accordance with its previously determined sensitivity of 55 ± 1 mV/pH. Approximately 400 s after the pH adjustment, nitrogen bubbling was reintroduced into the solution. Notably, while the platinum electrode exhibited a corresponding change in potential, indicative of its responsiveness to redox fluctuations, no such change was observed for the Ir PS+CP sample. Once the air was reintroduced into the solution, a potential shift was once again recorded for the platinum electrode, while the Ir PS+CP surface maintained its steady electrochemical response. This whole process is summarized in Figure 15.

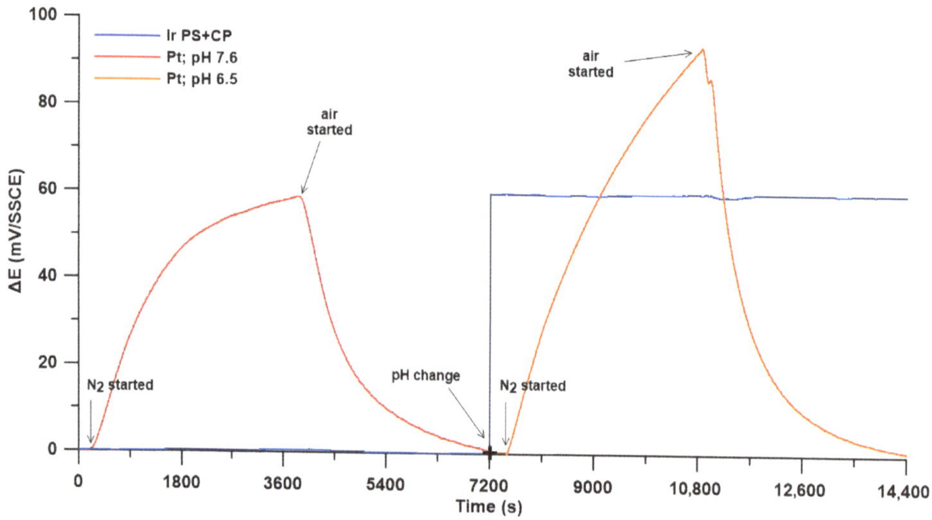

Figure 15. Effect of redox potential on the monitoring of pH change.

From the result of the EDS before and after the exposition (Table 5) of the Ir PS+CP sample, it was proven that there was no dissolution of IrO_2 into the solution during the pH changes. The fluoride listed in the table was a residue from the nanotube preparation that did not affect the sample response. This finding emphasizes the robustness and stability of the IrO_2-based pH sensor and confirms its suitability for prolonged exposure to varying pH conditions without compromising its structural integrity or performance. The absence of IrO_2 dissolution further supports confidence in the sensor's reliability and longevity, offering valuable insights for its potential applications in diverse environments and scenarios requiring precise pH monitoring capabilities. At the same time, when compared to the work of Liu et al. [17] where both iridium and ruthenium oxides were deposited on a polished titanium substrate, simpler fabrication was achieved with comparable results. At the same time, the preparation of the nanostructure resulted in a much larger real surface area compared to the geometric one, and hence the gain of a larger pH-sensitive surface. The same conclusion was reached in the work by Mingels et al. [19], but here, the iridium oxides were deposited on a platinum electrode with a gold interlayer. The novel potentiostatic process used here for depositing IrO_2 on a titanium substrate without a gold interlayer reduces the fabrication cost without losing the positive properties.

Table 5. Results of EDS analysis (wt.%) of Ir PS+CP before and after exposition in the whole pH range.

	Ti	Al	V	Ir	F
Ir PS+CP before	87.4 ± 1.3	6.1 ± 0.4	4.1 ± 1.4	1.3 ± 0.3	1.1 ± 0.1
Ir PS+CP after	87.3 ± 0.3	6.1 ± 0.3	3.9 ± 0.1	1.3 ± 0.1	1.4 ± 0.2

4. Conclusions

In this work, nanostructured TiAlV alloy samples were successfully prepared. According to the results of the analyses, reproducible potentiostatic deposition of Ir particles was achieved. During oxidation by cyclic polarization in the physiological solution, more stable iridium oxides were obtained on the surface of the Ir PS+CP sample. Based on the SIM and TEM-EDS data, it was found that IrO_2 particles were concentrated on the surface of the nanotubes but also distributed to the bottom of the nanotubes.

Due to the possibility of using the surface immediately after production, our prepared surface is suitable as a pH electrode. During our experiments, the sensitivity of the surface was shown to be 37 ± 3 mV/pH when the surface was used without the storage solution. If the measurement lasted 5 min, the sensitivity was already 43 ± 3 mV/pH. By keeping the sample in the storage solution for 40 h or five days, the surface sensitivity was already 55 ± 1 mV/pH. The electrochemical response in microvolume was measured on another prepared surface which was kept in the physiological solution at pH 7.6 before the experiment. Again, the pH change was successfully detected with the Nernstian sensitivity of 55 ± 1 mV/pH. It was further shown that the mixed surface of TiO_2 and IrO_2 did not respond to the change in the redox potential of the environment. Our experiments showed that the most accurate measurements were obtained when the sensor was kept in a storage solution and readings were taken at least 90 s after immersion. However, it was also possible to use the sensor without a storage medium. This surface with its properties would be suitable for use in medicine, especially for the rapid detection of inflammation in the surrounding area of the implant, when the punctate would be tested.

Author Contributions: Conceptualization, J.F. and A.M.; methodology, J.F. and J.J.; validation, J.F., V.H. and A.M.; formal analysis, J.J., K.K., A.M. and L.K.; investigation, J.J., V.H., A.M. and L.K.; resources, J.J., K.K., A.M. and L.K.; data curation, J.J., V.H. and A.M.; writing—original draft preparation, J.J. and J.F.; writing—review and editing, J.J., V.H., A.M. and J.F.; visualization, J.J.; supervision, J.F. and A.M.; project administration, J.F.; funding acquisition, J.F. All authors have read and agreed to the published version of the manuscript.

Funding: This research was funded by the Czech Health Research Council, grant number NU20-06-00424.

Institutional Review Board Statement: Not applicable.

Informed Consent Statement: Not applicable.

Data Availability Statement: Data are contained within the article.

Conflicts of Interest: The authors declare no conflicts of interest.

References

1. Tomšík, E.; Gunár, K.; Krunclová, T.; Ivanko, I.; Trousil, J.; Fojt, J.; Hybášek, V.; Daniel, M.; Sepitka, J.; Judl, T.; et al. Development of Smart Sensing Film with Nonbiofouling Properties for Potentiometric Detection of Local pH Changes Caused by Bacterial and Yeast Infections around Orthopedic Implants. *Adv. Mater. Interfaces* **2023**, *10*, 2201878. [CrossRef]
2. Busscher, H.J.; van der Mei, H.C.; Subbiahdoss, G.; Jutte, P.C.; van den Dungen, J.J.; Zaat, S.A.; Schultz, M.J.; Grainger, D.W. Biomaterial-associated infection: Locating the finish line in the race for the surface. *Sci. Transl. Med.* **2012**, *4*, 153rv110. [CrossRef] [PubMed]
3. Zhu, Y.; Weng, X.; Zhang, J.; Mao, J. Protective effect of additional cathelicidin antimicrobial peptide PR-39 on prosthetic-joint infections. *J. Orthop. Surg.* **2023**, *31*, 10225536231175237. [CrossRef] [PubMed]
4. Parvizi, J.; Tan, T.L.; Goswami, K.; Higuera, C.; Della Valle, C.; Chen, A.F.; Shohat, N. The 2018 Definition of Periprosthetic Hip and Knee Infection: An Evidence-Based and Validated Criteria. *J. Arthroplast.* **2018**, *33*, 1309–1314.e1302. [CrossRef] [PubMed]
5. Trampuz, A.; Zimmerli, W. Diagnosis and treatment of infections associated with fracture-fixation devices. *Injury* **2006**, *37* (Suppl. S2), S59–S66. [CrossRef] [PubMed]
6. Veletić, M.; Apu, E.H.; Simić, M.; Bergsland, J.; Balasingham, I.; Contag, C.H.; Ashammakhi, N. Implants with Sensing Capabilities. *Chem. Rev.* **2022**, *122*, 16329–16363. [CrossRef] [PubMed]
7. Judl, T.; Popelka, S.; Tomšík, E.; Hrubý, M.; Daniel, M.; Fojt, J.; Melicherčík, P.; Landor, I.; Jahoda, D. Acidity Is an Excellent Marker of Infection in Hip and Knee Arthroplasty. *J. Clin. Med.* **2024**, *13*, 688. [CrossRef] [PubMed]
8. Fisher, C.R.; Masters, T.L.; Johnson, S.; Greenwood-Quaintance, K.E.; Chia, N.; Abdel, M.P.; Patel, R. Comparative transcriptomic analysis of Staphylococcus epidermidis associated with periprosthetic joint infection under in vivo and in vitro conditions. *Int. J. Med. Microbiol.* **2024**, *315*, 151620. [CrossRef] [PubMed]
9. Rotstein, O.D.; Nasmith, P.E.; Grinstein, S. The Bacteroides by-product succinic acid inhibits neutrophil respiratory burst by reducing intracellular pH. *Infect. Immun.* **1987**, *55*, 864–870. [CrossRef]
10. Rotstein, O.D.; Pruett, T.L.; Fiegel, V.D.; Nelson, R.D.; Simmons, R.L. Succinic acid, a metabolic by-product of Bacteroides species, inhibits polymorphonuclear leukocyte function. *Infect. Immun.* **1985**, *48*, 402–408. [CrossRef]
11. Shahrestani, S.; Ismail, M.C.; Kakooei, S.; Beheshti, M.; Zabihiazadboni, M.; Zavareh, M.A. Iridium Oxide pH Sensor Based on Stainless Steel Wire for pH Mapping on Metal Surface. *IOP Conf. Ser. Mater. Sci. Eng.* **2018**, *328*, 012014. [CrossRef]
12. Widmer, A.F. New Developments in Diagnosis and Treatment of Infection in Orthopedic Implants. *Clin. Infect. Dis.* **2001**, *33*, S94–S106. [CrossRef] [PubMed]
13. McLister, A.; McHugh, J.; Cundell, J.; Davis, J. New Developments in Smart Bandage Technologies for Wound Diagnostics. *Adv. Mater.* **2016**, *28*, 5732–5737. [CrossRef] [PubMed]
14. Gray, M.; Meehan, J.; Ward, C.; Langdon, S.P.; Kunkler, I.H.; Murray, A.; Argyle, D. Implantable biosensors and their contribution to the future of precision medicine. *Vet. J.* **2018**, *239*, 21–29. [CrossRef] [PubMed]
15. Singewald, T.D.; Traxler, I.; Schimo-Aichhorn, G.; Hild, S.; Valtiner, M. Versatile, low-cost, non-toxic potentiometric pH-sensors based on niobium. *Sens. Bio-Sens. Res.* **2022**, *35*, 100478. [CrossRef]
16. Taheri, M.; Deen, I.A.; Packirisamy, M.; Deen, M.J. Metal oxide -based electrical/electrochemical sensors for health monitoring systems. *TrAC Trends Anal. Chem.* **2024**, *171*, 117509. [CrossRef]
17. Liu, B.; Zhang, J. A ruthenium oxide and iridium oxide coated titanium electrode for pH measurement. *RSC Adv.* **2020**, *10*, 25952–25957. [CrossRef] [PubMed]
18. Marzouk, S.A.M. Improved Electrodeposited Iridium Oxide pH Sensor Fabricated on Etched Titanium Substrates. *Anal. Chem.* **2003**, *75*, 1258–1266. [CrossRef] [PubMed]
19. Mingels, R.H.G.; Kalsi, S.; Cheong, Y.; Morgan, H. Iridium and Ruthenium oxide miniature pH sensors: Long-term performance. *Sens. Actuators B Chem.* **2019**, *297*, 126779. [CrossRef]
20. Prats-Alfonso, E.; Abad, L.; Casan-Pastor, N.; Gonzalo-Ruiz, J.; Baldrich, E. Iridium oxide pH sensor for biomedical applications. Case urea-urease in real urine samples. *Biosens. Bioelectron.* **2013**, *39*, 163–169. [CrossRef]
21. Cruz, A.M.; Abad, L.; Carretero, N.M.; Moral-Vico, J.; Fraxedas, J.; Lozano, P.; Subías, G.; Padial, V.; Carballo, M.; Collazos-Castro, J.E.; et al. Iridium Oxohydroxide, a Significant Member in the Family of Iridium Oxides. Stoichiometry, Characterization, and Implications in Bioelectrodes. *J. Phys. Chem. C* **2012**, *116*, 5155–5168. [CrossRef]
22. Zhao, R.; Xu, M.; Wang, J.; Chen, G. A pH sensor based on the TiO_2 nanotube array modified Ti electrode. *Electrochim. Acta* **2010**, *55*, 5647–5651. [CrossRef]
23. Awad, N.K.; Edwards, S.L.; Morsi, Y.S. A review of TiO_2 NTs on Ti metal: Electrochemical synthesis, functionalization and potential use as bone implants. *Mater. Sci. Eng. C Mater. Biol. Appl.* **2017**, *76*, 1401–1412. [CrossRef] [PubMed]
24. Benea, L.; Ravoiu, A.; Neaga, V.; Axente, E.R. Using Applied Electrochemistry to Obtain Nanoporous TiO_2 Films on Ti_6Al_4V Implant Alloys and Their Preclinical In Vitro Characterization in Biological Solutions. *Coatings* **2023**, *13*, 614. [CrossRef]
25. Machackova, N.; Jiru, J.; Hybasek, V.; Fojt, J. A Ru/RuO2-Doped TiO2 Nanotubes as pH Sensors for Biomedical Applications: The Effect of the Amount and Oxidation of Deposited Ru on the Electrochemical Response. *Materials* **2021**, *14*, 7912. [CrossRef]
26. Moravec, H.; Vandrovcova, M.; Chotova, K.; Fojt, J.; Pruchova, E.; Joska, L.; Bacakova, L. Cell interaction with modified nanotubes formed on titanium alloy Ti-6Al-4V. *Mater. Sci. Eng. C* **2016**, *65*, 313–322. [CrossRef]
27. Fojt, J.; Průchová, E.; Hybášek, V. Electrochemical impedance response of the nanostructured Ti–6Al–4V surface in the presence of S. aureus and E. coli. *J. Appl. Electrochem.* **2023**, *53*, 2153–2167. [CrossRef]

28. Saharudin, K.A.; Sreekantan, S.; Aziz, S.N.Q.A.A.; Hazan, R.; Lai, C.W.; Mydin, R.B.S.M.N.; Mat, I. Surface Modification and Bioactivity of Anodic Ti6Al4V Alloy. *J. Nanosci. Nanotechnol.* **2013**, *13*, 1696–1705. [CrossRef]
29. Kim, T.Y.; Yang, S. Fabrication method and characterization of electrodeposited and heat-treated iridium oxide films for pH sensing. *Sens. Actuators B Chem.* **2014**, *196*, 31–38. [CrossRef]
30. Dong, L.; Qiu, Y.; Gu, Y.; Wang, Y.; Li, L. Preparation of an iridium oxide micro pH electrode and its response rate under dynamic stray current interference. *Microchem. J.* **2024**, *196*, 109631. [CrossRef]
31. Jiang, P.; Zhang, Y.; Hu, R.; Shi, B.; Zhang, L.; Huang, Q.; Yang, Y.; Tang, P.; Lin, C. Advanced surface engineering of titanium materials for biomedical applications: From static modification to dynamic responsive regulation. *Bioact. Mater.* **2023**, *27*, 15–57. [CrossRef]
32. Manjakkal, L.; Szwagierczak, D.; Dahiya, R. Metal oxides based electrochemical pH sensors: Current progress and future perspectives. *Prog. Mater. Sci.* **2020**, *109*, 100635. [CrossRef]
33. Jang, H.; Lee, J. Iridium oxide fabrication and application: A review. *J. Energy Chem.* **2020**, *46*, 152–172. [CrossRef]

chemosensors

MDPI

Communication

Miniaturized Iridium Oxide Microwire pH Sensor for Biofluid Sensing

Khengdauliu Chawang [1], Sen Bing [1], Ki Yong Kwon [2] and J.-C. Chiao [1,*]

[1] Electrical and Computer Engineering, Southern Methodist University, Dallas, TX 75205, USA; kchawang@smu.edu (K.C.); sbing@smu.edu (S.B.)
[2] Plexon, Dallas, TX 75206, USA; ki@plexon.com
* Correspondence: jchiao@smu.edu

Abstract: pH regulation in human biofluids is a crucial step for disease diagnosis and health monitoring. Traditional pH sensors are limited by their bulky size in wearable systems, and fragile glass tips require frequent calibration, thus limiting their use in continuous monitoring. Flexible sensors, particularly those utilizing microwires and thread-based substrates, present advantages for small sample analysis, including natural breathability and suitability for bandage or textile integration. This study examines iridium oxide and silver–silver chloride coated on thin gold wires, fabricated using sol–gel and dip-coating processes known for their simplicity. The flexible microwires demonstrated promising pH performance from a study of their pH characteristics, sensitivity, hysteresis, and potential drift. Electrodes tested in microwells allowed for small sample volumes and localized pH measurement in a controlled environment. Additional integration into fabrics for sweat sensing in wearables highlighted their potential for continuous, real-time health monitoring applications.

Keywords: microwire; pH sensor; flexible; iridium oxide; biofluid

Citation: Chawang, K.; Bing, S.; Kwon, K.Y.; Chiao, J.-C. Miniaturized Iridium Oxide Microwire pH Sensor for Biofluid Sensing. *Chemosensors* **2024**, *12*, 168. https://doi.org/10.3390/chemosensors12080168

Received: 8 July 2024
Revised: 18 August 2024
Accepted: 20 August 2024
Published: 22 August 2024

1. Introduction

Human biofluid exhibits numerous biomarkers for disease diagnosis and for tracking health status. Biofluids maintain strict pH regulation, and minor changes can indicate the early onset of diseases. Real-time data collection and analysis can significantly improve patient outcomes through early detection and intervention strategies. An instance is the early detection of sepsis infection, which can be alerted by monitoring pH changes in the blood or bodily fluids [1,2]. Sweat, tears, saliva, and urine are options for the detection of biomarkers for specific diseases [3–6]. Continuous sweat production can be achieved through iontophoresis and chemical methods [7,8], allowing for real-time skin monitoring within a tiny amount of sweat. Skin pH assessment can indicate the deterioration or healing of the tissues. Typically, healthy skin displays a mildly acidic pH of 5 [9], while wounded skin tends to be more alkaline, nearing a pH of 8 [10]. In clinical environments, portable handheld or table-top pH sensors featuring flat-tip designs are employed for monitoring skin health. However, the fragile glass tip necessitates skilled users to operate and regular calibration, and prevents the continuous monitoring of skin chemistry. The amount of fluid needed by the sensing electrodes is not achievable with the use of iontophoresis. This prohibits tracking the continuous time-variant pH changes that provide critical physiological information.

Flexible sensors offer an excellent solution for real-time health monitoring. Flexible polymeric substrates with electrodes, being created through laser engraving, lithography, and printing techniques, have been demonstrated [11–13]. Although planar polymeric substrates have laid the groundwork for flexible pH sensing, they face limitations when it comes to analyzing tiny samples, such as those in cell analysis, and in terms of substrate breathability. Microwire and thread-based pH sensors may offer a great alternative. Multiplexed pH measurements using microwire electrodes were demonstrated for cell growth study in cell culture [14]. Thread electrodes integrated into textile substrates provide natural breathability during skin sensing. The use of thread-based pH sensors allows for sensing on various diagnostic platforms, such as bandages [15], and knitted into textiles [16]. Thread-based pH sensors require an initial step of metallization to create a conductive surface, and then the pH-sensitive material is coated. Metal microwires do not require the metallization step during the fabrication process. The pH-sensitive material can be deposited by sputtering, electrodeposition, or sol–gel processes [17–19]. Among these, the sol–gel process stands out for its relative simplicity. Researchers can reproduce high-quality thin films without sophisticated deposition hardware. Additionally, the surface homogeneity and thickness can be controlled during the coating step [20]. This work utilizes the sol–gel process to deposit iridium oxide (IrO_x) on a gold bonding wire. The biocompatibility and high sensitivity features of IrOx are crucial for biosensing, thus it has been a preferred coating material for implantable bioelectrodes [21]. Silver–silver chloride (Ag/AgCl) is widely used as a reference electrode material for stability, miniaturization, and fabrication ease [22].

Our group previously reported on IrO_x thin films coated on planar polyimide substrates [23]. This work explored IrO_x films conformally coated on 25 μm gold microwires. To complete the fabrication of the miniaturized sensing platform, Ag/AgCl films were deposited onto the microwires. The surface geometry of a microwire provided a three-dimensional sensing area beneficial for studying tiny samples. This was demonstrated by testing the electrodes immersed in microwells. Additionally, the microwires were weaved into fabric to illustrate wearable textile applications where sensing was effectively conducted using small sample volumes of solution. pH characteristics such as sensitivities, hysteresis, and drift were analyzed.

2. Materials and Methods

Figure 1 illustrates the device configuration. The preparation of the sol–gel solution, the dip-coating process, and the heat treatment were adapted from previous work [24]. After coating $IrCl_3$ on a gold (Au) microwire, heat treatment at 325 °C facilitated oxide growth. The IrO_x layer formed on the wire surface as illustrated in Figure 1a. The scanning electron microscope (SEM) image in Figure 1b shows an electrode diameter of 25 μm with a layer of IrO_x. A SEM image in Figure 2a shows the surface detail and successful IrO_x growth. Figure 2b shows the elemental peaks for Ir, O, and Au by energy dispersive spectroscopy (EDS) (JEOL IT500HR Field Emission).

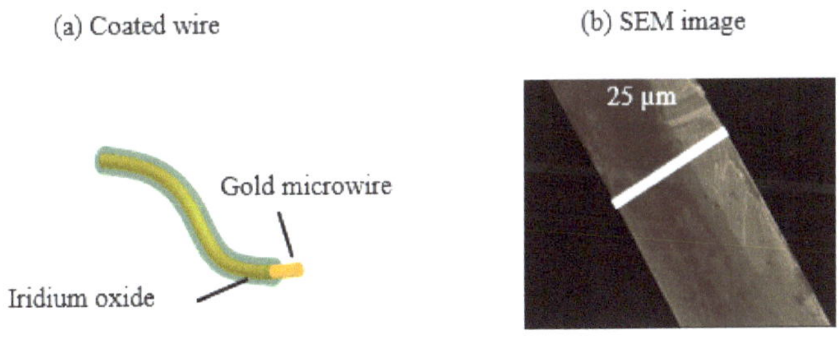

Figure 1. (**a**) Iridium oxide (IrO_x) film coated on gold microwire. (**b**) A SEM image showing the 25 μm coated microwire.

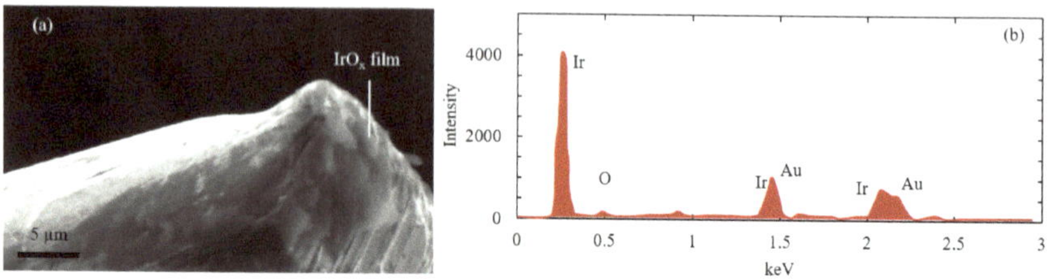

Figure 2. (**a**) A SEM image shows successful IrO_x growth. (**b**) EDS peaks indicate the elemental composition.

A potentiostat (CH Instruments, Bee Cave, TX, USA) was used for open-circuit potential (OCP) and cyclic voltammetry (CV) measurements. Phosphate-buffered saline (PBS) 1X (Fisher, Waltham, MA, USA), with a pH of 7.4 and a salt concentration (0.137 M) close to that of human body fluids, was chosen for its biological relevance [25]. Artificial sweat (Reagents, Charlotte, NC, USA) with an initial pH of 4.7 was adjusted from pH 4 to 9 using hydrochloric acid (HCl) and sodium hydroxide (NaOH). A commercial pH meter (Apera, Columbus, OH, USA) was used to verify the pH levels of the adjusted sweat solutions. A commercial glass rod (Basi, West Lafayette, IN, USA) Ag/AgCl reference electrode was used for tests in buffer pH 4, 7, and 10 solutions (Fisher, Manchester, NH, USA). Due to the large size of the commercial electrode, 50 mL solutions were prepared for verifying their pH values. The Ag/AgCl microwire was prepared by coating gold (Au) bonding wires with commercial ink (CHI., CA, USA). The 25 µm diameter microwires with IrO_x and Ag/AgCl enabled sensing in volumes as small as 0.2 mL. To test the microwire in these small volumes, 3-D printed microwells with a diameter of 10 mm and a 0.2 mL capacity were created. To further demonstrate wearable applications, the microwires were assembled in fabrics and tested with small solution volumes of 0.2 mL. Three IrO_x-Ag/AgCl microwire pairs were weaved into a fabric pad with a size of 5×5 cm^2. Each microwire pair was connected to a single input channel of a data acquisition card (NI, Austin, TX, USA). Each channel consists of two ports for the IrO_x working electrode and the Ag/AgCl reference electrode. This electrical interface connecting the electrode pairs embedded in the pad allows for three different pH-level solutions of 4, 6, and 9 to be dripped onto the pad and for their local potentials to be simultaneously recorded.

3. Results and Discussion

3.1. pH-Sensing Performance by Open-Circuit Potential Measurement

Figure 3a shows the OCP responses and hysteresis study across two test cycles using the IrO_x microwire vs. a commercial Ag/AgCl electrode. This was to verify solely the IrO_x performance using the commercial Ag/AgCl electrode as a standard reference. Each cycle involved testing in the sequence of pH 4-7-10-7-4. A 50 mL beaker filled with solutions was used for all of the tests. The electrodes underwent cleaning in deionized (DI) water between tests. Previously, hysteresis (dV) was defined and reported for IrO_x films on planar substrates [24]. The dV was characterized as the standard deviation between stabilized potentials at identical pH values in the test solutions, as indicated in Figure 3a. The correlation between pH and dV at pH 4, 7, and 10 was utilized to compute the pH fluctuation (dpH). The dV values of ±1.9, ±11, and ±0.3 mV corresponded to pH variations (dpH) of ±0.03, ±0.2, and ±0.06 pH, respectively, as depicted in Figure 3b. The highest dpH of ±0.2 pH occurred at pH 7, in contrast to both acidic and alkaline environments. The "±" sign here indicates the error range from its average output value. The difference in ionic concentrations of pH 10 and pH 7 likely caused high dV while returning from pH 10 to pH 7 during the cyclic testing. Alkaline pH 10 predominantly contains OH^- ions compared to neutral pH 7, with equal amounts of OH^- and H^+ ions. OH^- ions attached to the oxide states and tended to be more adherent on the surface even with deionized water cleaning to the electrode. Another study also reported that higher hysteresis in a neutral pH 7 solution in cyclic tests, compared to acidic or alkaline levels, existed in field effect transistor sensors [26]. The correlation between pH and potential was utilized for determining sensitivity. A sensitivity of −54 mV/pH for the IrO_x microwire electrode was comparable to the previously demonstrated IrO_x planar films at −51 mV/pH [24].

Ag/AgCl microwire was then used to replace the commercial Ag/AgCl electrode. Figure 4a shows the OCP responses from the IrO_x vs. Ag/AgCl microwire pair and (b) illustrates a sensitivity of −55 mV/pH. Figure 4b indicates the highest hysteresis (dV) of ±4.3 mV at pH 7, resulting in a pH variation (dpH) of ±0.07. The higher hysteresis for pH 7 was again a similar phenomenon, as mentioned before. From the OCP measurements, the IrO_x vs. Ag/AgCl microwires produced a lower pH variation of ±0.07 compared to ±0.2 by the commercial Ag/AgCl electrode. This is probably due to more residues on a larger surface area of the commercial Ag/AgCl electrode than the limited surface of the microwire.

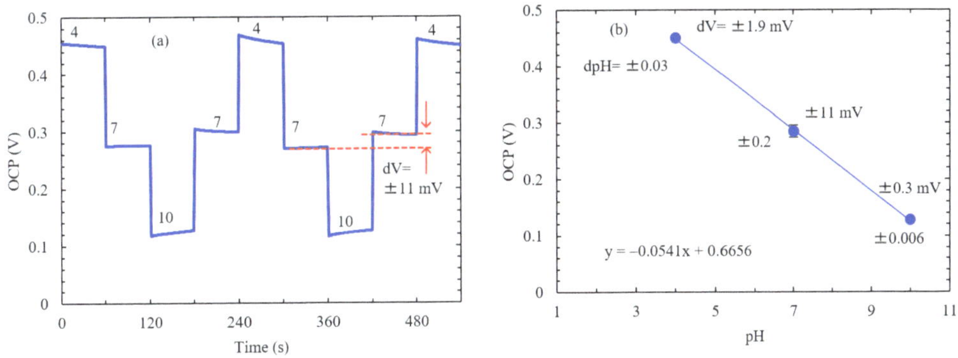

Figure 3. (**a**) Open-circuit potential (OCP) response of IrO_x microwire vs. a commercial Ag/AgCl electrode. The dotted lines indicate hysteresis (dV) at the identical pH level. (**b**) The sensitivity plot with hysteresis (dV) and corresponding pH variation (dpH) at three pH levels.

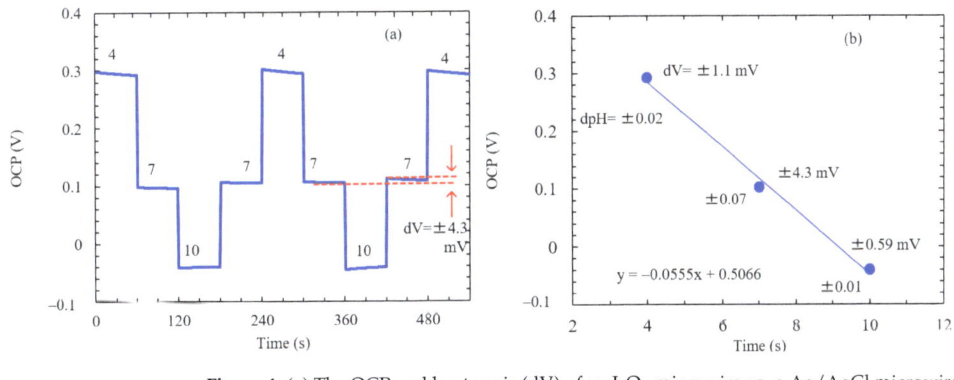

Figure 4. (**a**) The OCP and hysteresis (dV) of an IrO_x microwire vs. a Ag/AgCl microwire. (**b**) The sensitivity plot with hysteresis (dV) and corresponding pH variation (dpH) at three pH levels.

3.2. pH-Sensing Performance by Cyclic Voltammetry

Cyclic voltammetry is a versatile technique widely used to study the oxidative/reductive processes indicated by the peak potentials. To validate the functionality of the small microwires, a sensitivity study from CV was performed. The sensitivities were obtained from the CV profiles using the peak potential and pH relationship. CV was performed in a three-electrode system (CHI., USA) with an IrO_x microwire as the working electrode, commercial Ag/AgCl or microwire Ag/AgCl as the reference electrode, and platinum as the counter electrode. The tests were performed in the same buffer solutions in the potential range of −0.4 to +0.8 V at a scan rate of 30 mV/s.

The pH performance was compared between the IrO_x microwire vs. the commercial Ag/AgCl and IrO_x vs. Ag/AgCl microwires electrode pairs. For the IrO_x microwire vs. the commercial Ag/AgCl electrode, Figure 5a shows a sensitivity of −66.7 mV/pH, calculated from the peak potentials in the CV curves and pH values. A similar sensitivity of −63.7 mV/pH for the IrO_x vs. Ag/AgCl microwire pair is shown in Figure 5b. The increased sensitivity observed in the CV tests in contrast to the OCP measurements is ascribed to the differences in the oxide states. In CV scans, the oxide composition undergoes alterations owing to redox reactions. Conversely, in OCP measurements, the high input impedance of the instrument prevents substantial changes in the oxide states [27]. Nonetheless, the OCP and CV measurements demonstrated that IrO_x microwire performance was comparable between against a commercial reference electrode and the Ag/AgCl microwire made in our lab.

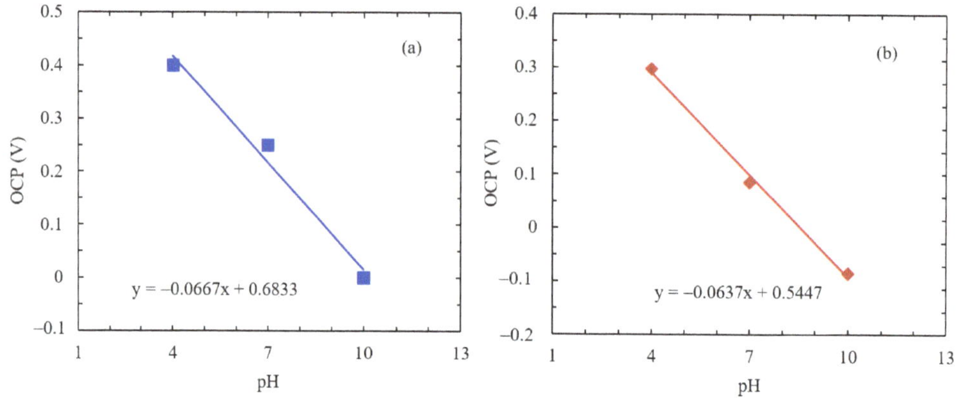

Figure 5. Sensitivities determined from cyclic voltammetry (CV) peak potentials and pH relationship for two different electrode-pair configurations: (**a**) an IrO_x microwire vs. a commercial Ag/AgCl electrode. (**b**) IrO_x vs. Ag/AgCl microwires.

3.3. Potential Drift Study

To study the potential drift behavior, Figure 6a shows an IrO_x microwire vs. a commercial Ag/AgCl electrode in 50 mL solution. Figure 6b shows the IrOx vs. Ag/AgCl microwires dipped in microwells filled with 0.2 mL solution in each well. The larger volume of 50 mL was used to accommodate the size of the commercial Ag/AgCl electrode. PBS 1× solution of pH 7.4 and a salt concentration of 0.137 M, close to human body fluid, was used in all stages of the drift studies.

Potential drift (V′) was previously defined as the slow non-random change in the output potential and was measured as the change between the initial and settled potentials [24]. Our group previously demonstrated a reset technique on planar IrO_x films that could improve the drifting issue [28]. Figure 6c,d show how the reset method applied to the IrO_x microwire vs. a commercial Ag/AgCl electrode, and the IrO_x vs. Ag/AgCl microwires, respectively, can improve the drift phenomena. The experimental protocol for both scenarios was the same: after the recording, they continued for 1.5 h with the pairs in solution. The blue-shaded areas show the potential drift ranges. Then, an electrical reset was applied to the electrode pairs, indicated by the yellow lines. The reset was performed by applying an external voltage of 0.2 V across the working and reference electrodes. The 0.2 V reset voltage was chosen because it has been shown to produce a stable OCP with low drifts compared to other reset values, demonstrated in previous work for IrO_x [28]. After 2 min, the voltage was turned off and the recording wires were reconnected to the electrode pairs to record the potentials for another 1.5 h. The red-shaded areas show the improved performance of output potentials with less drifting. The drift (V′) improvements after the reset for an IrO_x microwire vs. a commercial Ag/AgCl electrode in Figure 6c and the IrO_x vs. Ag/AgCl microwire pair in Figure 6d were changed from 11.8 mV to 4.2 mV and from 16.5 mV to 2.2 mV, reducing drifting by 64% and 87%, respectively. The potential drift improvement was accompanied by a faster settling time towards their stable potential values. The change in the output potential after reset can be attributed to the charge exchange processes at the OH^--rich (IrO_x) surface previously observed for the planar electrode [28]. This study confirmed that the voltage-reset technique applied to planar electrodes can also improve the performance of the microwire electrodes [28].

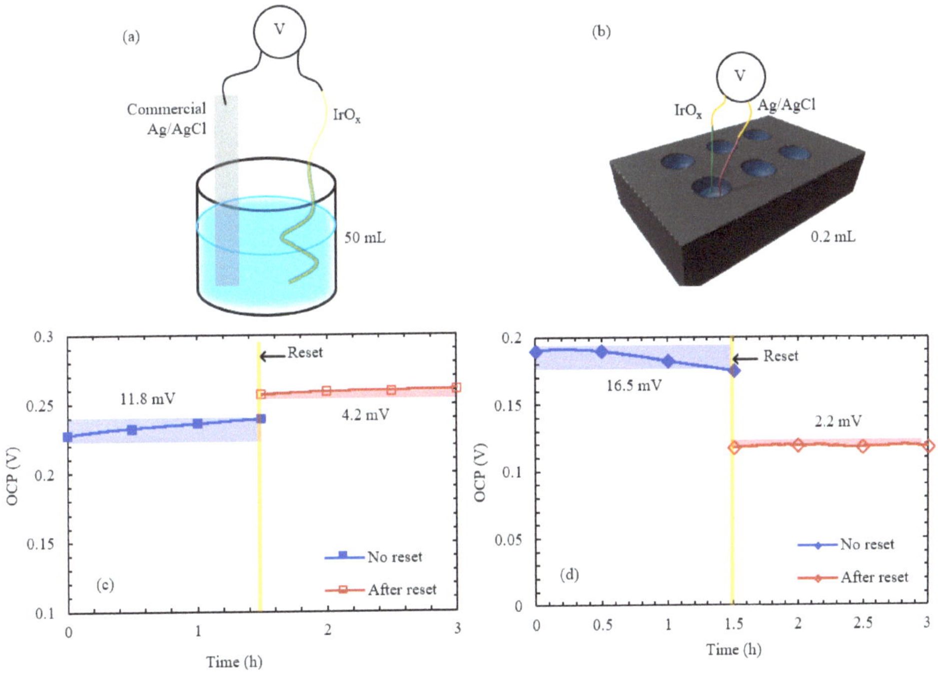

Figure 6. The test setup for the drift study among the two different electrode configurations includes the following: (**a**) an IrO_x microwire vs. a commercial Ag/AgCl electrode in a beaker with 50 mL of solution; (**b**) IrO_x vs. Ag/AgCl microwires in a microwell filled with 0.2 mL of solution. The potential drift (V′) analysis using the voltage reset technique for (**a**,**b**) electrode configurations is shown in (**c**,**d**), respectively. The yellow bar indicates the reset period of 2 min. The shaded areas of blue and red indicate the maximum drifting ranges.

3.4. Microwires in Small-Volume Sensing

Artificial sweat with an initial pH of 4.7 was used to create new solutions, with their pH levels adjusted to range from 4 to 9 using HCl and NaOH. A pH meter verified the adjusted pH levels. A pipette was used to drop in or draw out solutions or water from the microwells. Deionized water was used to clean the microwells and electrodes between tests to remove residues. A single IrO_x vs. Ag/AgCl microwire pair was immersed in the microwell and continuously tested from pH 4 to 9. Figure 7a shows the OCP step responses using the same microwire pair. The sensitivity of −57.4 mV/pH, shown in Figure 7b for the microwire, was comparable with the previous planar film design, which produced a sensitivity of −59 mV/pH, tested in artificial sweat [29].

To demonstrate potential wearable applications, three pairs of IrO_x vs. Ag/AgCl microwire pairs were weaved into a 5 × 5 cm^2 meshed fabric pad, as illustrated in Figure 8a. The meshed fabric was a stretchable cloth material for breathability that was important for long-term wound dressing. Each pair was connected to the data acquisition (DaQ) card. Each pair utilized a single channel. Solutions with pH 4, 6, and 9 were dripped on the fabric pad simultaneously with pipettes. The zoomed-in drawing in Figure 8a illustrates the pipette dripping a small volume, 0.2 mL, of artificial sweat solution onto the fabric pad containing the microwires. Figure 8b shows that the electrodes produce a sensitivity of −56 mV/pH across the three pH levels tested. These experiments demonstrated small-volume in situ sensing in two different scenarios.

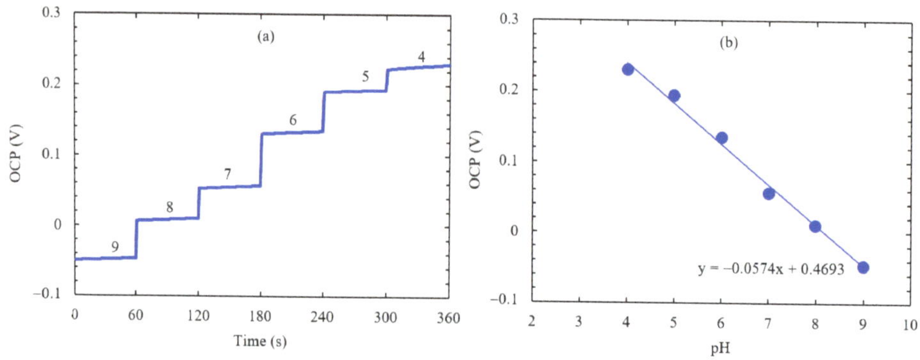

Figure 7. (**a**) OCP step response and (**b**) sensitivity of an IrO_x vs. Ag/AgCl microwire pair tested in microwells filled with 0.2 mL artificial sweat adjusted to pH 4–9.

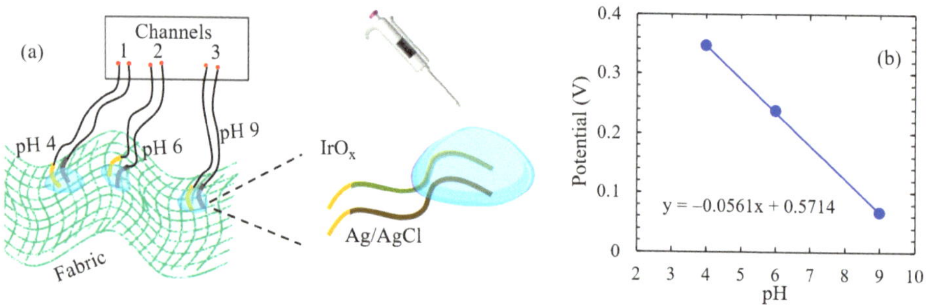

Figure 8. (**a**) Three IrO_x vs. Ag/AgCl microwire pairs weaved into the meshed fabric pad for wearable pH-sensing applications. The detailed drawing shows the application of the artificial sweat solution on the fabric area containing the microwires. (**b**) The sensitivity from the three pairs tested at pH 4, 6, and 9.

4. Conclusions

This study investigated the development and pH-sensing characteristics of microwire-based pH sensors. Our group previously demonstrated IrO_x based pH sensors built on planar surfaces. This manuscript makes a significant advancement by creating a much smaller, wire-shaped 25 μm IrO_x pH sensor. To complete the sensor system, a Ag/AgCl reference electrode was built utilizing the microwire. Sol–gel and dip coating processes were utilized to produce IrO_x and Ag/AgCl films on gold microwires.

Electroanalytical tests, such as open-circuit potential and cyclic voltammetry, indicated that IrO_x and Ag/AgCl microwire pairs exhibited performance comparable to their planar and commercial counterparts. When tested in control solutions, such as commercial buffers and adjusted pH solutions in PBS and artificial sweat, the microwire pairs showed near-Nernstian sensitivity and low hysteresis. Improved drifting was achieved after the voltage reset was applied and was also similar to the planar films. The reset technique enables long-term stability without frequently removing the sensor from the solution for calibration. The small-volume sensing ability was demonstrated using wells and embedding the microwires into 5×5 cm^2 fabrics.

The flexible microwire with a diameter of 25 μm enables localized pH measurements. The miniaturized configuration offers distinct advantages for both wearable and micro-volume sensing applications. Integrating these sensors into fabrics embedded in a bandage presents an approach for wearable monitoring systems. Their compact size also supports minimally invasive applications for real-time sensing in wound dressing. Moreover, the capacity to operate in low sample volumes is also beneficial for microfluidic systems where sample availability is limited. These attributes collectively position the microwire pH sensor as a promising tool for advancing biomedical research and clinical diagnostics. Future applications involve inserting these microwires into biological cells to interface directly with cellular environments without disrupting normal cellular functions.

Author Contributions: Conceptualization, J.-C.C. and K.C.; methodology, K.C.; software, K.C.; validation, K.C., S.B. and K.Y.K.; formal analysis, K.C.; investigation, S.B. and K.C.; resources, K.Y.K.; data curation, K.C.; writing—original draft preparation, K.C.; writing—review and editing, J.-C.C. and K.C.; visualization, S.B.; supervision, J.-C.C.; project administration, J.-C.C.; funding acquisition, J.-C.C. All authors have read and agreed to the published version of the manuscript.

Funding: This work was partially funded by the School of Engineering at Southern Methodist University, Mary and Richard Templeton Endowment.

Institutional Review Board Statement: Not applicable

Informed Consent Statement: Not applicable.

Data Availability Statement: The raw data supporting the conclusions of this article will be made available by the authors on request.

Conflicts of Interest: Ki Yong Kwon was employed by the company Plexon. All authors declare that the research was conducted in the absence of any commercial or financial relationships that could be construed as a potential conflict of interest.

References

1. Husabø, G.; Nilsen, R.M.; Flaatten, H.; Solligård, E.; Frich, J.C.; Bondevik, G.T.; Braut, G.S.; Walshe, K.; Harthug, S.; Hovlid, E. Early diagnosis of sepsis in emergency departments, time to treatment, and association with mortality: An observational study. *PLoS ONE* **2020**, *15*, e0227652. [CrossRef]
2. Ganesh, K.; Sharma, R.N.; Varghese, J.; Pillai, M. A profile of metabolic acidosis in patients with sepsis in an Intensive Care Unit setting. *Int. J. Crit. Illn. Inj. Sci.* **2016**, *6*, 178–181. [CrossRef] [PubMed]
3. Kaya, T.; Liu, G.; Ho, J.; Yelamarthi, K.; Miller, K.; Edwards, J.; Stannard, A. Wearable sweat sensors: Background and current trends. *Electroanalysis* **2019**, *31*, 411–421. [CrossRef]
4. Yang, X.; Pan, X.; Blyth, J.; Lowe, C.R. Towards the real-time monitoring of glucose in tear fluid: Holographic glucose sensors with reduced interference from lactate and pH. *Biosens. Bioelectron.* **2008**, *23*, 899–905. [CrossRef] [PubMed]
5. Zuliani, C.; Matzeu, G.; Diamond, D. A potentiometric disposable sensor strip for measuring pH in saliva. *Electrochim. Acta* **2014**, *132*, 292–296. [CrossRef]
6. Prats-Alfonso, E.; Abad, L.; Casañ-Pastor, N.; Gonzalo-Ruiz, J.; Baldrich, E. Iridium oxide pH sensor for biomedical applications. Case urea–urease in real urine samples. *Biosens. Bioelectron.* **2013**, *39*, 163–169. [CrossRef]
7. Bariya, M.; Nyein, H.Y.Y.; Javey, A. Wearable sweat sensors. *Nat. Electron.* **2018**, *1*, 160–171. [CrossRef]
8. Choi, J.; Ghaffari, R.; Baker, L.B.; Rogers, J.A. Skin-interfaced systems for sweat collection and analytics. *Sci. Adv.* **2018**, *4*, eaar3921. [CrossRef]
9. Fluhr, J.W.; Elias, P.M. Stratum corneum pH: Formation and function of the 'acid mantle'. *Exog. Dermatol.* **2002**, *1*, 163–175. [CrossRef]
10. Percival, S.L.; McCarty, S.; Hunt, J.A.; Woods, E.J. The effects of pH on wound healing, biofilms, and antimicrobial efficacy. *Wound Repair Regen.* **2014**, *22*, 174–186. [CrossRef]
11. Yang, Y.; Song, Y.; Bo, X.; Min, J.; Pak, O.S.; Zhu, L.; Wang, M.; Tu, J.; Kogan, A.; Zhang, H. A laser-engraved wearable sensor for sensitive detection of uric acid and tyrosine in sweat. *Nat. Biotechnol.* **2020**, *38*, 217–224. [CrossRef] [PubMed]
12. Ma, D.; Chon, S.; Cho, S.; Lee, Y.; Yoo, M.; Kim, D.; Lee, D.Y.; Lim, J.K. A novel photolithographic method for fabrication of flexible micro-patterned glucose sensors. *J. Electroanal. Chem.* **2020**, *876*, 114720. [CrossRef]
13. Dominiczak, J.; Krzemiński, J.; Wojcieszek, J.; Baraniecki, D.; Budny, F.; Wojciechowska, I.; Walter, P.; Pepłowski, A.; Górski, Ł.; Jakubowska, M. Aerosol-jet-printed potentiometric pH sensor for sweat measurements in smart patches. *Sens. Bio-Sens. Res.* **2024**, *43*, 100636. [CrossRef]

14. Dabaghi, M.; Saraei, N.; Xu, G.; Chandiramohan, A.; Yeung, J.; Nguyen, J.P.; Vukmirovic, M.; Selvaganapathy, P.R.; Hirota, J.A. PHAIR: A biosensor for pH measurement in air–liquid interface cell culture. *Sci. Rep.* **2021**, *11*, 3477. [CrossRef]
15. Terse-Thakoor, T.; Punjiya, M.; Matharu, Z.; Lyu, B.; Ahmad, M.; Giles, G.E.; Owyeung, R.; Alaimo, F.; Baghini, M.S.; Brunyé, T.T. Thread-based multiplexed sensor patch for real-time sweat monitoring. *npj Flex. Electron.* **2020**, *4*, 18. [CrossRef]
16. Jia, J.; Xu, C.; Pan, S.; Xia, S.; Wei, P.; Noh, H.Y.; Zhang, P.; Jiang, X. Conductive thread-based textile sensor for continuous perspiration level monitoring. *Sensors* **2018**, *18*, 3775. [CrossRef]
17. Wang, M.; Yao, S.; Madou, M. A long-term stable iridium oxide pH electrode. *Sens. Actuators B Chem.* **2002**, *81*, 313–315. [CrossRef]
18. Shahrestani, S.; Ismail, M.C.; Kakooei, S.; Beheshti, M.; Zabihiazadboni, M.; Zavareh, M.A. *Iridium Oxide pH Sensor Based on Stainless Steel Wire for pH Mapping on Metal Surface*; IOP Conference Series: Materials Science and Engineering; IOP Publishing: Bristol, UK, 2018; Volume 328, p. 012014.
19. Trovato, V.; Teblum, E.; Kostikov, Y.; Pedrana, A.; Re, V.; Nessim, G.D.; Rosace, G. Sol-gel approach to incorporate millimeter-long carbon nanotubes into fabrics for the development of electrical-conductive textiles. *Mater. Chem. Phys.* **2020**, *240*, 122218. [CrossRef]
20. Faustini, M.; Louis, B.; Albouy, P.A.; Kuemmel, M.; Grosso, D. Preparation of sol–gel films by dip-coating in extreme conditions. *J. Phys. Chem. C* **2010**, *114*, 7637–7645. [CrossRef]
21. Lee, I.; Whang, C.; Choi, K.; Choo, M.; Lee, Y. Characterization of iridium film as a stimulating neural electrode. *Biomaterials* **2002**, *23*, 2375–2380. [CrossRef]
22. Shinwari, M.W.; Zhitomirsky, D.; Deen, I.A.; Selvaganapathy, P.R.; Deen, M.J.; Landheer, D. Microfabricated reference electrodes and their biosensing applications. *Sensors* **2010**, *10*, 1679–1715. [CrossRef] [PubMed]
23. Chawang, K.; Bing, S.; Chiao, J. Printable and flexible iridium oxide-based pH sensor by a roll-to-roll process. *Chemosensors* **2023**, *11*, 267. [CrossRef]
24. Huang, W.; Cao, H.; Deb, S.; Chiao, M.; Chiao, J.-C. A Flexible pH Sensor Based on the Iridium Oxide Sensing Film. *Sens. Actuators A Phys.* **2011**, *169*, 1–11. [CrossRef]
25. Hopkins, E.; Sanvictores, T.; Sharma, S. *Physiology, Acid Base Balance*; StatPearls Publishing: Treasure Island, FL, USA, 2018.
26. Shin, P.; Ryssel, H. pH Sensing Properties of ISFETs with LPCVD Silicon Nitride Sensitive-Gate. *J. Electr. Eng. Inf. Sci.* **1997**, *2*, 82–87.
27. Olthuis, W.; Robben, M.; Bergveld, P.; Bos, M.; Van der Linden, W.E. pH Sensor Properties of Electrochemically Grown Iridium Oxide. *Sens. Actuators B Chem.* **1990**, *2*, 247–256. [CrossRef]
28. Chawang, K.; Xia, X.; Huang, M.-H.; Bing, S.; Benomar, M.; Doan, G.; Cao, H.; Chiao, J.-C. Improvement on Potential Drifting for Sol–Gel and Electroplated Iridium Oxide-Based pH-Sensing Films. *IEEE Sens. Lett.* **2023**, *7*, 2001504. [CrossRef]
29. Yang, X.; Chawang, K.; Chiao, J.-C. Wearable Iridium Oxide pH Sensors for Sweat pH Measurements. In Proceedings of the IEEE Sensors Conference, Montreal, QC, Canada, 27–30 October 2019.

chemosensors

MDPI

Article

Effects of Viscosity and Salt Interference for Planar Iridium Oxide and Silver Chloride pH Sensing Electrodes on Flexible Substrate

Khengdauliu Chawang, Sen Bing and Jung-Chih Chiao *

Electrical and Computer Engineering, Southern Methodist University, Dallas, TX 75205, USA
* Correspondence: jchiao@smu.edu

Abstract: The equivalency of pH measurements between aqueous and non-aqueous or viscous solutions is of great interest in biomedical applications as well for processing food and pharmaceuticals. Commercial glass-type electrodes have practical limitations, such as bulky sizes and membrane clogging in viscous environments. In this study, planar and flexible electrochemical pH sensors with iridium oxide as the sensing film have been developed by sol-gel and oxidation processes. A reference electrode was prepared by screen printing Ag/AgCl ink on the same polyimide substrate. The small form factors of the planar flexible electrodes provide an advantage in small volume or conformal surface measurements. Cyclic voltammetry was performed in different pH solutions. The electrode originally produced a response of −70.1 mV/pH at room temperature in aqueous solutions. The sensitivities were reduced when salt was added into the buffer solutions, although output potentials were increased. Sensing performances in a wide range of viscous conditions with various concentrations of added salt have been analyzed to study their effects on pH-sensing responses. Suitable calibration techniques using aqueous buffer solutions were studied for output potentials and their respective pH readings in viscous salt-added solutions. The mechanisms affecting output potentials are explained and results matched well for two different thickening agents. Specificity to pH changes measured by the planar IrO_x-Ag/AgCl pH electrodes showed how the potential-pH calibration should consider the interference effect of salt. The viscosity effects on pH reading errors became more pronounced as solution viscosity increased. Comparisons of pH readings to those from a commercial glass-bodied pH meter indicated that the planar electrodes provided predictable pH deviations that were confined to a limited range. The planar IrO_x-Ag/AgCl electrodes on flexible polyimide substrates have mostly been demonstrated with aqueous solutions in various diagnostic and environmental monitoring applications. This work provides more insights into pH sensing performance when the fluid is viscous and contains salt, which often is the case in biomedical and food-processing applications.

Keywords: pH sensor; salt interference; viscosity; flexible; calibration

Citation: Chawang, K.; Bing, S.; Chiao, J.-C. Effects of Viscosity and Salt Interference for Planar Iridium Oxide and Silver Chloride pH Sensing Electrodes on Flexible Substrate. *Chemosensors* **2022**, *10*, 371. https://doi.org/10.3390/chemosensors10090371

Academic Editor: Tae Geun Kim

Received: 22 August 2022
Accepted: 14 September 2022
Published: 17 September 2022

1. Introduction

Measuring the pH of viscous and non-aqueous solutions is of great importance in applied chemistry applications [1–3] such as in processing food [4,5], cosmetics [6,7], and pharmaceuticals [8,9]. The pH sensing mechanism involved in the aqueous solution is well established; however, studies in non-aqueous or viscous solutions have been limited. Some common issues faced using conventional glass-rod pH sensors are slow responses, frequent calibration, and glass membrane clogging and corrosion in the inner electrolyte, which limits their use in non-aqueous solutions. K. Izutsu et al. demonstrated slow responses of glass pH electrodes in a non-aqueous solvent tested in an acid-based titration method [2]. For reliable pH measurements, J. F. Artiola reported the need for frequent two-/three-point calibrations with periodic glass membrane soaking in a standard solution to remove organic salts [10]. The literature also suggests that the dissociation of functional groups on glass body surfaces may decrease Nernstian responses [11]. The traditional bulky glass-rod shaped pH sensors cannot be used for sensing a small volume of liquid or on planar surfaces. Owing to their features of miniaturization, integration with electronics, and the elimination of membranes, planar solid-state films have been proposed for pH sensing electrode materials [12–14].

Among metal oxides, iridium oxide (IrO_x) based films have been studied to exhibit higher Nernstian responses [15], reproducible super-Nernstian responses within wide pH ranges in aqueous solutions [16], and faster responses in non-aqueous solutions compared to commercial pH meters [2]. Stable pH-sensing performance after exposure to high temperatures and their biocompatibility feature was reported in the literature [17,18]. IrO_x films have been studied for use in implantable devices and cochlear prostheses [19–21]. Fabrication methods for iridium oxide films by radio-frequency (RF) sputtering [22,23], electroplating [24–26], chemical bath [27,28], and sol-gel deposition with oxidation have been conducted [29–31]. These methods are chosen according to the device and production requirements. The sol-gel method provides advantages for mass production, fabrication simplicity, and lower costs. Deposition through dipping in sol-gel solutions has been used in industrial manufacturing processes. It can be applied to a wide variety of substrates. Iridium oxide (IrO_x) based pH sensors on flexible substrates made by the sol-gel method have been demonstrated in the following ways: with Nernstian responses [32], in an array configuration [33], integrated with other chemical sensors [34], and for food-quality and medical implant applications [35–37]. In these demonstrations, planar IrO_x working electrodes integrated with Ag/AgCl reference electrodes were calibrated and tested in aqueous conditions with and without interfering salt effects.

Salt addition in an aqueous pH solution was demonstrated to have interfering effects on the potential and pH outputs previously demonstrated using planar electrodes in our group [34]. In this study, we explored the effects of ionic interferences from salt (NaCl), and viscosity on planar IrO_x and Ag/AgCl electrode potential outputs. Aqueous pH solutions were used as the calibration standard to estimate the pH values of viscous salt-added solutions. Solutions with different pH values, NaCl concentrations, and viscosities, which varied by adding thickening agents, were prepared. In each configuration of the solution, three pairs of planar IrO_x and Ag/AgCl electrodes were used to investigate performance variations.

We proposed planar pH electrodes under the consideration of the practical constraints faced when using glass-rod membrane-based sensors in viscous solutions. Preliminary results of using the planar electrodes were submitted and accepted to present at the 2022 IEEE Sensors Conference [38]. In this study, details of the investigation and experimental results for the effects of salt and viscosity on potential and pH outputs are reported.

Planar pH sensors were prepared, each consisting of an IrO_x working electrode (WE) and an Ag/AgCl reference electrode (RE) on a biocompatible, inert, and flexible polyimide substrate [38]. IrO_x was deposited as the H^+ ion sensing film by the sol-gel and oxidation processes. Ag/AgCl film was deposited by screen printing on the flexible substrate. The monolithic processes allowed miniaturization without the need for a vacuum environment. A commercial Ag/AgCl glass-rod electrode was used for calibration in aqueous pH buffer solutions, but could not be reliably used in viscosity experiments because of its dependency on ion migration to remain stable [10]. In a non-aqueous interface, the liquid junction potential may become large and unreliable. The planar electrodes, on the contrary, allow direct surface contact with the viscous solution samples.

2. Materials and Tools

Technical grade Iridium (IV) Chloride ($IrCl_4$) hydrate powder (Sigma Aldrich, Saint Louis, MO, USA) was used to prepare the sol-gel solution for WE. Ethyl Alcohol 200 Proof (Supelco, Bellefonte, PA, USA) was used during the preparation of the sol-gel solution. Glacial acetic acid (Fisher Scientific, Waltham, MA, USA) with 99.7% concentration was used as the reagent. To prepare the reference electrode, Ag/AgCl paste (011464, ALS, Kanagawa, Japan) was used. Electrical connections between electrodes and the interface were made of copper wires and adhesive silver epoxy (8331D, MG Chemicals, USA).

Different viscosities of the solution were achieved by adding thickening agents into pH solutions. Lab graded starch powder $(C_6H_{10}O_5)_n$ (Carolina Biological Supply Company, Burlington, NC, USA), and agar powder $C_{14}H_{24}O_9$ (Seaweed Solution, Trondheim, Norway), which is a and natural gelatin were used as the thickening agents. DI water with a conductivity of 0.055 μS/cm at room temperature was used for experiments. Hydrochloric acid (HCl) (LabChem, Zelienople, PA, USA) with a concentration of 50% was mixed with DI water to adjust the pH level to 4. The reason for the preparation of the custom-made pH 4 solution with HCl was to investigate interference from other ionic elements because commercial buffer solutions contained various chemicals. Similarly, custom-made pH 6 and pH 8 solutions were prepared with HCl and sodium hydroxide (NaOH) (Sigma Aldrich, MO, USA) in DI water. A commercial pH meter (A1311, Apera Instruments, Columbus, OH, USA) and pH/conductivity benchtop multiparameter (Orion Star A215, Thermo Fisher Scientific, Waltham, MA, USA) were used to adjust and verify the pH levels in aqueous solutions. The same benchtop multiparameter was used to measure the conductivity of aqueous solutions. Commercial buffer pH 4, 6, and 8 solutions (Fisher Scientific, MA, USA) were also used for experiments. Lab-grade sodium chloride (NaCl) (Macron, PA, USA) was added to the pH solutions after the thickening agents, which allowed for the study of ionic interferences to the sensing electrodes in viscous environments. Different viscous samples were prepared at the same time and stored in 50-mL plastic storage vials. Multiple viscometers were used including a modified Ostwald viscometer (CANNON Instrument Company, PA, USA), a rheometer (HR-3, TA Instruments, MN, USA), and a digital viscometer (KUNHEWUHUA, China). These meters together covered a wider range of viscosity. A data acquisition card (National Instruments 6201, USA) was used for signal recording.

3. Experiments

3.1. Electrode Preparation

Both the working electrode (WE) and reference electrode (RE) were prepared on a flexible polyimide substrate with a 125-μm thickness (500HN, Dupont, Newark, DE, USA). The WE consisted of two metal layers: gold (Au) and chromium (Cr), both deposited by electron-beam evaporation. Cr was used as an adhesion layer with a thickness of 50 nm. Au as the base metal layer had a thickness of 200 nm. IrO_x serving as the sensing film was deposited on Au by the sol-gel dipping process. The sol-gel solution was prepared by mixing 1 g of anhydrous $IrCl_4$ with 42 mL of 200 Proof alcohol, and 10 mL of 80% concentrated acetic acid was later added. Using a magnetic stirrer set at 500 rpm, the solution was mixed for 2 h at 25 °C before use.

The substrate with base metal was coated by dipping and was dried at 50 °C. The dipping and drying processes were repeated three times. Then, samples were heated for 4 h at 325 °C [39–41]. To prepare RE, a layer of Ag/AgCl paste was applied to the substrate and dried at 120 °C. Both WE and RE films were cut into strips of 2 mm × 15 mm and fixed with a spacing distance of 3 mm between the two electrodes.

3.2. Viscous Sample Preparation

Starch solutions with different concentrations (wt%) were prepared in beakers. Three types of solutions were used for tests at pH 4, 6, and 8. The solution for pH 4 was prepared by mixing DI water and HCl, and the solutions for pH 6 and 8 were commercial buffer solutions. The solution was heated in a beaker and starch was added at a boiling temperature (95 °C). Magnetic stirring was performed at 500 rpm for 20 min until aggregation of starch powders disappeared in the solutions.

After the starch powder was completely dissolved in the solution, salt was added, and stirring continued for 5 min in the hot solution. Similarly, agar-based viscous solutions were also prepared. A total of 183 different variations of viscous samples, including 165 starch-based and 18 agar-based solutions, were prepared. Each test vial contained a 50 mL solution with a specific viscosity, pH value, and salt concentration. Output potential recordings were performed at room temperature.

4. Results and Discussion

4.1. Cyclic Voltammetry Analysis

Electrochemical studies on the polyimide-based WE and RE were investigated using cyclic voltammetry (CV) analysis performed in aqueous pH solutions. Figure 1a shows the CV comparison in pH 4 solution without salt and with 1-M NaCl salt. The broader curves during cathodic and anodic scans were due to higher currents generated with conducting salt ions. Figure 1b compares CV curves of pH 4, 6, and 8 with 1-M NaCl salt added. The pH 4 solution only contained HCl, while the buffer pH 6 solution had sodium hydroxide (NaOH) and dihydrogen potassium phosphate (KH_2PO_4) groups; thiomersal ($C_9H_9HgNaO_2S$), NaOH, and KH_2PO_4 groups were in the pH 8 buffer solution. The additional conducting ions in the buffer pH 6 and pH 8 solutions generated higher currents and produced broader voltammograms. The CV analysis provided certainty that such planar electrodes could be used to generate sufficient potentials and currents in various pH solutions.

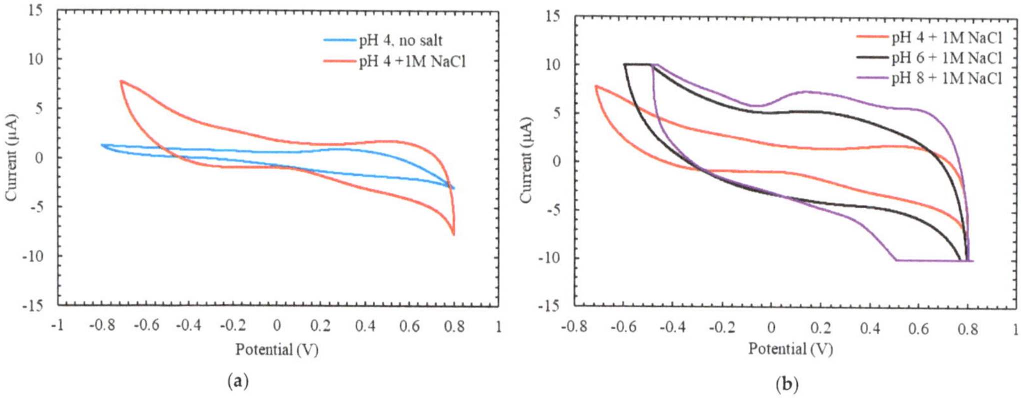

Figure 1. Cyclic voltammetry (CV) comparison of (**a**) pH 4 aqueous solution without salt (0 M) and 1-M NaCl salt added; and (**b**) pH 4, pH 6, and pH 8 aqueous solutions with 1-M NaCl salt added.

4.2. Nernstian Response in Aqueous Solutions

The ion exchanges on the working electrode (WE) surfaces created a potential difference against the reference electrode (RE) across the solution volume. Under ideal conditions without ionic or hydrodynamic interferences at room temperature, the Nernstian response in an aqueous solution should be around −59 mV/pH, as demonstrated in previous works [31–33].

Figure 2 compares the pH potential responses with no salt, 0.1-M, and 1-M NaCl salt added in aqueous pH 4, 6, and 8 solutions. A super-Nernstian response of −70.1 mV/pH was measured without salt added in the buffer solutions. Redox mechanisms for IrO_x producing Nernstian or super-Nernstian responses were explained in the literature [41–45]. Compared to our previous results, a possible explanation for the super-Nernstian response was that a controlled fabrication process was implemented. The WEs were prepared by multiple repeated steps of dip-coating and drying processes to achieve uniform thickness. The REs were prepared by screen printing commercial Ag/AgCl ink on the substrate instead of electroplating [38]. The sensitivity in the no-salt case decreased to near-Nernstian responses of −53.3 mV/pH when 0.1-M salt was added, and −47.3 mV/pH when 1-M NaCl was added. Increased output potentials were due to increased solution conductivities by salt ions. However, the output potential increased more in the alkaline solution as soon as salt was added. The decreased sensitivity validated that interference from salt ions existed.

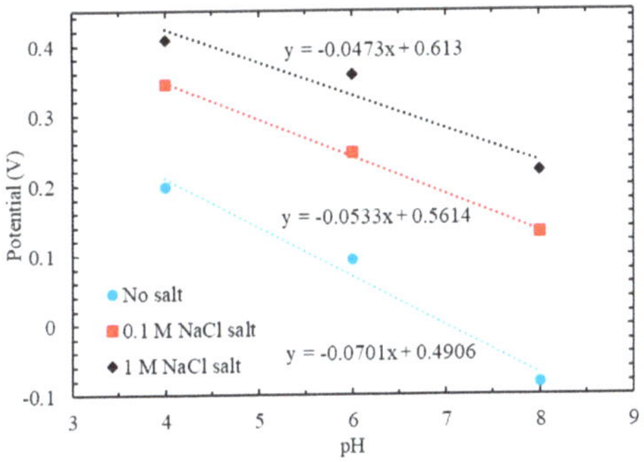

Figure 2. Sensitivities for aqueous no-salt (0 M), 0.1-M, and 1-M NaCl salt concentrations, calibrated at pH 4, 6, and 8 solutions.

To validate the effect of salt interference, the same solutions were measured by a glass-bodied pH meter. Figure 3a–c show pH and conductivity changes as salt was added to the pH 4, 6, and 8 solutions. The pH and conductivity were measured by a commercial pH/conductivity meter (Orion, A215). The addition of salt increased solution conductivity and made the solution seem more acidic. This effect decreased the output pH values calculated by the calibration curve obtained with standard calibration buffers. At 0.1-M salt concentrations, the pH values shifted to 4.01, 5.81, and 7.8 from their respective values of 4, 6, and 8. At 1-M salt concentrations, the values became 3.88, 5.2, and 7.4.

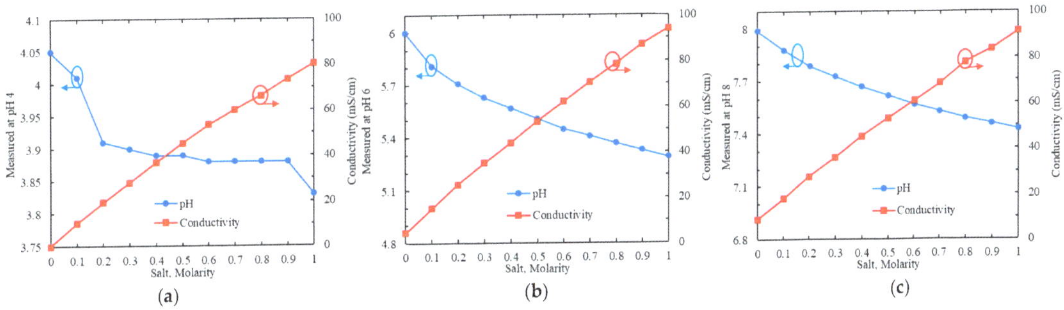

Figure 3. Measured pH and conductivity changes after 0.1–1 M NaCl salt addition in (**a**) pH 4, (**b**) pH 6, and (**c**) pH 8 aqueous solutions.

4.3. Viscosity Readings

Different percentages by weight, listed in Table 1, of starch and agar as thickening agents were prepared with no salt or salt added to achieve certain viscosities. Although the concentrations of the starch by weight varied from 1% to 4%, and 0.05% to 0.2% for agar, the viscosities varied quite significantly. Three different types of viscometers were used to validate the fluid viscosity because of their respective detection limitations.

An Ostwald viscometer was used to measure the viscosity of the starch sample with the lowest concentration of 1%. A rheometer for the concentration ranging from 1.0% to 2.0%, and a digital viscometer for 1.8% to 4.0% were used. For starch concentrations of 1.8% and 2.0%, both meters were used to cross-validate their viscosity values. The digital viscometer was used to measure agar samples since the viscosities were high. Agar solution with a concentration beyond 0.2% could not be measured by viscometers because it became a gel-like solid.

Table 1. Viscosity readings.

Thickening Agent	wt%	Viscosity (cP)
Starch	1	1.17
	1.4	2.39
	1.8	24
	2	47
	3	135
	4	820
Agar	0.05	54
	0.1	360
	0.2	2553

4.4. Experiments with Starch as the Thickening Agent

All experiments were performed by immersing the electrode area into the viscous solution samples. Electrodes were tested at pH 4, 6, and 8 solutions mixed with different NaCl concentrations (0–1 M with/without starch or agar). Output potential was recorded for 17 s and the electrodes were cleaned in deionized (DI) water for 30 s to remove residue from surfaces.

The custom-made pH 4 solution had a conductivity of 120 µS/cm. For weak acid and alkaline, the conductivities of custom-made pH 6 and 8 solutions mixing DI water with HCl or NaOH were 47 and 27 µS/cm, respectively. Unstable readings were observed for pH 6 and pH 8 cases. To overcome the issue, likely due to low conductivity, commercial buffer solutions at pH 6 and 8 with conductivities of 4726 and 8460 µS/cm were used instead. As mentioned previously during the CV analysis, various conducting ions were present in buffer pH 6 and 8 solutions, and it was expected that the additional ions in the solutions caused the measured voltage across the electrodes to decrease. Amylose in starch counteracted to decrease the solution resistance with additional electrons. The effects of amylose and matrices negated each other, so the potential decreased with smaller steps as the starch concentration increased. At lesser H^+ concentrations in pH 6 and 8, the effect from matrices to constrain flows of ions was less significant. Thus, the role of amylose became more pronounced in decreasing resistance in the solution, and potential increased with bigger steps as the starch concentration increased (purple arrow).

might create interfering effects.

Figure 4 shows the measured potentials in the starch-based measurements with (a) no salt and (b) 1 M NaCl salt added. The outputs in time stayed stable without significant drifts. Some fluctuation in signals might be due to liquid motions in beakers. Although other salt concentrations were also tested, only no-salt and 1-M salt cases are shown for illustration purposes. The five colors indicate 0%, 1%, 2%, 3%, and 4% starch added, with viscosities of 0.89, 1.17, 47, 135, and 820 cP, respectively.

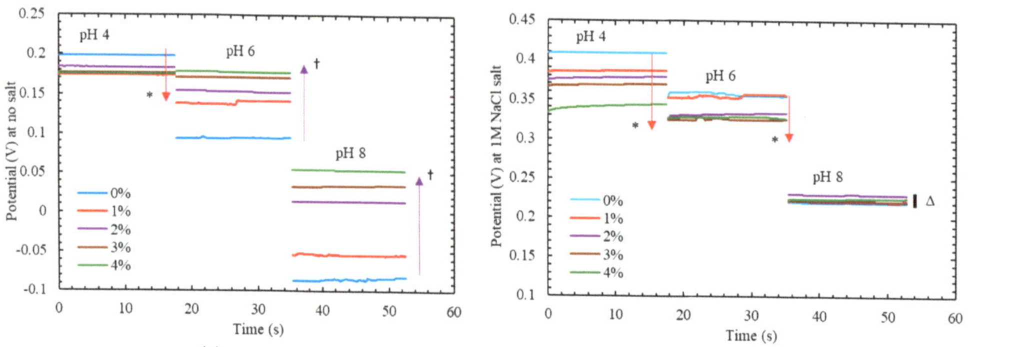

(a) (b)

Figure 4. Output potentials in viscous starch-based solutions at pH 4, 6, and 8 for (**a**) no salt added and (**b**) 1 M salt concentration. The symbols * and † indicate a monotonic decrease and increase as viscosity increases. The symbol Δ indicates no drastic change in output potentials as viscosity increases.

Figure 4. Output potentials in viscous starch-based solutions at pH 4, 6, and 8 for (**a**) no salt added and (**b**) 1 M salt concentration. The symbols * and † indicate a monotonic decrease and increase as viscosity increases. The symbol Δ indicates no drastic change in output potentials as viscosity increases.

When salt was added, free flows of conductive ions significantly increased, and potentials at pH 4, 6, and 8 increased with significant steps. The matrices of amylopectin could constrain the ionic movements of salt. Compared to amylose, the salt provided more impact on conductivity increases, thus the amylose effect became less obvious. In return,

The arrows in Figure 4a,b indicate the decreasing (red) or increasing (purple) trends in potentials as starch concentration increases. When starch is heated above 90 °C, granules swell and release ionic species known as branched amylopectin and linear amylose into the solution [46,47]. Branched amylopectin is responsible for aggregation and creates viscous matrices, while linear amylose increases solution conductivity [48]. Electrochemical impedance spectroscopy (EIS) showed that, after starch gelatinization, electrical resistance in the viscous matrix was created by amylopectin which was an insoluble material [46]. In the no salt added case in Figure 4a, pH 4 has a higher H^+ concentration with the free flow of electrons from the redox equilibrium. The viscous matrices in high ionic concentration played a significant role and increased the electrical resistance that constrained the movements of free ions. The potential drop in the solution increased, which caused the measured voltage across the electrodes to decrease. Amylose in starch counteracted to decrease the solution resistance with additional electrons. The effects of amylose and matrices negated each other, so the potential decreased with smaller steps as the starch concentration increased. At lesser H^+ concentrations in pH 6 and 8, the effect from matrices to constrain flows of ions was less significant. Thus, the role of amylose became more pronounced in decreasing resistance in the solution, and potential increased with bigger steps as the starch concentration increased (purple arrow).

When salt was added, free flows of conductive ions significantly increased, and potentials at pH 4, 6, and 8 increased with significant steps. The matrices of amylopectin could constrain the ionic movements of salt. Compared to amylose, the salt provided more impact on conductivity increases, thus the amylose effect became less obvious. In return, the potential decreases at pH 4 (red arrow, Figure 4b) became more prominent as the viscosity increased. Compared to Figure 4a, the electrical resistance in the solution increased more when its viscosity increased. For the same reason, the constraining effect of matrices became more on free salt ions that overpowered the increased conductivity by amylose. As a result, Figure 4b shows that there was almost no change of potential at pH 8 for different viscosities. The negating effect was more obvious for pH 6 solutions comparing the opposite directions of the purple and red arrows in Figure 4a,b. When there was no salt added, the effect from amylose made the potential increase with increasing viscosities as conductivities increased. However, the abundant ions from adding salt were constrained by amylopectin matrices, which allows for the potential decrease when increasing viscosities. The potential decreases are shown in two steps in Figure 4b, instead of the multiple steps for pH 6 and 8 solutions in Figure 4a, because the negating effects were not linear or equal. It should also be noted that the viscosities for 1% and 2% starch concentrations were 1.17 and 47 cP, respectively, with a 40x increase, while the increase was 2.9 and 6 times from 47 to 135 cP and from 135 to 820 cP for the starch concentration increasing from 2% to 3% and 3% to 4%, respectively.

Figure 5a–c shows the comprehensive data of output potentials with different salt concentrations and starch percentages (at different viscosities) in solutions. Three electrode pairs were tested under the same protocol from no salt (0 M) to 1-M salt concentration with a step change of 0.1 M before they were switched to another viscosity. The decreasing trends of potentials when the viscosity increased at pH 4 and the increasing trends at pH 8 validated the results in Figure 4. For pH 6 without salt added, the potential increased when the viscosity increased. With salt added, the potential decreased when viscosity increased. The upward arrow at pH 4, 6, and 8 shows output potential monotonically increases as more salt is added to the aqueous or viscous solutions. The data matched well with those in Figure 4. From the aqueous solutions to cases with starch added, the potential differences were much larger at pH 6 and 8 because amylose provides more electrical currents in the solutions. Generally speaking, when NaCl was added, even as low as the 0.1 M concentration, the effects from the starch molecule matrices, which reduced conductivity, were not as effective as the increasing conductivity effect by adding salt ions. Potential changes had smaller steps for the salt-added cases when viscosities changed, compared to the potential jumps from the no-salt cases.

matched well with those in Figure 4. From the aqueous solutions to cases with starch added, the potential differences were much larger at pH 6 and 8 because amylose provides more electrical currents in the solutions. Generally speaking, when NaCl was added, even as low as the 0.1 M concentration, the effects from the starch molecule matrices, which reduced conductivity, were not as effective as the increasing conductivity effect by adding salt ions. Potential changes had smaller steps for the salt-added cases when viscosities changed, compared to the potential jumps from the no-salt cases.

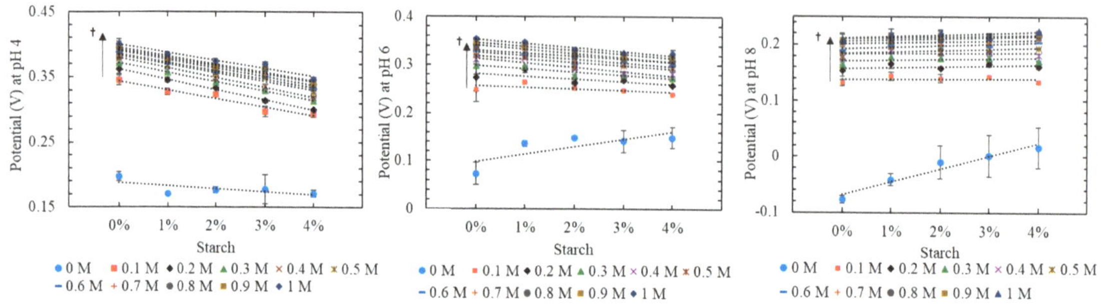

(a) (b) (c)

Figure 5. Measured output potential using starch as a thickening agent from 0–4% and salt 0–1 M added to (**a**) pH 4, (**b**) pH 6, and (**c**) pH 8 solutions. Error bars at every pH are the standard deviation of three sensors (n = 3). The † symbol indicates potential increases monotonically when 0.1–1 M NaCl salt is added.

The error bars in the plots of Figure 5 indicate the standard deviations measured by three sensors separately. The standard deviations in general were higher in solutions with higher viscosities when no salt was added to the solutions, particularly for the pH 8 cases. This might be due to the ions from amylose in viscous solutions, which were not uniform in the beakers, and the electrodes not experiencing more pronounced electrical current variations.

4.5. Experiments with Agar as the Thickening Agent

The experimental setup and procedure using agar powder as the thickening agent were similar to the starch-based experiments. Solutions were prepared with the custom-made pH 4 solution, and commercial pH 6 and 8 buffer solutions were also similar to the starch-based solutions. Each pH level had salt concentrations of 0 and 1 M, and three different viscosity levels.

Figure 6a,b show the results for agar-based measurements in which the three colors indicate 0%, 0.2%, and 0.6% agar added. The viscosity for 0.2% agar was measured as 2553 cP. When more than 0.2% agar was added to the solution, the viscometer could not receive a reading because the solution became gel-like. The electrodes could still be inserted into the gel, but it was obvious that they experienced mechanical resistance. Measurement results in the (a) no-salt condition are compared to those of (b) 1 M salt concentrations in Figure 6. Without added salt, a decreasing trend (red arrow) of potentials at pH 4 and increasing trends (purple arrow) at pH 6 and 8 were observed, as in Figure 6a. With 1 M salt added, the potentials stayed roughly in the similar levels at pH 4 and 8. The overall trend also showed decreased potentials when the viscosity increased. The trends for both no-salt and salt-added cases agreed with the starch-based cases in Figure 4. Although there is no report of ionic species contributed by agar, literature reported the conductivity of agar prepared with DI water was 73 µS/cm at 0.89% concentration [49]. The matrices and conductivity effects thus play similar roles to those in the starch-based solutions.

At pH 4 (red arrow, Figure 6a), the viscous matrices restricted flows of ions and increased the electrical resistance in the solution, which led to decreased output potentials. The potentials remained similar for 0.2% and 0.6% agar concentrations because these solutions have become very viscous. Most of the free-moving ions were constrained by the matrices. In Figure 6b, at pH 6 with salt added, there was the same phenomenon of a step drop of potential from 0% to 0.2% agar concentration, but no drop from 0.2% to 0.4% because most of the free ions from salt have been constrained at 0.2%. Increased potentials were observed (purple arrows in Figure 6a) at pH 6 and 8. Without added salt, low currents in low H^+ concentrations increased the conductivities with Agar, which reduced the potential drop in the solution and increased the potential between electrodes.

rations have become very viscous. Most of the free moving ions were constrained by the matrices. In Figure 6b, at pH 6 with salt added, there was the same phenomenon of a step drop of potential from 0% to 0.2% agar concentration, but no drop from 0.2% to 0.4% because most of the free ions from salt have been constrained at 0.2%. Increased potentials were observed (purple arrows in Figure 6a) at pH 6 and 8. Without added salt, low currents in low H^+ concentrations increased the conductivities with Agar, which reduced the potential drop in the solution and increased the potential between electrodes.

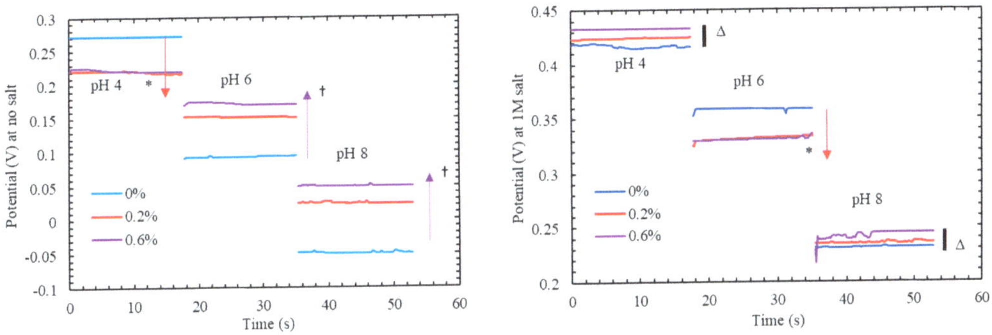

(a) (b)

Figure 6. Output potentials in viscous agar solutions at pH 4, 6, and 8 for (**a**) no salt and (**b**) 1-M NaCl salt concentration. The symbols * and † indicate a monotonic decrease and increase as viscosity increases. The symbol Δ indicates no drastic change in output potential as viscosity increases.

When 1-M salt was added, the conductivities were dominated by the salt ions. Thus the additional conductivity from Agar of 73 µS/cm [49] did not play a significant role, which was much lower compared to 86.7 mS/cm from salt. The trend of pH 8 in Figure 6b with increasing viscosities was similar to the one in Figure 4b. The increasing trends at pH 4 and 8 were observed, but with smaller changes. Due to the high viscosity already at 0.2% agar concentration, conductivities in the solutions did not vary much.

In all tests, conductivity in the solution has been observed to play a role. The viscous matrix created by starch or agar change the potential output depending on initial ionic concentrations. The multifaceted mechanisms of conductivity and viscosity seem to counteract their effects in high ionic concentrations.

4.6. pH Calibration

The aforementioned measurements showed changes in output potentials under the influences of viscosity and salt concentrations. To illustrate the viscosity effect on pH reading, 2% and 4% viscous starch-based solutions, corresponding to viscosities 47 and 820 cP, were considered. With salt added, it was clear from previous results that the potentials could be significantly increased due to higher conductivities. Thus, pH reading errors may rise differently with respect to salt contents in solutions. We considered two scenarios for using three-point calibration curves at (1) unknown or (2) known salt concentrations in aqueous solutions. The pH values in solutions with different salt concentrations and viscosities were then calculated from the calibration curves.

Figure 7 demonstrates erroneous pH values when the calibration was conducted without knowing salt concentrations. The potential-pH relationship was first found in aqueous pH 4, 6, and 8 solutions without added salt. A sensitivity slope of −68.4 mV/pH was found and was used to convert the potentials generated in the viscous solutions (2% and 4% starch) to pH values. The initial pH values of 4, 6, and 8 were calculated as 2.2, 3.3, and 4.96 at 0.1-M salt cases, and 1.48, 2.03, and 3.76 at 1-M salt cases. The error bar in each salt concentration indicates the pH deviations at 2% and 4% starch concentrations from the aqueous conditions. The calibration clearly generated significant pH errors with the existence of salt and viscosity when the sensor calibration curve was produced without knowing salt concentrations.

The second calibration method was conducted at a known salt concentration. The three-point calibration was conducted at aqueous pH 4, 6, and 8 with known salt concentrations. For each salt concentration between 0.1 and 1 M, an individual sensitivity slope was generated. Then potentials in viscous starch solutions with different salt concentrations were converted to pH values. Figure 8 shows the results. The pH values calculated for the viscous and salt-added pH 6 and 8 cases significantly improved with more accurate pH values. For example, the pH values for the pH 6 cases were 5.8 and 6.05 at 2% and 4% starch, respectively, at the 0.1-M salt concentration. Similarly, the pH values for the pH 8 cases were 7.9 and 8.01 at 2% and 4% starch, respectively, at the 0.1-M salt concentration. For comparison, the pH values measured by a commercial pH meter (A1311, Apera) for the pH 6 cases were 5.76 and 5.66 at 2% and 4% starch, respectively, at the 0.1-M salt concentration. For the pH 8 cases, the commercial pH meter readings were 7.32 and 7.79 at 2% and 4% starch at the 0.1-M salt concentration, respectively. Improvements in pH errors were also observed at other salt concentrations in Figure 8.

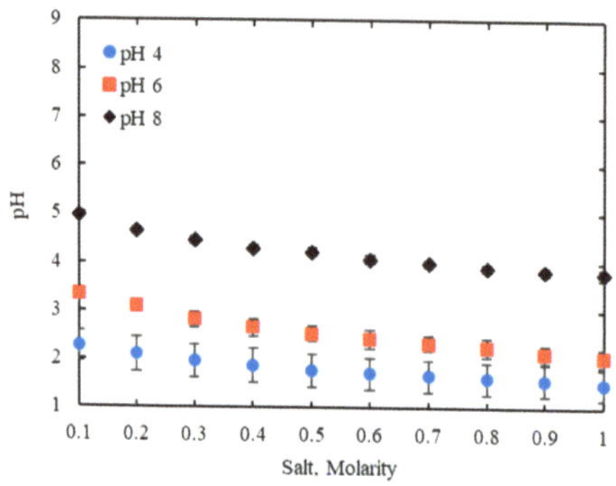

Figure 7. The pH values were calculated by the three-point calibration slope at aqueous pH 4, 6, and 8 with no salt or starch added (0 M). The error bar indicates the pH range for the viscous solutions with 2% or 4% starch added.

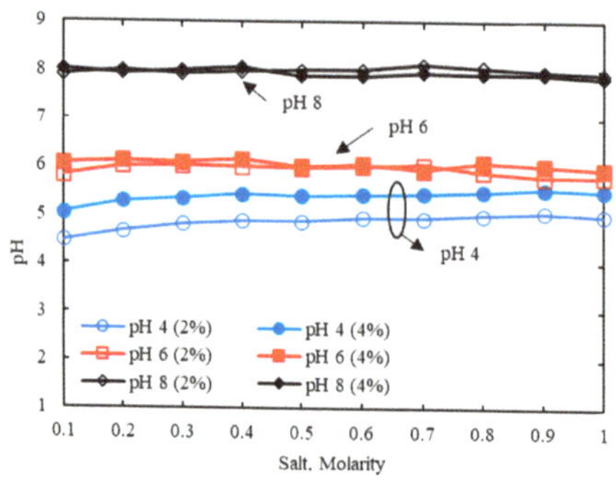

Figure 8. The pH values were calculated by the three-point calibrations conducted with aqueous pH 4, 6, and 8 solutions at individual NaCl concentrations. The viscosities of the 2% and 4% starch-added solutions correspond to 47 and 820 cP, respectively.

The effect of viscosity on the pH calibration error was particularly higher at pH 4, as indicated in Figure 8. The significant role of amylopectin matrices by starch in the high H^+ concentration created a higher potential difference. Since more ions could be constrained, the solution conductivity was lower compared to its aqueous case. As a result, higher pH errors occurred as viscosity increased. For example, the initial pH values of pH 4 changed to 4.46 and 5.04 at 2% and 4% starch, respectively, at the 0.1-M salt concentration. The larger pH errors at the more viscous solutions across all salt concentrations manifested the amylopectin matrix effect. Using a commercial pH meter, the results showed pH of 3.35 and 4.34 at 2% and 4% starch, respectively, for the 0.1-M salt case.

4.7. Specificity to pH

Based on the output potentials and calibrated pH levels in the different solutions, specificity to pH was performed to estimate the changes generated under ionic interferences. Potential deviation (ΔV) and pH deviation (ΔpH) were analyzed; ΔV in Figure 9 is the potential output difference between the aqueous and viscous solutions for a specific viscosity. There are five groups of viscosity data: four groups for starch from 1.17–820 cP and one group for agar at 2553 cP because the viscosity of 0.6% agar could not be measured because the solution became a gel-like solid (indicated by the † symbol).

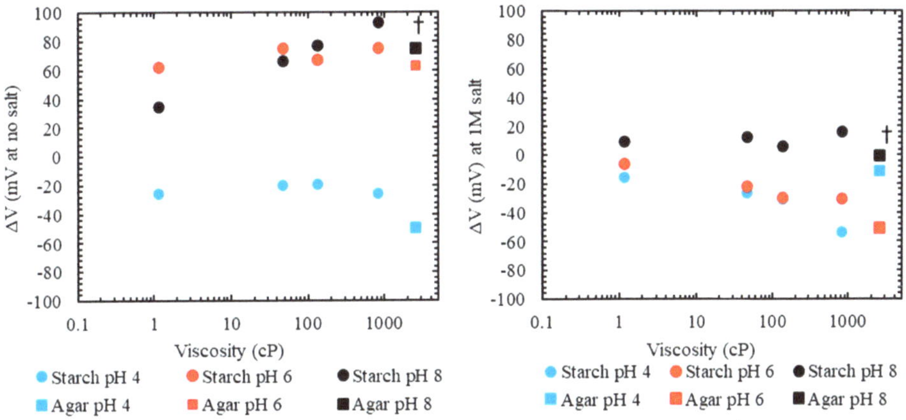

(a) (b)

Figure 9. Potential change (ΔV) for viscous solutions with starch or agar added, at (**a**) no salt and (**b**) 1-M salt concentrations. Symbol †: only one group of ΔV data for agar at pH 4, 6, and 8.

Figure 9a shows the ΔV of starch and agar-added solutions at pH 4 (blue), pH 6 (red), and pH 8 (black) without salt added. Data above the zero point at the *y*-axis mean potential increases with increased viscosity, and data below the zero point mean potential decreases. The trends in output potential changes were previously indicated by red and purple arrows in Figures 4 and 6. Lower H^+ concentrations at pH 6 and 8 showed higher ΔV, whereas pH 4 remained relatively similar with increased viscosities. For the 1-M salt added cases, the pH 4 solutions experienced higher resistance from the amylopectin, which created higher ΔV, as shown in Figure 9b. Reduced ΔVs for pH 6 and 8 were due to the dominating conductivities of the added salt ions in low H^+ concentrations. Thus, potential deviations from viscous matrices were reduced. Overall, Figure 9 shows that ΔV tends to become higher with an increased viscosity if the conductivity is low in the solution.

Table 2 lists the worst ΔV for the starch cases (with concentrations of 1%, 2%, 3%, and 4% by weight) at no-salt, 0.1-M, and 1-M salt concentrations. The worst ΔV is defined as the highest potential deviation from the aqueous case due to viscosity. The values for no-salt and 1-M salt can be seen in Figure 9. Additional values for the 0.1-M salt cases were added in Table 2 to compare the ΔV with the 1-M case. The positive or negative signs indicate potential increase or decrease with an increasing viscosity, respectively. It was evident that ΔV became higher as viscosity increased, and ΔV became smaller once the salt was added. Similar ΔV values for adding salt of 0.1 M and 1 M showed smaller changes in potential outputs because the conductivities in the solutions were dominated by salt.

The pH deviation (ΔpH) is defined as the pH difference between the aqueous and viscous solutions. Aqueous solutions at pH 4, 6, and 8 at a specific salt concentration were used as the calibration standard. The pH values for the viscous solutions were calculated from their respective calibration slopes. Figure 10a shows the pH deviations measured by the IrO_x-Ag/AgCl electrode pairs for the 2% starch solutions that have a viscosity of

47 cP at different salt concentrations. The pH deviations with salt concentrations between 0.1 and 1 M were within the ranges of (0.3–0.6), (0.06–0.5), and (0.03–0.2) from their correct values of pH 4, 6, and 8, respectively. Figure 10b shows ΔpH measured by a commercial pH meter in the same conditions. The deviations were more predictable and confined to certain ranges for the IrO_x-Ag/AgCl electrode pair sensor with different salt concentrations. The commercial meter produced wider variations of errors, particularly for the alkaline cases.

Table 2. Worst ΔV (mV) for starch cases.

	1.17 cP	47 cP	135 cP	820 cP
No salt	62.1	74.7	77.5	92.9
0.1 M	−17.9	−21	−47.6	−52
1 M	−16	−26.5	−30.8	−53.6

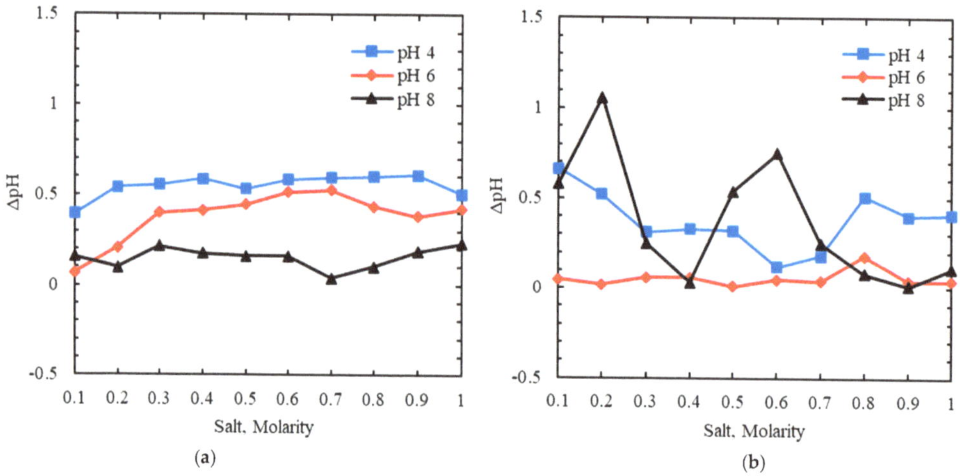

Figure 10. The pH deviation (ΔpH) at the 2% starch solutions was measured by (**a**) the IrOx-Ag/AgCl pH sensor electrodes and (**b**) a commercial pH meter when 0.1–1 M NaCl was added.

Figure 11a shows the 4% starch cases with a viscosity of 820 cP. The pH deviations with the salt concentrations between 0.1 and 1 M were within the ranges of (0.9–1.1), (0.1–0.6), and (0.04–0.3) from their correct values of pH 4, 6, and 8, respectively. Figure 11b shows the results measured by a commercial meter. The planar electrodes once again provided better predictability for the pH values while the commercial meter obtained more diverse deviations. It is obvious that the pH deviation ranges became larger for more viscous solutions because the calibration slopes were conducted in respective aqueous solutions.

Although the deviation values in Figure 11a were higher, they were more confined within a range compared to Figure 11b. The diverse ΔpH detected by the commercial meter shown in Figures 10b and 11b for pH 8 were due to low H^+ concentrations and likely membrane clogging in the highly viscous environment.

In a typical human body, sodium concentration is around 0.1 M [50,51], and biofluids such as saliva, serum, and whole blood viscosities are in the range of 1–5 cP [52,53]. It is clear the pH deviations will not be as large as those shown in this paper when using our planar pH sensors in biomedical applications. This study covered wide ranges of viscosity and salt concentrations in order to study potential interference and calibration issues. Such issues may occur in the cases of food processing and preparation in which viscosities depend on processing methods. For example, the same ketchup can have different viscosities in different yield stresses during extrapolation, initiated flow, squeezing flow, or vane methods. The different viscosities may produce varied pH readings in different steps of processing for the same ketchup. The viscosities of common food such as honey and sauce range between 23–578 cP [54,55]. Nectar-like and honey-like consistency food products added with a variety of thickening agents range between 51–350 and 351–1750 cP, respectively [56]. It was noted in [56] that there is variability in viscosity measurements within the same product line for thickening various liquids. Therefore, one would expect pH reading errors along the same product line when the solutions change viscosities. In this study, the viscosities covered the ranges of interest for the cases in [56].

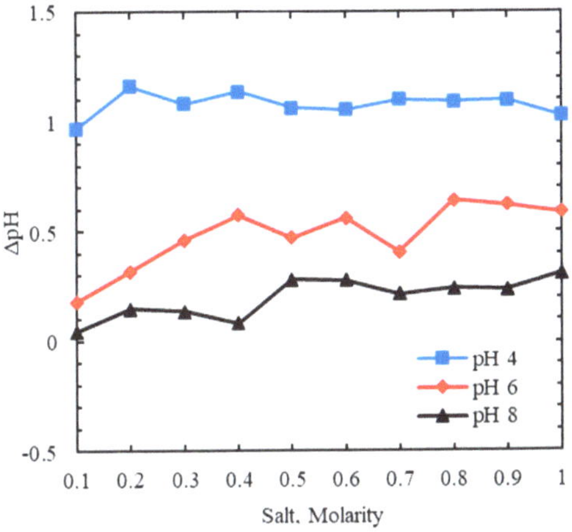

(a)

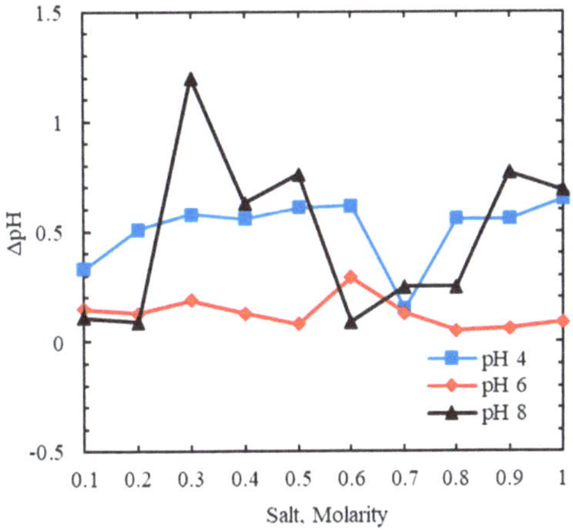

(b)

Figure 11. pH deviation (ΔpH) at the 4% starch concentration, measured by (**a**) the IrOx-Ag/AgCl pH sensor electrodes and (**b**) a commercial pH meter when 0.1–1 M NaCl was added.

5. Conclusions

This work investigated the effects of viscosity and salt interference on pH reading. A pair of planar IrO_x-Ag/AgCl electrodes deposited on a flexible polyimide substrate were used. The deposition of IrO_x was conducted with sol-gel dipping and oxidation processes, while Ag/AgCl was screen printed. The cyclic voltammetry (CV) analysis was conducted at different pH levels. Super Nernstian response was achieved in aqueous solutions. Changes in Nernstian responses with salt additions showed that conductive ions created interference effects and potential shifts. The outcome showed that sensitivity was reduced.

The effects of viscosity and salt conductivity on sensor output potentials were studied. Viscous solutions were prepared with starch and agar as thickening agents. Different concentrations of salt were added to the aqueous and viscous solutions. Both factors were studied simultaneously and counteracting effects on potential outputs were found. The trends of potential shifts were examined and matched well for two different thickening agents. Two calibration methods for the potential-pH relationship were analyzed. It was clear that the calibration should be conducted with known salt concentrations to avoid large pH errors in solutions with high salt concentrations or high viscosity. Specificity to pH performance was investigated at two different viscosities with different salt concentrations and pH levels. The planar electrodes showed that the pH deviations were more predictable and confined within limited ranges compared to those measured with a commercial pH meter. It was also shown the pH deviations increased when viscosity increased. The results demonstrated potential uses of the planar flexible IrO_x-Ag/AgCl pH sensors in aqueous and viscous solutions, as well as the related issues of viscosity and salt interference. This investigation was conducted with wide ranges of viscosities and salt concentrations. It covers potential applications in diagnostics of human bodily fluids as well as cosmetic, food, and pharmaceutical industries. The planar, flexible, and miniature features of the IrO_x-Ag/AgCl electrodes can operate without the concern for membrane clogging in viscous solutions. Thus, it can enable new applications.

Author Contributions: Conceptualization, J.-C.C.; methodology, J.-C.C. and K.C.; software, K.C.; validation, K.C. and S.B.; formal analysis, K.C.; investigation, K.C.; resources, J.-C.C.; data curation, K.C.; writing—original draft preparation, K.C.; writing—review and editing, K.C., S.B. and J.-C.C.; visualization, K.C.; supervision, J.-C.C.; project administration, J.-C.C.; funding acquisition, J.-C.C. All authors have read and agreed to the published version of the manuscript.

Funding: This research is partially funded by the Mary and Richard Templeton Centennial Chair endowment.

Institutional Review Board Statement: Not applicable.

Informed Consent Statement: Not applicable.

Data Availability Statement: No public data disclosure.

Acknowledgments: The authors sincerely appreciate the facility and financial support from the Lyle School of Engineering and the generosity of Mary and Richard Templeton.

Conflicts of Interest: The authors declare no conflict of interest.

References

1. Frant, M.S. How to Measure pH in Mixed and Nonaqueous solutions. *Today's Chem. Work* **1995**, 39–42.
2. Izutsu, K.; Yamamoto, H. Response of an Iridium Oxide pH-Sensor in Nonaqueous Solutions. Comparison with Other pH-Sensors. *Anal. Sci.* **1996**, *12*, 905–909. [CrossRef]
3. Pandey, P.C.; Singh, G. Tetraphenylborate doped polyaniline based novel pH sensor and solid-state urea biosensor. *Talanta* **2001**, *55*, 773–782. [CrossRef]
4. Samaranayake, C.P.; Sastry, S.K. In-situ pH measurement of selected liquid foods under high pressure. *Innov. Food Sci. Emerg. Technol.* **2013**, *17*, 22–26. [CrossRef]
5. Karastogianni, S.; Girousi, S.; Sotiropoulos, S. pH: Principles and measurement. *Encycl. Food Health* **2016**, *4*, 333–338.
6. Woolf, A.D.; Woolf, A.; Shaw, J.S. Nail primer cosmetics: Correlations between product pH and adequacy of labeling. *J. Toxicol. Clin.Toxicol.* **1999**, *37*, 827–832. [CrossRef]
7. Liudmyla, K.; Olena, C.; Nadiia, S. Chemical properties of Helix aspersa mucus as a component of cosmetics and pharmaceutical products. *Mater. Today Proc.* **2022**, *62*, 7650–7653. [CrossRef]
8. Traple, M.A.L.; Saviano, A.M.; Francisco, F.L.; Lourenço, F.R. Measurement uncertainty in pharmaceutical analysis and its application. *J. Pharm. Anal.* **2014**, *4*, 1–5. [CrossRef]
9. Mäder, K. Pharmaceutical applications of in vivo EPR. *Phys. Med. Biol.* **1998**, *43*, 1931. [CrossRef]
10. Artiola, J.F. Environmental chemical properties and processes. In *Environmental Monitoring and Characterization*; Elsevier Science: Amsterdam, The Netherlands, 2004; pp. 241–261.
11. Kurzweil, P. Metal Oxides and Ion-Exchanging Surfaces as pH Sensors in Liquids: State-of-the-Art and Outlook. *Sensors* **2009**, *9*, 4955–4985. [CrossRef]
12. Noh, J.; Park, S.; Boo, H.; Kim, H.C.; Chung, T.D. Nanoporous Platinum Solid-state Reference Electrode with Layer-by-layer Polyelectrolyte Junction for pH Sensing Chip. *Lab. A Chip* **2011**, *11*, 664–671. [CrossRef] [PubMed]
13. Suzuki, H.; Hiratsuka, A.; Sasaki, S.; Karube, I. Problems Associated with the Thin-film Ag/AgCl Reference Electrode and a Novel Structure with Improved Durability. *Sens. Actuators. B Chem.* **1998**, *46*, 104–113. [CrossRef]
14. Kim, T.Y.; Hong, S.A.; Yang, S. A Solid-State Thin-Film Ag/AgCl Reference Electrode Coated with Graphene Oxide and its Use in a pH Sensor. *Sensors* **2015**, *15*, 6469–6482. [CrossRef] [PubMed]
15. Manjakkal, L.; Szwagierczak, D.; Dahiya, R. Metal oxides based electrochemical pH sensors: Current progress and future perspectives. *Prog. Mater. Sci.* **2020**, *109*, 100635. [CrossRef]
16. Baur, J.E.; Spaine, T.W. Electrochemical deposition of iridium (IV) oxide from alkaline solutions of iridium (III) oxide. *J. Electroanal Chem.* **1998**, *443*, 208–216. [CrossRef]
17. Kreider, K. Iridium Oxide Thin-Film Stability in High-temperature Corrosive Solutions. *Sensors Actuators B Chem.* **1991**, *5*, 165–169. [CrossRef]
18. Jang, H.; Lee, J. Iridium Oxide Fabrication and Application: A Review. *J. Energy Chem.* **2020**, *46*, 152–172. [CrossRef]
19. Bahrami, H.; Mirbozorgi, S.A.; Ameli, R.; Rusch, L.A.; Gosselin, B. Flexible, polarization-diverse UWB antennas for implantable neural recording systems. *IEEE Trans. Biomed. Circuits Syst.* **2015**, *10*, 38–48. [CrossRef]

20. Sun, Y.; Lacour, S.P.; Brooks, R.A.; Rushton, N.; Fawcett, J.; Cameron, R.E. Assessment of the biocompatibility of photosensitive polyimide for implantable medical device use. *J. Biomed. Mater. Res. Part A Off. J. Soc. Biomater. Jpn. Soc. Biomater. Aust. Soc. Biomater. Korean Soc. Biomater.* **2009**, *90*, 648–655. [CrossRef]
21. Shamma-Donoghue, S.A.; May, G.A.; Cotter, N.E.; White, R.L.; Simmons, F.B. Thin-film multielectrode arrays for a cochlear prosthesis. *IEEE Trans. Electron Devices* **1982**, *29*, 136–144. [CrossRef]
22. Klein, J.D.; Clauson, S.L.; Cogan, S.F. The influence of substrate bias on the morphology and charge capacity of rf-sputtered iridium oxide films. *J. Mater. Res.* **1989**, *4*, 1505–1510. [CrossRef]
23. Liu, D.; Yu, S.; Son, S.; Joo, S. Electrochemical performance of iridium oxide thin film for supercapacitor prepared by radio frequency magnetron sputtering method. *Ecs Trans.* **2008**, *16*, 103. [CrossRef]
24. Han, M.; McCreery, D.B. A new chronic neural probe with electroplated iridium oxide microelectrodes. In Proceedings of the 30th Annual International Conference of the IEEE Engineering in Medicine and Biology Society, Vancouver, BC, Canada, 20–25 August 2008; pp. 4220–4221.
25. Pikulski, M.; Gorski, W. Iridium-based electrocatalytic systems for the determination of insulin. *Anal. Chem.* **2000**, *72*, 2696–2702. [CrossRef] [PubMed]
26. Ohsaka, T.; Matsubara, Y.; Hirano, K.; Ohishi, T. Electroplating of iridium–cobalt alloy. *Trans. IMF* **2007**, *85*, 265–269. [CrossRef]
27. Chen, P.; Chen, Y.; Huang, C. Free-standing iridium oxide nanotube array for neural interface electrode applications. *Mater. Lett.* **2018**, *221*, 293–295. [CrossRef]
28. Chen, Y.; Chung, T.; Wu, P.; Chen, P. A cost-effective fabrication of iridium oxide films as biocompatible electrostimulation electrodes for neural interface applications. *J. Alloy. Compd.* **2017**, *692*, 339–345. [CrossRef]
29. Chawang, K.; Chou, S.; Bing, S.; Wu, P.; Chiao, J. Characterization of pH Sensors Based on Iridium Oxide and Gold Encapsulated Polypropylene Membranes. In Proceedings of the IEEE Sensors Conference, Sydney, Australia, 31 October–4 November 2021.
30. Marsh, P.; Huerta, M.; Le, T.; Yang, X.; Chiao, J.-C.; Cao, H. Wireless Iridium Oxide-Based pH Sensing Systems. In Proceedings of the IEEE Sensors Conference, New Delhi, India, 28–31 October 2018.
31. Nguyen, C.M.; Rao, S.; Yang, X.; Dubey, S.; Mays, J.; Cao, H.; Chiao, J.-C. Sol-gel Deposition of Iridium Oxide for Biomedical Micro-devices. *Sensors* **2015**, *15*, 4212–4228. [CrossRef] [PubMed]
32. Yang, X.; Chawang, K.; Chiao, J.-C. Wearable Iridium Oxide pH Sensors for Sweat pH Measurements. In Proceedings of the IEEE Sensors Conference, Montreal, QC, Canada, 28–31 October 2019.
33. Nguyen, C.M.; Huang, W.; Rao, S.; Cao, H.; Tata, U.; Chiao, M.; Chiao, J.-C. Sol-Gel Iridium Oxide-Based pH Sensor Array on Flexible Polyimide Substrate. *IEEE Sens. J.* **2013**, *13*, 3857–3864. [CrossRef]
34. Yang, X.; Chiao, J.-C. Integrated pH and Sodium Sensor Array Based on Iridium Oxide Film. In Proceedings of the IEEE Sensors Conference, New Delhi, India, 28–31 October 2018.
35. Huang, W.; Deb, S.; Seo, Y.; Rao, S.; Chiao, M.; Chiao, J.C. A Passive Radio-Frequency pH-Sensing Tag for Wireless Food-Quality Monitoring. *IEEE Sens. J.* **2011**, *12*, 487–495. [CrossRef]
36. Cao, H.; Rao, S.; Tang, S.; Tibbals, H.F.; Spechler, S.; Chiao, J. Batteryless Implantable Dual-Sensor Capsule for Esophageal Reflux Monitoring. *Gastrointest. Endosc.* **2013**, *77*, 649–653. [CrossRef]
37. Deb, S.; Tang, S.; Abell, T.L.; McLawhorn, T.; Huang, W.; Lahr, C.; To, S.F.; Easter, J.; Chiao, J. Development of Innovative Techniques for the Endoscopic Implantation and Securing of a Novel, Wireless, Miniature Gastrostimulator (with videos). *Gastrointest. Endosc.* **2012**, *76*, 179–184. [CrossRef] [PubMed]
38. Chawang, K.; Bing, S.; Chiao, J.-C. Investigation of pH Sensing in Viscous Salt-added Solution by Iridium Oxide Film. In Proceedings of the IEEE Sensors Conference, Dallas, TX, USA, 30 October–2 November 2022.
39. Huang, W.; Cao, H.; Deb, S.; Chiao, M.; Chiao, J.-C. A Flexible pH Sensor Based on the Iridium Oxide Sensing Film. *Sens. Actuators A Phys.* **2011**, *169*, 1–11. [CrossRef]
40. Yang, X.; Fu, T.; Kota, P.K.; Tjia, M.; Nguyen, C.M.; Chiao, J.-C. Lactate Sensors on Flexible Substrates. *Biosensors* **2016**, *6*, 48. [CrossRef] [PubMed]
41. Osaka, A.; Takatsuna, T.; Miura, Y. Iridium oxide films via sol-gel processing. *J. Non Cryst. Solids* **1994**, *178*, 313–319. [CrossRef]
42. Olthuis, W.; Robben, M.; Bergveld, P.; Bos, M.; van der Linden, W.E. pH Sensor Properties of Electrochemically Grown Iridium Oxide. *Sensors Actuators B Chem.* **1990**, *2*, 247–256. [CrossRef]
43. Chou, S.C.; Hsieh, Y.C.; Cheang, W.H.; Sun, B.Y.; Chu, C.Y.; Chen, S.Y.; Chiao, J.C.; Wu, P.W. A flexible IrO2 membrane for pH sensing. *Sci. Rep.* **2022**, *12*, 11712. [CrossRef]
44. Wang, G.; Zhou, M.; Goettel, J.T.; Schrobilgen, G.J.; Su, J.; Li, J. Identification of an iridium-containing compound with a formal oxidation state of IX. *Nature* **2014**, *514*, 475–477. [CrossRef]
45. Bause, S.; Decker, M.; Gerlach, F.; Näther, J.; Köster, F.; Neubauer, P.; Vonau, W. Development of an iridium-based pH sensor for bioanalytical applications. *J. Solid State Electrochem.* **2018**, *22*, 51–60. [CrossRef]
46. Hernandez-Jaimes, C.; Lobato-Calleros, C.; Sosa, E.; Bello-Pérez, L.A.; Vernon-Carter, E.J.; Alvarez-Ramirez, J. Electrochemical Characterization of Gelatinized Starch Dispersions: Voltammetry and Electrochemical Impedance Spectroscopy on Platinum Surface. *Carbohydr. Polym.* **2015**, *124*, 8–16. [CrossRef]
47. Rajeswari, A.; Gopi, S.; Christy, E.J.S.; Jayaraj, K.; Pius, A. Current research on the blends of chitosan as new biomaterials. In *Handbook of Chitin and Chitosan*; Elsevier: Amsterdam, The Netherlands, 2020; pp. 247–283.

48. Bementa, E.; Rajan, M.A.J.; Gnanadass, E.S. Effect of Prolonged Duration of Gelatinization in Starch and Incorporation with Potassium Iodide on the Enhancement of Ionic Conductivity. *Polym.-Plast. Technol. Eng.* **2017**, *56*, 1632–1645. [CrossRef]
49. Bennett, D. NaCl Doping and the Conductivity of Agar Phantoms. *Mater. Sci. Eng. C* **2011**, *31*, 494–498. [CrossRef]
50. Strazzullo, P.; Leclercq, C. Sodium1. *Adv. Nutr. Bethesda Md.* **2014**, *5*, 188–190. [CrossRef] [PubMed]
51. Pohl, H.R.; Wheeler, J.S.; Murray, H.E. Sodium and potassium in health and disease. In *Interrelations between Essential Metal Ions and Human Diseases*; Springer: Berlin/Heidelberg, Germany, 2013; pp. 29–47.
52. International Association for Dental Research. Available online: https://iadr.abstractarchives.com/abstract/18iags-2954851/relationship-of-saliva-viscosity-and-ph-at-resting-and-stimulation (accessed on 7 September 2022).
53. Piagnerelli, M.; Boudjeltia, K.Z.; Vanhaeverbeek, M.; Vincent, L.-J. Physiological Reviews. In *Applied Physiology in Intensive Care Medicine*, 2nd ed.; Pinsky, M.R., Brochard, L., Mancebo, J., Hedenstierna, G., Eds.; Springer: Berlin/Heidelberg, Germany, 2006; pp. 273–282.
54. Yanniotis, S.; Skaltsi, S.; Karaburnioti, S. Effect of moisture content on the viscosity of honey at different temperatures. *J. Food Eng.* **2006**, *72*, 372–377. [CrossRef]
55. Kim, H. The mineral contents, viscosity and sensory characteristics of demi-glace sauce according to the varying quantity of omija added. *J. Korean Soc. Food Cult.* **2004**, *19*, 667–677.
56. Garcia, J.M.; Chambers, E.; Matta, Z.; Clark, M. Viscosity measurements of nectar-and honey-thick liquids: Product, liquid, and time comparisons. *Dysphagia* **2005**, *20*, 325–335. [CrossRef]

chemosensors

Communication

High-Sensitivity pH Sensor Based on Coplanar Gate AlGaN/GaN Metal-Oxide-Semiconductor High Electron Mobility Transistor

Seong-Kun Cho and Won-Ju Cho *

Department of Electronic Materials Engineering, Kwangwoon University, 20 Gwangun-ro, Nowon-gu, Seoul 01897, Korea; whtjdrms98@gmail.com
* Correspondence: chowj@kw.ac.kr; Tel.: +82-2-940-5163

Abstract: The sensitivity of conventional ion-sensitive field-effect transistors is limited to the Nernst limit (59.14 mV/pH). In this study, we developed a pH sensor platform based on a coplanar gate AlGaN/GaN metal-oxide-semiconductor (MOS) high electron mobility transistor (HEMT) using the resistive coupling effect to overcome the Nernst limit. For resistive coupling, a coplanar gate comprising a control gate (CG) and a sensing gate (SG) was designed. We investigated the amplification of the pH sensitivity with the change in the magnitude of a resistance connected in series to each CG and SG via Silvaco TCAD simulations. In addition, a disposable extended gate was applied as a cost-effective sensor platform that helped prevent damages due to direct exposure of the AlGaN/GaN MOS HEMT to chemical solutions. The pH sensor based on the coplanar gate AlGaN/GaN MOS HEMT exhibited a pH sensitivity considerably higher than the Nernst limit, dependent on the ratio of the series resistance connected to the CG and SG, as well as excellent reliability and stability with non-ideal behavior. The pH sensor developed in this study is expected to be readily integrated with wide transmission bandwidth, high temperature, and high-power electronics as a highly sensitive biosensor platform.

Keywords: AlGaN/GaN MOS HEMT; coplanar gate; resistive coupling; pH sensor

Citation: Cho, S.-K.; Cho, W.-J. High-Sensitivity pH Sensor Based on Coplanar Gate AlGaN/GaN Metal-Oxide-Semiconductor High Electron Mobility Transistor. *Chemosensors* **2021**, *9*, 42. https://doi.org/10.3390/chemosensors9030042

Academic Editor: Pietro Salvo

Received: 28 January 2021
Accepted: 24 February 2021
Published: 25 February 2021

1. Introduction

Developments in big data, artificial intelligence, deep learning, and internet of things in the fourth industrial revolution have led to increased human-machine interactions and, consequently, the requirement of the miniaturization and functionalization of electronic devices. In particular, biosensors are being actively studied as a core technology for human-machine interactions. Field-effect transistor (FET)-based biosensors offer various advantageous properties, including a short response time and accurate detection, and they can be mass produced and miniaturized [1,2]. However, conventional ion-sensitive FETs (ISFETs) have a sensitivity limit of 59.14 mV/pH, called the Nernst limit, which hinders their commercialization [3,4]. Double gate structure ISFETs based on silicon-on-insulator are drawing attention because they can overcome the Nernst limit via asymmetric capacitive coupling of the upper and lower gate oxides [5–8]. In addition, in our previous work, we studied a silicon-based device that amplifies the sensitivity through capacitive coupling between the coplanar gate and the floating gate, and here we investigated the capacitive coupling effect according to the coplanar gate area and the corresponding sensitivity amplification [9]. However, owing to the limitations of the physical characteristics of silicon, these transistors cannot easily be integrated with wide transmission bandwidths and at high temperatures. Meanwhile, compound semiconductor-based AlGaN/GaN high electron mobility transistors (HEMTs) exhibit low on-resistance and rapid switching speeds owing to the high-mobility two-dimensional electron gas induced by the discontinuity of the conduction band in the AlGaN/GaN heterostructure and its piezoelectric polarization [10]. In addition, AlGaN/GaN HEMT-based sensors have shown potential for application for the detection of DNA, antigens, glucose, cellular responses, and gas; further, these sensors can be monolithically integrated with signal processing and amplification circuits or radio frequency (RF) signal transmission circuits, which are convenient for reading data and receiving remote signals [11–18]. Nevertheless, the leakage current of AlGaN/GaN HEMTs due to the trap-assisted tunneling and surface state is crucial. In order to attain low power consumption and high conversion efficiency, high breakdown voltage, leakage suppression, and high on/off current ratio are crucial. Meanwhile, the metal-oxide-semiconductor (MOS) structure can achieve a high breakdown voltage and on/off current ratio because the gate insulator effectively prevents the gate leakage and suppresses the surface leakage [19,20].

In this study, we developed a pH sensor based on an AlGaN/GaN MOS HEMT using the resistive coupling of a coplanar gate structure comprising a control gate (CG) and a sensing gate (SG). Previous AlGaN/GaN MOS HEMT-based ISFETs can easily be damaged, because the HEMT device in these is directly exposed to a chemical solution. To prevent this damage, we designed an AlGaN/GaN MOS HEMT transducer unit that converts biochemical signals into electrical signals and an extended gate (EG) sensing unit that is directly exposed to the pH solution. This separate structure of the transducer and sensing units fundamentally prevents damage to the expensive HEMT and provides a cost-effective sensor platform by easily replacing the damaged low-cost sensing unit. The variation in the sensitivity of the pH sensor was evaluated according to the ratio of the series resistance connected to the CG (R_{CG}) and SG (R_{SG}) of the AlGaN/GaN MOS HEMT transducer, and an amplified sensitivity that exceeded the Nernst limit was obtained. In addition, non-ideal behaviors, such as hysteresis and drift effects, were investigated to verify the stability and reliability of the sensor.

2. Materials and Methods

The coplanar gate AlGaN/GaN MOS HEMT transducer unit was fabricated as follows. AlGaN/GaN heterostructures were grown in metal organic chemical vapor deposition reactors on (0001) sapphire substrates. A 25 nm low-temperature GaN nucleation layer was grown on the sapphire substrate, followed by the successive growth of a 1 μm high-resistance GaN layer and 1.8 μm unintentionally doped GaN epitaxial layer. Subsequently, a 12 nm $Al_{0.25}GaN$ layer was grown. Mesa separation was realized by etching a 500 nm AlGaN/GaN layer, where the active region was defined by BCl_3- and Cl_2- based inductively coupled plasma etching using a photoresist (PR) mask. A 500 nm SiO_2 field oxide layer was deposited via RF magnetron sputtering for a flat surface to avoid breaking of the gate metals at the steep steps in the mesa region. After removing the PR mask and SiO_2 layer from the active region, the AlGaN/GaN heterostructure was cleaned using an ammonium sulfide solution with excess sulfur [$(NH_4)_2S_x$, 40% sulfur] to remove the native oxide and prevent the formation of oxide. Subsequently, a 5 nm SiO_2 layer and 10 nm Ta_2O_5 layer were deposited as gate insulators via RF magnetron sputtering. The source and drain electrodes of Ti/Al/Ni/Au (=20/100/25/50 nm) were formed via e-beam evaporator and lift-off, followed by rapid thermal treatment at 800 °C for 30 s in N_2 ambient. To form series resistors between the HEMT gate (on channel) and the coplanar gates (CG and SG), a 50 nm indium-tin-oxide (ITO) film with a sheet resistance of 1.7×10^3 Ω/sq was deposited. We designed resistors with a width of 2 μm and lengths of 10, 20, and 30 μm and obtained resistances of 17, 34, and 51 kΩ, respectively. For the HEMT gate and coplanar gate electrodes, Ni/Au (=25/100 nm) was deposited via e-beam evaporation and lift-off. Meanwhile, to fabricate the EG sensing unit, a 150 nm ITO conductive layer and a 50 nm SnO_2 sensing membrane were subsequently deposited on glass substrates via RF magnetron sputtering. Subsequently, a polydimethylsiloxane reservoir was placed on top. Figure 1a,b shows schematics of the coplanar gate AlGaN/GaN MOS HEMT and EG, respectively.

We finally constructed the pH sensor by connecting the EG to the SG of the AlGaN/GaN MOS HEMT with an electrical cable (Model 5342, Pomana electronics, Everett, WA, USA) to transfer the potential of the pH solution to the SG. We performed all electrical measurements of device characteristics using an Agilent 4156B precision semiconductor parameter analyzer (Agilent Technologies, Santa Clara, CA, USA). Also, these measurements were performed in a shielded dark box to prevent external elements such as noise and light. In particular, a commercial Ag/AgCl reference electrode (Horiba 2086A-06T, Kyoto, Japan) composed of ceramic-plug junction and internal solution saturated with KCl and AgCl was used for pH sensing.

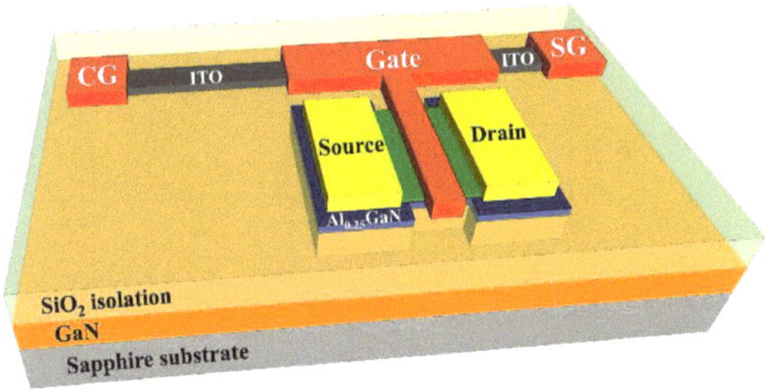

(a)

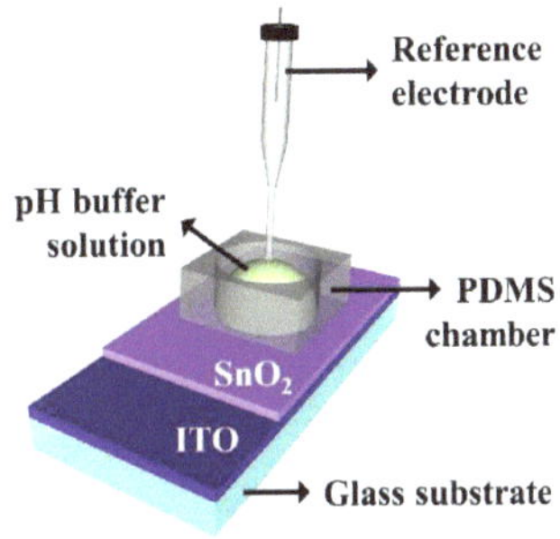

(b)

Figure 1. Schematic of (**a**) coplanar gate AlGaN/GaN metal-oxide-semiconductor high electron mobility transistor (MOS HEMT) transducer unit, and (**b**) extended gate (EG) sensing unit.

3. Results

3.1. Silvaco TCAD Simulations

The voltages applied to the CG and SG are denoted as V_{CG} and V_{SG}, respectively. The total resistance (R_T) of a coplanar gate can be expressed as $R_T = R_{CG} + R_{SG}$, where R_{CG} and R_{SG} are the resistances of the CG and SG, respectively. Then, the gate voltage (V_{FG}) of the MOS HEMT is expressed as in Equation (1), and the voltage between CG and SG can be calculated using Equation (2). Eventually, the potential change in SG (ΔV_{SG}) is amplified by a factor of R_{CG}/R_{SG}, resulting in a voltage change in the CG (ΔV_{CG}), which implies that a significantly small change in potential in the SG is amplified by the resistive coupling effect and can be detected in the CG.

$$V_{FG} = \frac{R_{SG}}{R_T} V_{CG} + \frac{R_{CG}}{R_T} V_{SG}, \tag{1}$$

$$V_{CG} = \frac{R_T}{R_{SG}} V_{FG} - \frac{R_{CG}}{R_{SG}} V_{SG}, \tag{2}$$

$$\therefore \Delta V_{CG} \propto \frac{R_{CG}}{R_{SG}} \Delta V_{SG}, \tag{3}$$

Figure 2b shows Silvaco TCAD simulations for the transfer characteristic curves of the AlGaN/GaN MOS HEMT with R_{CG}:R_{SG} = 2:1. The Silvaco TCAD simulation proposed in this study was to evaluate the signal amplification capability by resistive coupling, and was performed as follows: (1) Atals simulation for MOS HEMT device with conmob (specify concentration dependent mobility model), fldmob (specify lateral electric field-dependent model), SRH (specifies Shockley-Read-Hall recombination). (2) Mixedmode module in Atlas for circuit simulation for resistive coupling between CG and SG. (3) Tunneling conduction such as band-to-band tunneling, Fowler-Nordheim tunneling, direct quantum tunneling, trap assist tunneling, phonon assist electron tunneling, Schottky tunneling, etc. were ignored. As a result, using Equation (3), when ΔV_{SG} is 2 V, ΔV_{CG} is calculated to be 4 V, and the resistance ratio R_{CG}/R_{SG} is increased to 2. Figure 2c shows the dependence of the amplification factor ($\Delta V_{CG}/\Delta V_{SG}$) considering ratios R_{CG}:R_{SG} = 1:2, 1:1, 2:1, and 3:1, where ΔV_{CG} is determined at a drain current of 1 nA. The amplification coefficients at ratios R_{CG}:R_{SG} = 1:2, 1:1, 2:1, and 3:1 were found to be 0.5, 1, 2, and 3, respectively.

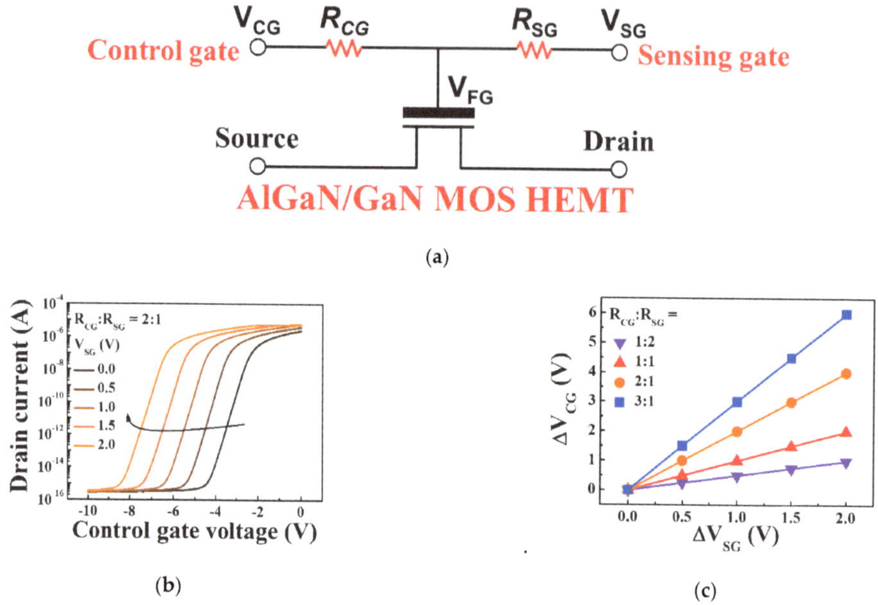

Figure 2. (**a**) Simplified equivalent circuit of the pH sensor based on the coplanar gate AlGaN/GaN MOS HEMT proposed herein. Silvaco TCAD simulation results of (**b**) transfer characteristic curves shift as a function of ΔV_{SG}, and (**c**) dependence of the amplification factor ($\Delta V_{CG}/\Delta V_{SG}$) on R_{CG}:R_{SG} of the pH sensor.

3.2. Electrical Characteristics of Coplanar Gate AlGaN/GaN MOS HEMT

Figure 3a shows the transfer characteristic curves of the fabricated coplanar gate AlGaN/GaN MOS HEMT. Here, the threshold voltage is −0.99 V, electron concentration is 7.47×10^{12} cm^{-2}, electron mobility is 2835.9 cm^2/V s, on/off current ratio is 5.55×10^8, and subthreshold swing is 79.08 mV/dec. A low gate leakage current was observed because the SiO_2/Ta_2O_5 stacked gate insulator effectively blocked the gate leakage and suppressed the surface leakage. In addition, the output characteristic curves in Figure 3b indicate a high drain drive current due to the high electron concentration and mobility of the AlGaN/GaN MOS HEMT.

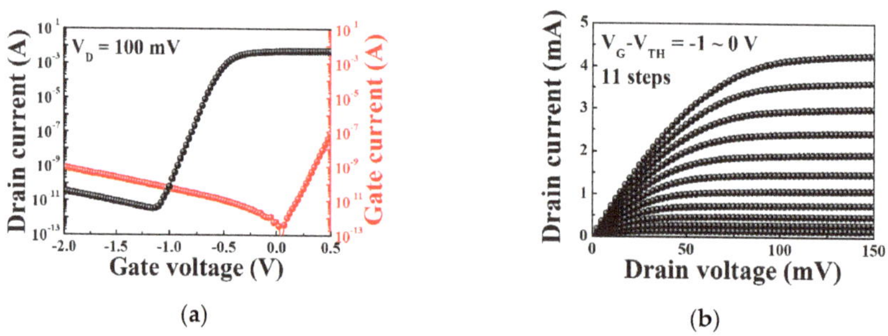

Figure 3. (**a**) Transfer characteristics and (**b**) output characteristic curves of the fabricated AlGaN/GaN MOS HEMT.

3.3. pH Sensing Characteristics of Coplanar Gate AlGaN/GaN MOS HEMT

There is a Nernst limit for pH sensing properties that cannot exceed 59.14 mV/pH according to site-binding theory. In the site-binding theory, the ability to detect ions depends entirely on the surface potential (ψ), which is summarized in the Equation (4): [21,22]

$$\psi = 2.303\frac{kT}{q}\frac{\beta}{\beta+1}(pH_{pzc} - pH), \tag{4}$$

where k is Boltzmann constant, T is the absolute temperature, q is the elementary charge, pH_{pzc} is the pH at the point of zero charge, and β is a parameter that denotes the chemical sensitivity of the sensing membrane. The ψ depends on the properties of the sensing membrane and the pH level of the electrolyte. Thus, the shift of the threshold voltage (ΔV_{TH}) for the transistor is determined by $\Delta\psi$. Based on the simulation result, when resistive coupling is introduced, ΔV_{TH} becomes R_{CG}/R_{SG} times $\Delta\psi$, which can amplify the sensitivity.

Figure 4a–c shows the transfer characteristic curves of the pH sensor for the ratios R_{CG}:R_{SG} = 1:1, 2:1, and 3:1, respectively. We measured the sensing properties using buffer solutions with the following pH concentrations: 3.07, 4.08, 5.99, 6.95, 8.97, and 9.87. From Figure 4, it can be seen that as the pH value increases, the transfer characteristic curve shifts in the positive voltage direction. Figure 4d shows a plot of ΔV_{CG} as a function of the pH value. Here, ΔV_{CG} is defined as the shift of the V_{CG} extracted from the drain current of 1 nA, corresponding to the pH sensitivity as a function of the concentration of the buffer solution. For R_{CG}:R_{SG} = 1:1, the pH sensitivity is 56.63 mV/pH, which is lower than the Nernst limit of 59.14 mV/pH. However, for R_{CG}:R_{SG} = 2:1 and 3:1, the pH sensitivities are 112.17 and 167.71 mV/pH, respectively, which are amplified by a factor of R_{CG}/R_{SG} and exceed the Nernst limit. In addition, the linear fitting line for extracting the sensitivity in Figure 4d showed more than 99.8% linearity in R_{CG}:R_{SG} = 1:1, 2:1, and 3:1. Figure 4d also shows the deviation for 20 measurements at each pH value, indicating reliable sensing performance. This means that there is no Debye screening length limit for the various pH buffer solutions used in this study.

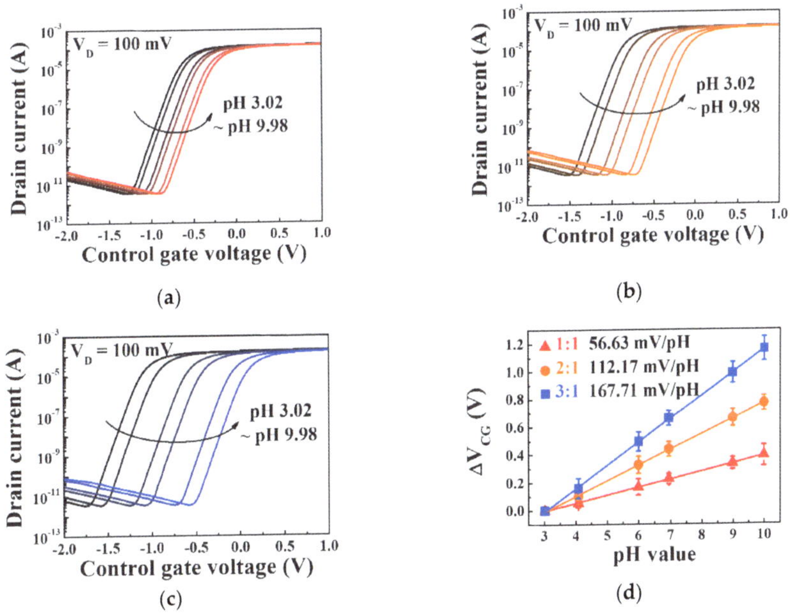

Figure 4. Transfer characteristic curves of the pH sensor based on a coplanar gate AlGaN/GaN MOS HEMT for R_{CG}:R_{SG} = (**a**) 1:1, (**b**) 2:1, and (**c**) 3:1. (**d**) ΔV_{CG} as a function of the pH value.

Reliability-related hysteresis effects are due to the presence of slowly reacting OH-sites on the surface of the sensing membrane and the transport of certain species in the bulk of the sensing membrane [23–25]. Meanwhile, stability-related drift effects are due to the hopping and/or trap-limited transport of OH-related species from the electrolyte and defects present in the sensing membrane [26,27]. These non-ideal behaviors limit the accuracy of the sensor. Figure 5a shows the hysteresis characteristics when the pH is changed as follows: 7 → 10 → 7 → 4 → 7. It can be seen that, for R_{CG}:R_{SG} = 1:1, 2:1, and 3:1, the hysteresis voltages are 1.8, 3.5, and 7.1 mV, respectively. Figure 5b shows the drift rates upon exposure to a pH 7 buffer solution for 10 h. It can be seen that, for R_{CG}:R_{SG} = 1:1, 2:1, and 3:1, the drift rates are 0.4, 0.6, and 0.7 mV/h, respectively.

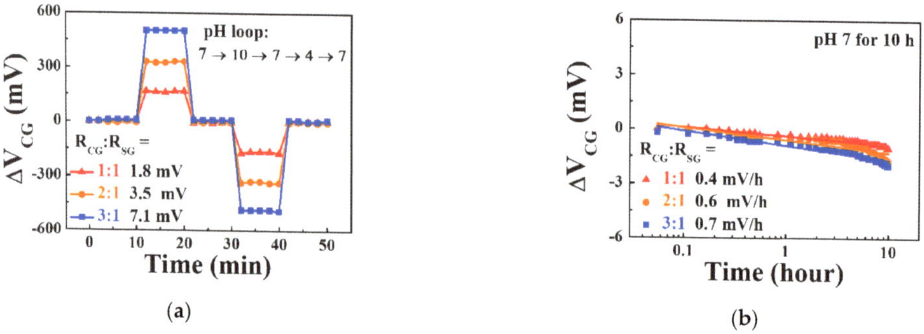

Figure 5. (**a**) Hysteresis effect and (**b**) drift effect of the pH sensor based on a coplanar gate AlGaN/GaN MOS HEMT for R_{CG}:R_{SG} = 1:1, 2:1, and 3:1.

4. Conclusions

We developed a pH sensor based on a coplanar gate AlGaN/GaN MOS HEMT with increased sensitivity using the resistive coupling effect. Through Silvaco TCAD simulations, the pH sensitivity amplification corresponding to the ratio of the series resistance connected to the CG and SG (R_{CG}/R_{SG}) is predicted. The pH sensitivity of the fabricated sensors increases proportionally with R_{CG}/R_{SG}. In particular, for R_{CG}:R_{SG} = 2:1 and 3:1, the sensitivity was 112.17 and 167.71 mV/pH, respectively, which are greater than the Nernst limit. We also verified the stability and reliability of the sensor by evaluating its non-ideal behaviors, such as the hysteresis and drift effects. In addition, by applying the disposable EG sensing unit, damage to the AlGaN/GaN MOS HEMT transducer unit due to chemical solutions was prevented. The results of this study indicate that the pH sensor developed has high performance, stability, and reliability and is also disposable. Thus, since the sensor proposed in this study is based on a HEMT device, it is expected to be suitable for integration with wide transmission bandwidth, high temperature, speed and frequency power electronics. In addition, the pH sensor proposed in this study detects the potential change of the sensing membrane using EG. Therefore, when the coplanar gate AlGaN/GaN MOS HEMT proposed in this study is introduced into the subsequent study on EG, various biological events such as enzyme-substrate reactions, antigen-antibody binding, and nucleic acid hybridization can be detected with high sensitivity.

Author Contributions: S.-K.C.: conceptualization, formal analysis, methodology, investigation, data curation, visualization, software, resources, writing—original draft. W.-J.C.: conceptualization, methodology, investigation, resources, formal analysis, funding acquisition, supervision, validation, writing—review and editing. All authors have read and agreed to the published version of the manuscript.

Funding: The present research has been conducted by the Research Grant of Kwangwoon University in 2020, and this work was supported by the National Research Foundation of Korea (NRF) grant funded by the Korea government (MSIT) (No. 2020R1A2C1007586). This research was funded and conducted under the Competency Development Program for Industry Specialists of the Korean Ministry of Trade, Industry and Energy (MOTIE), operated by Korea Institute for Advancement of Technology (KIAT) (No. P0002397, HRD program for Industrial Convergence of Wearable Smart Devices).

Conflicts of Interest: The authors declare no conflict of interest.

References

1. Bergveld, P. Development of an Ion-Sensitive Solid-State Device for Neurophysiological Measurements. *IEEE Trans. Biomed. Eng.* **1970**, *17*, 70–71. [CrossRef]
2. Bergveld, P. Thirty years of ISFETOLOGY: What happened in the past 30 years and what may happen in the next 30 years. *Sens. Actuators B Chem.* **2003**, *88*, 1–20. [CrossRef]
3. Knopfmacher, O.; Tarasov, A.; Fu, W.; Wipf, M.; Niesen, B.; Calame, M.; Schoönenberger, C. Nernst Limit in Dual-Gated Si-Nanowire FET Sensors. *Nano Lett.* **2010**, *10*, 2268–2274. [CrossRef]
4. Spijkman, M.; Smits, E.C.P.; Cillessen, J.F.M.; Biscarini, F.; Blom, P.W.M.; De Leeuw, D.M. Beyond the Nernst-limit with dual-gate ZnO ion-sensitive field-effect transistors. *Appl. Phys. Lett.* **2011**, *98*, 043502. [CrossRef]
5. Spijkman, M.-J.; Myny, K.; Smits, E.C.P.; Heremans, P.; Blom, P.W.M.; De Leeuw, D.M. Dual-Gate Thin-Film Transistors, Integrated Circuits and Sensors. *Adv. Mater.* **2011**, *23*, 3231–3242. [CrossRef] [PubMed]
6. Lim, H.K.; Fossum, J.G. Threshold voltage of thin-film silicon-on-insulator (SOI) MOSFET's. *IEEE Trans. Electron. Devices* **1983**, *30*, 1244–1251.
7. Jang, H.J.; Cho, W.J. Performance enhancement of capacitive-coupling dual-gate ion-sensitive field-effect transistor in ultra-thin-body. *Sci. Rep.* **2014**, *4*, 5284. [CrossRef]
8. Lee, I.-K.; Lee, K.H.; Lee, S.; Cho, W.-J. Microwave Annealing Effect for Highly Reliable Biosensor: Dual-Gate Ion-Sensitive Field-Effect Transistor Using Amorphous InGaZnO Thin-Film Transistor. *ACS Appl. Mater. Interfaces* **2014**, *6*, 22680–22686. [CrossRef]
9. Jeon, J.-H.; Cho, W.-J. Ultrasensitive Coplanar Dual-Gate ISFETs for Point-of-Care Biomedical Applications. *ACS Omega* **2020**, *5*, 12809–12815. [CrossRef]
10. Ambacher, O.; Smart, J.; Shealy, J.R.; Weimann, N.G.; Chu, K.; Murphy, M.; Schaff, W.J.; Eastman, L.F.; Dimitrov, R.; Wittmer, L.; et al. Two-dimensional electron gases induced by spontaneous and piezoe-lectric polarization charges in N-and Ga-face AlGaN/GaN heterostructures. *J. Appl. Phys.* **1999**, *85*, 1999. [CrossRef]
11. Sarangadharan, I.; Regmi, A.; Chen, Y.-W.; Hsu, C.-P.; Chen, P.-C.; Chang, W.-H.; Lee, G.-Y.; Chyi, J.-I.; Shiesh, S.-C.; Lee, G.-B.; et al. High sensitivity cardiac troponin I detection in physiological environment using AlGaN/GaN High Electron Mobility Transistor (HEMT) Biosensors. *Biosens. Bioelectron.* **2018**, *100*, 282–289. [CrossRef] [PubMed]
12. Espinosa, N.; Schwarz, S.U.; Cimalla, V.; Ambacher, O. Detection of different target-DNA concentrations with highly sensitive AlGaN/GaN high electron mobility transistors. *Sens. Actuators B Chem.* **2015**, *210*, 633–639. [CrossRef]
13. Li, J.-D.; Cheng, J.-J.; Miao, B.; Wei, X.-W.; Xie, J.; Zhang, J.-C.; Zhang, Z.-Q.; Li, H.-W.; Wu, D.-M. Label free electrical detection of prostate specific antigen with millimeter grade biomolecule-gated AlGaN/GaN high electron mobility transistors. *Microsyst. Technol.* **2014**, *21*, 1489–1494. [CrossRef]
14. Li, J.-D.; Cheng, J.-J.; Miao, B.; Wei, X.-W.; Xie, J.; Zhang, J.-C.; Zhang, Z.-Q.; Wu, D.-M. Detection of prostate-specific antigen with biomolecule-gated AlGaN/GaN high electron mobility transistors. *J. Micromech. Microeng.* **2014**, *24*, 75023. [CrossRef]
15. Lee, C.T.; Chiu, Y.S. Photoelectrochemical passivated ZnO-based nanorod structured glucose biosensors using gate-recessed AlGaN/GaN ion-sensitive field-effect-transistors. *Sens. Actuators B Chem.* **2015**, *210*, 756–761. [CrossRef]
16. Chu, B.H.; Kang, B.S.; Hung, S.C.; Chen, K.H.; Ren, F.; Sciullo, A.; Gila, B.P.; Pearton, S.J. Aluminum Gallium Nitride (GaN)/GaN High Electron Mobility Transistor-Based Sensors for Glucose Detection in Exhaled Breath Condensate. *J. Diabetes Sci. Technol.* **2010**, *4*, 171–179. [CrossRef]
17. Pulikkathodi, A.K.; Sarangadharan, I.; Hsu, C.-P.; Chen, Y.-H.; Hung, L.-Y.; Lee, G.-Y.; Chyi, J.-I.; Lee, G.-B.; Wang, Y.-L. Enumeration of circulating tumor cells and investigation of cellular responses using aptamer-immobilized AlGaN/GaN high electron mobility transistor sensor array. *Sens. Actuators B Chem.* **2018**, *257*, 96–104. [CrossRef]
18. Jung, S.; Baik, K.H.; Ren, F.; Pearton, S.J.; Jang, S. Pt-AlGaN/GaN hydrogen sensor with water-blocking PMMA layer. *IEEE Electron Device Lett.* **2017**, *38*, 657–660. [CrossRef]
19. Lee, C.S.; Liao, Y.H.; Chou, B.Y.; Liu, H.Y.; Hsu, W.C. Composite HfO_2/Al_2O_3-dielectric AlGaAs/InGaAs MOS-HEMTs by using RF sputtering/ozone water oxidation. *Superlattices Microstruct.* **2014**, *72*, 194–203. [CrossRef]
20. Seok, O.; Ahn, W.; Han, M.-K.; Ha, M.-W. High on/off current ratio AlGaN/GaN MOS-HEMTs employing RF-sputtered HfO_2 gate insulators. *Semicond. Sci. Technol.* **2013**, *28*, 025001. [CrossRef]
21. Pan, T.-M.; Huang, M.-D.; Lin, C.-W.; Wu, M.-H. Development of high-κ $HoTiO_3$ sensing membrane for pH detection and glucose biosensing. *Sens. Actuators B Chem.* **2010**, *144*, 139–145. [CrossRef]
22. Chou, J.-C.; Liao, L.P. Study on pH at the point of zero charge of TiO_2 pH ion-sensitive field effect transistor made by the sputtering method. *Thin Solid Films* **2005**, *476*, 157–161. [CrossRef]
23. Chang, S.P.; Yang, T.H. Sensing performance of EGFET pH sensors with CuO nanowires fabricated on glass substrate. *Int. J. Electrochem. Sci.* **2012**, *7*, 5020–5027.
24. Gandhi, O.P.; Chen, J.-Y.; Riazi, A. Currents Induced in a Human Being for Plane-Wave Exposure Conditions 0–50 MHz and for RF Sealers. *IEEE Trans. Biomed. Eng.* **1986**, *33*, 757–767. [CrossRef] [PubMed]
25. Bousse, L.; Mostarshed, S.; van der Schoot, B.; De Rooij, N.F. Comparison of the hysteresis of Ta_2O_5 and Si_3N_4 pH-sensing insulators. *Sens. Actuators B Chem.* **1994**, *17*, 157–164. [CrossRef]
26. Bousse, L.; Bergveld, P. The role of buried OH sites in the response mechanism of inorganic-gate pH-sensitive ISFETs. *Sens. Actuators* **1984**, *6*, 65–78. [CrossRef]
27. Jamasb, S.; Collins, S.; Smith, R.L. A physical model for drift in pH ISFETs. *Sens. Actuators B Chem.* **1998**, *49*, 146–155. [CrossRef]

chemosensors

MDPI

Article

Fully Transparent and Highly Sensitive pH Sensor Based on an a-IGZO Thin-Film Transistor with Coplanar Dual-Gate on Flexible Polyimide Substrates

Tae-Hwan Hyun and Won-Ju Cho *

Department of Electronic Materials Engineering, Kwangwoon University, Seoul 139-701, Republic of Korea
* Correspondence: chowj@kw.ac.kr; Tel.: +82-2-940-5163

Abstract: In this paper, we propose a fully transparent and flexible high-performance pH sensor based on an amorphous indium gallium zinc oxide (a-IGZO) thin-film transistor (TFT) transducer with a coplanar dual-gate structure on polyimide substrates. The proposed pH sensor system features a transducer unit consisting of a floating gate (FG), sensing gate (SG), and control gate (CG) on a polyimide (PI), and an extended gate (EG) sensing unit on a separate glass substrate. We designed a capacitive coupling between (SG) and (CG) through the FG of an a-IGZO TFT transducer to contribute to sensitivity amplification. The capacitance ratio (C_{SG}/C_{CG}) increases linearly with the area ratio; therefore, the amplification ratio of the pH sensitivity was easily controlled using the area ratio of SG/CG. The proposed sensor system improved the pH sensitivity by up to 359.28 mV/pH (C_{SG}/C_{CG} = 6.16) at room temperature (300 K), which is significantly larger than the Nernstian limit of 59.14 mV/pH. In addition, the non-ideal behavior, including hysteresis and drift effects, was evaluated to ensure stability and reliability. The amplification of sensitivity based on capacitive coupling was much higher than the increase in the hysteresis voltage and drift rate. Furthermore, we verified the flexibility of the a-IGZO coplanar dual-gate TFT transducer through a bending test, and the electrical properties were maintained without mechanical damage, even after repeated bending. Therefore, the proposed fully transparent and highly sensitive a-IGZO coplanar dual-gate TFT-based pH sensor could be a promising wearable and portable high-performance chemical sensor platform.

Keywords: transparent; flexible; PI substrate; a-IGZO; coplanar dual-gate; capacitive coupling; pH sensor; FET

Citation: Hyun, T.-H.; Cho, W.-J. Fully Transparent and Highly Sensitive pH Sensor Based on an a-IGZO Thin-Film Transistor with Coplanar Dual-Gate on Flexible Polyimide Substrates. *Chemosensors* **2023**, *11*, 46. https://doi.org/10.3390/chemosensors11010046

Academic Editors: Pietro Salvo and Luigi Campanella

Received: 22 November 2022
Revised: 22 December 2022
Accepted: 3 January 2023
Published: 4 January 2023

1. Introduction

Recently, research interest in chemical sensors has increased owing to the increased interest in medical care worldwide, and these sensors can detect signals of small amounts of chemicals or biomolecules. Chemical sensors can be applied to various fields such as food manufacturing, environmental conditioning, and biological monitoring (blood, sweat, urine). Accordingly, many types of chemical sensors for detecting pH, viruses, proteins, and chemicals have been reported [1–3]. Among them, the field-effect transistor (FET)-type sensor platform has attracted considerable interest owing to its excellent features, such as fast response, label-free detection, compatibility with CMOS technology, and easy signal processing [4–6]. In a study by Bergveld, the author considered FET-based sensors as ion-sensitive FETs (ISFETs) [7]. However, because ISFETs detect signals from chemicals through direct contact with the gate dielectric sensing membrane, there is a possibility of its degradation by chemicals. To avoid these reliability-related issues, extended-gate FET (EGFET)-type sensor platforms were introduced [8]. The EGFET consists of two separate parts: a transducer unit and sensing unit. Various high-performance sensing membrane materials have been developed by applying an extended gate that is electrically connected to the gate electrode of the FET to prevent degradation caused by chemical damage [9–11]. However, the most significant limitation hindering the commercialization of

FET-based sensors is the physical sensitivity limit of 59.14 mV/pH at 300 K, which is known as the Nernstian limit [12,13]. Among the many approaches reported in the literature, FET-based sensors with dual-gate structures can overcome the Nernstian limit through the self-amplification of the capacitive coupling between the two gate electrodes [14–18]. Furthermore, interest in portable and wearable sensors as promising next-generation sensor platforms have continued to increase [19–24]. Transparent and flexible sensors can be applied to wearable or portable sensor systems, which results in the point-of care (POC) or real-time monitoring of wound, skin, sweat, and blood that is difficult to achieve with conventional rigid sensor systems. Amorphous oxide semiconductors (AOSs) are widely applied transparent material for transparent TFTs due to their transparency, ease of processing, and high electron mobility [25]. Many studies have been conducted on FET-based sensors fabricated on flexible substrates, including polyimide (PI), polyethylene naphthalate (PEN), and polyethylene terephthalate (PET) [26–30]. In particular, PI is a desirable material for transparent and flexible substrates because it is suitable for CMOS technology owing to its excellent thermal, chemical, and mechanical properties [31,32].

In this study, we propose a fully transparent high-performance coplanar dual-gate thin-film transistor (TFT)-based pH sensor on a flexible PI substrate. We used amorphous indium gallium zinc oxide (a-IGZO) channel layers, indium tin oxide (ITO) source/drain (S/D), and ITO gate electrodes to obtain fully transparent optical properties. An amorphous oxide semiconductor material with high transmittance in the visible light range was used to obtain transparent optical properties [25,33]. We also fabricated an extended gate with an SnO_2 sensing membrane, which ensured excellent sensing properties close to the theoretical Nernstian limit with an acid and base affinity constant of 2.5×10^6 and 1.1×10^{-5}, respectively [18]. Our pH sensor system consists of an a-IGZO coplanar dual-gate TFT transducer and SnO_2 extended-gate (EG) sensing units. In the proposed pH sensor system, we designed a capacitive coupling between the sensing gate (SG) and the control gate (CG) via the floating gate (FG) of the a-IGZO transducer to improve its sensitivity amplification. In particular, the SG and CG were located on the same plane on the gate insulating film, but the FG was located below the a-IGZO FET channel and was electrically separated from the CG and SG by a gate insulating film. The capacitance ratio (C_{SG}/C_{CG}) changes according to the combination of the areas of CG and SG, indicating that the proposed sensor is a self-amplifiable chemical sensor platform with tunable sensitivity. This tunable sensitivity is a beneficial feature of the capacitive coupling-based coplanar dual-gate structure pH sensor that cannot be achieved with a dual-gate structure consisting of a top- and bottom-gate. The C_{SG} and C_{CG} values of top- and bottom-gate structure pH sensors are fixed because they are determined by the pattern size of the channel layer. However, the C_{SG} and C_{CG} values of the proposed coplanar dual-gate structure pH sensors are controlled by the gate electrode pattern sizes because the biases in each gate's electrodes are applied to the channel layer via the FG. Therefore, the proposed coplanar dual-gate pH sensor based on capacitive coupling has the advantage of tunable sensitivity over various conventional chemical sensors. In addition, in comparison to conventional SOI substrate-based dual-gate structure pH sensors, the proposed pH sensor provides various advantages in its material, process, and device design. We also evaluated the non-ideal behavior, such as the hysteresis and drift effects, to ensure its stability and reliability. To ensure its flexibility, the mechanical and electrical stabilities of the a-IGZO coplanar dual-gate TFT transducer on a flexible PI substrate were determined through repeated bending tests.

2. Materials and Methods

Figure 1 shows a schematic illustration of the fabricated a-IGZO coplanar dual-gate TFT transducer and SnO_2 EG sensing units. To construct the pH sensor system, we connected the two units using an electric cable, as indicated by the dotted line. Specifically, the gate electrode of the transducer unit was electrically connected to the conductive layer of the sensing unit to apply the chemical potential of the sensing membrane to the gate electrode. The fully transparent and flexible coplanar dual-gate TFTs were fabricated to prepare the transducer unit. The transducer unit was fabricated using the following procedure. We prepared 6-μm-thick PI films on 1.5 cm × 1.5 cm size glass plates covered with a 100/100 nm thick SiN_x/SiO_2 adhesive layer. The PI substrates were wet cleaned, for 10 min each, using a standard solvent cleaning process with deionized water (DI) and 2-propyl-alcohol (IPA) in an ultrasonic bath. The substrates were then dried in an oven at 100 °C for 1 h to evaporate the residual solvent and moisture. Subsequently, a 300-nm-thick ITO layer, 100-nm-thick SiO_2 layer, and 50-nm-thick a-IGZO layer were sequentially deposited on the FG, gate insulating film, and channel layer, respectively. The active regions of the TFTs were formed by photolithography and a lift-off process of the a-IGZO layer. The channel width/length ratio of the patterned IGZO channel layer was 80/120 μm. Subsequently, an ITO film with a thickness of 150 nm was deposited, and coplanar dual-gate electrodes (SG, CG) and S/D electrodes were simultaneously formed by a lift-off process. In particular, the CG electrodes patterned with various sizes contributed to achieving various amplification ratios. Finally, the a-IGZO coplanar dual-gate TFT fabricated on the PI substrate was annealed at 250 °C in O_2 ambient for 30 min (PMA). We also prepared the EG sensing unit using the following procedure. A 300-nm-thick ITO conductive layer was deposited as an electrode on a cleaned 1.5 cm × 2.5 cm glass substrate, followed by a 50 nm thick SnO_2 layer as a sensing membrane. Finally, a 0.6 cm inner diameter polydimethylsiloxane (PDMS) reservoir was attached to the SnO_2 sensing membrane to accommodate the electrolyte solution. The ITO, SiO_2, a-IGZO, and SnO_2 layers used for the a-IGZO TFT and SnO_2 EG fabrication were deposited using an RF magnetron sputtering system.

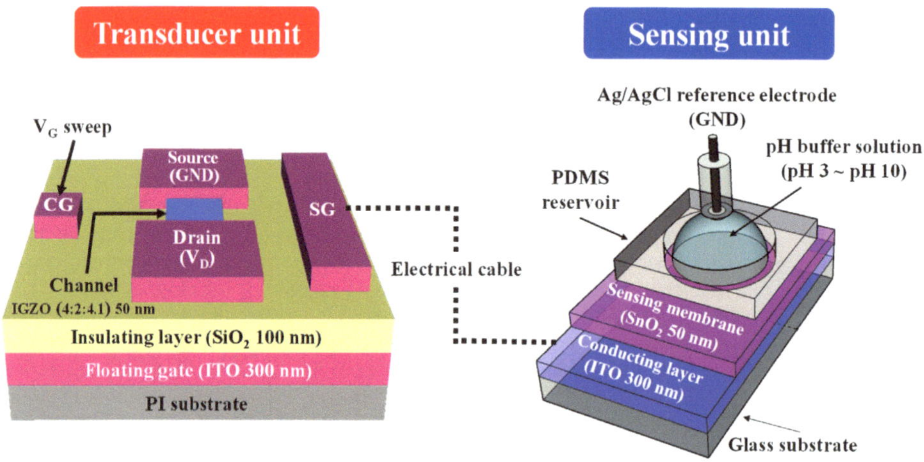

Figure 1. Schematic illustration of an a-IGZO coplanar dual-gate TFT transducer and SnO_2 EG sensing units. The dotted line represents the electrical connection between the two units. Reference electrode is connected to the ground unit of measurement instrument.

Figure 2a,b show photographs of the prepared transparent and flexible a-IGZO coplanar dual-gate TFT transducer and SnO_2 EG sensing units, respectively. Figure 2c shows the optical transmittance spectra of the PI substrate and fabricated a-IGZO coplanar dual-gate TFT transducer unit. The inset shows a photograph of the transparent transducer. The average transmittance of the transducer unit was 76.96% under visible light (wavelength 550–800 nm), whereas that of the PI film was 88.59%.

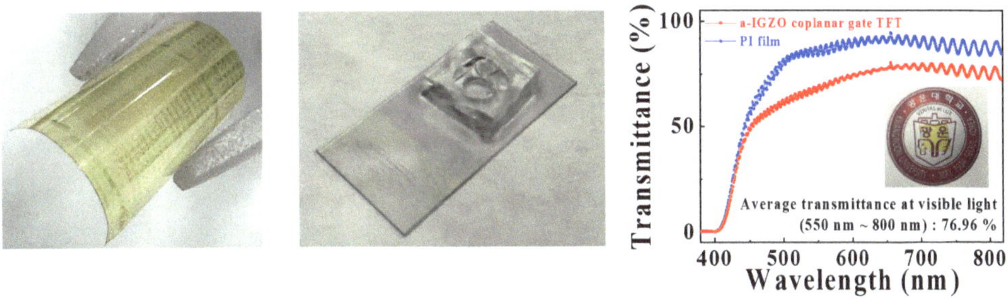

(a) (b) (c)

Figure 2. Photographs of the fabricated (**a**) transducer unit and (**b**) sensing unit. (**c**) Optical transmittance spectra of the PI film and a-IGZO coplanar dual-gate TFT transducer unit. The inset shows a photograph of the fabricated transparent transducer unit. The average transmittance at visible light (wavelength 550–800 nm) of the device is 76.96 %.

The capacitance–voltage (C–V) characteristics were measured using an Agilent 4284A Precision LCR meter (Agilent Technologies, Santa Clara, CA, USA). All the electrical characteristics of the a-IGZO TFTs and pH sensor platforms were characterized using an Agilent 4156 B Precision Semiconductor Parameter Analyzer (Agilent Technologies) in a dark box to eliminate noise or light. A pH buffer solution (pH 3.0, 4.0, 6.0, 7.0, 9.0, 10.0) and a commercial Ag/AgCl reference electrode (Horiba 2086A-06T, Kyoto, Japan) were prepared for pH sensing.

3. Results

3.1. C–V Characteristics of the Coplanar Dual-Gate

Figure 3a shows an optical microscope image of a coplanar dual-gate TFT. The SG was designed to have a fixed size of 90×420 μm^2, whereas the CG had various sizes of 90×420 μm^2, 80×200 μm^2, 80×110 μm^2, and 60×70 μm^2. The measured C–V curves for the various CG sizes are shown in Figure 3b. The capacitances of the CGs with the four dimensions specified above were 2.1 pF, 4.54 pF, 6.67 pF, and 12.65 pF, respectively. Meanwhile, the capacitance of the SG was 12.94 pF, which is almost identical to that of the CG of the same size. The relationship between the gate area and capacitance is shown in Figure 3c; the figure shows that the capacitance increased linearly with an increase in the area. The inset shows the relationship between the C_{SG}/C_{CG} and the gate area ratio (A_{SG}/A_{CG}), which indicates that the C_{SG}/C_{CG} is linearly proportional to the A_{SG}/A_{CG}. Therefore, by adjusting the area of the SG and CG, we can easily control the C_{SG}/C_{CG}, which is similar to the amplification ratio in capacitive coupling.

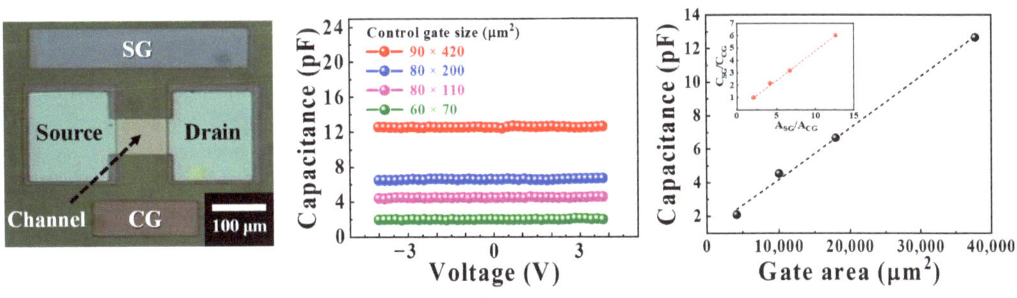

(a) (b) (c)

Figure 3. (**a**) Optical microscope image, (**b**) capacitance–voltage curve, and (**c**) capacitance–gate area relationship of the a-IGZO coplanar dual-gate TFT. The inset shows the relationship between C_{SG}/C_{CG} and A_{SG}/A_{CG}.

3.2. DC Bias Coupling Test of the a-IGZO Coplanar Dual-Gate TFT

Figure 4a shows the schematic illustration of the electrical equivalent circuit of an a-IGZO coplanar dual-gate TFT, depicting a simplified model in which the parasitic capacitance components are ignored. The CG where the gate voltage sweeps and the SG where the electrochemical potential of the pH buffer solution is biased are capacitively connected via an electrically isolated FG. In this case, the voltages of the coplanar gates (V_{CG} and V_{SG}) are capacitively coupled to FG (V_{FG}), as expressed in Equation (1). The relationship between V_{CG} and V_{SG} can then be expressed as given in Equation (2). Consequently, the change in the potential of the SG (ΔV_{SG}) can be amplified as the capacitance ratio of the C_{SG}/C_{CG} by capacitive coupling, as expressed in Equation (3). In addition, the ratio of the sensing gate capacitance to the control gate capacitance can modify the relationship between ΔV_{SG} and ΔV_{CG}.

$$V_{FG} = \frac{C_{CG}}{C_{SG}+C_{CG}}V_{CG} + \frac{C_{SG}}{C_{SG}+C_{CG}}V_{SG} \tag{1}$$

$$V_{CG} = \frac{C_{SG}+C_{CG}}{C_{CG}}V_{FG} - \frac{C_{SG}}{C_{CG}}V_{SG} \tag{2}$$

$$\therefore \Delta V_{CG} \propto \frac{C_{SG}}{C_{CG}}\Delta V_{SG} \tag{3}$$

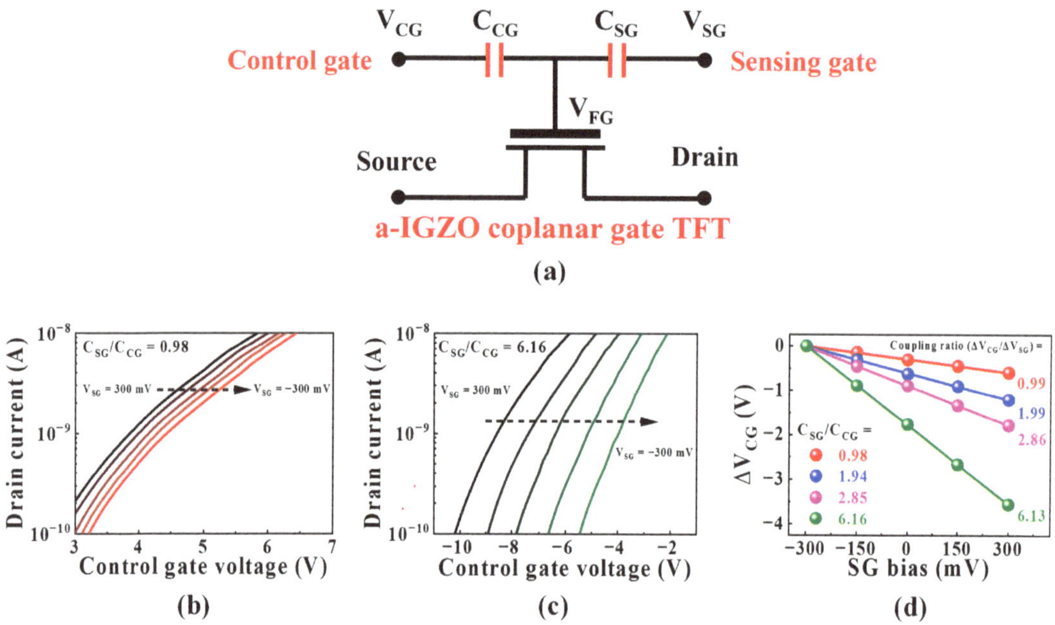

Figure 4. (**a**) Simplified schematic illustration of the electrical equivalent circuit of an a-IGZO coplanar dual-gate TFT. Shifts in the transfer characteristic curve for amplification factor of (**b**) 0.98 and (**c**) 6.16 when SG bias (V_{SG}) is varied from +300 mV to −300 mV at intervals of 150 mV. (**d**) $\Delta V_{CG}/\Delta V_{SG}$ for various values of C_{SG}/C_{CG} obtained at I_{Read} = 1 nA.

Prior to the pH sensing measurements, a DC bias-coupling test was conducted to verify the amplification factor because of capacitive coupling. When a DC bias voltage is applied to the SG, the threshold voltage of the CG shifts according to the magnitude of the SG bias. The shifts in the transfer characteristic curve for C_{SG}/C_{CG} of 0.98 and 6.16 are shown in Figure 4b,c, respectively. When the SG bias was varied between +300 mV and

−300 mV, at intervals of 150 mV, a decrease in the drain current and a rightward shift in the transfer characteristic curve were observed in response to a decrease in the V_{SG}. Figure 4d shows the amplification factor ($\Delta V_{CG}/\Delta V_{SG}$) for various C_{SG}/C_{CG} values extracted at a read drain current (I_{Read}) of 1 nA. It can be observed that there is a linear proportional relationship between $\Delta V_{CG}/\Delta V_{SG}$ and C_{SG}/C_{CG}. For the C_{SG}/C_{CG} values of 0.98, 1.94, 2.85, and 6.16, the values of $\Delta V_{CG}/\Delta V_{SG}$ were 0.99, 1.99, 2.86 and 6.13, respectively. Therefore, we demonstrated that ΔV_{SG} can be amplified by the amplification factor.

Table 1 lists the values of $\Delta V_{CG}/\Delta V_{SG}$ obtained from the DC bias test for various amplification factors.

Table 1. Amplification factors ($\Delta V_{CG}/\Delta V_{SG}$) obtained from the DC bias test of the a-IGZO coplanar dual-gate TFT.

C_{SG}/C_{CG}	$\Delta V_{CG}/\Delta V_{SG}$	R^2 (%)
0.98	0.99	99.93
1.94	1.99	99.95
2.85	2.86	99.99
6.16	6.13	99.98

3.3. *pH Sensing Characteristics of the a-IGZO Coplanar Dual-Gate TFT pH sensor*

The pH response of the FET-type chemical sensor can be explained by combining the Gouy–Chapman–Stern (GCS) theory and the site-binding model (SBM) [34–36]. According to the GCS theory, an electric double layer is created at the interface between the sensing membrane and the electrolyte solution. In addition, the surface potential (ψ) of the corresponding interface in the SBM is a critical parameter for the ion-sensing capability, which is summarized in Equation (4) [37,38]:

$$2.303\left(pH_{pzc} - pH\right) = \beta\psi + \sinh^{-1}\left[\frac{\sigma_0}{2q(K_b/K_a)^{1/2}N_s}\right] - \ln\left(1 - \frac{\sigma_0}{qN_s}\right) \tag{4}$$

where k is the Boltzmann constant, T is the temperature of the Kelvin system, q is the elementary charge, β is the dimensionless chemical sensitivity of the sensing membrane, pH_{pzc} is the pH at which the net charge of the surface is zero, σ_0 is the charge density, and the N_s is the total number of the sites per unit area; ψ varies depending on the chemical properties of the sensing membrane and the pH of the electrolyte. The values of the pH_{pzc} and β of SnO_2 that we adopt as sensing membrane are 5.6 and 58.6, respectively. According to this model, the sensing characteristics of the FET-type chemical sensors are determined using $\Delta\psi$. However, in this model, the sensitivity of the conventional single-gate FET-type pH sensor cannot exceed the physical limit of ~59.14 mV/pH at 300 K, which is known as the Nernstian limit. To overcome this fatal drawback, we introduced a coplanar dual-gate structure based on capacitive coupling, which amplifies the small potential change in the SG and makes it detectable in the CG.

Figure 5a,b show the transfer characteristic curves of the a-IGZO coplanar dual-gate TFT pH sensor for C_{SG}/C_{CG} values of 0.98 and 6.16, respectively, indicating that they shift with the pH. The sensing properties were measured with a pH buffer solution of pH 3 to 10 at 300 K. The practical pH sensitivities of various C_{SG}/C_{CG} values obtained at I_{Read} = 1 nA are shown in Figure 5c. The pH sensitivities for C_{SG}/C_{CG} values of 0.98, 1.94, 2.85, and 6.16 were 57.77 mV/pH, 116.4 mV/pH, 174.38 mV/pH, and 359.28 mV/pH, respectively. The pH sensitivity without capacitive coupling (C_{SG}/C_{CG} = 1) was 58.29 mV/pH. It is noteworthy that the proposed sensor exhibited high-performance pH sensing properties that far exceeded the Nernstian limit without additional amplification circuits. This is because the capacitive coupling of the coplanar dual-gate structure enables self-amplification in practical pH sensing operations.

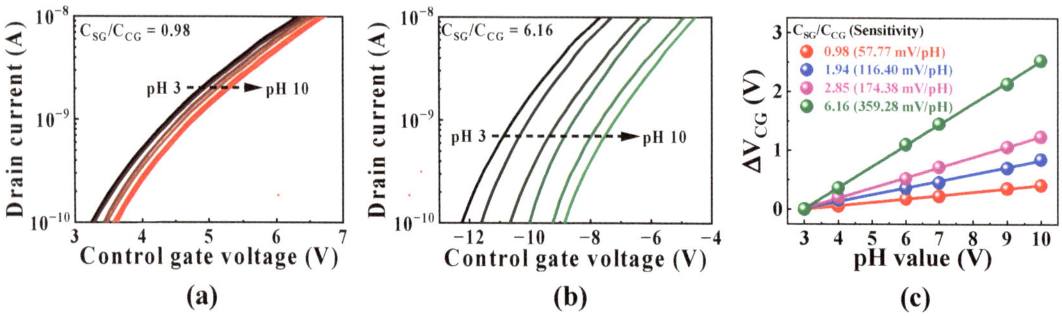

Figure 5. Transfer characteristics curves of the a-IGZO coplanar dual-gate TFT pH sensor in various pH buffer solutions for C_{SG}/C_{CG} of (**a**) 0.98 and (**b**) 6.16. (**c**) pH sensitivity for various C_{SG}/C_{CG} values (I_{Read} = 1 nA).

3.4. Non-Ideal Behavior of the a-IGZO Coplanar Gate TFT pH Sensor

In addition to sensitivity, stability and reliability are important performance indicators of chemical sensors. To verify whether the proposed sensor system ensures repetitive sensing operation over a relatively short period and long period, we measured the hysteresis and drift effects. The hysteresis and drift effects are typical non-ideal behaviors that prevent accurate detection by sensors. The hysteresis effect often arises from the reaction between electrolyte ions (H^+ or OH^-) and the surface, or from the slow transport of ionic species in the sensing membrane bulk [39]. The drift effect is caused by the permeation of ionic species in the electrolyte or by defects in the sensing membrane through hopping or trap-limited transport [40,41]. The hysteresis effect was measured for a total of 50 min by changing the pH of the buffer solution at 300 K in the order of 7→10→7→4→7. Then, the hysteresis voltage (V_H) was extracted from the ΔV_{CG} difference between the start and end points of the pH loop. Figure 6a shows the V_H for various C_{SG}/C_{CG} values; the V_H values were 5.29 mV, 9.13 mV, 13.83 mV, and 20.54 mV for C_{SG}/C_{CG} values of 0.98, 1.94, 2.85, and 6.16, respectively. The drift rate (R_{drift}) was determined by immersion in a pH 7 buffer solution at 300 K for 10 h. Figure 6b shows the drift rates; the R_{drift} values were 7.84 mV/h, 16.71 mV/h, 32.21 mV/h, and 65.08 mV/h for C_{SG}/C_{CG} values of 0.98, 1.94, 2.85 and 6.16, respectively. As capacitive coupling amplifies the surface potential of the sensing membrane connected to the SG, both the V_H and R_{drift} increase with an increase in the C_{SG}/C_{CG}. However, it can be observed that the increments in V_H and R_{drift} are smaller than that of sensitivity.

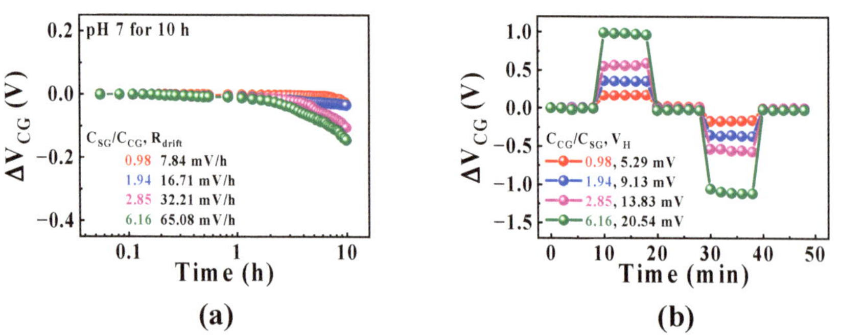

Figure 6. Non-ideal behavior of the a-IGZO coplanar dual-gate TFT pH sensor. (**a**) Hysteresis and (**b**) drift effects for various C_{SG}/C_{CG} values.

Table 2 summarizes the pH sensing properties of the proposed a-IGZO coplanar dual-gate TFT pH sensor. It can be observed that the increments in the V_H and R_{drift} with an increase in the C_{SG}/C_{CG} are less than 9.1% and 18.5% that of sensitivity, respectively. Therefore, the proposed a-IGZO coplanar gate TFT pH sensor is a stable and reliable chemical sensor platform with a sensitivity high above the Nernstian limit.

Table 2. pH sensing characteristics of the a-IGZO coplanar dual-gate TFT pH sensor.

C_{SG}/C_{CG}	Sensitivity (mV/pH)	$\Delta V_{CG}/\Delta V_{SG}$	V_H (mV)	R_{drift} (mV/h)	V_H to Sensitivity (%)	R_{drift} to Sensitivity (%)
0.98	57.77	0.99	5.29	7.84	9.1	13.3
1.94	116.4	1.99	9.13	16.71	7.8	14.4
2.85	174.38	2.99	13.83	32.21	7.9	18.5
6.16	359.28	6.16	20.54	65.08	5.7	18.1

3.5. Bending Test of the a-IGZO Coplanar Dual Gate TFT

In the sensor system, the flexibility of the sensor must be evaluated by the sensing characteristics after repeated bending operation. For flexible chemical sensor platform applications, it is necessary to maintain the sensing characteristics, including sensitivity and the amplification factor, without a significant degradation of the electrical properties, even after repeated bending operations. As a result of the mechanical stress accompanying the deformation of the flexible PI substrate, various parts of the TFT device, such as the gate insulating film, channels, electrodes, or their interfaces, may undergo irreversible mechanical damage [42].

Figure 7a shows a TFT transducer unit bent to a diameter of 3 mm using a vernier caliper. The inset shows an optical microscope image after the bending test, that is, 500 bending cycles to a diameter of 3 mm. Compared with the sample before the bending test, there were no recognizable defects in the a-IGZO channel, S/D electrode, or coplanar gates, indicating that there was no mechanical damage caused by the bending. Therefore, the bending tests verified the flexibility and mechanical strength of the a-IGZO coplanar dual-gate TFT on the PI substrate. Bending not only affects the optical and mechanical properties, but also the electrical characteristics [42,43]. In FET-type sensor systems in which the electrical characteristics directly affect the sensing characteristics, poor electrical characteristics lead to the deterioration of the sensing characteristics. Therefore, to verify the flexible characteristics of the -IGZO coplanar dual-gate TFT on the PI substrate, we measured the pH sensing characteristics after repeated bending tests. Figure 7b,c show the transfer curves after 500 bending cycles to a diameter of 3 mm for C_{SG}/C_{CG} values of 0.94 and 6.16, respectively. Figure 7d shows the pH sensitivity for various C_{SG}/C_{CG} values after the repeated bending tests. The pH sensitivity obtained from the bending test, using the same samples, slightly decreased from 57.77 mV/pH, 116.4 mV/pH, 174.38 mV/pH, and 359.28 mV/pH before bending to 56.45 mV/pH, 113.6 mV/pH, 165.45 mV/pH, and 345.09 mV/pH after bending, respectively. In addition, slight changes in the $\Delta V_{CG}/\Delta V_{SG}$, from 0.99, 1.99, 2.99, and 6.16 to 0.97, 1.95, 2.84, and 5.92, were observed before and after bending, respectively. It is considered that the slight decrease in the sensitivity and $\Delta V_{CG}/\Delta V_{SG}$ after the bending test is caused by repeated mechanical stresses on the device. However, the decrease in sensitivity is almost negligible, up to 5.1%, and the bent device still gives a high pH sensing performance, above the Nernstian limit, as summarized in Table 3. Accordingly, we conclude that the a-IGZO coplanar dual-gate TFT on a PI substrate is a suitable flexible chemical sensor system that gives a high sensing performance, even after repeated bending tests.

bending, respectively. It is considered that the slight decrease in the sensitivity and $\Delta V_{CG}/\Delta V_{SG}$ after the bending test is caused by repeated mechanical stresses on the device. However, the decrease in sensitivity is almost negligible, up to 5.1%, and the bent device still gives a high pH sensing performance, above the Nernstian limit, as summarized in Table 3. Accordingly, we conclude that the a-IGZO coplanar dual-gate TFT on a PI substrate is a suitable flexible chemical sensor system that gives a high sensing performance, even after repeated bending tests.

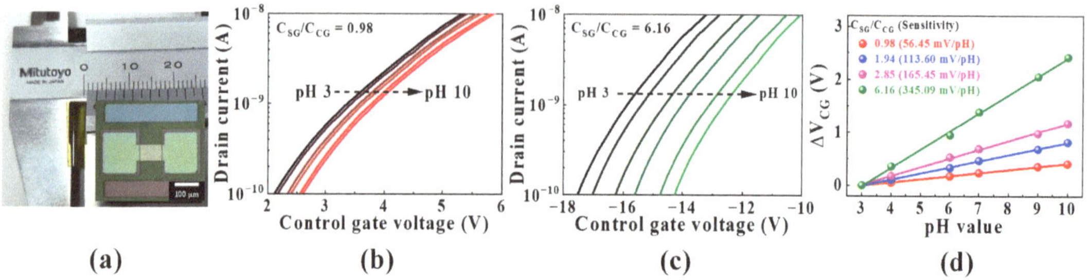

Figure 7. (**a**) Image of an a-IGZO coplanar dual-gate TFT converter unit on a PI substrate bent to a diameter of 3 mm using vernier calipers. The inset shows an optical microscope image obtained after the bending test. Transfer characteristic curves after 500 bending cycles to a diameter of 3 mm for C_{SG}/C_{CG} values of (**b**) 0.98 and (**c**) 6.16. (**d**) pH sensitivity for various C_{SG}/C_{CG} values after repeated bending tests (I_{Read} = 1 nA).

Table 3. PH-sensing characteristics obtained from the bending test of the a-IGZO coplanar dual-gate TFT.

C_{SG}/C_{CG}	Sensitivity before Bending (mV/pH)	Bending Cycles (Times)	Sensitivity after Bending (mV/pH)	Decrease Rate of Sensitivity after Bending (%)
0.98	58.77	100	57.42	2.3
		200	57.28	2.5
		300	56.98	3.0
		400	56.52	3.8
		500	56.45	3.9
1.94	116.40	100	116.08	0.3
		200	115.89	0.4
		300	114.77	1.4
		400	113.73	2.3
		500	113.60	2.4
2.85	174.38	100	173.37	0.6
		200	172.90	0.8
		300	169.46	2.8
		400	168.99	3.1
		500	165.45	5.1
6.16	359.28	100	358.02	0.3
		200	356.69	0.7
		300	351.19	2.3
		400	349.58	2.7
		500	345.09	3.8

4. Conclusions

We investigated a fully transparent and flexible high-performance pH sensor based on an a-IGZO TFT transducer with a coplanar dual-gate structure on a PI substrate. The proposed pH sensor system was constructed by electrically connecting a sensing unit and an a-IGZO TFT transducer unit prepared on different substrates to protect the transducer from chemical damage. The transducer unit consists of an ITO FG, SG electrodes, CG, a-IGZO TFT channel, and ITO S/D electrodes on a flexible PI substrate, which are all transparent materials. The EG sensing unit was prepared on a separate glass substrate. In the proposed pH sensor system, we designed a capacitive coupling between the SG and CG through the FG of the a-IGZO TFT transducer to contribute to sensitivity amplification. We conducted a DC bias-coupling test and found that the C_{SG}/C_{CG} ratio increased linearly with the area ratio of the SG to CG (A_{SG}/A_{CG}) and determined the sensitivity amplification. We measured the potentials of various buffer solutions using a pH sensor composed of an a-IGZO TFT transducer unit and a SnO_2 EG sensing unit and found that the practical pH sensitivity was amplified in a linear ratio to C_{SG}/C_{CG}. The amplification ratio could be determined by the area ratio of the SG to CG, which increased the pH sensitivity to 359.28 mV/pH at a C_{SG}/C_{CG} value of 6.16; this value is significantly larger than the Nernstian limit of 59.14 mV/pH at room temperature (300 K). In addition, we evaluated the stability and reliability by measuring the non-ideal behaviors, including the hysteresis and drift effects. The amplification of sensitivity with an increase in the C_{SG}/C_{CG} ratio was much larger than the increase in the hysteresis voltage and drift rate, indicating that our proposed a-IGZO coplanar dual-gate TFT pH sensor is a stable and reliable high-sensitivity FET-based chemical sensor platform. Finally, the flexibility of the a-IGZO coplanar dual-gate TFT converter was evaluated via a bending test, and the electrical properties were maintained without mechanical damage, even after 500 bending cycles, to a diameter of 3 mm. Therefore, the fully transparent and highly sensitive IGZO coplanar dual-gate TFT-based pH sensor proposed in this study can be applied to wearable and portable high-performance chemical sensor platforms.

Author Contributions: T.-H.H.: conceptualization, formal analysis, investigation, methodology, data curation, visualization, and writing—original draft. W.-J.C.: conceptualization, methodology, investigation, resources, formal analysis, funding acquisition, supervision, validation, and writing—review and editing. All authors have read and agreed to the published version of the manuscript.

Funding: This study was funded by the National Research Foundation of Korea (NRF) grant funded by the Korea government (MSIT), grant number 2020R1A2C1007586.

Institutional Review Board Statement: Not applicable.

Informed Consent Statement: Not applicable.

Data Availability Statement: Not applicable.

Acknowledgments: The present research has been conducted by the Research Grant of Kwangwoon University in 2022 and by the Excellent research support project of Kwangwoon University in 2022.

Conflicts of Interest: The authors declare no conflict of interest.

References

1. Singh, A.K.; Pandey, A.; Chakrabarti, P. Fabrication, characterization, and application of CuO nano wires as electrode for ammonia sensing in aqueous environment using extended gate-FET. *IEEE Sens. J.* **2021**, *21*, 5779–5786. [CrossRef]
2. Capua, L.; Sprunger, Y.; Elettro, H.; Risch, F.; Grammoustianou, A.; Midahuen, R.; Ernst, T.; Barraud, S.; Gill, R.; Ionescu, A.M. Label-free C-reactive protein Si nanowire FET sensor arrays with super-Nernstian back-gate operation. *IEEE Trans. Electron Devices* **2022**, *69*, 2159–2165. [CrossRef]
3. Dolai, S.; Tabib-Azar, M. Zika virus field effect transistor. *IEEE Sens. J.* **2021**, *21*, 4122–4128. [CrossRef]
4. Bausells, J.; Carrabina, J.; Errachid, A.; Merlos, A. Ion-sensitive field-effect transistors fabricated in a commercial CMOS technology. *Sens. Actuators B Chem.* **1999**, *57*, 56–62. [CrossRef]
5. Moser, N.; Lande, T.S.; Toumazou, C.; Georgiou, P. ISFETs in CMOS and emergent trends in instrumentation: A review. *IEEE Sens. J.* **2016**, *16*, 6496–6514. [CrossRef]
6. Olthuis, W.; Robben, M.A.M.; Bergveld, P.; Bos, M.; van der Linden, W.E. pH sensor properties of electrochemically grown iridium oxide. *Sens. Actuators B Chem.* **1990**, *2*, 247–256. [CrossRef]
7. Bergveld, P. Development of an ion-sensitive solid-state device for neurophysiological measurements. *IEEE Trans. Biomed. Eng.* **1970**, *BME-17*, 70–71. [CrossRef]
8. van der Spiegel, J.; Lauks, I.; Chan, P.; Babic, D. The extended gate chemically sensitive field effect transistor as multi-species microprobe. *Sens. Actuators* **1983**, *4*, 291–298. [CrossRef]
9. Rao, L.; Wang, P.; Qian, Y.; Zhou, G.; Nötzel, R. Comparison of the extended gate field-effect transistor with direct potentiometric sensing for super-Nernstian InN/InGaN quantum dots. *ACS Omega* **2020**, *5*, 32800–32805. [CrossRef] [PubMed]
10. Xue, M.; Mackin, C.; Weng, W.-H.; Zhu, J.; Luo, Y.; Luo, S.-X.L.; Lu, A.-Y.; Hempel, M.; McVay, E.; Kong, J.; et al. Integrated biosensor platform based on graphene transistor arrays for real-time high-accuracy ion sensing. *Nat. Commun.* **2022**, *13*, 5064. [CrossRef]
11. Chen, C.-P.; Ganguly, A.; Lu, C.-Y.; Chen, T.-Y.; Kuo, C.-C.; Chen, R.-S.; Tu, W.-H.; Fischer, W.B.; Chen, K.-H.; Chen, L.-C. Ultrasensitive in situ label-free DNA detection using a GaN nanowire-based extended-gate field-effect-transistor sensor. *Anal. Chem.* **2011**, *83*, 1938–1943. [CrossRef]

12. Spijkman, M.; Smits, E.C.P.; Cillessen, J.F.M.; Biscarini, F.; Blom, P.W.M.; de Leeuw, D.M. Beyond the Nernst-limit with dual-gate ZnO ion-sensitive field-effect transistors. *Appl. Phys. Lett.* **2011**, *98*, 043502. [CrossRef]
13. Knopfmacher, O.; Tarasov, A.; Fu, W.; Wipf, M.; Niesen, B.; Calame, M.; Schönenberger, C. Nernst limit in dual-gated Si-nanowire FET sensors. *Nano Lett.* **2010**, *10*, 2268–2274. [CrossRef]
14. Spijkman, M.-J.; Myny, K.; Smits, E.C.P.; Heremans, P.; Blom, P.W.M.; de Leeuw, D.M. Dual-gate thin-film transistors, integrated circuits and sensors. *Adv. Mater.* **2011**, *23*, 3231–3242. [CrossRef] [PubMed]
15. Lee, I.-K.; Lee, K.H.; Lee, S.; Cho, W.-J. Microwave annealing effect for highly reliable biosensor: Dual-gate ion-sensitive field-effect transistor using amorphous InGaZnO thin-film transistor. *ACS Appl. Mater. Interfaces* **2014**, *6*, 22680–22686. [CrossRef] [PubMed]
16. Cheng, Q.; Wang, M.; Tao, M.; Yin, R.; Li, Y.; Yang, N.; Xu, W.; Gao, C.; Hao, Y.; Yang, Z. Planar dual gate GaN HEMT cascode amplifier as a voltage readout PH sensor with high and tunable sensitivities. *IEEE Electron Device Lett.* **2020**, *41*, 485–488. [CrossRef]
17. Jang, H.-J.; Gu, J.-G.; Cho, W.-J. Sensitivity enhancement of amorphous InGaZnO thin film transistor based extended gate field-effect transistors with dual-gate operation. *Sens. Actuators B Chem.* **2013**, *181*, 880–884. [CrossRef]
18. Cho, S.-K.; Cho, W.-J. Ultra-high sensitivity pH-sensors using silicon nanowire channel dual-gate field-effect transistors fabricated by electrospun polyvinylpyrrolidone nanofibers pattern template transfer. *Sens. Actuators B Chem.* **2021**, *326*, 128835. [CrossRef]
19. Costa, J.C.; Pouryazdan, A.; Panidi, J.; Spina, F.; Anthopoulos, T.D.; Liedke, M.O.; Schneider, C.; Wagner, A.; Münzenrieder, N. Flexible IGZO TFTs and their suitability for space applications. *IEEE J. Electron Devices Soc.* **2019**, *7*, 1182–1190. [CrossRef]
20. Segev-Bar, M.; Haick, H. Flexible sensors based on nanoparticles. *ACS Nano* **2013**, *7*, 8366–8378. [CrossRef]
21. Hu, B.; Chen, W.; Zhou, J. High performance flexible sensor based on inorganic nanomaterials. *Sens. Actuators B Chem.* **2013**, *176*, 522–533. [CrossRef]
22. Xu, K.; Lu, Y.; Takei, K. Multifunctional skin-inspired flexible sensor systems for wearable electronics. *Adv. Mater. Technol.* **2019**, *4*, 1800628. [CrossRef]
23. Nag, A.; Mukhopadhyay, S.C.; Kosel, J. Wearable flexible Sensors: A review. *IEEE Sens. J.* **2017**, *17*, 3949–3960. [CrossRef]
24. Nakata, S.; Arie, T.; Akita, S.; Takei, K. Wearable, flexible, and multifunctional healthcare device with an ISFET chemical sensor for simultaneous sweat PH and skin temperature monitoring. *ACS Sens.* **2017**, *2*, 443–448. [CrossRef]
25. Kamiya, T.; Hosono, H. Material characteristics can applications of transparent amorphous oxide semiconductor. *NPG Asia Mater.* **2010**, *2*, 15–22. [CrossRef]
26. Han, S.-T.; Peng, H.; Sun, Q.; Venkatesh, S.; Chung, K.-S.; Lau, S.C.; Zhou, Y.; Roy, V.A.L. An overview of the development of flexible sensors. *Adv. Mater.* **2017**, *29*, 1700375. [CrossRef]
27. Vilela, D.; Romeo, A.; Sánchez, S. Flexible sensors for biomedical technology. *Lab A Chip* **2016**, *16*, 402–408. [CrossRef]
28. Jakob, M.H.; Gutsch, S.; Chatelle, C.; Krishnaraja, A.; Fahlteich, J.; Weber, W.; Zacharias, M. Flexible thin film PH sensor based on low-temperature atomic layer deposition. *Phys. Status Solidi-Rapid Res. Lett.* **2017**, *11*, 1700123. [CrossRef]
29. Mansouri Majd, S.; Salimi, A. Ultrasensitive flexible FET-type aptasensor for CA 125 cancer marker detection based on carboxylated multiwalled carbon nanotubes immobilized onto reduced graphene oxide film. *Anal. Chim. Acta* **2018**, *1000*, 273–282. [CrossRef]
30. Park, S.J.; Kwon, O.S.; Lee, S.H.; Song, H.S.; Park, T.H.; Jang, J. Ultrasensitive flexible graphene based field-effect transistor (FET)-type bioelectronic nose. *Nano Lett.* **2012**, *12*, 5082–5090. [CrossRef]
31. Singh, K.; Her, J.-L.; Lou, B.-S.; Pang, S.-T.; Pan, T.-M. An extended-gate FET-based pH sensor with an InZnxOy membrane fabricated on a flexible polyimide substrate at room temperature. *IEEE Electron Device Lett.* **2019**, *40*, 804–807. [CrossRef]
32. Gao, X.; Lin, L.; Liu, Y.; Huang, X. LTPS TFT process on polyimide substrate for flexible AMOLED. *J. Disp. Technol.* **2015**, *11*, 666–669. [CrossRef]
33. Kamiya, T.; Nomura, K.; Hosono, H. Present status of amorphous In–Ga–Zn–O thin-film transistors. *Sci. Technol. Adv. Mater.* **2010**, *11*, 044305. [CrossRef]
34. Chen, S.; Bomer, J.G.; Carlen, E.T.; van den Berg, A. Al2O3/silicon nanoISFET with near ideal Nernstian response. *Nano Lett.* **2011**, *11*, 2334–2341. [CrossRef]
35. Yates, D.E.; Levine, S.; Healy, T.W. Site-binding model of the electrical double layer at the oxide/water interface. *J. Chem. Soc. Faraday Trans. 1* **1974**, *70*, 1807–1818. [CrossRef]
36. Tarasov, A.; Wipf, M.; Bedner, K.; Kurz, J.; Fu, W.; Guzenko, V.A.; Knopfmacher, O.; Stoop, R.L.; Calame, M.; Schönenberger, C. True reference nanosensor realized with silicon nanowires. *Langmuir* **2012**, *28*, 9899–9905. [CrossRef] [PubMed]
37. Landheer, D.; Aers, G.; McKinnon, W.R.; Deen, M.J.; Ranuares, J.C. Model for the field effect from layers of biological macromolecules on the gates of metal-oxide semiconductor transistors. *J. Appl. Phys* **2005**, *98*, 044701. [CrossRef]
38. Chou, J.-C.; Liao, L.P. Study on pH at the point of zero charge of TiO2 pH ion-sensitive field effect transistor made by the sputtering method. *Thin Solid Film.* **2005**, *476*, 157–161. [CrossRef]
39. Tsai, C.-N.; Chou, J.-C.; Sun, T.-P.; Hsiung, S.-K. Study on the sensing characteristics and hysteresis effect of the tin oxide pH electrode. *Sens. Actuators B Chem* **2005**, *108*, 877–882. [CrossRef]
40. Jamasb, S.; Collins, S.; Smith, R.L. A physical model for drift in pH ISFETs. *Sens. Actuators B Chem.* **1998**, *49*, 146–155. [CrossRef]
41. Bousse, L.; Bergveld, P. The role of buried OH sites in the response mechanism of inorganic-gate pH-sensitive ISFETs. *Sens. Actuators* **1984**, *6*, 65–78. [CrossRef]

42. Cheng, J.; Wu, L.; Du, X.-W.; Jin, Q.-H.; Zhao, J.-L.; Xu, Y.-S. Flexible solution-gated graphene field effect transistor for electrophysiological recording. *J. Microelectromech Syst.* **2014**, *23*, 1311–1317. [CrossRef]
43. Shin, K.; Xiong, W.; Cho, C.Y.; Cleavelin, C.R.; Schulz, T.; Schruefer, K.; Patruno, P.; Smith, L.; Liu, T.-J.K. Study of bending-induced strain effects on MuGFET performance. *IEEE Electron Device Lett.* **2006**, *27*, 671–673. [CrossRef]

chemosensors

MDPI

Article

High-Performance Bidirectional Chemical Sensor Platform Using Double-Gate Ion-Sensitive Field-Effect Transistor with Microwave-Assisted Ni-Silicide Schottky-Barrier Source/Drain

Yeong-Ung Kim and Won-Ju Cho *

Department of Electronic Materials Engineering, Kwangwoon University, Gwangun-ro 20, Nowon-gu, Seoul 01897, Korea; herokim5163@gmail.com
* Correspondence: chowj@kw.ac.kr; Tel.: +82-2-940-5163

Abstract: This study proposes a bidirectional chemical sensor platform using ambipolar double-gate ion-sensitive field-effect transistors (ISFET) with microwave-assisted Ni-silicide Schottky-barrier (SB) source and drain (S/D) on a fully depleted silicon-on-insulator (FDSOI) substrate. The microwave-assisted Ni-silicide SB S/D offer bidirectional turn-on characteristics for both p- and n-type channel operations. The *p*- and *n*-type operations are characterized by high noise resistance as well as improved mobility and excellent drift performance, respectively. These features enable sensing regardless of the gate voltage polarity, thus contributing to the use of detection channels based on various target substances, such as cells, antigen-antibodies, DNA, and RNA. Additionally, the capacitive coupling effect existing between the top and bottom gates help achieve self-amplified pH sensitivity exceeding the Nernst limit of 59.14 mV/pH without any additional amplification circuitry. The ambipolar FET sensor performance was evaluated for bidirectional electrical characteristics, pH detection in the single-gate and double-gate modes, and reliability in continuous and repetitive operations. Considering the excellent characteristics confirmed through evaluation, the proposed ambipolar chemical sensor platform is expected to be applicable to various fields including biosensors. And through linkage with subsequent studies, various medical applications and precision detector operations for specific markers will be possible.

Keywords: Ni-silicide; Schottky-barrier source and drain; ambipolar conductance characteristics; microwave irradiation (MWI); ion-sensitive field-effect transistor (ISFET); capacitive coupling

Citation: Kim, Y.-U.; Cho, W.-J. High-Performance Bidirectional Chemical Sensor Platform Using Double-Gate Ion-Sensitive Field-Effect Transistor with Microwave-Assisted Ni-Silicide Schottky-Barrier Source/Drain. *Chemosensors* **2022**, *10*, 122. https://doi.org/10.3390/chemosensors10040122

Academic Editor: Pietro Salvo

Received: 10 February 2022
Accepted: 22 March 2022
Published: 24 March 2022

1. Introduction

Owing to the frequent incidence of global pandemics, medical care is garnering extensive attention among researchers. Hence, sensors based on various detection technologies are being introduced, such as antibody-coated virus sensors, label-free electrochemical sensors, and field-effect transistor (FET)-based biosensors [1–3]. Biosensors have great potential for detecting disease markers and microorganisms in clinical diagnoses and point-of-care (PoC) detection, and have therefore been used for monitoring in biomedical, environmental, industrial, and agricultural applications. Among the different types of biosensing systems available, the FET-type biosensor is one of the most attractive electrical biosensors. Several FET-based biosensors have been developed to study the biomolecular interactions that drive biological responses of in vitro and in vivo systems [4,5]. Furthermore, ion-sensitive FETs (ISFETs) are the most common type of biosensors and are highly advantageous owing to their fast measurement capabilities, easy operation with a small amount of the sample, label-free sensing, low cost owing to CMOS process compatibility, and compact or portable instrumentation [6,7]. However, despite these advantages, ISFETs cannot be commercialized owing to their relatively low sensitivity because of the Nernst limitation (~59.14 mV/pH at room temperature) [6,8]. To overcome this issue, many studies have attempted to increase the sensitivities of ISFETs [9,10]. In particular, the double-gate (DG) sensing mode of ISFETs can amplify sensitivity through capacitive coupling considering the capacitances of the gate electrodes positioned above and below the thin-film channel without additional amplifying circuitry [11]. Meanwhile, the increase in series resistance and capacitance of the source/drain (S/D) junction of the ISFET due to miniaturization and integration of biomedical sensors can limit rapid biosignal responses, which are important for understanding S/D formation in early diagnosis and improved therapy in the future [12,13]. Typically, S/D formation is achieved via high-dose ion implantation and high-temperature activation annealing processes. However, this method is complicated and can cause defects owing to lattice damages caused by the implanted ions [14,15]. Conversely, the Schottky-barrier (SB) S/D using metal-based silicide has a simpler process compared to conventional impurity doping and can hence reduce the series resistance further [14]. Among the transition metal silicides, Ni-silicide exhibits a low resistance, low process temperature, low Si consumption, and significantly lower resistance degradation in a narrow line. Additionally, it can achieve ambipolar switching behaviors because it is a midgap metal that is used as in standard metal contacts for both *n*- and *p*-type MOSFETs [16,17]. The ambipolar switching characteristic has two operating regions that are turned on depending on the polarity of the gate bias, which is advantageous in many applications [18–20].

Therefore, in this study, we propose a bidirectional chemical sensor platform based on an ambipolar DG ISFET and a separated extended-gate (EG) structure. Considering that *p*-type MOSFETs have almost two orders of magnitude lower noise than *n*-type MOSFETs, *p*-type ISFETs are preferable for reducing noise. Conversely, *n*-type ISFETs exhibit higher mobility and improved drift performances [21,22]. The proposed ambipolar ISFET has a unique advantage: it can operate as a *p*-type device when high noise immunity is required and as an *n*-type device when improved mobility or drift performance is required. Furthermore, to achieve excellent ambipolar switching properties, Ni-silicide was formed using high-efficiency microwave irradiation (MWI) with a low thermal budget [23,24]. Thus, we successfully implemented a bidirectional chemical sensor platform capable of pH sensing for both positive and negative gate voltage (V_G) polarities. Additionally, we used the EG structure to prevent direct contact between the gate dielectric of the ISFET and analytes to improve immunity to nonideal effects and the lifetime of the transducer. The transducer performance was further evaluated by measuring the transfer and output characteristics of the fabricated DG ambipolar ISFET. We also evaluated the pH sensitivity in the conventional single-gate (SG) and proposed DG sensing modes to amplify sensitivity by itself without needing additional amplification circuitry. The nonideal effects, such as hysteresis and drift, were evaluated to verify the sensor stability during repetitive and continuous operations. With its simple process and bidirectional operation, the proposed chemical sensor platform is expected to be a promising candidate for detecting specific markers in various applications.

2. Materials and Methods

2.1. Fabrication of the Ambipolar DG ISFET and EG Unit

The proposed ambipolar DG ISFETs were fabricated on a silicon-on-insulator (SOI) substrate comprising a 100-nm-thick *p*-type top Si layer doped with 1×10^{15} cm^{-3} and a 200-nm-thick buried oxide layer. After removing the surface contaminants of the SOI substrate through RCA cleaning, the upper Si layer was etched with 2.38 wt.% tetramethylammonium hydroxide (TMAH) solution to prepare a 60-nm-thick, fully depleted SOI substrate. The active area of a FET transducer of length (L) and width (W) 20 μm and 10 μm, respectively, was defined by photolithography and plasma reactive ion etching (RIE) using SF_6 gas.

After depositing a 150-nm-thick Ni layer using an electron beam evaporator, Ni was selectively left in the S/D region through a lift-off process. Then, using a 600 W MWI source for 2 min in an N_2 ambient, Ni-silicide, which is a key material for bipolar switching, was formed [25]. Unreacted Ni was removed using a sulfur peroxide mixture (SPM) for 10 min.

As the top-gate insulator, a 70-nm-thick SiO_2 layer was deposited using an RF magnetron sputtering system. A 150-nm-thick Al film was deposited with an e-beam evaporator, followed by photolithography and wet etching by H_3PO_4 to form a top-gate electrode. The buried oxide layer of the SOI substrate serves as the bottom-gate insulator of the FET. Additionally, a 150-nm-thick Al film was deposited on the backside of the SOI substrate to form the bottom-gate electrode. After the local opening of the S/D contact for the holes in the top gate-oxide layer, a forming gas (H_2:N_2 = 5%:95%) annealing was performed for 30 min at 450 °C to improve the electrical characteristics of the ISFET.

Meanwhile, a separate EG unit was prepared to prevent deterioration owing to direct contact between the buffer solution and gate insulator of the FET transducer, which was fabricated on a 1.5 × 3 cm^2 glass substrate. A 300-nm-thick indium tin oxide (ITO) conductive layer and a 50-nm-thick SnO_2 sensing film were sequentially deposited on the glass substrate using an RF magnetron sputtering system. This layer acts as a receptor for the surface potential change and transfers the applied potential change. After the formation of the sensing layer, polydimethylsiloxane (PDMS) reservoir was attached with silicone glue for injecting the pH buffer solution. Figure 1 shows the schematic of the ambipolar DG ISFET and EG unit.

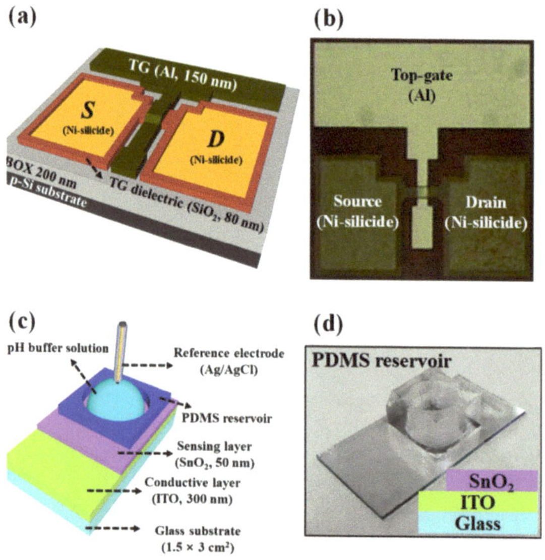

Figure 1. (**a**) Schematic and (**b**) microscopic image of ambipolar DG ISFET with Ni-silicide SB S/Ds on an SOI substrate. (**c**) Schematic and (**d**) photograph of the fabricated EG unit.

2.2. Characterization of the Fabricated Ambipolar DG ISFET

We evaluated the electrical properties of the fabricated bidirectional chemical sensor platform using an Agilent 4156B precision semiconductor parameter analyzer. The device was placed in a dark box to exclude optical and electrical interferences from the external environment. For the pH response analysis, a commercial ceramic-plug junction-type Ag/AgCl electrode (Horiba 2080-06T) with an internal solution AgCl-saturated KCl electrolyte was used as the reference electrode. In this experiment, we defined a reference voltage (V_{REF}) at the I_D of 100 pA (read current, I_R) to quantify the amount of transfer characteristics (V_G–I_D) curve shift depending on the pH value of the buffer solution. Furthermore, the hysteresis and drift effects were measured, which indicated reliability degradation from repetitive and continuous operations. The hysteresis width voltage (V_H) was determined from the V_{REF} difference between the initial and final pH states in the loop path of pH 7 → 4 → 7 → 10 → 7. Additionally, the drift effect monitored the modulation of V_{REF} when the SnO_2 sensing membrane was exposed to a pH 7 buffer solution for 10 h.

2.3. Signal Amplification of the Ambipolar DG ISFET

For an ISFET with a DG structure, the sensing operation can be performed in either the SG mode, which measures the pH (or biosignal) of the target analyte using the top-gate electrode (reference electrode), or the DG mode, which measures the signal of the target analyte with the bottom-gate electrode. In the SG sensing mode, the bottom-gate electrode is grounded, and V_G is swept for the top-gate electrode. At this time, the EG unit and reference electrode are located at the top-gate electrode, as shown in Figure 2a. Therefore, the shift of the V_G–I_D curve is determined only by the surface potential of the sensing membrane. The modulation of V_{REF} in the top-gate operation (ΔV_{REF}^{T}), which quantitatively represents the degree of V_G–I_D curve shift, can be expressed as

$$\Delta V_{REF}^{T} = -\Delta\varphi \tag{1}$$

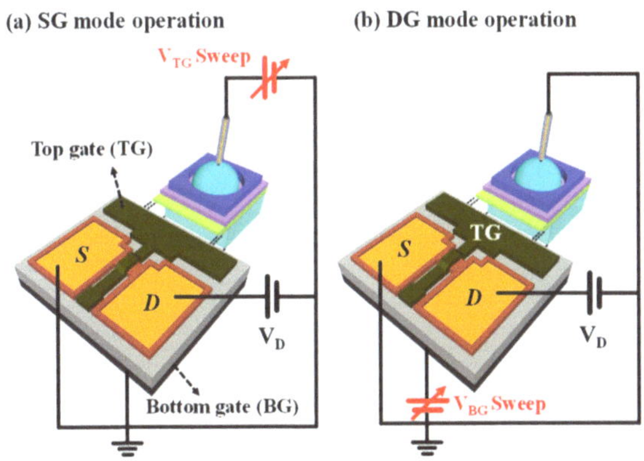

Figure 2. Simplified schematics of the DG ISFET sensor platform.

The pH sensitivity is limited to 59.14 mV/pH at room temperature owing to the site binding theory [8,26]. Therefore, the detection accuracy will be poor if the difference in concentrations is not significant. Similarly, in the DG sensing mode, the EG unit and reference electrode are connected to the top-gate electrode, and V_G is swept for the bottom-gate electrode, as shown in Figure 2b. In this case, the pH sensitivity may exceed the sensitivity limit owing to the capacitive coupling effect between the top- and bottom-gate dielectric layers and is expressed as

$$\Delta V_{REF}^{B} = -\frac{C_{tox}}{C_{box}}\Delta\varphi = \frac{C_{tox}}{C_{box}}\Delta V_{REF}^{T} \tag{2}$$

where the modulation of V_{REF} in the bottom-gate operation (ΔV_{REF}^{B}) is amplified by the capacitance ratio of the top to bottom gate dielectric layers [9,10,27]. C_{tox} and C_{box} are the top- and bottom-gate dielectric capacitances, respectively. ΔV_{REF}^{B} and ΔV_{REF}^{T} are the reference voltage modulations of the top and bottom gates, respectively. In this study, considering that the top- and bottom-gate dielectrics are the same as SiO_2, Equation (2) can be expressed according to the thickness ratio of the top- to bottom-gate dielectrics as

$$\Delta V_{REF}^{B} = -\frac{t_{box}}{t_{tox}}\Delta V_{REF}^{T} \tag{3}$$

where t_{tox} and t_{box} are the thicknesses of the top- and bottom-gate dielectrics, respectively.

3. Results and Discussion

3.1. Evaluation of the Ni-Silicide S/D Region for Ambipolar DG ISFET

Ni-silicide is a typical midgap metal used in both NMOS and PMOS FETs, which enables ambipolar switching operations. In this study, a Ni-silicide SB S/D was formed with a low thermal budget through high-efficiency energy transfer of microwaves to implement an ambipolar DG ISFET with excellent electrical characteristics. Furthermore, to determine the optimal Ni-silicide formation conditions by MWI, we evaluated the electrical properties and crystallinity of the Ni-silicide layer based on microwave power. Figure 3a shows the sheet resistance (Rs) of the Ni-silicide layer under various microwave power conditions measured using a four-point probe. The MWI process for Ni silicidation was performed in an N_2 ambient for 2 min, and the Rs was measured before and after removing the unreacted Ni using an SPM. Before SPM etching, because of the remaining Ni layer, all samples exhibited low Rs. Conversely, in the absence of MWI, the unreacted Ni was completely eliminated after SPM etching, which resulted in high Rs. However, above 250 W, the Rs decreases significantly after treatment with SPM and Ni-silicide formation is realized, whereas at 500 W, the decrease in Rs is almost saturated. It is worth noting that 600 W of MWI exhibits the lowest Rs. Therefore, 600 W was determined as the optimal condition in terms of resistance, thermal budget, and power consumption. Figure 3b shows the crystallinity of the characterized Ni-silicide layer by X-ray diffraction (XRD) analysis. The XRD pattern of the as-deposited Ni film showed peaks at (111) and (200) corresponding to pure Ni crystals. Conversely, after the silicidation process, several other peaks were observed. At a low microwave power of 250 W, a Ni-silicide peak corresponding to (310) was observed, indicating silicide formation. Meanwhile, several strong peaks corresponding to (211), (220), (310), (221), and (301) appeared above 500 W. Accordingly, we verified that Si and Ni reacted to form high-quality Ni-silicide through MWI treatment exceeding 500 W [28,29]. From the above results, an MWI of 600 W, which provides the lowest Rs and good crystallinity, was used for the Ni-silicide SB S/D formation in the fabrication of the ambipolar DG ISFET. Figure 3c shows the characterization of Schottky contact of S/D. The characteristic evaluation was performed through Schottky diodes. In the case of as-deposited Ni film, the rectification characteristics of the diodes are hardly exhibited due to the low on-current and high leakage current due to interfacial defects between the Si and unreacted Ni layers. On the other hand, in the Ni silicide junction diodes in which the silicide reaction occurred by 600 W MWI process, clear rectification characteristics of high on-current and low leakage current were observed.

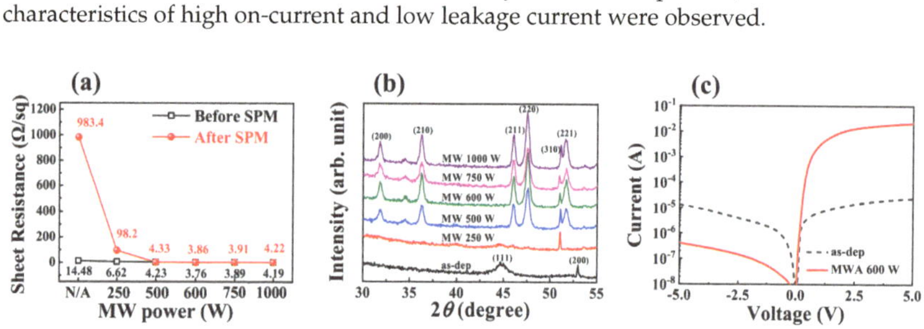

Figure 3. (**a**) Sheet resistance (Rs) and (**b**) XRD patterns of the microwave-assisted Ni-silicides formed under various MWI powers. (**c**) The characterization of Schottky contact of S/D.

3.2. Electrical Characteristics of the Ambipolar DG ISFET with Ni-Silicide SB S/D

The electrical characteristics of the FET transducer play an essential role in the pH sensing operation; hence, we measured the transfer characteristic (V_G–I_D) curves during the top- and bottom-gate operations to verify these characteristics. Figure 4 shows the electrical characteristics of the proposed device. Figure 4a shows the V_G–I_D curves for the top- and bottom-gate operations in the *p*- and *n*-regions, respectively. The drain voltages were −50 mV and 50 mV for the *p*- and *n*-regions, respectively. It is seen that the fabricated device exhibits distinct ambipolar conduction behaviors for the top- and bottom-gate operations. Furthermore, changing the polarity of the gate bias can cause the FET to behave as a *p*-type or an *n*-type channel. Additionally, the top gate allows a larger drive current compared to the bottom gate considering that the top-gate oxide is thinner than the bottom-gate oxide. Figure 4b shows the output characteristics (V_D–I_D) curves for the top- and bottom-gate operations in the *p*- and *n*-regions, respectively. For both the top- and bottom-gate operations, the drain current (I_D) exhibited a pinch-off characteristic that increased linearly in the low drain voltage (V_D) region before gradually saturating in the high V_D region. Based on these transfer and output characteristics, we can confirm that the ambipolar DG FET transducer with Ni-silicide SB S/D was well-formed by the 600 W MWI process.

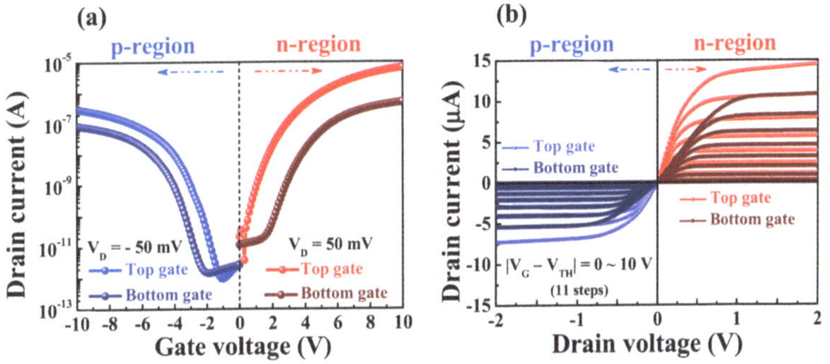

Figure 4. Transfer characteristic (V_G–I_D) curves of the bipolar DG ISFETs operated by the top- or bottom-gate voltages in the (**a**) *p* and *n* regions. (**b**) Output characteristic (V_D–I_D) curves of the bipolar DG ISFETs operated by the top- or bottom-gate voltages in the *p* and *n* regions, respectively.

3.3. pH Sensing Performance of the Ambipolar DG ISFET

Generally, pH sensitivity is determined by the changes in the surface potential ($\Delta\varphi$) per unit pH value, where φ is defined by the ion concentration in the pH buffer solution, given as [8,26]:

$$\varphi = 2.303\frac{kT}{q}\left(\frac{\beta}{\beta+1}\right)(pH_{pzc} - pH) \quad (4)$$

where k is the Boltzmann constant, T is the absolute temperature, q is the elementary charge, pH_{pzc} is the pH value at the point of zero charge, and β is the chemical sensitivity of the gate dielectric. Figure 5 shows the V_G–I_D curves of the ambipolar DG ISFET for different pH buffer solutions. The pH values of the buffer solutions used for the measurements were 3.07, 4.08, 5.99, 6.95, 8.97, and 9.87. Figure 5a,b show the sensing properties of the *p* and *n* regions in the SG sensing mode, respectively. In both cases, as the pH of the buffer solution increased, the transfer characteristic curve shifted toward the positive V_G direction. Figure 5c,d show the shift of the V_G–I_D curve in the DG sensing mode. In this case, despite the same sensing membrane, the shift of the transfer curve is larger than that in the SG sensing mode. This indicates that the pH sensitivity is highly dependent on the transducer performance and quality of the sensing membrane. Since the DG ISFET is capacitively coupled between the top- and bottom-gate electrodes, it can self-amplify the sensitivity according to Equation (3) without the need for any additional amplification circuitry.

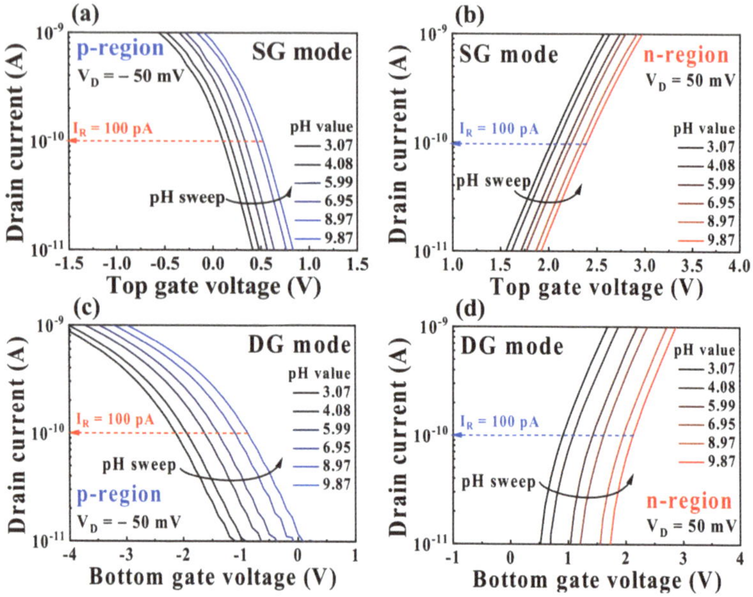

Figure 5. Transfer characteristic (V_G–I_D) curves of the ambipolar DG ISFET for different pH buffer solutions. SG mode sensing in the (**a**) *p* region and (**b**) *n* region. DG mode sensing in the (**c**) *p* region and (**d**) *n* region.

Figure 6 shows the shift in the reference voltage (ΔV_{REF}) for the *p* and *n* regions as a function of the pH value. Herein, V_{REF} was determined at a constant read drain current (I_R) of 100 pA. The SG-mode pH sensitivities of the ambipolar sensor were 58.8 and 58.7 mV/pH in the *p* and *n* regions, respectively. According to the site coupling theory, the ISFET exhibits a low sensitivity of 59.14 mV/pH, whereas the proposed ambipolar sensor exhibits a sensitivity close to the Nernst limit in the SG sensing mode. Conversely, in the DG sensing mode, the pH sensitivities increased by approximately 3 times compared to those in the SG sensing mode. The extracted pH sensitivities were 177.5 and 175.0 m V/pH for the *p* and *n* regions, respectively; this magnification is almost identical to the potential increase by the amplification factor, t_{tox}/t_{box}, in Equation (3).

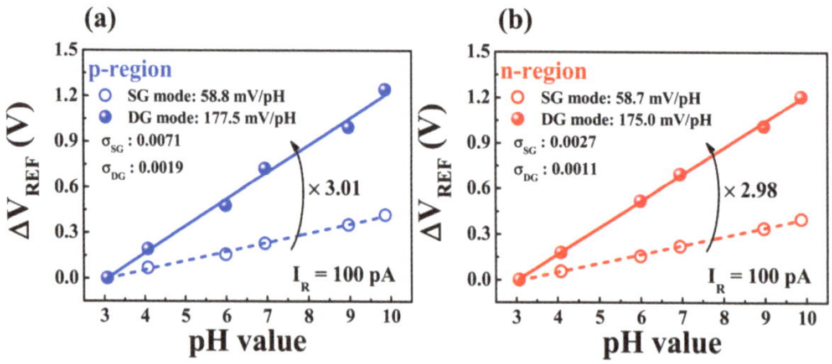

Figure 6. Reference voltage shift (ΔV_{REF}) in the (**a**) *p*- and (**b**) *n*-region operations as a function of the pH value. The symbols and lines represent the experimental data and linear fits, respectively.

3.4. Reliability and Stability of the Ambipolar DG ISFET

As both the pH sensitivity and reliability of the fabricated ambipolar ISFETs must be ensured for repeated and continuous operations, we evaluated the hysteresis and drift effects to derive the reliability in the SG and DG sensing modes. Figure 7 shows the hysteresis width voltage (V_H) measurements when the sensing membrane undergoes a gradual change in the pH buffer solution. The sensing membrane of the EG unit comprises hydroxy groups, which are capable of capturing or releasing hydrogen ions. Although the surface of the sensing membrane is capable of fast bonding to hydrogen ions, the bulk exhibits slower bonding. This results in the degradation of reliability during the repetitive operation, called the hysteresis effect [30–32]. Measurements were conducted for a total of 50 min at intervals of 2 min and 10 min per buffer solution. The V_H values obtained in the pH loop of $7 \rightarrow 4 \rightarrow 7 \rightarrow 10 \rightarrow 7$ were 5.8 mV and 4.2 mV for the *p* and *n* regions, respectively, for the SG sensing mode and 8.6 mV and 5.1 mV, respectively, for the DG sensing mode.

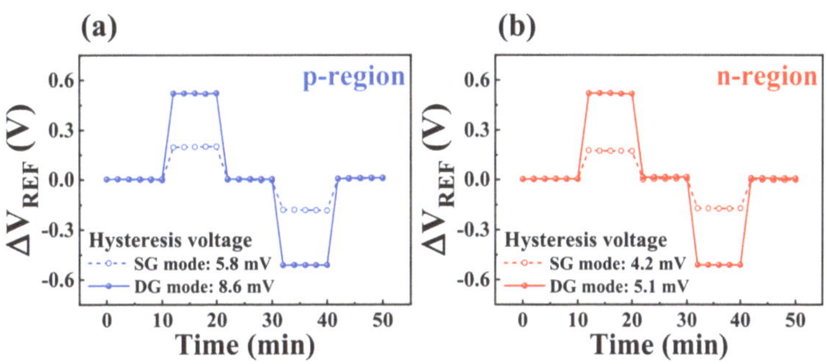

Figure 7. Hysteresis effect of the ambipolar DG ISFET in the (**a**) *p* region and (**b**) *n* region for three different buffer solutions with pH values of 4, 7, and 10. The open and closed circles represent ΔV_{REF} in the SG and DG sensing modes, respectively.

During continuous operation for a long period, a hydration layer is formed on the surface of the sensing film, which diffuses ions of the electrolyte into the sensing membrane to alter its properties [30,33]. This results in a drift phenomenon accompanied by significant noise, which is considered a critical obstacle. Figure 8 shows the drift effect measurement when the EG is immersed in a pH 7 buffer solution for 10 h. The drift rates were 23.0 and 14.1 mV/h in the *p* and *n* regions, respectively, for the SG sensing mode and 30.2 and 18.3 mV/h, respectively, for the DG sensing mode. Although the differences in the drift rates were insignificant, the *n*-type operation was better compared to the *p*-type operation. As in the case of the hysteresis voltage, the drift rates of the SG and DG modes were much smaller than the sensitivities. The ratios of drift rate to sensitivity were 39.12% and 24.02% in the SG mode and 17.01% and 10.46% in the DG mode, respectively. These results mean that changing from the SG to DG mode increases the sensitivity by 201.87% and 198.13% and drift rates by only 31.3% and 29.79%, respectively. This indicates that in both the *p* and *n* channels, the stability and reliability are higher in the DG mode than in the SG mode owing to high sensitivity.

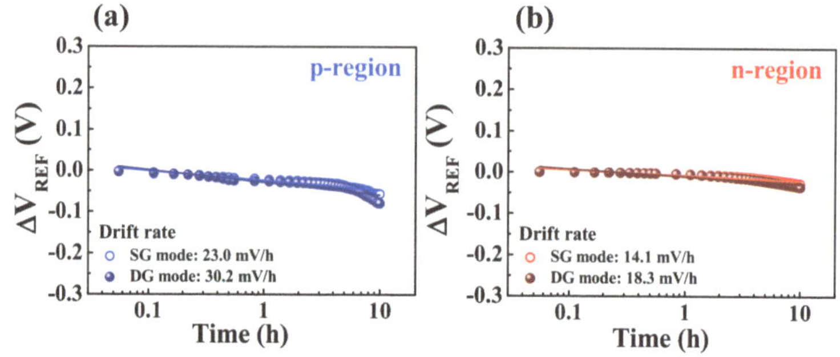

Figure 8. Drift effects of the ambipolar DG ISFET in the (**a**) *p* region and (**b**) *n* region when the sensing membrane is immersed in a pH 7 buffer for 10 h. ΔV_{REF} was monitored in the SG and DG sensing modes.

Table 1 summarizes the pH sensing performances of the proposed ambipolar DG ISFETs. The fabricated device exhibits amplified sensitivity depending on the capacitance ratio of the top- to bottom-gate electrodes. Specifically, it is worth noting that the increase in nonideal behavior is significantly lower than the amplification of sensitivity, which indicates that the proposed bidirectional chemical sensor platform is capable of amplification with guaranteed reliability and stability, along with operating as a *p*-type device with good noise resistance or an *n*-type device with good mobility and drift performance.

Table 1. pH sensing performance of the fabricated bipolar DG ISFETs.

		Sensitivity (mV/pH)	V_H (mV)	Drift Rate (mV/h)
p-region	SG mode	58.8	5.8	23.0
	DG mode	177.5	8.6	30.2
n-region	SG mode	58.7	4.2	14.1
	DG mode	175.0	5.1	18.3

Finally, the ISFET inverter is considered as a potential application of the proposed bipolar ISFET. However, there are some applications where analog details are not required, and simply determining whether the chemical concentration or pH is above or below a certain threshold is sufficient. In particular, for DNA sequencing where a "yes"/"no" answer to the pH change is sufficient, an ISFET inverter can determine whether a chain extension reaction has occurred [34,35]. Fabricating a typical ISFET inverter requires a CMOS process where *n*-type and *p*-type dopants are implanted and activated. This process is complex and time-consuming, resulting in high fabrication costs. However, the proposed ambipolar ISFET does not require doping of the *n*-type and *p*-type impurities and is easy to implement in a CMOS-like inverter with simple silicide formation. Therefore, the CMOS-like inverter based on the proposed ambipolar ISFET may be a promising candidate for DNA logic gates.

4. Conclusions

This study proposes a high-performance bidirectional chemical sensor platform based on ambipolar DG ISFETs with microwave-assisted Ni-silicide SB S/D and a separative EG unit. The proposed device exhibits ambipolar switching and excellent electrical characteristics as a transducer. Furthermore, the fabricated bipolar chemical sensor platform exhibits pH sensitivities of 58.8 and 58.7 mV/pH in the *p*- and *n*-regions, respectively, in the SG sensing mode compared to the theoretical limit of sensitivity at room temperature. Conversely, we achieved excellent pH sensitivities of 177.5 and 175.0 mV/pH in the *p*- and *n*-regions, respectively, in the DG sensing mode, which is approximately three times that achieved in the SG sensing mode, owing to the capacitive coupling effect between the top and bottom gates. The V_H values were 5.8 mV and 4.2 mV in the *p*- and *n*-regions, respectively, in the SG sensing mode and 8.6 mV and 5.1 mV, respectively, in the DG sensing mode. Additionally, the drift rates were 23.0 mV/h and 14.1 mV/h in the *p*- and *n*-regions, respectively, in the SG sensing mode and 30.2 mV/h and 18.3 mV/h, respectively, in the DG sensing mode. Although the DG mode has larger V_H and drift rates compared to the SG mode, the increase in nonideal effects was found to be relatively small compared to the sensitivity amplification ratio. Thus, the application of Ni-silicide S/D and DG structures enables bidirectional sensing operation of the *p*- and *n*-channels depending on the purpose of use while ensuring high sensitivity and stability. In addition to these advantages, the proposed bidirectional chemical sensor platform, which has a simple process and bidirectional operation capability, is expected to be applicable in various fields including biosensors. Furthermore, when subsequent study containing bio-selective elements is introduced, it will be possible to implement a biological event monitoring system that can detect regardless of the polarity of the target substance and secure high sensitivity.

Author Contributions: Y.-U.K.: conceptualization, formal analysis, methodology, investigation, data curation, visualization, software, resources, and writing—original draft. W.-J.C.: conceptualization, methodology, investigation, resources, formal analysis, funding acquisition, supervision, validation, and writing—review and editing. All authors have read and agreed to the published version of the manuscript.

Funding: This work was supported by the National Research Foundation of Korea (NRF) grant funded by the Korean government (MSIT) (No. 2020R1A2C1007586).

Institutional Review Board Statement: Not applicable.

Informed Consent Statement: Not applicable.

Data Availability Statement: Not applicable.

Acknowledgments: The present Research has been conducted by the Research Grant of Kwangwoon University in 2022 and the Excellent research support project of Kwangwoon University in 2022.

Conflicts of Interest: The authors declare no conflict of interest.

References

1. Ndiaye, M.; Oyewobi, S.S.; Abu-Mahfouz, A.M.; Hancke, G.P.; Kurien, A.M.; Djouani, K. IoT in the wake of COVID-19: A survey on contributions, challenges and evolution. *IEEE Access* **2020**, *8*, 186821–186839. [PubMed]
2. Giwan, S.; Geonhee, L.; Mi, J.K.; Seung-Hwa, B.; Minsuk, C.; Keun, B.K.; Chang-Seop, L.; Sangmi, J.; Deaui, P.; Hong, G.K.; et al. Rapid detection of COVID-19 causative virus (SARS-CoV-2) in human nasopharyngeal swab specimens using a field-effect transistor-based biosensor. *ACS Nano* **2020**, *14*, 5135–5142.
3. Abdulhadee, Y.; Umaporn, P.; Sirirat, R.; Nattiya, H.; Orawon, C.; Sudkate, C. Paper-based electrochemical biosensor for diagnosing COVID-19: Detection of SARS-CoV-2 antibodies and antigen. *Biosens. Bioelectron.* **2021**, *176*, 112912.
4. Kim, S.; Ahn, J.H.; Park, T.J.; Lee, S.Y.; Choi, Y.K. A biomolecular detection method based on charge pumping in a nanogap embedded field-effect-transistor biosensor. *Appl. Phys. Lett.* **2009**, *94*, 243903.
5. Tsai, C.-C.; Chiang, P.-L.; Sun, C.-J.; Lin, T.-W.; Tsai, M.-H.; Chang, Y.-C.; Chen, Y.-T. Surface potential variations on a silicon nanowire transistor in biomolecular modification and detection. *Nanotechnology* **2011**, *22*, 135503.
6. Bergveld, P. Thirty years of ISFETOLOGY: What happened in the past 30 years and what may happen in the next 30 years. *Sens. Actuators B Chem.* **2003**, *88*, 1–20.
7. Sadighbayan, D.; Hasanzadeh, M.; Ghafar-Zadeh, E. Biosensing based on field-effect transistors (FET): Recent progress and challenges. *Trends Anal. Chem.* **2020**, *133*, 116067.
8. Bergveld, P. Development of an ion-sensitive solid-state device for neurophysiological measurements. *IEEE. Trans. Biomed. Eng.* **1970**, *1*, 70–71.
9. Li, B.-R.; Chen, C.-W.; Yang, W.-L.; Lin, T.-Y.; Pan, C.-Y.; Chen, Y.-T. Biomolecular recognition with a sensitivity-enhanced nanowire transistor biosensor. *Biosens. Bioelectron.* **2013**, *45*, 252–259.
10. Cho, S.-K.; Cho, W.-J. High-sensitivity pH sensor based on coplanar gate AlGaN/GaN metal-oxide-semiconductor high electron mobility transistor. *Chemosensors* **2021**, *9*, 4213.
11. Jang, H.J.; Cho, W.J. Performance enhancement of capacitive-coupling dual-gate ion-sensitive field-effect transistor in ultra-thin-body. *Sci. Rep.* **2014**, *4*, 1–8.

12. Ioannidis, E.G.; Theodorou, C.G.; Haendler, S.; Josse, E.; Dimitriadis, C.A.; Ghibaudo, G. Impact of source–drain series resistance on drain current mismatch in advanced fully depleted SOI n-MOSFETs. *IEEE Electron. Device Lett.* **2015**, *36*, 433–435.
13. Rahhal, L.; Bajolet, A.; Diouf, C.; Cros, A.; Rosa, J.; Planes, N.; Ghibaudo, G. New methodology for drain current local variability characterization using Y function method. In Proceedings of the 2013 IEEE International Conference on Microelectronic Test Structures (ICMTS), Osaka, Japan, 25–28 March 2013; pp. 99–103.
14. Dubois, E.; Larrieu, G. Low Schottky barrier source/drain for advanced MOS architecture: Device design and material considerations. *Solid State Electron. Lett.* **2002**, *46*, 997–1004.
15. Hwang, H.; Lee, D.-H.; Hwang, J.M. Degradation of MOSFETs drive current due to halo ion implantation. *Tech. Dig. Int. Electron. Devices Meet.* **1996**, *4*, 567–570.
16. Won-Jae, L.; Do-Woo, K.; Soon-Young, O.; Yong-Jin, K.; Ying-Ying, Z.; Zhun, Z.; Shi-Guang, L.; Soon-Yen, J.; In-Sik, H.; Tae-Kyu, G. Work function variation of nickel silicide using an ytterbium buffer layer for Schottky barrier metal-oxide-semiconductor field-effect transistors. *Int. J. Appl. Phys.* **2007**, *101*, 103710.
17. Seger, J.; Hellström, P.E.; Lu, J.; Malm, B.G.; von Haartman, M.; Östling, M.; Zhang, S.L. Lateral encroachment of Ni-silicides in the source/drain regions on ultrathin silicon-on-insulator. *Appl. Phys. Lett.* **2005**, *86*, 253507.
18. Xu, Z.; Zi, W.; Ruxin, S.; Yadan, Z.; Lunan, Z.; Di, X.; Lizhen, H.; Lifeng, C. High performance gas sensors with dual response based on organic ambipolar transistors. *J. Mater. Chem. C* **2021**, *9*, 1584–1592.
19. Ajay, S.; Rakhi, N.; Manoj, S.; Mridula, G. Ambipolar behaviour of tunnel field effect transistor (TFET) as an advantage for biosensing applications. In *Physics of Semiconductor Devices*; Springer: Berlin/Heidelberg, Germany, 2014; pp. 171–174.
20. Yuan, L.; Guo, Z.; Hailong, Z.; Zheng, L.; Rui, C.; Yang, X.; Vincent, G.; Yu, H.; Xiangfeng, D. Ambipolar barristors for reconfigurable logic circuits. *Nano Lett.* **2017**, *17*, 1448–1454.
21. Jakobson, C.G.; Nemirovsky, Y. 1/f noise in ion sensitive field effect transistors from subthreshold to saturation. *IEEE Trans. Electron. Devices* **1999**, *46*, 259–261.
22. Panahi, A.; Sadighbayan, D.; Forouhi, S.; Ghafar-Zadeh, E. Recent advances of field-effect transistor technology for infectious diseases. *Biosensors* **2021**, *11*, 103.
23. Schmitt, A.L.; Higgins, J.M.; Szczech, J.R.; Jin, S. Synthesis and applications of metal silicide nanowires. *J. Mater. Chem.* **2010**, *20*, 223–235.
24. Byon, K.; Tham, D.; Fischer, J.E.; Johnson, A.T. Systematic study of contact annealing: Ambipolar silicon nanowire transistor with improved performance. *Appl. Phys. Lett.* **2007**, *90*, 143513.
25. Min, J.G.; Lee, D.H.; Kim, Y.U.; Cho, W.J. Implementation of ambipolar polysilicon thin-film transistors with nickel silicide Schottky junctions by low-thermal-budget microwave annealing. *Nanomaterials* **2022**, *12*, 628.
26. Yates, D.E.; Levine, S.; Healy, T.W. Site-binding model of the electrical double layer at the oxide/water interface. *J. Chem. Soc. Faraday Trans.* **1974**, *70*, 1807–1818.
27. Jang, H.J.; Cho, W.J. Fabrication of high-performance fully depleted silicon-on-insulator based dual-gate ion-sensitive field-effect transistor beyond the Nernstian limit. *Appl. Phys. Lett.* **2012**, *100*, 073701.
28. Bhaskaran, M.; Sriram, S.; Mitchell, D.R.G.; Short, K.T.; Holland, A.S.; Mitchell, A. Effect of multi-layered bottom electrodes on the orientation of strontium-doped lead zirconate titanate thin films. *Micron* **2008**, *40*, 11–14.
29. Wen-Li, C.; Chung-Hua, C.; Jui-Yuan, C.; Chun-Wei, H.; Yu-Ting, H.; Kuo-Chang, L.; Cheng-Lun, H.; Ping-Hung, Y.; Wen-Wei, W. Single-crystalline δ-Ni2Si nanowires with excellent physical properties. *Nanoscale Res. Lett.* **2013**, *8*, 1–5.
30. Chou, J.C.; Wang, Y.F. Preparation and study on the drift and hysteresis properties of the tin oxide gate ISFET by the sol–gel method. *Sens. Actuators B Chem.* **2002**, *86*, 58–62.
31. Bousse, L.; Mostarshed, S.; van der Schoot, B.; De Rooij, N.F. Comparison of the hysteresis of Ta2O5 and Si3N4 pH-sensing insulators. *Sens. Actuators B Chem.* **1994**, *17*, 157–164.
32. Bousse, L.; Bergveld, P. The role of buried OH sites in the response mechanism of inorganic-gate pH-sensitive ISFETs. *Sens. Actuators* **1984**, *6*, 65–78.
33. Chou, J.C.; Hsiao, C.N. The hysteresis and drift effect of hydrogenated amorphous silicon for ISFET. *Sens. Actuators B Chem.* **2000**, *66*, 181–183.
34. Al-Ahdal, A.; Toumazou, C. ISFET-based chemical switch. *IEEE Sens. J.* **2011**, *12*, 1140–1146.
35. Wong, W.; Shepherd, L.; Georgiou, P.; Toumazou, C. Towards ISFET based DNA logic for rapid nucleic acid detection. *Sensors* **2009**, *8*, 1451–1454.

chemosensors

Article

Enhancement of Ion-Sensitive Field-Effect Transistors through Sol-Gel Processed Lead Zirconate Titanate Ferroelectric Film Integration and Coplanar Gate Sensing Paradigm

Dong-Gyun Mah [1], Seong-Moo Oh [2], Jongwan Jung [2] and Won-Ju Cho [1,*]

[1] Department of Electronic Materials Engineering, Kwangwoon University, Gwangun-ro 20, Nowon-gu, Seoul 01897, Republic of Korea; madong13@kw.ac.kr
[2] Department of Nanotechnology and Advanced Materials Engineering, Sejong University, Seoul 05006, Republic of Korea; 17011353@sju.ac.kr (S.-M.O.)
* Correspondence: chowj@kw.ac.kr; Tel.: +82-2-940-5163

Abstract: To facilitate the utility of field effect transistor (FET)-type sensors, achieving sensitivity enhancement beyond the Nernst limit is crucial. Thus, this study proposed a novel approach for the development of ferroelectric FETs (FeFETs) using lead zirconate titanate (PZT) ferroelectric films integrated with indium–tungsten oxide (IWO) channels synthesized via a cost-effective sol-gel process. The electrical properties of PZT-IWO FeFET devices were significantly enhanced through the strategic implementation of PZT film treatment by employing intentional annealing procedures. Consequently, key performance metrics, including the transfer curve on/off ratio and subthreshold swings, were improved. Moreover, unprecedented electrical stability was realized by eliminating the hysteresis effect during double sweeps. By leveraging a single-gate configuration as an FeFET transformation element, extended-gate (EG) detection methodologies for pH sensing were explored, thereby introducing a pioneering dimension to sensor architecture. A measurement paradigm inspired by plane gate work was adopted, and the proposed device exhibited significant resistive coupling, consequently surpassing the sensitivity thresholds of conventional ion-sensitive field-effect transistors. This achievement represents a substantial paradigm shift in the landscape of ion-sensing methodologies, surpassing the established Nernst limit (59.14 mV/pH). Furthermore, this study advances FeFET technology and paves the way for the realization of highly sensitive and reliable ion sensing modalities.

Keywords: ferroelectric field-effect transistors (FeFETs); ion-sensitive field-effect transistors (ISFETs); lead zirconate titanate (PZT) ferroelectric films; extended fate ion-sensitive field-effect transistors (EG-ISFETs); methodologies for pH sensing; sensitivity enhancement; resistive coupling

Citation: Mah, D.-G.; Oh, S.-M.; Jung, J.; Cho, W.-J. Enhancement of Ion-Sensitive Field-Effect Transistors through Sol-Gel Processed Lead Zirconate Titanate Ferroelectric Film Integration and Coplanar Gate Sensing Paradigm. *Chemosensors* **2024**, *12*, 134. https://doi.org/10.3390/chemosensors12070134

Received: 31 May 2024
Revised: 2 July 2024
Accepted: 8 July 2024
Published: 9 July 2024

1. Introduction

The continuous development of industries emphasizes the importance of ion sensitivity measurement, particularly pH measurement, in various fields, such as biology, medicine, and environmental monitoring [1–3]. Consequently, several studies have focused on ion-sensitive field-effect transistor (ISFET) technology, which holds potential for label-free detection, high sensitivity, fast response time, and compatibility with complementary metal-oxide-semiconductor (CMOS) processes [4–6]. In typical ISFETs, the transducer unit and the sensing membrane are integrated into a single platform. However, frequent sensitivity measurements degrade the reliability of the transducer unit, leading to the proposal of the extended gate ion-sensitive field-effect transistor (EG-ISFET), which separates the transducer and the sensing membrane [7–9]. Recently, extensive research has been conducted on the transducer, sensing membrane, and overall sensor platform structure to improve the sensing characteristics and process costs of EG-ISFETs [10–13]. However, studies that comprehensively address the performance enhancement of both the transducer and the sensor platform structure are not widely reported. Therefore, in this study, we fabricated a cost-effective transducer and evaluated its electrical performance and reliability. Additionally, we introduced a resistive coupling method to replace the previously studied vertical dual-gate and coplanar gate structures to achieve sensitivity exceeding the Nernst limit [14–17].

Initially, we focused on the enhancement of the characteristics of the transducer unit to drive innovative advancements in EG-ISFET technology. As part of this approach, we fabricated ferroelectric field-effect transistors (FeFETs) incorporating lead zirconate titanate (PZT) ferroelectric films and indium–tungsten oxide (IWO) channels. The PZT films were synthesized through a cost-effective sol-gel method [18–21]. PZT films exhibit high residual polarization, low coercive voltage, and compatibility with silicon processes within integrated circuits. Further, integration with amorphous oxide semiconductor channels facilitates the implementation of low-power consumption and excellent on/off performance characteristics [22,23]. However, PZT films exhibit counterclockwise hysteresis in the transfer curve owing to typical polarization characteristics, which can result in reliability and reproducibility issues when applied to ISFET platforms [24,25]. Consequently, PZT films are subjected to thermal treatment to form a thin trapping layer approximately 2 nm thick [26–28]. Traditional dual sweeps exhibit polarization and charge trapping characteristics in counterclockwise and clockwise directions, respectively [29]. By balancing these two characteristics, we optimized thin-film transistors (TFTs) for the transducer unit, consequently achieving outstanding performance metrics: a hysteresis voltage of less than 10 mV, subthreshold swing (SS) of 88.75 mV/dec, and a current on/off ($I_{on/off}$) ratio of approximately 10^9. The performance metrics achieved are noteworthy when compared to bottom gate structure transistors based on PZT thin films, particularly in terms of μ_{FE}, SS, and $I_{on/off}$ ratio. To facilitate the use of the proposed device as a transducer in the sensor platform, preliminary DC bias tests were conducted to observe threshold voltage changes in the transfer curve [30]. Excellent consistency exceeding 99% across various read currents (I_{REF}) was confirmed.

Consequently, in this study's EG-ISFET platform, a single-gate configuration PZT-FeFET was employed as the transducer unit, with an SnO_2 sensing membrane. A sensitivity similar to the Nernst limit was realized in pH sensitivity measurements [31,32]. Further, we adopted an innovative resistive coupling method inspired by the plane gate in the sensor platform [33]. Through adjustments to the resistance ratio between the control gate (CG) resistance (R_{CG}) and the sensing gate (SG) resistance (R_{SG}), we achieved a high sensitivity of 287.2 mV/pH, which surpassed the traditional Nernst limit of 59.14 mV/pH. Moreover, to ensure the reliability of the proposed platform, we verified the measurement sensitivity, hysteresis, and drift effects according to pH value [34,35]. In addition, direct comparisons and evaluations with results obtained using the existing single-gate-based EG-ISFET method for the case of $R_{CG}/R_{SG} = 1$ were conducted.

Thus, our innovative approach, which ingeniously engineered the polarization and trapping characteristics, achieved unprecedented electrical performance improvements in FeFET technology and clearly overcame the limitations of the sensing mechanism imposed by the polarization characteristics being studied [36–38]. Furthermore, the successful implementation of sensitivity amplification via integration with the self-resistive coupling sensing platform demonstrated the feasibility of creating an efficient single-gate structure platform, surpassing the existing complex and high-cost transducer fabrication methods.

2. Materials and Methods

2.1. Material Specifications

The fabricated EG-ISFET comprised transducer and sensing units. For the fabrication of the sensing unit, glass substrates (7059 glass; Corning Inc., New York, NY, USA), indium–tin oxide (ITO) sputter target (purity $\geq$ 99.99%, THIFINE Co., Ltd., Incheon, Republic of Korea), SnO_2 sputter target (purity $\geq$ 99.99%, THIFINE Co., Ltd.), polydimethylsiloxane (PDMS; Sylgard 184 silicon elastomer; Dow Corning, Midland, MI, USA), and Ag/AgCl electrode (Horiba 2080-06T; Kyoto, Japan) were utilized. Further, sensitivity to pH buffer solution (Samchun Chemical, Pyeongtaek, Republic of Korea) was measured. To fabricate the transducer unit, the following materials were employed: indium–tungsten oxide (IWO) sputter target (purity $\geq$ 99.99%, THIFINE Co., Ltd.), ITO sputter target (purity $\geq$ 99.99%, THIFINE Co., Ltd.), lead acetate trihydrate ($Pb(CH_3CO_2)_2 \cdot 3H_2O$), Ti-isopropoxide and Zr-propoxide (Zr:Ti = 52:48), Pt (purity $\geq$ 99.99%, THIFINE Co., Ltd.), and Ti (purity $\geq$ 99.99%, THIFINE Co., Ltd.), SiO_2 (purity $\geq$ 99.99%, THIFINE Co., Ltd.).

2.2. Formation of PZT Thin Film and Fabrication Process of FeFET

The fabrication process flow of the device, as illustrated in Figure 1, is as follows. A PZT thin film was fabricated on a Pt/Ti/SiO_2/Si substrate using the sol-gel method. The PZT thin film was deposited via spin-coating at a rotation speed of 3000 rpm for 30 s. Subsequently, to promote evaporation and combustion of volatile substances and minimize film cracking, it was dried on a hot plate at 150 °C for 1 min, followed by annealing at 450 °C for 7 min and then at 650 °C for 3 min. To achieve the desired thickness, the annealing processes were repeated six times, yielding a 300 nm thick PZT film. Further, for film crystallization and intentional trap layer formation, furnace annealing of the PZT film was conducted at 650 °C for 30 min. Thereafter, an IWO channel layer with dimensions of 60 µm width × 120 µm length was deposited to a thickness of 20 nm using RF sputtering. Given the high conductivity of IWO, it is essential to engineer the off-current characteristics. Therefore, optimization was achieved through a comprehensive parameter evaluation extracted from the transfer curves of PZT-IWO FeFETs with IWO channel thicknesses of 20, 35, and 50 nm. Subsequently, ITO was deposited to a thickness of 100 nm and consequently patterned using a lift-off process to form source and drain electrodes measuring 120 µm × 150 µm. To enhance the electrical properties of the IWO semiconductor, the deposited film was annealed in a 300 °C oxygen environment for 30 min. Considering the predominant conductivity of indium in the IWO material, an engineered optimal channel thickness was determined, and the annealing conditions were optimized to enhance electrical characteristics such as the $I_{on/off}$ and SS.

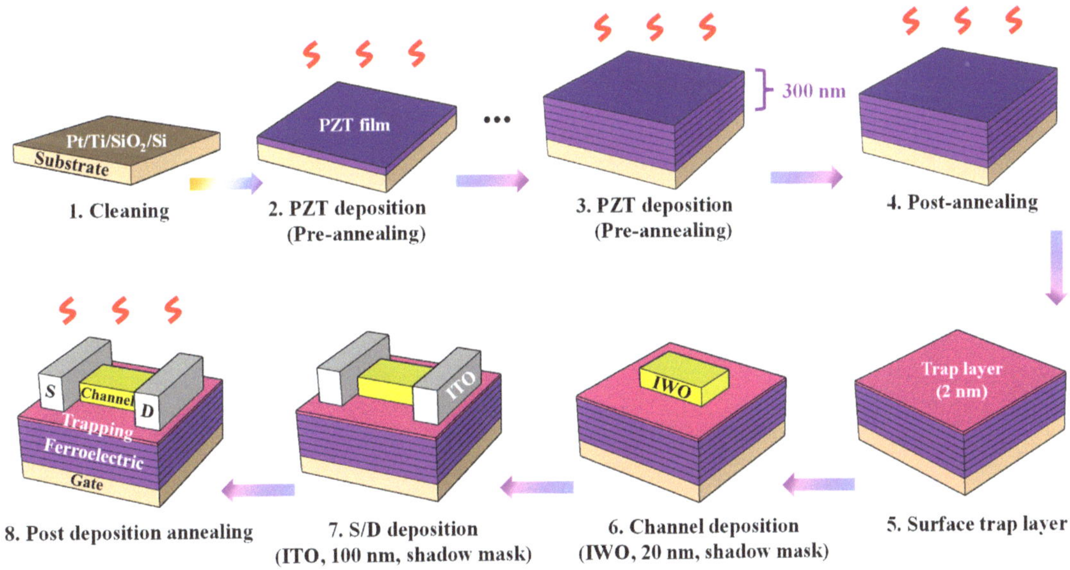

Figure 1. Formation of thin film and trap layer through thermal treatment of sol-gel PZT, and fabrication process of FeFET.

2.3. Fabrication of pH Selective EG Sensing Unit

This study addressed the limitations of sensing platforms reliant solely on a single device, such as traditional ISFETs, particularly the issue of reduced transistor lifespan resulting from repeated measurements. To circumvent this challenge, a separate sensing unit (EG) and transducer (FeFET) were fabricated [39]. The EG was designed to relay applied potential changes, arising from variations in surface ion concentration, to the transducer gate. This was achieved via the deposition of a 300 nm thick ITO layer onto a glass substrate measuring 1.5 cm × 3 cm. Subsequently, a 50 nm thick SnO_2 layer (sensing layer) was deposited onto the ITO layer to serve as a receptor for detecting surface potential changes corresponding to variations in the pH buffer solution. In addition, to facilitate contact with the pH buffer solution, a polydimethylsiloxane (PDMS) reservoir was affixed atop the sensing layer. Finally, a reference electrode (Ag/AgCl) was positioned approximately 3 mm away from the sensing layer within the PDMS reservoir for immersion in the buffer solution.

2.4. Design of a Self-Resistive Coupling Circuit for Sensitivity Amplification

Single-gate based EG-ISFETs are plagued by the limitations of the Nernst limit, which has prompted significant research efforts aimed at enhancing the sensitivity performance. Consequently, methods to improve device structure, such as the introduction of vertical dual-gate and coplanar gate structures, have emerged, shifting focus beyond the transducer's electrical operation excellence alone. However, the complexity and cost of platform fabrication, coupled with inefficiencies in sensitivity adjustment, have resulted in new challenges. These challenges were addressed by designing an innovative self-resistive coupling circuit. The sensitivity amplification mechanism was similar to co-planar gate type sensor platforms using the CG and SG concepts. However, we introduced a simple and efficient approach by incorporating resistors into the EG-ISFET sensor platform. The surface potential change resulting from the adsorption and desorption of hydrogen ions in the sensing unit, connected to R_{SG}, along with the operating voltage supplied by Agilent 4156B precision semiconductor parameter analyzer (Hewlett-Packard Co., Palo Alto, CA, USA) connected to R_{CG}, was capacitively coupled to FG (V_{FG}), which was applied to the gate voltage of FeFET. Specifically, artificial adjustment according to the ratio of the control gate resistor (R_{CG}) to the sensing gate resistor (R_{SG}) was performed. Consequently, the stability and superior sensitivity amplification of the innovative sensor platform were successfully validated through pH sensitivity measurements at various resistance ratios.

2.5. Device Characterization

The deposition thickness of all materials during the fabrication processes of the transducer and sensing units was measured using the Dektak XT Bruker stylus profiler (Bruker, Hamburg, Germany). The electrical characteristics of the transducer and its operational performance as a sensor platform were evaluated using a precision semiconductor parameter analyzer (Agilent 4156B), in conjunction with a DC power supply. In addition, measurements were conducted within a dark box to minimize external interference factors such as light and noise. Finally, sensitivity measurements were performed for pH 3, 4, 6, 7, 9, and 10 buffer solutions.

3. Results and Discussion

3.1. Electrical Characteristics and Mechanism of PZT-FeFET

The inherent operational characteristics of the transducer unit within the proposed EG-ISFET sensor-based self-resistive coupling platform are considered pivotal factors. Therefore, the semiconductor characteristics of the FeFET, encompassing both transfer and output characteristics, must be assessed. Specifically, enhancements in the $I_{on/off}$, SS, and mobility facilitate swift responses to variations in pH and ion concentration, thereby improving the real-time monitoring. Moreover, the absence of hysteresis in the dual-sweep transfer curve operation ensures the reliability of the semiconductor device for repetitive sensing measurements.

As depicted earlier in Figure 1, intentional polarity and trap characteristics interaction were induced via the formation of a trapping layer on the surface of the PZT thin film through conventional thermal annealing (CTA) processes at elevated temperatures. Figure 2a illustrates the double-sweep transfer curve of the PZT-FeFET, revealing exemplary transistor behavior characterized by a threshold voltage (V_{th}) of 0.6 V, SS of 85.75 mV/dec, $I_{on/off}$ of 1.91×10^8, and mobility of 15.87 $cm^2V^{-1}s^{-1}$. The bottom Pt serves as the gate, with voltage applied via the Agilent 4156B, while the drain voltage was set to 1 V.

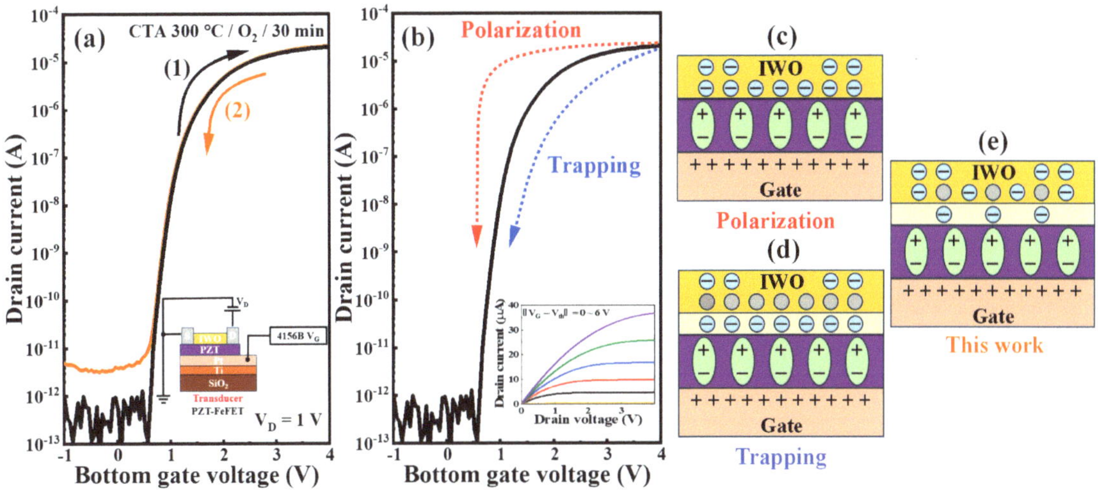

Figure 2. Transfer characteristics (V_G–I_D) curves of the PZT FeFET: (**a**) double sweep, and (**b**) direction of hysteresis with respect to polarization and trapping. The inset in (**b**) illustrates the output characteristics of the PZT-FeFET. The mechanisms, depicting the relative magnitudes of polarization and trapping, are as follows: (**c**) polarization dominance, (**d**) trapping dominance, and (**e**) interaction between polarization and trapping.

In Figure 2b, the salient features of the double-sweep curve are depicted under conditions of polarization and charge trap dominance. In general, anticlockwise hysteresis is observed during the backward sweep owing to the dielectric properties of the PZT thin film caused by polarization. In contrast, clockwise hysteresis occurs when electrons are trapped at the interface between the channel and the insulating layer owing to the presence of a trapping layer. During the dynamic phase where the gate voltage sweeps from its maximum (Max. V_g) to 0 V, the polarization characteristics (Figure 2c) and trapping behaviors (Figure 2d) synergize. This yields a device with minimal hysteresis, as evidenced by the 10 mV hysteresis value depicted in Figure 2e. Further, the inset in (b) presents the output curve characteristics of the PZT-FeFET, measured in 11 steps by varying $|V_G–V_{th}|$ as 0–6 V.

Table 1 presents the performance parameters of PZT-based FETs for the devices fabricated in this study and those from earlier studies [29,40–43]. To enable a comprehensive comparison, an analysis of the parameters extracted from the transfer curves was conducted. First, the hysteresis window (V_H) was derived from the transfer curve upon the application of a gate voltage (V_g) through a double sweep. Without considering the anticlockwise and clockwise directions that appeared according to the dominant characteristics of polarization and trapping, the absolute value of the V_H was summarized. The V_{th} was extracted using the constant-current method, employing a predetermined drain current (I_{ds}) value that considered the channel's width and length. In addition, the SS was extracted within the range where the I_{ds} value shifted from 10^{-10} to 10^{-9} A. The field-effect mobility (μ_{FE}) was derived using the equation $[I_{ds}/V_g] \times [L/(WC_{ox}V_d)]$, where W and L represent the width and length of the channel, respectively, and C_{ox} denotes the capacitance per area extracted from the metal-PZT-metal capacitor. Finally, the $I_{on/off}$ was calculated based on the maximum I_{on} value at the highest V_g and the minimum I_{off} value at approximately V_g = 0.5 V.

Table 1. Comparison of transfer curve parameters for PZT film-based FeFETs.

Channel	V_H (V)	V_{th} (V)	SS (mV/dec)	μ_{FE} ($cm^2V^{-1}s^{-1}$)	$I_{on/off}$ (A/A)	Ref.
IGZO	4	1.2, 2.6	1250	1.5, 3	1.5×10^6	[40]
IGZO	0.5	-	330	10.23	9.5×10^6	[41]
MoS_2	0.01	0.2~0.4	85.9	10.01	~10^8	[29]
ITO	1.2	-	88	-	10^5	[42]
IGZO	3.5	-	520	14.2	5.6×10^7	[43]
IWO	0.01	0.8	85.75	15.87	1.91×10^8	This work

Figure 3a presents a comparison of the V_H and μ_{FE}. V_H was widely distributed across 0.01–4 V. This indicated that the proposed device exhibited relatively high stability at the interface between the PZT thin film and the channel as well as in terms of the operating characteristics of the channel. In terms of μ_{FE}, the excellent conductivity of IWO resulted in the devices exhibiting a mobility that was 5.86 $cm^2V^{-1}s^{-1}$ higher than similar devices with comparable hysteresis window values. Figure 3b presents a comparison of the SS and on/off ratio. Notably, among devices with excellent SS values ranging between 85 and 90 (mV/dec), the fabricated device demonstrated the best performance characteristics.

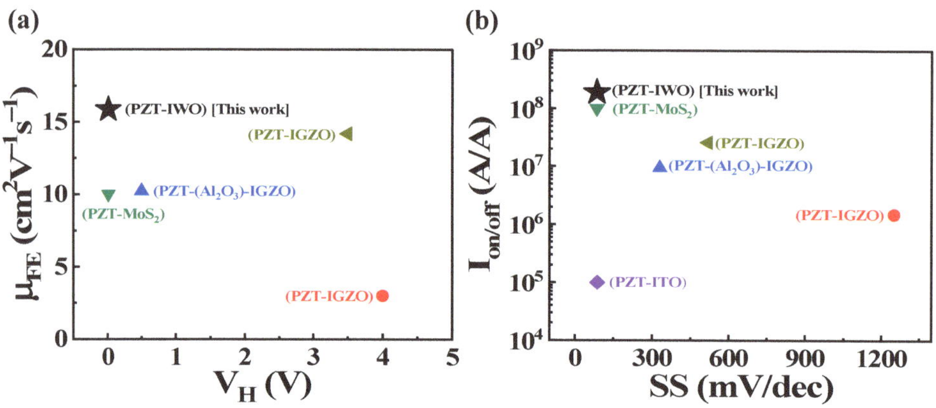

Figure 3. Comparison of electrical characteristics parameters of PZT film-based FeFETs: (**a**) field-effect mobility (μ_{FE}) and hysteresis window (V_H) and (**b**) current on/off ratio ($I_{on/off}$) and subthreshold swing (SS). PZT-IGZO (bronze) [43] (Teng, W.; Bao, S.Y.; Hu, Y.Q.; Deng, X.; Guan, Z.; Chen, B.B.; Zhong, N.; Xiang, P.H. 2024). PZT-(Al_2O_3)-IGZO [41] (Jo, Y.; Lee, J.Y.; Park, E.; Kim, H.S.; Choi, H.J.; Mun, S.; Kim, Y.; Hur, S.; Yoon, J.H.; Jang, J.S.; et al. 2023). PZT-MoS_2 [29] (Zhou, C.; Wang, X.; Raju, S.; Lin, Z.; Villaroman, D.; Huang, B.; Chan, H.L.W.; Chan, M.; Chai, Y. 2015). PZT-IGZO (red) [40] (Besleaga, C.; Radu, R.; Balescu, L.M.; Stancu, V.; Costas, A.; Dumitru, V.; Stan, G.; Pintilie, L. 2019). PZT-ITO [42] (Tue, P.T.; Miyasako, T.; Trinh, B.N.Q.; Li, J.; Tokumitsu, E.; Shimoda, T. 2010).

3.2. DC Bias Test for the Application of PZT-FeFET as Transducer

Variation in the pH value and DC bias voltage can change the potential difference applied to the gate of the transducer. This indicates the existence of a fundamental mechanism that affects the V_{th} of the transfer curve [44–47]. Therefore, prior to integrating the PZT-FeFET into the SG-based EG-ISFET sensor platform, measurements were conducted to assess the performance of the transducer through a DC bias test. Figure 4a illustrates a schematic of the DC bias test. An Agilent 4156B was used to apply the voltage, and the effective voltage applied to the gate was adjusted by varying the magnitude of the DC bias in the power supply. Specifically, to evaluate the comprehensive performance of the device's operating voltage range, we applied a wide range of gate voltages from −3 V to 6 V, which exceeded the gate voltage sweeping range shown in Figure 2a. Figure 4b presents the characteristics of 15 steps of the transfer curve measured by varying the DC bias voltage from 0.7 V to −0.7 V in 0.1 V increments during the voltage sweep. Figure 4d–f present enlarged data of transfer curves at various I_{REF} levels (0.1, 1, and 10 nA). To verify the accuracy and consistency of the I_{REF} in three cases owing to changes in the DC bias voltage, the change in the reference voltage was measured based on the voltage at a DC bias voltage of 0.7 V. Considering the fitted data presented in Figure 4c, the slopes for I_{REF} of 0.1, 1, and 10 nA were 0.92, 0.89, and 0.91, respectively, with good linearity of 99.91%, 99.87%, and 99.75%, respectively. Considering the potential interference from the power supply and external resistance elements between measurements, the results of the DC bias test demonstrated sufficient reliability when compared to the ideal slope value of 1. Thus, the DC bias test confirmed the capability of measuring subtle sensitivity changes owing to potential differences in the manufactured device.

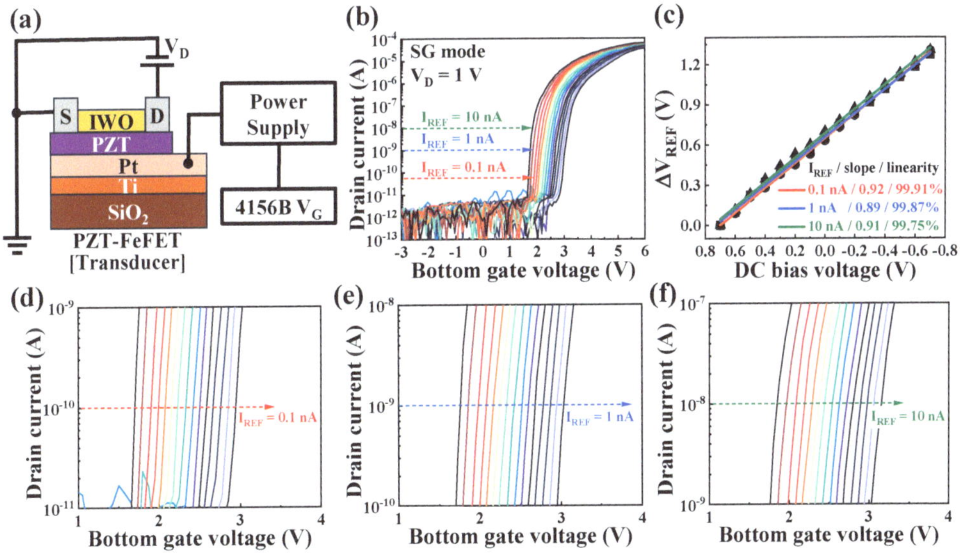

Figure 4. (**a**) Schematic illustrating the DC bias test using PZT-FeFET. (**b**) Transfer curve spanning the gate voltage range from −3 V to 6 V (with DC bias varying from 0.7 V to −0.7 V in 0.1 V steps across 15 increments). (**c**) Variation in reference voltage according to DC bias at I_{REF} levels (0.1, 1, and 10 nA). Enlarged transfer curves corresponding to the I_{REF} levels: (**d**) I_{REF} = 0.1 nA, (**e**) I_{REF} = 1 nA, and (**f**) I_{REF} = 10 nA.

3.3. SG-Based EG-ISFET for pH Sensing

The validation of the suitability of using PZT-FeFET as a transducer underwent several stages, culminating in the sensitivity measurements based on ion concentration variations using an actual sensor platform. Figure 5a shows the proposed sensor platform, which was implemented in an EG-ISFET type structure featuring separate transducer and sensing units. Figure 5b shows the characteristics of the transfer curve of the transducer as the pH value varied from 3 to 10. These characteristics, indicating off and currents below 10 pA and above 10 μA, respectively, within the gate voltage sweep range (−1 V to 3 V), demonstrated sufficient reliability compared to both the operational characteristics of the FeFET and those obtained from the DC bias test. Consequently, the reference voltage shift (ΔV_{REF}) value was fitted based on variations in the pH values, as shown in the enlarged data presented in Figure 5b, as depicted in Figure 5c. The results yielded an R^2 value of 99.79 and a sensitivity of 57.2 mV/pH at an I_{REF} of 1 nA. Compared to the well-known Nernst limit of 59.14 mV/pH, a slight sensitivity difference of 1.94 mV/pH was observed, affirming the excellent sensitivity characteristics of the proposed sensor platform despite its separation into a transducer and sensing unit.

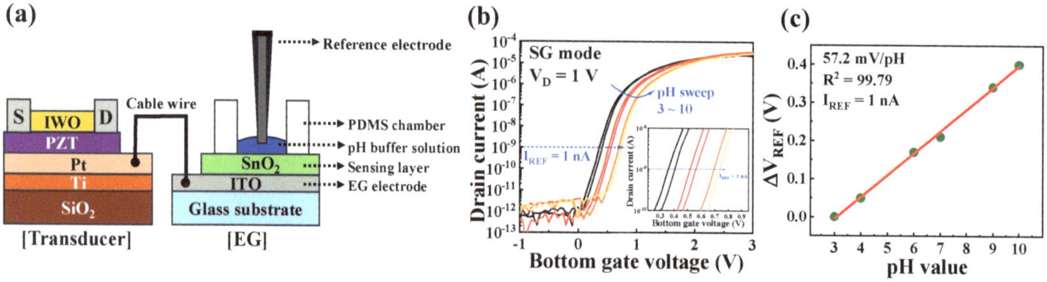

Figure 5. (**a**) Schematic of a PZT single-gate FeFET transducer with SnO_2 EG sensing units. (**b**) V_G–I_D curves of the EG-ISFET for different pH buffer solutions. The inset in (**b**) highlights data pertaining to the I_{REF} region. (**c**) Reference voltage shift (ΔV_{REF}) as a function of pH value.

3.4. Sensitivity Amplification by Self-Resistive Coupling

This study implemented an innovative sensor platform that mimicked the functionality of a co-planar gate-TFT type transducer using a self-resistive coupling approach. Figure 6a presents the equivalent circuit illustrating the resistive coupling phenomenon. Through precise adjustments to the ratio of R_{CG} to R_{SG} using resistors, we evaluated the amplification ratios for three scenarios: 1:1, 3:1, and 5:1. Initially, R_{CG} and R_{SG} were connected in series, and the total resistance ($R_{CG} + R_{SG}$) is denoted as R_T in Equation (1). The operational voltage V_{CG} of the transducer and the surface potential change V_{SG} induced by the sensing unit were capacitively coupled to the V_{FG} of the transducer gate for operation. Consequently, V_{FG} can be expressed as shown in Equation (2), and V_{CG} as shown in Equation (3). This indicated that the potential change (ΔV_{SG}) was amplified by the ratio of R_{CG}/R_{SG}, which changed the CG voltage (ΔV_{CG}), as indicated in Equation (4). Thus, minor potential changes at the sensing gate (SG) can be enhanced through the resistive coupling effect and detected at the control gate (CG). Figure 6b summarizes the ΔV_{REF} across the transfer curve depicted in Figure 6c–e concerning pH values. It was confirmed that ΔV_{REF} was amplified with sensitivities of 58.7, 130.9, and 287.2 mV/pH, corresponding to changes in the R_{CG}:R_{SG} ratio, respectively. Consequently, the sensitivity increased by 2.23, 2.19 times, and up to 4.89 times, respectively, depending on the resistive ratio, thereby demonstrating maximum amplification. Moreover, the linearity of ΔV_{REF} for each resistive coupling ratio exhibited reliability values of 99.18%, 99.74%, and 99.72%. Compared to the sensitivity of 57.2 mV/pH for the single-gate EG-ISEFET shown in Figure 5c, the value of 58.7 mV/pH for the 1:1 ratio confirmed the positive potential of the proposed resistive coupling platform.

$$R_T = R_{CG} + R_{SG} \tag{1}$$

$$V_{FG} = \frac{R_{SG}}{R_T} V_{CG} + \frac{R_{CG}}{R_T} V_{SG} \tag{2}$$

$$V_{CG} = \frac{R_T}{R_{SG}} V_{FG} - \frac{R_{CG}}{R_{SG}} V_{SG} \tag{3}$$

$$\therefore \Delta V_{CG} \propto \frac{R_{CG}}{R_{SG}} \Delta V_{SG} \tag{4}$$

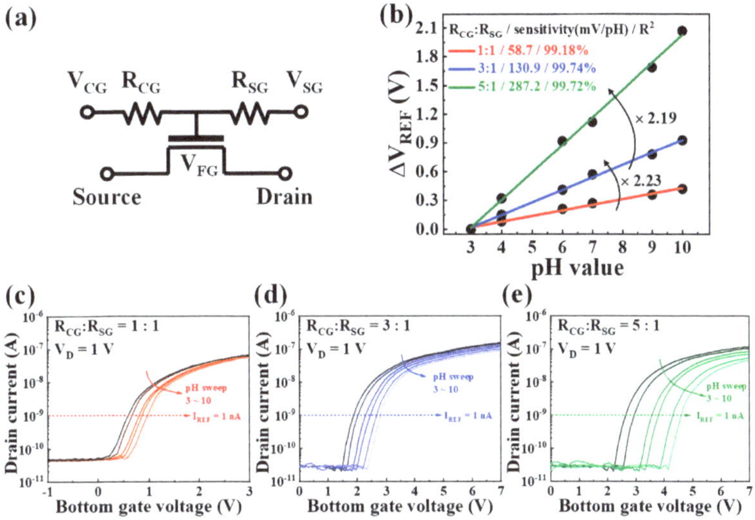

Figure 6. (**a**) Schematic of the equivalent circuit demonstrating the resistive coupling effect. (**b**) Amplification of pH sensitivity corresponding to various R_{CG}:R_{SG} ratios. V_G–I_D curves illustrating the resistive coupling effect for different pH buffer solutions: (**c**) 1:1, (**d**) 3:1, and (**e**) 5:1.

3.5. Non-Ideal Effects of the Proposed Sensor Platform

The integration of the extended-gate (EG) concept into the sensing platform highlights the necessity of verifying the stability and reliability of chemical or physical reactions on the sensing membrane. This verification was achieved through the evaluation of hysteresis and drift effects [48–50]. Hysteresis measurements were conducted 35 times at 2-min intervals along a designated pH loop: 7, 10, 7, 4, and 7, with the hysteresis voltage defined as the difference between the reference voltages at the first and last pH 7 buffer solution. Further, drift measurements were performed during a 10-h exposure to a pH 7 buffer solution. The measured drift rate (R_D) was the variation in the reference voltage per hour caused by ions penetrating the sensing membrane over a sufficient period, based on mechanisms such as hopping or trap-limited transport [51]. Initially, the hysteresis and drift effects of the traditional SG-based EG-ISFET (SG mode) and the proposed resistive coupling EG-ISFET (RC mode) were compared directly. As shown in Figure 7a, the V_H for the two types of platforms was recorded at 5.8 and 6.1 mV, respectively. The difference of 0.3 mV was considered within an acceptable range for sensor platform application. Figure 7c shows the values of drift voltage variation, with the R_D in SG and RC modes exhibiting similar characteristics of 4.24 and 4.48 mV/h, respectively. Subsequently, the hysteresis and drift effects for the RC mode with R_{CG}:R_{SG} ratios of 1:1, 3:1, and 5:1 were compared. As depicted in Figure 7b, the hysteresis voltages were 6.1, 10.1, and 15.8 mV for 1:1, 3:1, and 5:1, respectively, indicating that the values increased according to the resistive ratio. Consequently, as shown in Figure 7d, the drift rates were 4.48, 6.82, 9.19 mV for 1:1, 3:1, and 5:1, respectively, similarly illustrating that the values varied as per the resistive ratio.

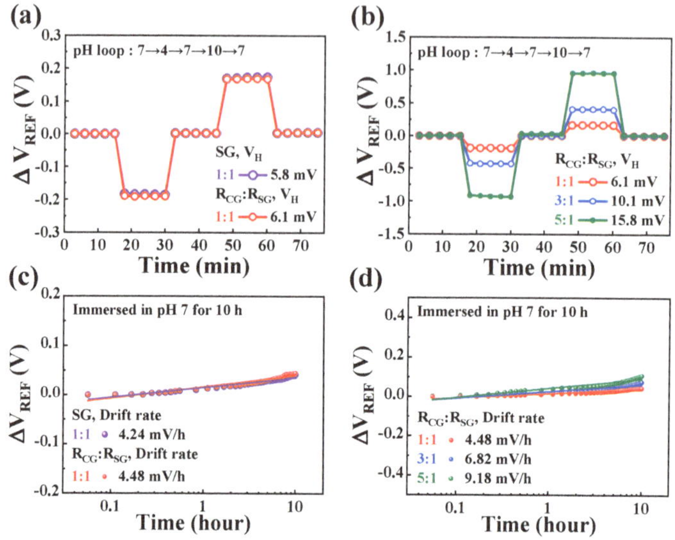

Figure 7. Comparison of non-ideal effects between single-gate (SG) mode and resistivity coupling (RC) mode. Hysteresis and drift behaviors depicted for: (**a**,**c**) SG mode and R_{CG}:R_{SG} = 1:1, (**b**,**d**) various R_{CG}:R_{SG} ratios.

Table 2 summarizes the pH sensing performance of both the SG and RC modes. At an amplification ratio of ×1, both modes exhibited similar values for hysteresis voltage, drift rate, and sensitivity, thereby ensuring the reliability of the proposed sensor platform. In addition, in the RC mode, as the amplification ratio increased from ×1 to ×5, the non-ideal operation parameters (hysteresis voltage, drift rate) and sensitivity increased by 2.59, 2.05, and 4.89 times, respectively. Consequently, the amplification of sensitivity was superior to the increase in non-ideal operations.

Table 2. pH sensing characteristics of SG mode and resistivity coupling mode.

Amplification Ratio (R_{CG}/R_{SG})	Hysteresis Voltage (mV)	Drift Rate (mV/h)	Sensitivity (mV/pH)
×1 (SG mode)	5.8	4.24	57.2
×1 (RC mode)	6.1	4.48	58.7
×3 (RC mode)	10.1	6.82	130.9
×5 (RC mode)	15.8	9.18	287.2

4. Conclusions

This study aimed to realize comprehensive advancements in two key aspects: the fabrication of transducers with excellent electrical properties and the establishment of an innovative sensor platform. Initially, we proposed a pioneering methodology to develop FeFETs via the integration of PZT ferroelectric films with IWO channels. Leveraging cost-effective sol-gel processing, we substantially augmented the electrical characteristics of PZT-IWO FeFET devices. Key performance parameters such as the on/off ratio (85.75) and subthreshold swing (1.91 × 10^8) of the transfer curve were notably enhanced. Moreover, our successful annealing process effectively mitigated the hysteresis effects during double sweeps, thereby ensuring incomparable electrical stability. To apply the proposed FeFET to a sensor platform, we conducted DC bias tests to ascertain reliability, consequently achieving an exemplary sensitivity of 57.2 mV/pH in the EG-ISFET structure. This indicated an improvement over conventional ISFET methodologies. Moreover, by harnessing a single gate EG-ISFET, introducing an innovative sensor architecture based on resistive coupling for pH detection, and fine-tuning the resistive ratio between the R_{CG} and the R_{SG}, a remarkable sensitivity enhancement of 287.2 mV/pH was achieved. Consequently, this study demonstrates significant advancements in FeFET technology and offers promising applications across different domains. By integrating FeFET technology into a self-resistive coupling sensing platform, precise pH measurements can be enabled. Overcoming selectivity potential issues, such as the cross-sensitivity of the SnO_2 sensing membrane to other ions, would provide solutions to various challenges in healthcare and environmental monitoring. Additionally, for the successful commercialization and practical implementation of this integrated approach, future research focusing on structural improvements such as miniaturization and packaging is necessary. In conclusion, the results of this study, based on the FeFET technology in the enhanced platform, have the potential to establish it as a crucial component in advanced ion detection systems.

Author Contributions: D.-G.M.: conceptualization, formal analysis, methodology, investigation, data curation, visualization, resources, and writing—original draft. S.-M.O.: investigation, resources. J.J.: investigation, resources, writing—review and editing. W.-J.C.: conceptualization, methodology, investigation, resources, formal analysis, funding acquisition, supervision, validation, and writing—review and editing. All authors have read and agreed to the published version of the manuscript.

Funding: This work was supported by the Korea Institute for Advancement of Technology (KIAT) grant funded by the Korean government (MOTIE) (Grant No. P0020967, The Competency Development Program for Industry Specialist).

Institutional Review Board Statement: Not applicable.

Informed Consent Statement: Not applicable.

Data Availability Statement: The original contributions presented in the study are included in the article, further inquiries can be directed to the corresponding author.

Acknowledgments: The present research has been conducted by the Research Grant of Kwangwoon University in 2024 and the Excellent Research Support Project of Kwangwoon University in 2024.

Conflicts of Interest: The authors declare no conflicts of interest.

References

1. Capua, L.; Sprunger, Y.; Elettro, H.; Risch, F.; Grammoustianou, A.; Midahuen, R.; Ernst, T.; Barraud, S.; Gill, R.; Ionescu, A.M. Label-Free C-Reactive Protein Si Nanowire FET Sensor Arrays with Super-Nernstian Back-Gate Operation. *IEEE Trans. Electron. Devices* **2022**, *69*, 2159–2165. [CrossRef]
2. Dolai, S.; Tabib-Azar, M. *Zika virus* field effect transistor. *IEEE Sens. J.* **2021**, *21*, 4122–4128. [CrossRef]
3. Singh, A.K.; Pandey, A.; Chakrabarti, P. Fabrication, characterization, and application of CuO nano wires as electrode for ammonia sensing in aqueous environment using extended gate-FET. *IEEE Sens. J.* **2021**, *21*, 5779–5786. [CrossRef]
4. Bausells, J.; Carrabina, J.; Errachid, A.; Merlos, A. Ion-sensitive field-effect transistors fabricated in a commercial CMOS technology. *Sens. Actuators B* **1999**, *57*, 56–62. [CrossRef]
5. Olthuis, W.; Robben, M.A.M.; Bergveld, P.; Bos, M.; van der Linden, W.E. pH sensor properties of electrochemically grown iridium oxide. *Sens. Actuators B* **1990**, *2*, 247–256. [CrossRef]
6. Moser, N.; Lande, T.S.; Toumazou, C.; Georgiou, P. ISFETs in CMOS and emergent trends in instrumentation: A review. *IEEE Sens. J.* **2016**, *16*, 6496–6514. [CrossRef]
7. Huang, X.; Yu, H.; Liu, X.; Jiang, Y.; Yan, M.; Wu, D.; Dual-Mode, A. A Dual-Mode Large-arrayed CMOS ISFET sensor for accurate and high-throughput PH sensing in biomedical diagnosis. *IEEE Trans. Biomed. Eng.* **2015**, *62*, 2224–2233. [CrossRef]
8. Pijanowska, D.G.; Torbicz, W. PH-ISFET based urea biosensor. *Sens. Actuators B* **1997**, *44*, 370–376. [CrossRef]
9. Chodavarapu, V.P.; Titus, A.H.; Cartwright, A.N. CMOS ISFET microsystem for biomedical applications. In Proceedings of the 2005 IEEE International Reliability Physics Symposium, Irvine, CA, USA, 30 October 2005; IEEE Publications: Piscataway, NJ, USA, 2005; Volume 2005, p. 4.
10. Rezaee, A.; Carrabina, J. Dual-gate organic thin-film transistor and multiplexer chips for the next generation of flexible EG-ISFET sensor chips. *Sensors* **2023**, *23*, 6577. [CrossRef]
11. Pan, T.M.; Lin, L.A.; Ding, H.Y.; Her, J.L.; Pang, S.T. A simple and highly sensitive flexible sensor with extended-gate field-effect transistor for epinephrine detection utilizing InZnSnO sensing films. *Talanta* **2024**, *275*, 126178. [CrossRef]

12. Chen, C.H.; Liu, S.B.; Chang, S.P. Performance Improvement of a $ZnGa_2O_4$ Extended-Gate Field-Effect Transistor pH Sensor. *ACS Omega* **2024**, *9*, 15304–15310. [CrossRef] [PubMed]
13. Gubanova, O.; Poletaev, A.; Komarova, N.; Grudtsov, V.; Ryazantsev, D.; Shustinskiy, M.; Shibalov, M.; Kuznetsov, A. A novel extended gate ISFET design for biosensing application compatible with standard CMOS. *Mater. Sci. Semicond. Process* **2024**, *177*, 108387. [CrossRef]
14. Harb, A.; Istanbullu, M. High sensitive PH sensor with graphene based dual-gate ISFET. *J. Nano. Sci. Adv. Mater.* **2023**, *2*, 19–24.
15. Zain, A.S.M.; Dinar, A.M.; Salehuddin, F.; Hazura, H.; Hanim, A.R.; Idris, S.K.; Hamid, A.M.A. Beyond Nernst sensitivity of ion sensitive field effect transistor based on ultra-thin body box FDSOI. *J. Phys. Conf. Ser.* **2020**, *1502*, 012048. [CrossRef]
16. Kim, Y.-U.; Cho, W.-J. Self-sensitivity amplifiable dual-gate ion-sensitive field-effect transistor based on a high-k engineered dielectric layer. *Jpn. J. Appl. Phys.* **2023**, *62*, SC1056. [CrossRef]
17. Hyun, T.H.; Cho, W.J. High-performance FET-based dopamine-sensitive biosensor platform based on SOI substrate. *Biosensors* **2023**, *13*, 516. [CrossRef]
18. Tan, G.; Kweon, S.H.; Kanno, I. Piezoelectric properties of epitaxial $Pb(Zr Ti)O_3$ thin films grown on Si substrates by the sol–gel method. *Thin Solid. Film.* **2023**, *764*, 139612. [CrossRef]
19. Zhang, Q.; Whatmore, R.W. Sol-gel PZT and Mn-doped PZT thin films for pyroelectric applications. *J. Phys. D Appl. Phys.* **2001**, *34*, 2296–2301. [CrossRef]
20. Kweon, S.H.; Kanayama, Y.; Tan, G.; Koganezawa, T.; Kanno, I. In-situ study on piezoelectric responses of sol-gel derived epitaxial $Pb[Zr,Ti]O_3$ thin films on Si substrate. *J. Eur. Ceram.* **2024**, *44*, 3887–3894. [CrossRef]
21. Weng, L.; Bao, X.; Sagoe-Crentsil, K. Effect of acetylacetone on the preparation of PZT materials in sol-gel processing. *Mater. Sci. Eng. B* **2002**, *96*, 307–312. [CrossRef]
22. Wu, Z.; Jie, W.; Yang, Z.; Hao, J. Hybrid heterostructures and devices based on two-dimensional layers and wide bandgap materials. *Mater. Today Nano* **2020**, *12*, 100092. [CrossRef]
23. Pinto, R.M.R.; Gund, V.; Dias, R.A.; Nagaraja, K.K.; Vinayakumar, K.B. CMOS-integrated aluminum nitride MEMS: A review. *J. Microelectromech. Syst.* **2022**, *31*, 500–523. [CrossRef]
24. Pintilie, I.; Pasuk, I.; Ibanescu, G.A.; Negrea, R.; Chirila, C.; Vasile, E.; Pintilie, L. The impact of the $Pb(Zr,Ti)O_3$-ZnO interface quality on the hysteretic properties of a metal-ferroelectric-semiconductor structure. *J. Appl. Phys.* **2012**, *112*, 104103. [CrossRef]
25. Damjanovic, D. Contributions to the piezoelectric effect in ferroelectric single crystals and ceramics. *J. Am. Ceram. Soc.* **2005**, *88*, 2663–2676. [CrossRef]
26. Chen, H.M.; Lee, J.Y.M. Electron trapping process in ferroelectric lead–zirconate–titanate thin-film capacitors. *Appl. Phys. Lett.* **1998**, *73*, 309–311. [CrossRef]
27. Afanasjev, V.P.; Petrov, A.A.; Pronin, I.P.; Tarakanov, E.A.; Kaptelov, E.J.; Graul, J. Polarization and self-polarization in thin $PbZr_{1-x}Ti_xO_3$ (PZT) films. *J. Phys. Condens. Matter.* **2001**, *13*, 8755–8763. [CrossRef]
28. Chen, J.; Dun, G.; Hu, J.; Lin, Z.; Wang, Y.; Lu, T.; Li, P.; Wei, T.; Zhu, J.; Wang, J.; et al. Polarized tunneling transistor for ultrafast memory. *ACS Nano* **2023**, *17*, 12374–12382. [CrossRef] [PubMed]
29. Zhou, C.; Wang, X.; Raju, S.; Lin, Z.; Villaroman, D.; Huang, B.; Chan, H.L.W.; Chan, M.; Chai, Y. Low voltage and high ON/OFF ratio field-effect transistors based on CVD MoS_2 and ultra high-k gate dielectric PZT. *Nanoscale* **2015**, *7*, 8695–8700. [CrossRef] [PubMed]
30. Cho, S.K.; Cho, W.J. Ultra-high sensitivity pH-sensors using silicon nanowire channel dual-gate field-effect transistors fabricated by electrospun polyvinylpyrrolidone nanofibers pattern template transfer. *Sens. Actuators B* **2021**, *326*, 128835. [CrossRef]
31. Parizi, K.B.; Xu, X.; Pal, A.; Hu, X.; Wong, H.S.P. ISFET pH sensitivity: Counter-ions play a key role. *Sci. Rep.* **2017**, *7*, 41305. [CrossRef]
32. Garrido, J.A.; Härtl, A.; Kuch, S.; Stutzmann, M.; Williams, O.A.; Jackmann, R.B. pH sensors based on hydrogenated diamond surfaces. *Appl. Phys. Lett.* **2005**, *86*, 7. [CrossRef]
33. Cho, S.K.; Cho, W.J. High-sensitivity pH sensor based on coplanar gate AlGaN/GaN metal-oxide-semiconductor high electron mobility transistor. *Chemosensors* **2021**, *9*, 42. [CrossRef]
34. Chou, J.C.; Wang, Y.F. Preparation and study on the drift and hysteresis properties of the tin oxide gate ISFET by the sol-gel method. *Sens. Actuators B* **2002**, *86*, 58–62. [CrossRef]
35. Hossain, S.; Rahman, M.T. Micro-heater embedded ISFET pH sensor with high-k gate dielectrics for enhanced sensitivity. *Eng. Res. Express* **2023**, *5*, 035068. [CrossRef]
36. Xu, J.; Jiang, S.Y.; Zhang, M.; Zhu, H.; Chen, L.; Sun, Q.Q.; Zhang, D.W. Ferroelectric $HfZrO_x$-based MoS_2 negative capacitance transistor with ITO capping layers for steep-slope device application. *Appl. Phys. Lett.* **2018**, *112*, 103104. [CrossRef]
37. Islam Sakib, F.; Hasan, M.A.; Mohona, M.D.; Hossain, M. Negative capacitance dual-gated ISFETs as ultra-sensitive pH sensors. *ACS Omega* **2023**, *8*, 48756–48763. [CrossRef]
38. Xue, H.; Peng, Y.; Jing, Q.; Zhou, J.; Han, G.; Fu, W. Sensing with extended gate negative capacitance ferroelectric field-effect transistors. *Chip* **2024**, *3*, 100074. [CrossRef]
39. Chi, L.L.; Chou, J.C.; Chung, W.Y.; Sun, T.P.; Hsiung, S.K. Study on extended gate field effect transistor with tin oxide sensing membrane. *Mater. Chem. Phys.* **2000**, *63*, 19–23. [CrossRef]
40. Besleaga, C.; Radu, R.; Balescu, L.M.; Stancu, V.; Costas, A.; Dumitru, V.; Stan, G.; Pintilie, L. Ferroelectric field effect transistors based on PZT and IGZO. *IEEE J. Electron. Devices Soc.* **2019**, *7*, 268–275. [CrossRef]

41. Jo, Y.; Lee, J.Y.; Park, E.; Kim, H.S.; Choi, H.J.; Mun, S.; Kim, Y.; Hur, S.; Yoon, J.H.; Jang, J.S.; et al. Epitaxial PZT film-based ferroelectric field-effect transistors for artificial synapse. *ACS Appl. Electron. Mater.* **2023**, *5*, 4549–4555. [CrossRef]
42. Tue, P.T.; Miyasako, T.; Trinh, B.N.Q.; Li, J.; Tokumitsu, E.; Shimoda, T. Optimization of Pt and PZT films for ferroelectric-gate thin film transistors. *Ferroelectrics* **2010**, *405*, 281–291. [CrossRef]
43. Teng, W.; Bao, S.Y.; Hu, Y.Q.; Deng, X.; Guan, Z.; Chen, B.B.; Zhong, N.; Xiang, P.H. Ferroelectric controlled interfacial effect on the electronic properties of PZT gated IGZO channel thin-film transistors. *ACS Appl. Electron. Mater.* **2024**, *6*, 1063–1070. [CrossRef]
44. Bakker, E.; Pretsch, E.; Bühlmann, P. Selectivity of potentiometric ion sensors. *Anal. Chem.* **2000**, *72*, 1127–1133. [CrossRef]
45. Tran, T.N.T.; Qiu, S.; Chung, H.-J. Potassium ion selective electrode using polyaniline and matrix-supported ion-selective PVC membrane. *IEEE Sens. J.* **2018**, *18*, 9081–9087. [CrossRef]
46. Chen, X.-W.; Huang, S.-R.; Huang, N.-T. Dual Ion-Selective Membrane Deposited Ion-Sensitive Field-Effect Transistor Integrating a Whole Blood Processing Microchamber for In Situ Blood Ion Testing. *ACS Sens.* **2023**, *8*, 904–913. [CrossRef] [PubMed]
47. van den Berg, A.; Grisel, A.; Verney-Norberg, E. An ISFET-based calcium sensor using a photopolymerized polysiloxane membrane. *Sens. Actuators B* **1991**, *4*, 235–238. [CrossRef]
48. Chou, J.C.; Weng, C.Y. Sensitivity and hysteresis effect in Al_2O_3 gate pH-ISFET. *Mater. Chem. Phys.* **2001**, *71*, 120–124. [CrossRef]
49. Jamasb, S.; Collins, S.; Smith, R.L. A physical model for drift in pH ISFETs. *Sens. Actuators B* **1998**, *49*, 146–155. [CrossRef]
50. Chiang, J.L.; Jan, S.S.; Chou, J.C.; Chen, Y.C. Study on the temperature effect, hysteresis and drift of PH-ISFET devices based on amorphous tungsten oxide. *Sens. Actuators B* **2001**, *76*, 624–628. [CrossRef]
51. Bousse, L.; Bergveld, P. The role of buried OH sites in the response mechanism of inorganic-gate pH-sensitive ISFETs. *Sens. Actuators* **1984**, *6*, 65–78. [CrossRef]

chemosensors

MDPI

Article

Variability Assessment of the Performance of MoS_2-Based BioFETs

Juan Cuesta-Lopez *, Alejandro Toral-Lopez, Enrique G. Marin, Francisco G. Ruiz, Francisco Pasadas, Alberto Medina-Rull and Andres Godoy *

Departamento de Electrónica y Tecnología de Computadores, Facultad de Ciencias, Universidad de Granada, 18071 Granada, Spain

* Correspondence: jcuesta@ugr.es (J.C.-L.); agodoy@ugr.es (A.G.)

Abstract: Two-dimensional material (2DM)-based Field-Effect Transistors (FETs) have been postulated as a solid alternative for biosensing applications thanks to: (i) the possibility to enable chemical sensitivity by functionalization, (ii) an atomically thin active area which guarantees optimal electrostatic coupling between the sensing layer and the electronic active region, and (iii) their compatibility with large scale fabrication techniques. Although 2DM-based BioFETs have demonstrated notable sensing capabilities, other relevant aspects, such as the yield or device-to-device variability, will demand further evaluation in order to move them from lab-to-fab applications. Here, we focus on the latter aspect by analyzing the performance of MoS_2-based BioFETs for the detection of DNA molecules. In particular, we explore the impact of the randomized location and activation of the receptor molecules at the sensing interface on the device response. Several sensing interface configurations are implemented, so as to evaluate the sensitivity dependence on device-to-device variability.

Keywords: BioFET; MoS_2; DNA; device-to-device variability; numerical modeling

Citation: Cuesta-Lopez, J.; Toral-Lopez, A.; Marin, E.G.; Ruiz, F.G.; Pasadas, F.; Medina-Rull, A.; Godoy, A. Variability Assessment of the Performance of MoS_2-Based BioFETs. *Chemosensors* **2023**, *11*, 57. https://doi.org/10.3390/chemosensors11010057

Academic Editor: Tung-Ming Pan

Received: 23 November 2022
Revised: 23 December 2022
Accepted: 2 January 2023
Published: 10 January 2023

1. Introduction

In the context of the Internet of Things (IoT), the possibility to continuously monitor information from the environment and translate it into electrical signals, that can be easily processed and stored, has become a present and ubiquitous reality. Among the myriad of signals to be sensed, those connected with human healthcare, or in a more general perspective with bio-activity, occupy a prime position [1]. There exists, indeed, a relentless demand for the design and optimization of accurate and robust electronic bio-sensors able to gather and process information of varied sources in real-time.

In this scenario, Biomolecular Field-Effect Transistors (BioFETs) are expected to play a key role, prevailing over other sensing alternatives thanks to: (i) a quick and label-free operation [2], (ii) high sensitivity and selectivity [3,4], and (iii) the capability to measure in vivo [5] and even real-time [6] signals. Although the first demonstration of these devices dates back to the 1970s [7], BioFETs implementation over the last decades has been hindered by a complex and unsolved trade-off between increased sensitivity and compatibility with large scale fabrication. The sensitivity was, indeed, boosted in the beginnings of the century thanks to the use of nanowire structures [8–10], due to their optimized surface-to-volume ratio. However, their widespread deployment was compromised by a difficult mass-scale production and integration with planar technology [11]. The appearance of two-dimensional materials (2DMs) around two decades ago opened new possibilities to reach the best balance between the two facets [12,13]: 2DMs represent the optimal surface-to-volume ratio structural arrangement, while their integration with current thin-film fabrication processes is feasible [14,15]. Moreover, a large variety of 2DMs can be synthesized [16], showing insulating, metallic or semiconducting behavior while the weak inter-layer forces that hold them together can be harnessed to seamlessly stack them in so-called van-der-Waals heterostructures, opening a wide variety of possibilities to be explored [17].

In this roadmap for the development of 2DM-based BioFETs, computational tools are called to play a key role to: (i) better understand the physical mechanisms that govern their behavior, and (ii) serve as a guide for faster device optimization enabling the lab-to-fab gap closure. Several approaches have been exploited for the latter which are based either in TCAD [18–20] or ad-hoc simulation tools [21–25], that are however not able to deal with essential aspects of mass-scale production such as reproducibility and device-to-device variability. This essential step, mandatory to increase the technology readiness level and prepare the leap from research to manufacture, demands an early performance evaluation with computational tools capable to reproduce heterogeneous device architectures, scenarios and operations. To that end, we present an improved numerical tool for 2DM-based BioFETs [26] that enables a comprehensive evaluation of the sensor variability through the impact in the device response of an arbitrary receptor distribution along a sensing interface. The capabilities of the proposed approach will be exemplified with the study of MoS_2-based BioFETs aiming for the detection of DNA molecules. Different receptor distributions and activation configurations will be considered to assess the sensor electrical read-out robustness against variability. Devices employed for this study are assumed to be identical in their I–V characteristics, and the only source of variability is originated by the receptor molecules.

The rest of this work is organized as follows: Section 2 describes the simulation model, while the details of the simulated device and the results obtained for different receptor configurations are collected in Section 3. Finally, Section 4 sums up the main conclusions.

2. Materials and Methods

The implemented numerical tool self-consistently solves the Poisson and continuity equations in the sensor and its environment. The electrolyte corresponds to a Phosphate Buffer Saline (PBS) solution, whose regulatory chain of reactions—dependent on ionic strength and temperature—are described in [26], while the ion concentration is determined by the modified Boltzmann equation, including steric effects due to the finite size of the ions. This numerical tool has been previously validated with experimental results at the levels of the semiconductor device [27,28], the sensing interface [29], and the complete BioFET device [26]. DNA is selected as the analyte of interest due to their relevance in numerous biological processes, although this study could be extended to other biomolecules [29]. The DNA receptor and target sequences are modeled taking into consideration the individual shape and charge distribution of the DNA chains (Figure 1), what is by itself a notable step-forward in the modeling of DNA sensing, substantially upgrading conventional charge-box models [18]. In more detail, the receptor molecules (attached to the device sensing surface) are single-stranded DNA (ssDNA) molecules, whose complementary nucleotide sequence (target molecule) is to be detected, as in conventional DNA sensors. These molecules are assumed to be disposed along the device surface with a vertical orientation. Other molecules orientations with respect to the sensing surface, i.e., with a certain tilting angle, could be considered, although it has been demonstrated in earlier studies that they do not result in significant changes in the sensing layer surface potential [30]. Once the complementary ssDNA is captured, it forms a double-stranded (dsDNA) molecule with the receptor attached to the sensing interface (Figure 1a). The actual charge and size of the receptor and target molecules depend on the number of nucleotides in the sequences. For ssDNA molecules, each nucleotide contributes to $-1e$ charge (where e is the electron elementary charge) and has a length of $\sim$0.34 nm. For dsDNA molecules, the charge per nucleotide is $-2e$ and the length is kept unaltered [31] (Figure 1c). It is worth it to note that the ssDNA is flexible, while the dsDNA has a compact rigid structure, a feature also included in the model, where the electrolyte ions are allowed to permeate and mix with ssDNA molecules while they are kept out of the spatial region occupied by dsDNA molecules (Figure 1b).

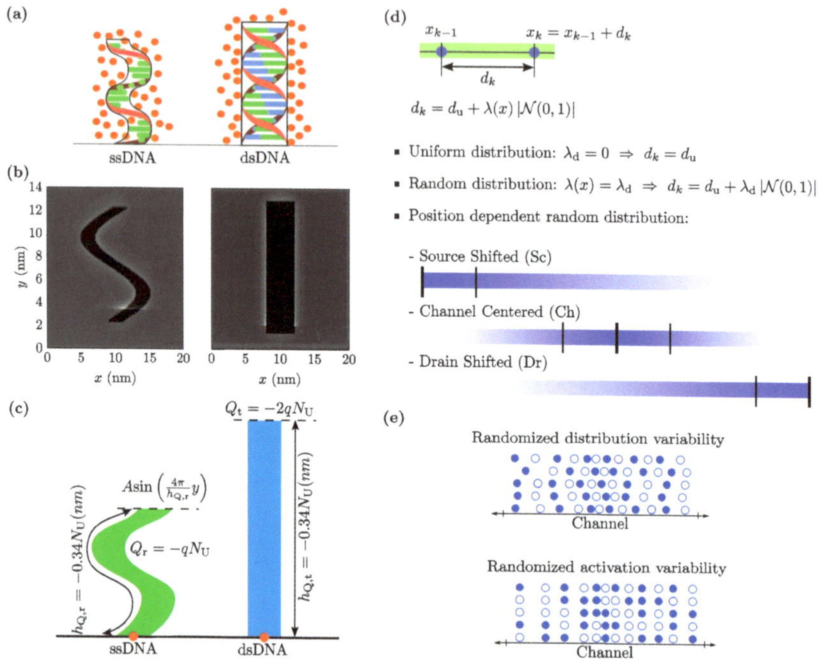

Figure 1. (**a**) Models for the ssDNA (receptor) and dsDNA (receptor-target complex) molecules. ssDNA model considers a sinusoid-shaped charged region where ions are allowed to enter, while the dsDNA model considers a box-shaped region where ions cannot penetrate. (**b**) Examples of simulated ion distributions regarding idle and activated DNA receptors. (**c**) Mathematical modeling of the length, shape and charge of ssDNA and dsDNA receptors. In both cases, the length of the molecule h_Q and its charge Q is scaled with the number of nucleotides N_U. (**d**) Location of receptors, controlled with the distance between them (for the k-th receptor, $d_k = x_k - x_{k-1}$). That distance is considered constant (uniform distribution) or defined by a half-normal distribution with a position dependent variance. This latter enables the definition of densely occupied regions at the sensing interface, allowing one to choose between the three distributions indicated by the blue bars. (**e**) Example of 5 channel-centered (Ch) distributions with $P = 50\%$, where filled and empty circles represent activated and deactivated receptors, respectively. Randomized distribution variability: considering a fixed activated receptor configuration (one occupied—one empty), several distributions are represented. Randomized activation variability: assuming a fixed receptor distribution, five activated receptor configurations are depicted.

The impact of the receptor allocation in the sensor variability is evaluated considering different spatial distributions and activations of the ssDNA chains according to two scenarios (Figure 1d): (i) a uniform spacing between receptors (d_u) along the sensing layer length (L_{Chn}), which is used as a reference; and (ii) a random distribution where the distance between nearby receptors (d_k) pivots around the uniform spacing as $d_k = d_u + \lambda(x)|\mathcal{N}(0,1)|$, where $|\mathcal{N}(0,1)|$ is a half-normal distribution with unity variance and $\lambda(x) = \lambda_d\left(1 - \exp\left[-\frac{1}{2}\left(\frac{x-\mu}{\alpha L_{Chn}}\right)^2\right]\right)$ is a factor determining the amplitude of this random position correction to the uniform spacing. Here, μ and α control the amplitude and variance of the random correction, which eventually resembles a Gaussian distribution.

Once the position of the receptors is determined, their state (idle or activated) is set according to an activation probability ($0\% \leq P \leq 100\%$). Thus, for a given number of receptor molecules (N) in the sensing layer, the number of activated receptors is given by $N_{act} = PN$. The assignment of activated receptors for a given P value can be set:

(i) uniformly, following a certain spatial frequency, or (ii) randomly, being the activation of each receptor equally probable. In this way, we can extend the study of different receptor spatial distributions (which roots in distinct fabricated devices) also to varying receptor activation configurations (which correspond to different operation conditions in a single device). Examples of both degrees of freedom are shown in Figure 1e.

3. Results

In order to exemplify the capabilities of the proposed approach to evaluate the impact of the variability of the sensing interface in the response of 2DM-based BioFETs, we have focused on a MoS_2-based DNA-sensor. More specifically, the structure is defined by a MoS_2 monolayer (0.65 nm-thick and 50 cm^2/Vs carrier mobility) with length $L_{semic} = 400$ nm, sandwiched between a $t_{box} = 20$ nm-thick SiO_2 layer, acting as a substrate, and a $t_{ox} = 10$ nm-thick SiO_2 top-oxide layer. Above this latter one, a sensing layer of $L_{Chn} = 300$ nm length is placed. For this channel length, and assuming low contact resistance, we can consider that the device performance is controlled by the channel properties rather than any contact effects, as probed by [32,33]. A number of $N = 12$ ssDNA receptors are allocated along the sensing interface, either uniformly or following the aforementioned random spatial distribution, with $\lambda_d = 0.5$ nm, $\mu = 50, 200, 350$ nm, and $\alpha = 1/8$. The electrolyte is defined by a PBS solution, with composition [NaCl] = 140 mM, [KCl] = 2.7 mM and [NaH_3PO_4] = 10 mM, regulating a pH of 7.4. An schematic depiction of the structure under study is shown in Figure 2.

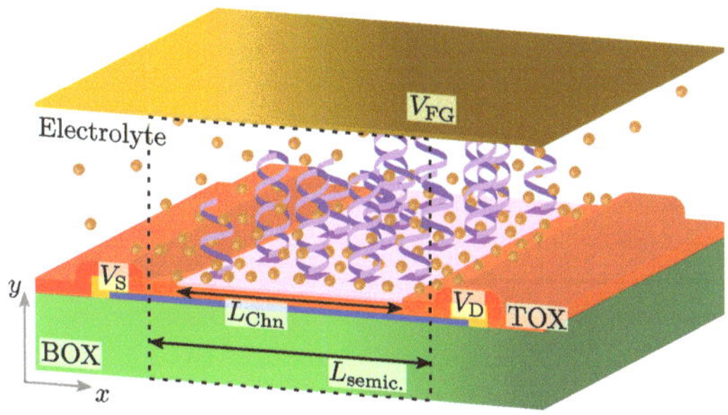

Figure 2. MoS_2-based DNA BioFET outline. The dashed line determines the simulated area.

The electrical operation of the device is studied for a drain-to-source voltage of $V_{DS} = 0.1$ V and a varying bias of the reference electrode immersed in the electrolyte (V_{FG}). From the drain-to-source current (I_{DS}), it is possible to define the device sensitivity (S) at a certain activation probability (P), as:

$$S(P) = I_{DS(0\%)} - I_{DS(P)}. \quad (1)$$

where $I_{DS(P)}$ stands for the output current calculated with an activation probability P. The $I_{DS} - V_{FG}$ and $S - V_{FG}$ electrical read-outs as a function of P are depicted in Figure 3 for the scenario where the receptors are uniformly distributed, for two different PBS concentrations. The increase in P results in a shift of the transfer curves (Figure 3b) towards positive gate biases, due to the electrostatic repulsion that the negatively charged dsDNA molecules produce in the 2DM electron charge. When the ionic strength of the electrolyte is reduced, i.e., when we move from a 1×PBS electrolyte to a 0.1×PBS electrolyte, the Debye length of the electrolyte ions increases from $\lambda_D = 0.74$ nm to $\lambda_D = 2.34$ nm. Thus, the capability of the electrolyte to screen the dsDNA charges is reduced. As a consequence, the impact of the charged dsDNA molecules in the 2DM channel is enhanced. These effects can be also appreciated in the sensitivity curves (Figure 3c), that notably saturate for high V_{FG} values, when the change in the carrier concentration in the channel (and therefore in I_{DS}) is no more significantly impacted by the dsDNA molecules. Note that sensitivity is usually referred in the literature as a value relative to the current in the absence of any binding event of the target antigen, i.e., $S_r(P) = \left[I_{DS(0\%)} - I_{DS(P)}\right] / I_{DS(0\%)}$ [34,35] (see Figure 3d). When defined in this way, $S(P)$ maximum values correspond to the operation of the BioFET in the sub-threshold regime, where the sensor current is very low ($I_{ds} \approx$ pA/μm, see Figure 3a). The BioFET variability response in this regime, in a realistic experimental scenario, would be blurred by the noise background or limited by the lower limit sensitivity of the employed measuring equipment. For this reason, in order to investigate the random activation and distribution of single receptors and their impact in the variability, the absolute difference of the sensor output currents was considered as the sensor sensitivity definition (see Figure 3c).

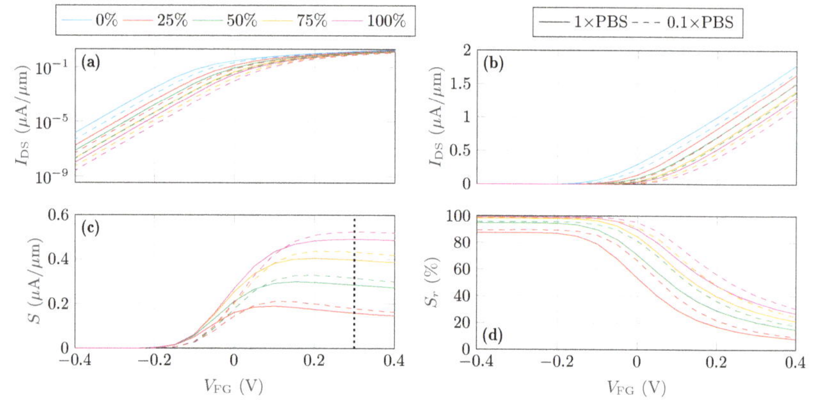

Figure 3. (**a**) Current density $I_{DS} - V_{FG}$ shown in logarithmic scale. (**b**) Current density $I_{DS} - V_{FG}$ shown in linear scale. (**c**) Absolute current sensitivity $S - V_{FG}$. (**d**) Relative current sensitivity $S_r - V_{FG}$. All results are calculated for different values of P with a uniform spatial distribution of receptors. Solid and dashed lines correspond to the 1×PBS electrolyte and 0.1×PBS electrolyte scenarios, respectively. The dotted line in (c) indicates the value $V_{FG} = 0.3$ V for which $S(P = 100\%)$ is maximum.

In order to compare the sensitivity in different distribution and activation scenarios, we set an operating bias point at $V_{FG} = 0.3$ V, which corresponds to the value where $S(P = 100\%)$ is maximum. At this operating point, we have analyzed two sources of device-to-device variability: (i) the receptors spatial distribution, that accounts for different receptor location configurations, and (ii) the receptors activation distribution, that considers different configurations to activate the receptor molecules for a fixed P value. Figure 1e illustrates both scenarios.

3.1. Randomized Distribution Variability

Three main scenarios were analysed for the randomized location of the receptors in the sensing layer: channel-centered (Ch), source-shifted (Sc) and drain-shifted (Dr). Ten different receptor distributions are generated for each of them. The three cases resulted in a similar behavior in terms of variability (see Appendix A Figure A1). Thus, for the sake of clarity, we have mostly focused on one of them: the channel centered. From the set of S values obtained in the different random distributions, we calculated the standard deviation σ_S and mean value $\langle S \rangle$ that characterize the set. Figure 4a shows $\langle S \rangle$ as a function of the activation probability (solid lines), with the different S values marked with symbols and σ_S plotted as a shadowed region around $\langle S \rangle$. The results are compared with those of a uniform distribution of receptors for two PBS concentrations (dashed lines). The differences between the uniform distribution and the mean behavior of the randomized distributions is more acute as P increases, with the randomized distribution outperforming the uniform distribution for all P values. Although the number of activated receptors is the same, when the receptors are randomly distributed with a higher concentration at the centre of the channel, they are closer to each other and the ions diluted in the electrolyte have a reduced screening impact. Moreover, σ_S increases with P due to the higher effective charge of the activated receptors. In fact, their remarkable influence in the drain current enhance the effect of the distribution randomness. As for the PBS, a lower value results in a higher Debye length, λ_D, which translates into a lower electrolyte screening and a higher S for all cases, but at the cost of a higher variability, σ_S.

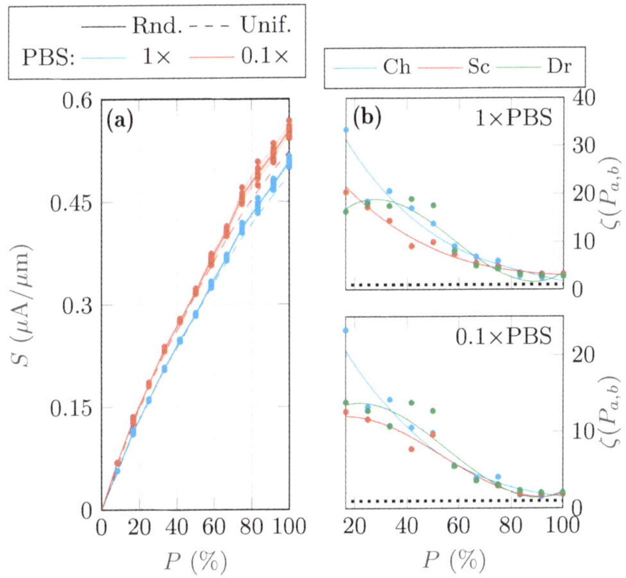

Figure 4. (**a**) Device sensitivity as a function of the activation percentage. Ten random distributions following a channel centered case are considered, fixing the configuration of activated receptors for each P value. Markers account for simulated data, while shaded areas depict σ_S and lines correspond to $\langle S \rangle$. Both 1×PBS (blue) and 0.1×PBS (red) concentrations were simulated. Results obtained for the uniform distribution are depicted by dashed lines. (**b**) $\zeta(P)$ ratio. Markers represent simulated data, while colours refer to channel-centered, source-shifted and drain-shifted distributions. Both PBS concentrations are separated in two graphics for the sake of clarity. The dotted lines indicate $\zeta = 1$. Fitting parameters are summarized in Appendix A Table A1.

One key criterion to characterize the role of the variability in the sensor response is the minimum change in the activation of the receptors that can be detected without being blurred by the variability of the response, i.e., by σ_S. If the difference between $\langle S(P_a) \rangle$ and $\langle S(P_b) \rangle$ (i.e., the sensitivity mean values for a set of distributions with two P values) is smaller than the addition of their standard deviations, $\sigma_S(P_a)$ and $\sigma_S(P_b)$, the activation probabilities P_a and P_b cannot be discerned in a reliable manner. The ratio of these two magnitudes provides a clear idea of the resolution of the sensor with respect to the variability:

$$\zeta(P_{a,b}) = \frac{|\langle S(P_a) \rangle - \langle S(P_b) \rangle|}{\sigma_S(P_a) + \sigma_S(P_b)}. \tag{2}$$

A value of ζ greater than 1 assures a bijective relation between sensitivity and activation percentage for these two values P_a and P_b. Of course, the closer P_a and P_b are, the lower the value of ζ. Here, we have studied the case where $P_b = P_a \pm \frac{1}{12}$. Figure 4b shows ζ for the three random distribution cases and the two values of PBS. ζ decreases with P as a consequence of the higher variability observed, but in all cases is substantially above 1 when $P < 80\%$. Only at very high activation percentages the ζ value approximates the limit. When comparing the different molecule distributions, i.e., source-, channel- and drain-centered, we observe that the channel-centered distribution yield a higher ζ value. This is due to the effect of the source and drain contact biases, which electrically attract the ions in the electrolyte closer to the sensing interface, rising the screening capability of the electrolyte in these areas and reducing the sensor sensitivity. The molecule receptors located in the middle of the channel, on the contrary, are surrounded by ions free of such electrostatic attraction.

3.2. *Randomized Activation Variability*

For a single fabricated device with a given distribution of N receptors, among which N_{act} of them are activated, there are $\binom{N}{N_{\text{act}}} = \frac{N!}{N_{\text{act}}!(N-N_{\text{act}})!}$ possible activation combinations, where $\binom{N}{N_{\text{act}}} = 1$ for both $P = 0\%$ and $P = 100\%$, and it is maximized for $N_{\text{act}} = N/2$. Among all possible combinations, twelve different activated receptors configurations for each value of P were selected (when possible). This activation variability was studied for the three distribution scenarios, namely, channel-centered, source-shifted and drain-shifted, and the results compared with a uniform distribution (see Appendix A Figure A2). Again, we considered $1\times$PBS and $0.1\times$PBS electrolytes for the sake of completeness.

Figure 5a compares $\langle S \rangle$ for the channel-centered scenario and the uniform receptor distribution, with each activation combination represented by a symbol and σ_S plotted as a shadowed region. The trend of $\langle S \rangle$ with P and the PBS concentration and its comparison with the uniform distribution cast similar conclusions as in the previous subsection. The behavior of σ_S, however, is conditioned by the actual possible combinations that N and N_{act} allow, being 0 at $P = 0\%$ and $P = 100\%$ and broadening at $N_{\text{act}} = N/2$. Finally, ζ has also been studied as a function of the activation probability (Figure 5b). While it keeps far from 1 at low and high P values, it evidences low values in the intermediate P range, when the amount of activation combinations is large. When comparing the behavior of ζ with the previous random distribution study (Figure 4b), it can be observed that the resolution of the sensor is compromised at different activation percentages in each case: only for very large activation percentages in the former case, and for intermediate values in the latter one, evidencing that the prevalence of the random receptor activation or their location in the variability of the sensor sensitivity depends on P.

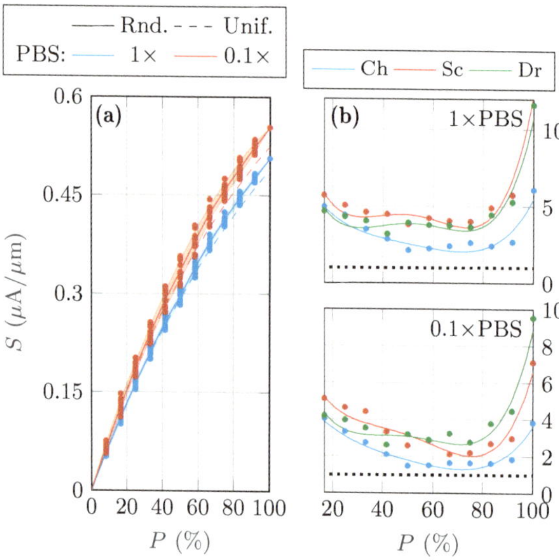

Figure 5. (**a**) Device sensitivity as a function of the activation percentage calculated for twelve randomized activated receptors configurations corresponding to the same channel-centered distribution.

Markers account for simulated data, while shaded areas depict σ_S and lines correspond to $\langle S \rangle$. Both $1\times$PBS (blue) and $0.1\times$PBS (red) concentrations were simulated. Results obtained for the uniform distribution are depicted by dashed lines. (**b**) $\zeta(P)$ ratio. Markers represent simulated data, while colours refer to channel-centered, source-shifted and drain-shifted distributions. Both PBS concentrations are separated in two graphics for the sake of clarity. The dotted lines indicate $\zeta = 1$. Fitting parameters are summarized in Appendix A Table A2.

4. Discussion

A comprehensive simulator has been developed able to study the variability in the response of 2DM-based BioFETs due to the randomization of both the location and the activation of the receptors at the solid–liquid interface. Other relevant sources of variability, such as interface defects or traps, are not investigated here, for the sake of brevity, although their role and the intertwined interaction with the distribution and the activation of molecules can play an important role to fully understand device-to-device technological variability. The presented solver is demonstrated to be a suitable theoretical tool to assess the impact of device-to-device variability and an excellent help to close the gap from lab-to-fab implementations. We have tested the capabilities of the proposed simulator studying 2D MoS_2-based BioFETs aiming for the detection of DNA molecules. To this end, we have implemented a purposely designed model for receptor molecules that handles the inherent features of the receptor ssDNA and the target dsDNA molecules, and their interaction with the surrounding electrolyte.

The device sensitivity S is impacted by these random changes in the sensing interface, showing noticeable scatter in its value when different distribution and activation scenarios are considered. In particular, S is observed to be higher when the receptors are randomly placed in the sensing interface (as compared to uniform equispaced locations). This is because the perfectly uniform distribution of the receptors is better screened by the electrolyte ions, reducing the electrostatic coupling of the target molecules and the channel. The increase in S with the randomization of the receptors positions, however, comes hand in hand with a variability in the sensor response σ_S, which is also enhanced when the PBS ionic strength is reduced. When random patterns of activation of the receptors are investigated, we observe again a higher S and higher σ_S for lower ionic strengths and for random receptor distributions. Finally, when the two sources of variability (receptor location and receptor activation) are compared, we observe that the resolution of the sensor is compromised at different activation probabilities P.

5. Conclusions

In summary, these results demonstrate the capability of the proposed model to consider the device variability in the study of 2DM-based BioFETs. This constitutes a relevant step forward in the state-of-the-art of numerical modeling, contributing to clarify device-to-device variations that could be experimentally expected due to changes in the receptor location as well as the single-device variations due to fluctuations in the receptor activation.

Author Contributions: J.C.-L. and A.G. are the corresponding authors. J.C.-L., E.G.M. and A.T.-L. developed the simulators. J.C.-L., E.G.M., F.G.R. and F.P. conceived the numerical experiments. J.C.-L., A.T.-L. and A.M.-R. conducted the simulations and data processing. E.G.M., F.G.R. and A.G., acquired funding. J.C.-L. and A.T.-L. wrote the original draft. All authors contributed to the revision and editing of the final manuscript. All authors have read and agreed to the published version of the manuscript.

Funding: This research was funded by the Spanish Government under the projects PID2020-116518GB-I00 and TED2021-129769B-I00 (MCIU/AEI/FEDER, UE); the FEDER/Junta de Andalucía under projects A-TIC-646-UGR20 and P20-00633; and the European Commission under Horizon 2020 projects WASP (No. 825213). F. Pasadas and A. Medina-Rull also acknowledge PAIDI 2020 grant 20804, and PTA grant PTA2020-018250-I, respectively. J. Cuesta-Lopez acknowledges the FPU program FPU019/05132, and A. Toral-Lopez the support of Plan Propio of Universidad de Granada.

Institutional Review Board Statement: Not applicable.

Informed Consent Statement: Not applicable.

Data Availability Statement: The data that support the findings of this study are available from the corresponding author upon reasonable request.

Conflicts of Interest: The authors declare no conflict of interest.

Abbreviations

The following abbreviations are used in this manuscript:

2DM	Two-dimensional material
FET	Field-effect transistor
IoT	Internet of Things
BioFET	Biomolecular field-effect transistor
PBS	Phosphate Buffer Saline
ssDNA	Single-stranded DNA
dsDNA	Double-stranded DNA

Appendix A

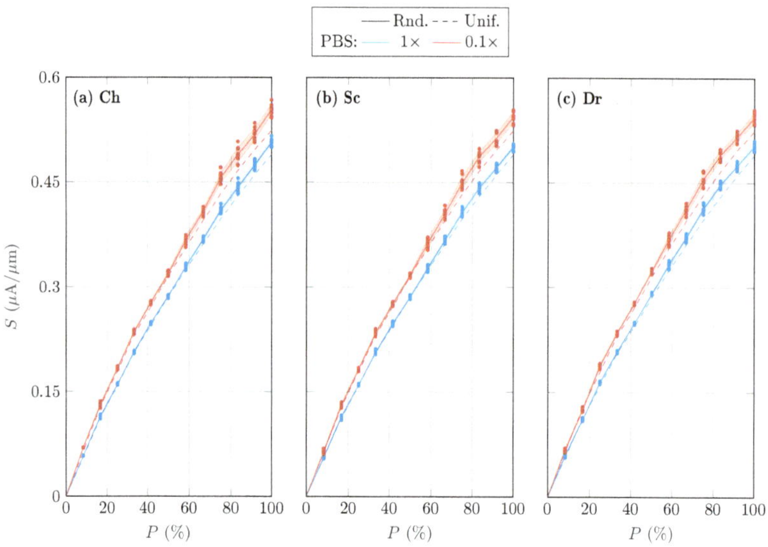

Figure A1. Device sensitivity as a function of the activation percentage. Ten random distributions following (**a**) channel-centered, (**b**) source-shifted and (**c**) drain-shifted cases are considered, fixing the configuration of activated receptors for each P value. Markers account for simulated data, while shaded areas depict σ_S and lines correspond to $\langle S \rangle$. Both 1×PBS (blue) and 0.1×PBS (red) concentrations were simulated. Results obtained for the uniform distribution are depicted by dashed lines.

Table A1. Parameters of the fitting curves $\zeta = a + bP + cP^2 + dP^3$ from Figure 4b.

PBS	$a(\cdot 10)$	$b(\cdot 10^{-1})$	$c(\cdot 10^{-2})$	$d(\cdot 10^{-5})$
Ch 1×PBS	4.87	−12.25	1.18	−4.25
Ch 0.1×PBS	3.09	−7.19	0.59	−1.69
Sc 1×PBS	3.27	−7.74	0.68	−2.11
Sc 0.1×PBS	1.02	2.26	−0.80	4.98
Dr 1×PBS	45.29	11.46	−2.69	1.53
Dr 0.1×PBS	91.48	4.56	−1.29	7.71

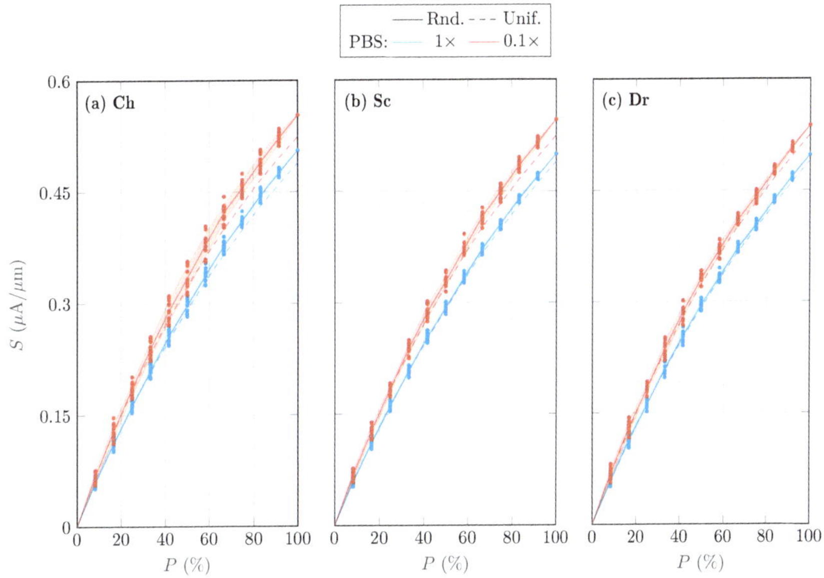

Figure A2. Device sensitivity as a function of the activation percentage calculated for twelve randomized activated receptors configurations corresponding to the same (**a**) channel-centered, (**b**) source-shifted or (**c**) drain-shifted distribution. Markers account for simulated data, while shaded areas depict σ_S and lines correspond to $\langle S \rangle$. Both 1×PBS (blue) and 0.1×PBS (red) concentrations were simulated. Results obtained for the uniform distribution are depicted by dashed lines.

Table A2. Parameters of the fitting curves $\zeta = a + bP + cP^2 + dP^3 + eP^4$ from Figure 5b.

PBS	$a(\cdot 10)$	$b(\cdot 10^{-1})$	$c(\cdot 10^{-2})$	$d(\cdot 10^{-4})$	$e(\cdot 10^{-6})$
Ch 1 × PBS	0.97	−4.46	1.16	−1.48	0.72
Ch 0.1 × PBS	0.68	−2.56	58.60	−0.73	0.37
Sc 1 × PBS	1.58	−10.60	3.48	−4.82	2.36
Sc 0.1 × PBS	1.05	−5.30	1.62	−2.29	1.16
Dr 1 × PBS	1.34	−9.10	2.98	−4.09	1.99
Dr 0.1 × PBS	1.06	−6.55	2.09	−2.89	1.43

References

1. Karthick, G.S.; Pankajavalli, P.B. A Review on Human Healthcare Internet of Things: A Technical Perspective. *SN Comput. Sci.* **2020**, *1*, 1–19. [CrossRef]
2. Li, Z.; Chen, Y.; Li, X.; Kamins, T.I.; Nauka, K.; Williams, R.S. Sequence-Specific Label-Free DNA Sensors Based on Silicon Nanowires. *Nano Lett.* **2004**, *4*, 245–247. [CrossRef]
3. Lee, J.; Dak, P.; Lee, Y.; Park, H.; Choi, W.; Alam, M.A.; Kim, S. Two-dimensional layered MoS_2 biosensors enable highly sensitive detection of biomolecules. *Sci. Rep.* **2014**, *4*, 7352. [CrossRef]
4. Stern, E.; Wagner, R.; Sigworth, F.J.; Breaker, R.; Fahmy, T.M.; Reed, M.A. Importance of the debye screening length on nanowire field effect transistor sensors. *Nano Lett.* **2007**, *7*, 3405–3409. [CrossRef]
5. Wang, L.; Wang, Y.; Wong, J.I.; Palacios, T.; Kong, J.; Yang, H.Y. Functionalized MoS_2 nanosheet-based field-effect biosensor for label-free sensitive detection of cancer marker proteins in solution. *Small* **2014**, *10*, 1101–1105. [CrossRef]
6. Cui, Y.; Wei, Q.; Park, H.; Lieber, C.M. Nanowire nanosensors for highly sensitive and selective detection of biological and chemical species. *Science* **2001**, *293*, 1289–1292. [CrossRef]
7. Bergveld, P. Development of an Ion-Sensitive Solid-State Device for Neurophysiological Measurements. *IEEE Trans. Biomed. Eng.* **1970**, *BME-17*, 70–71. [CrossRef] [PubMed]
8. Tian, R.; Regonda, S.; Gao, J.; Liu, Y.; Hu, W. Ultrasensitive protein detection using lithographically defined Si multi-nanowire field effect transistors. *Lab Chip* **2011**, *11*, 1952–1961. [CrossRef]
9. Patolsky, F.; Zheng, G.; Lieber, C.M. Fabrication of silicon nanowire devices for ultrasensitive, label-free, real-time detection of biological and chemical species. *Nat. Protoc.* **2006**, *1*, 1711–1724. [CrossRef] [PubMed]

10. Stern, E.; Klemic, J.F.; Routenberg, D.A.; Wyrembak, P.N.; Turner-Evans, D.B.; Hamilton, A.D.; LaVan, D.A.; Fahmy, T.M.; Reed, M.A. Label-free immunodetection with CMOS-compatible semiconducting nanowires. *Nature* **2007**, *445*, 519–522. [CrossRef] [PubMed]
11. Mao, S.; Chang, J.; Pu, H.; Lu, G.; He, Q.; Zhang, H.; Chen, J. Two-dimensional nanomaterial-based field-effect transistors for chemical and biological sensing. *Chem. Soc. Rev.* **2017**, *46*, 6872–6904. [CrossRef]
12. Hess, L.H.; Becker-Freyseng, C.; Wismer, M.S.; Blaschke, B.M.; Lottner, M.; Rolf, F.; Seifert, M.; Garrido, J.A. Electrical coupling between cells and graphene transistors. *Small* **2015**, *11*, 1703–1710. [CrossRef] [PubMed]
13. Sarkar, D.; Liu, W.; Xie, X.; Anselmo, A.C.; Mitragotri, S.; Banerjee, K. MoS_2 Field-Effect Transistor for Next-Generation Label-Free Biosensors. *ACS Nano* **2014**, *8*, 3992–4003. [CrossRef]
14. Li, D.; Wi, S.; Chen, M.; Ryu, B.; Liang, X. Nanoimprint-assisted shear exfoliation plus transfer printing for producing transition metal dichalcogenide heterostructures. *J. Vac. Sci. Technol. B Nanotechnol. Microelectron. Mater. Process. Meas. Phenom.* **2016**, *34*, 1–6. [CrossRef]
15. Quellmalz, A.; Wang, X.; Sawallich, S.; Uzlu, B.; Otto, M.; Wagner, S.; Wang, Z.; Prechtl, M.; Hartwig, O.; Luo, S.; et al. Large-area integration of two-dimensional materials and their heterostructures by wafer bonding. *Nat. Commun.* **2021**, *12*, 917. [CrossRef]
16. Gibney, E. 2D or not 2D. *Nature* **2015**, *522*, 274–276. [CrossRef]
17. Novoselov, K.S.; Mishchenko, A.; Carvalho, A.; Castro Neto, A.H. 2D materials and van der Waals heterostructures. *Science* **2016**, *353*, 461–473. [CrossRef] [PubMed]
18. Chung, I.Y.; Jang, H.; Lee, J.; Moon, H.; Seo, S.M.; Kim, D.H. Simulation study on discrete charge effects of SiNW biosensors according to bound target position using a 3D TCAD simulator. *Nanotechnology* **2012**, *23*, 065202. [CrossRef] [PubMed]
19. Bandiziol, A.; Palestri, P.; Pittino, F.; Esseni, D.; Selmi, L. A TCAD-Based Methodology to Model the Site-Binding Charge at ISFET/Electrolyte Interfaces. *IEEE Trans. Electron Devices* **2015**, *62*, 3379–3386. [CrossRef]
20. Passeri, D.; Morozzi, A.; Kanxheri, K.; Scorzoni, A. Numerical simulation of ISFET structures for biosensing devices with TCAD tools. *BioMed. Eng.* **2015**, *14*, S3. [CrossRef]
21. Nair, P.R.; Alam, M.A. Design considerations of silicon nanowire biosensors. *IEEE Trans. Electron Devices* **2007**, *54*, 3400–3408. [CrossRef]
22. Yang, J.; He, J.; Liu, F.; Zhang, L.; Liu, F.; Zhang, X.; Chan, M. A compact model of silicon-based nanowire MOSFETs for circuit simulation and design. *IEEE Trans. Electron Devices* **2008**, *55*, 2898–2906. [CrossRef]
23. Shadman, A.; Rahman, E.; Khosru, Q.D. Monolayer MoS_2 and WSe_2 Double Gate Field Effect Transistor as Super Nernst pH sensor and Nanobiosensor. *Sens.-Bio-Sens. Res.* **2016**, *11*, 45–51. [CrossRef]
24. Datta, K.; Shadman, A.; Rahman, E.; Khosru, Q.D.M. Trilayer TMDC Heterostructures for MOSFETs and Nanobiosensors. *J. Electron. Mater.* **2016**, *46*, 1248–1260. [CrossRef]
25. Yang, X.; Frensley, W.R.; Zhou, D.; Hu, W. Performance Analysis of Si Nanowire Biosensor by Numerical Modeling for Charge Sensing. *IEEE Trans. Nanotechnol.* **2012**, *11*, 501–512. [CrossRef]
26. Toral-Lopez, A.; Marin, E.G.; Gonzalez-Medina, J.M.; Romero, F.J.; Ruiz, F.G.; Morales, D.P.; Rodriguez, N.; Godoy, A. Assessment of three electrolyte-molecule electrostatic interaction models for 2D material based BioFETs. *Nanoscale Adv.* **2019**, *1*, 1077–1085. [CrossRef] [PubMed]
27. Toral-Lopez, A.; Pasadas, F.; Marin, E.G.; Medina-Rull, A.; Gonzalez-Medina, J.M.; Ruiz, F.J.G.; Jiménez, D.; Godoy, A. Multi-scale analysis of radio-frequency performance of 2D-material based field-effect transistors. *Nanoscale Adv.* **2021**, *3*, 2377–2382. [CrossRef]
28. Toral-Lopez, A.; Marin, E.G.; Medina, A.; Ruiz, F.G.; Rodriguez, N.; Godoy, A. GFET Asymmetric Transfer Response Analysis through Access Region Resistances. *Nanomaterials* **2019**, *9*, 1027. [CrossRef]
29. Toral-Lopez, A.; Kokh, D.B.; Marin, E.G.; Wade, R.C.; Godoy, A. Graphene BioFET sensors for SARS-CoV-2 detection: A multiscale simulation approach. *Nanoscale Adv.* **2022**, *4*, 3065–3072. [CrossRef]
30. Nishio, Y.; Uno, S.; Nakazato, K. Three-Dimensional Simulation of DNA Sensing by Ion-Sensitive Field-Effect Transistor: Optimization of DNA Position and Orientation. *Jpn. J. Appl. Phys.* **2013**, *52*, 04CL01. [CrossRef]
31. Poghossian, A.; Cherstvy, A.; Ingebrandt, S.; Offenhäusser, A.; Schöning, M.J. Possibilities and limitations of label-free detection of DNA hybridization with field-effect-based devices. *Sens. Actuators B Chem.* **2005**, *111–112*, 470–480. [CrossRef]
32. Daus, A.; Vaziri, S.; Chen, V.; Köroğlu, Ç.; Grady, R.W.; Bailey, C.S.; Lee, H.R.; Schauble, K.; Brenner, K.; Pop, E. High-performance flexible nanoscale transistors based on transition metal dichalcogenides. *Nat. Electron.* **2021**, *4*, 495–501. [CrossRef]
33. Kumar, A.; Schauble, K.; Neilson, K.M.; Tang, A.; Ramesh, P.; Wong, H.S.; Pop, E.; Saraswat, K. Sub-200 $\Omega \cdot \mu$m Alloyed Contacts to Synthetic Monolayer MoS_2. In Proceedings of the 2021 IEEE International Electron Devices Meeting (IEDM), San Francisco, CA, USA, 11–16 December 2021. [CrossRef]

34. Yan, X.; Gu, Y.; Li, C.; Tang, L.; Zheng, B.; Li, Y.; Zhang, Z.; Yang, M. Synergetic catalysis based on the proline tailed metalloporphyrin with graphene sheet as efficient mimetic enzyme for ultrasensitive electrochemical detection of dopamine. *Biosens. Bioelectron.* **2016**, *77*, 1032–1038. [CrossRef] [PubMed]
35. Gao, A.; Lu, N.; Wang, Y.; Dai, P.; Li, T.; Gao, X.; Wang, Y.; Fan, C. Enhanced sensing of nucleic acids with silicon nanowire field effect transistor biosensors. *Nano Lett.* **2012**, *12*, 5262–5268. [CrossRef] [PubMed]

chemosensors

MDPI

Article

Two-Dimensional Dy_2O_3-Pd-PDA/rGO Heterojunction Nanocomposite: Synergistic Effects of Hybridisation, UV Illumination and Relative Humidity on Hydrogen Gas Sensing

Hanie Hashtroudi [1,*], Aimin Yu [1], Saulius Juodkazis [1,2] and Mahnaz Shafiei [1,*]

1 School of Science, Computing and Engineering Technologies, Swinburne University of Technology, Melbourne 3122, Australia; aiminyu@swin.edu.au (A.Y.); ajuodkazis@swin.edu.au (S.J.)

2 World Research Hub Initiative (WRHI), School of Materials and Chemical Technology, Institute of Technology, 2-12-1, Ookayama, Meguro-Ku, Tokyo 152-8550, Japan

* Correspondence: hhashtroudi@swin.edu.au (H.H.); mshafiei@swin.edu.au (M.S.)

Abstract: A two-dimensional (2D) Dy_2O_3-Pd-PDA/rGO heterojunction nanocomposite has been synthesised and tested for hydrogen (H_2) gas sensing under various functioning conditions, including different H_2 concentrations (50 ppm up to 6000 ppm), relative humidity (up to 25 %RH) and working temperature (up to 200 °C). The material characterisation of Dy_2O_3-Pd-PDA/rGO nanocomposite performed using various techniques confirms uniform distribution of Pd NPs and 2D Dy_2O_3 nanostructures on multi-layered porous structure of PDA/rGO nanosheets (NSs) while forming a nanocomposite. Moreover, fundamental hydrogen sensing mechanisms, including the effect of UV illumination and relative humidity (%RH), are investigated. It is observed that the sensing performance is improved as the operating temperature increases from room temperature (RT = 30 °C) to the optimum temperature of 150 °C. The humidity effect investigation revealed a drastic enhancement in sensing parameters as the %RH increased up to 20%. The highest response was found to be 145.2% towards 5000 ppm H_2 at 150 °C and 20 %RH under UV illumination (365 nm). This work offers a highly sensitive and selective hydrogen sensor based on a novel 2D nanocomposite using an environmentally friendly and energy-saving synthesis approach, enabling us to detect hydrogen molecules experimentally down to 50 ppm.

Keywords: two-dimensional Dy_2O_3-Pd-PDA/rGO heterojunction nanocomposite; conductometric devices; hydrogen gas sensors; UV illumination; relative humidity; low operating temperature

Citation: Hashtroudi, H.; Yu, A.; Juodkazis, S.; Shafiei, M. Two-Dimensional Dy_2O_3-Pd-PDA/rGO Heterojunction Nanocomposite: Synergistic Effects of Hybridisation, UV Illumination and Relative Humidity on Hydrogen Gas Sensing. *Chemosensors* **2022**, *10*, 78. https://doi.org/10.3390/chemosensors10020078

Academic Editor: Pietro Salvo

Received: 7 January 2022
Accepted: 10 February 2022
Published: 14 February 2022

1. Introduction

Hydrogen (H_2) is a non-toxic, odourless, and colourless gas that can be used as a renewable energy source [1]. H_2 is abundant on earth in different molecular forms, including water and organic chemical compounds that contain hydrogen-carbon bonds such as hydrocarbons [2]. Since H_2 is a highly flammable and explosive gas, reliable and effective sensors must be used to strictly monitor any leakage during its production procedure, storage, transport, and usage [2]. To date, different 2D semiconducting materials, including metal oxides, transition metal dichalcogenides (TMDs), and graphene-based materials, have been used as hydrogen sensors due to their electrochemical and physical characteristics [3–6]. In general, 2D semiconductors have a high number of active sites, high surface to volume ratios, and exceptional optical and electrical properties that are frequently stimulated by a concentration of charge density near or on surfaces, resulting in enhanced sensing performance (i.e., high response, gas selectivity, fast response and recovery, and durability) [7–9]. Graphene-based materials are considered as promising hydrogen sensing candidates with low operating temperatures due to their excellent charge carrier mobility, high conductivity, and electrochemical stability [10]. However, their recovery is slow and mostly show poor selectivity due to the absence of a direct bandgap and few dangling bonds on the surface [11]. To overcome these issues, surface modification such as functionalisation with noble metal dopants (i.e., Pd, Au, and Pt) and hybridisation with metal oxides to develop a suitable composite sensing layer have been employed [12–15]. In addition, studies have shown that utilising UV radiation can significantly improve the H_2 sensing performance of graphene-based materials [16]. Functionalising the graphene-based materials, including the reduced graphene oxide (rGO) with noble metals, facilitates the H_2 molecules dissociation in the sensing layer and forms a dipole layer on the metal surface [13,17,18].

Thus, in this work we decided to combine Pd nanoparticles (NPs) with rGO nanosheets (NSs) to maximise the electro-catalytic activities [19]. Still, the spontaneous agglomeration of rGO NSs prepared by the traditional oxidation of graphite technique limits the anchor sites for Pd NPs growth. Another drawback of using the conventional methods to reduce the GO is using toxic chemicals, which produces hazardous residues that are harmful to the environment on an industrial scale [20]. Dopamine (DA) is an environmentally friendly chemical that has been used for the reduction and functionalisation of GO since 2010 [21]. The reduction process occurs when the DA starts self-polymerisation to form polydopamine (PDA) in the presence of GO oxygen functional groups in the weak alkaline pH environment, with the catechol groups undergoing oxidation until producing the quinon groups [21,22]. Therefore, to overcome all the disadvantages of traditional GO reduction methods, in this work, the GO has been reduced and functionalised by Pd NPs using a green approach with DA through a facile wet chemical process to achieve a high quality Pd-PDA/rGO nanocomposite [21].

Recently, rare earth metal oxides have also attracted considerable attention based on their thermal, physical, and chemical properties [23]. Among them, dysprosium oxide (Dy_2O_3) is a stable C-type rare-earth sesquioxide structured metal oxide that generally exists in two monoclinic and hexagonal structures [24]. It shows excellent physical and chemical properties based on its high surface to volume ratios, thermal stability, and, more importantly, by having a number of electrons in its 4f subshell [25]. Dy_2O_3 has been used in many applications, including pH sensors [25], magneto-optical memory materials [26], and gas storage [27]. He et al. [28] reported the catalytic behaviour of Dy_2O_3, which is based on its high hydrogen adsorption capability and being able to create new active sites through the hydrogen spill-over process. This proves that Dy_2O_3 might have remarkable electrochemical properties to be used as a hydrogen gas sensing layer. However, to date, there are no studies of the hydrogen sensing performance of any Dy_2O_3-based materials, which is considered the primary motivation for the present work. In this work, GO NSs are reduced by DA and functionalised with Pd NPs and 2D Dy_2O_3 nanostructures, making heterojunction Dy_2O_3-Pd-PDA/rGO nanocomposite as the final product. The synthesis process and results of surface characteristics and hydrogen gas sensing properties of the heterojunction Dy_2O_3-Pd-PDA/rGO nanocomposite is reported here for the first time to the best of the authors' knowledge. The experimental results confirm that the fabricated sensor is highly responsive to H_2 at very low concentrations with quick response and recovery, as well as long-term stability, which can be attributed to the excellent physical and chemical characteristics of the heterojunction nanocomposite.

2. Materials and Methods

2.1. Material Synthesis and Sensor Fabrication

Two-dimensional dysprosium oxide (Dy_2O_3) dispersion with the concentration of 0.15 mg/mL was obtained from 2D Semiconductors Inc, Scottsdale, AZ, USA, and the commercially produced graphene oxide (GO) powder was purchased from JCNANO INC Advanced Materials Supplier, Nanjing, China. The other used materials for the synthesis of Dy_2O_3-Pd-PDA/rGO nanocomposite, including dopamine hydrochloride (DA), sodium borohydride ($NaBH_4$), and palladium (II) chloride ($PdCl_2$), were purchased from Sigma-Aldrich (Sydney, NSW, Australia).

The schematic synthesis process of the Dy_2O_3-Pd-PDA/rGO nanocomposite is shown in Figure 1. Dy_2O_3-Pd-PDA/rGO nanocomposite was synthesised through the facile wet chemistry method in the following steps [21,22]. At first, 10 mg of GO NSs were sonicated in 30 mL of Tris buffer (pH 8.5) for 1 h, followed by adding 10 mg of DA and a further sonication to dissolve DA completely. The reaction mixture was left under vigorous shaking conditions for 24 h. This step aims to reduce the GO and functionalised it with PDA through the self-polymerisation process. The synthesised PDA/rGO nanocomposite was collected after centrifugation and washed several times with deionised water (DI). Afterwards, 1 mL of PDA/rGO (1 mg/mL) was added to a mixture comprised of 3 mL of DI water and 1 mL of $PdCl_2$ solution (5 mg/mL) and mixed thoroughly for 2 h. Then, 1 mL freshly made $NaBH_4$ solution (0.5 M) was added and stirred for 5 h. The Pd-PDA/rGO nanocomposite was collected via centrifugation and washed three times with DI water [22]. Afterwards, 0.5 mL of Pd-PDA/rGO dispersion (1 mg/mL) and 0.1 mL of Dy_2O_3 (0.15 mg/mL) was mixed for 4 h for the immobilization of Dy_2O_3 onto Pd-PDA/RGO nanocomposite. The final product of Dy_2O_3-Pd-PDA/rGO nanocomposite was collected using centrifugation and washed thoroughly three times by DI water. Then, 0.5 mL of DI was added to the final powder to make a dispersion. Subsequently, the dispersion was drop-casted onto a 10 mm × 6 mm gold interdigitated electrode fingers with a spacing of ~10 μm to make the final sensor device.

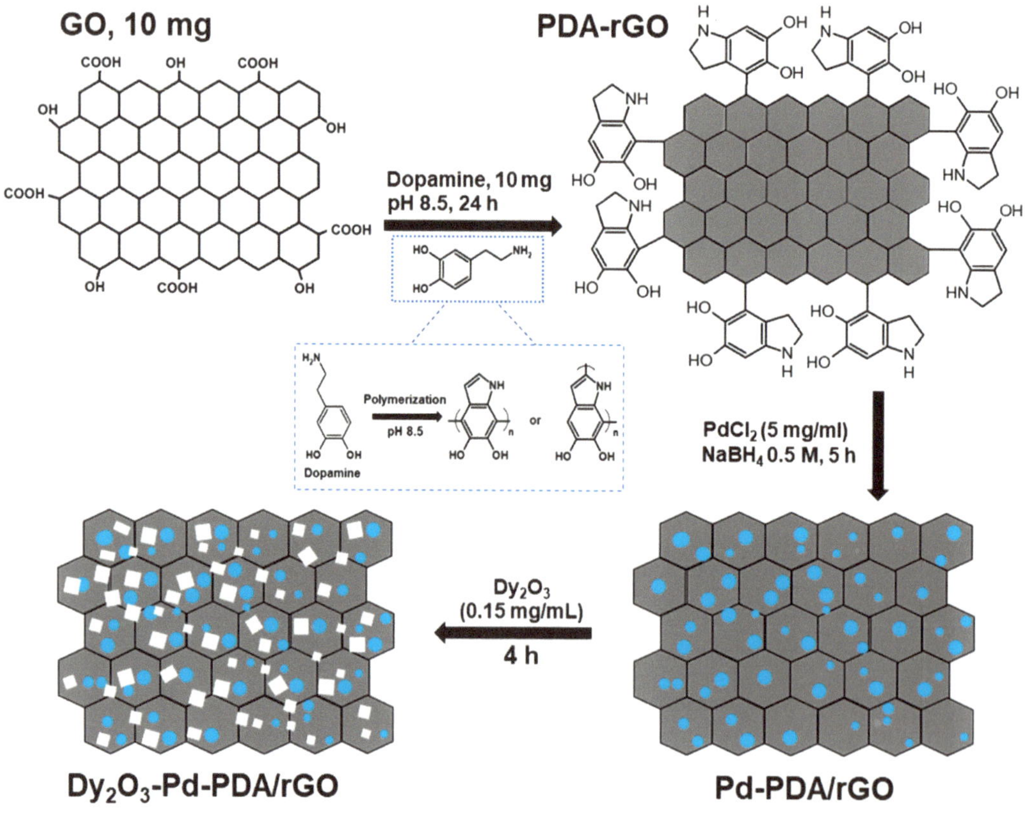

Figure 1. Schematic diagram of the synthesis process of the Dy_2O_3-Pd-PDA/rGO nanocomposite.

2.2. Material Characterisation

Different techniques were applied to analyse the structural and morphological characteristics of the synthesised Dy_2O_3-Pd-PDA/rGO nanocomposite. The surface morphology and elemental composition of the nanocomposite were investigated using a scanning electron microscope (EBL-SEM, Raith150 Two, Dortmund, Germany, at different magnifications, an energy-dispersive X-ray spectroscope (EDS, TESCAN MIRA3 FEG-SEM combined with Thermo Scientific UltraDry EDS) (Corporation Park, VIC, Australia), and a high-resolution transmission electron microscope (HRTEM, JEOL ARM200F' NeoARM'at 200 kV, Tokyo, Japan). The crystal phases of the samples were investigated using an X-ray diffractometer (D8-Advanced, Bruker, Germany) with Cu K_α and λ = 1.54 Å at 40 kV and 20 mA over a range of 5–95 degrees. A small amount of the dried Dy_2O_3-Pd-PDA/rGO nanocomposite powder was added to ethanol, then dispersed on a carbon-coated 300 mesh copper TEM grid and allowed to dry in air at RT for the HRTEM characterisation. The Raman spectra measurements of the sample were carried out using the Raman spectroscopy (Renishaw plc, Gloucestershire, UK) excited by a 514.5 nm laser.

2.3. Hydrogen Sensing Measurments

The gas sensing properties of the Dy_2O_3-Pd-PDA/rGO nanocomposite based conductometric sensor were investigated using a gas sensing system, as illustrated in Figure 2a. Figure 2b shows a photo of the IDTs used to fabricate the sensors. This system consists of a temperature and humidity controlled Linkam stage (T96, Linkam Scientific Instruments Ltd.) (Tadworth, UK), six mass flow controllers (GE50A MFCs) (Andover, MA, USA), to regulate the gas concentrations, a humidity generator, and a built-in heater (LNP96) (Tadworth, UK). A 365 nm UV LED (Harrison, NJ, USA), (M365D1 LED, with 8.9 $\mu W/mm^2$ power-driven with a current of 700 mA) was integrated on top of the stage with a distance of 1 cm from the sensor during the experiments [8]. Different operating temperatures from RT (30 °C) to 200 °C, relative humidity from 0 to ~25 %RH at 150 °C, and hydrogen concentration from 50 to 10,000 ppm were tested. The selectivity was determined by exposing the sensor to 50 ppm of different gases, including hydrogen, acetone, ammonia, and nitrogen dioxide.

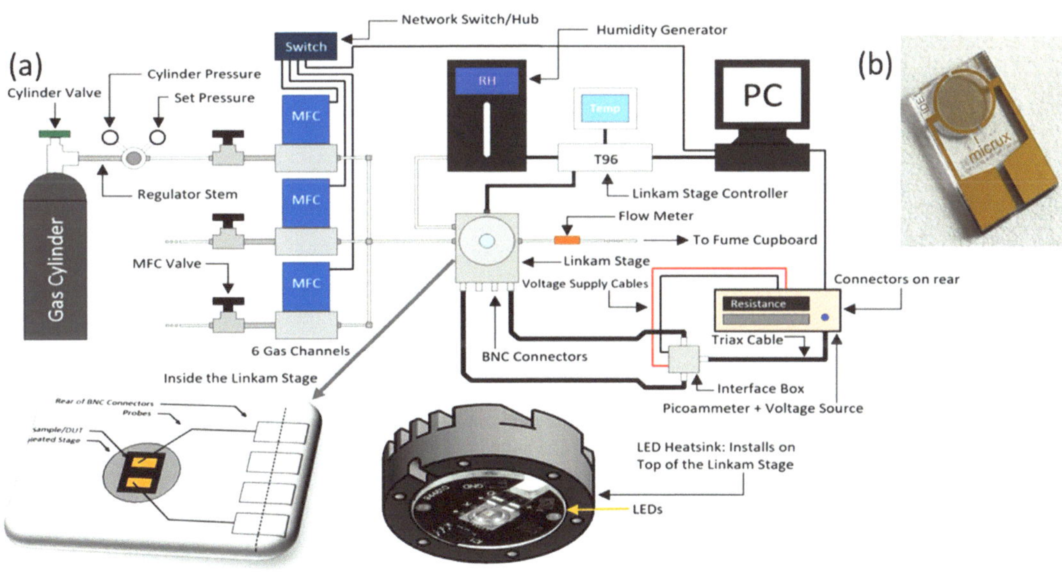

Figure 2. (**a**) Schematic diagram of the gas sensing system. Adapted with permission [8]. (**b**) A photo of the IDTs used for the fabricated sensors.

Hydrogen gas (0.1% and 1%) balanced in synthetic air with different concentrations at a uniform flow rate of 200 sccm were generated using dry synthetic air as a reference gas and the MFCs. Gas exposure time for each experiment was 15 min, followed by 2 h of purging with synthetic air. A Keithley Piccoammeter (model 6487) (Cleveland, OH, USA) periodically measured interval currents throughout the experiments. A change in sensor resistivity was measured upon exposure to the target gas at a bias voltage of 1 V and was correlated to the gas concentrations. The sensors response (R) towards a target gas was calculated as below:

$$R = \left(\frac{R_g}{R_a}\right) \times 100, \quad (1)$$

where R_a is the resistivity of the sensor in air and R_g is the resistivity in the target analyte. The response time is considered when the sensor reaches 90% of the response, and the recovery time is calculated when the sensor recovered 90% back to its baseline. In this work, three similar sensors were fabricated, and the sensing results were comparable. Each experiment was performed at least three times under the exact working conditions to confirm the repeatability and reliability of the sensing procedure.

3. Results and Discussion

This section may be divided by subheadings. It should provide a concise and precise description of the experimental results, their interpretation, as well as the experimental conclusions that can be drawn.

3.1. Material Characterisation

XRD pattern of Dy_2O_3-Pd-PDA/rGO nanocomposite was analysed to indicate its degree of crystallinity and is illustrated in Figure 3. A relatively broad peak at $2\theta = 26.6°$ corresponds to (002) graphitic plane of the partially restacked rGO nanosheets. The intralayer spacing between the rGO nanosheets was calculated to be 0.33 nm by the Bragg's low [29], as follows:

$$\lambda = 2d\sin(\theta), \quad (2)$$

where λ is the X-ray beam wavelength (0.154 nm), d is the distance between the rGO NSs, and θ is the diffraction angle. In the following, the graphene layers number in rGO NSs was determined to be <3 layers from the Scherrer Equation [30], as below:

$$X = \frac{K\lambda}{\beta.\cos(\theta)}, \quad (3)$$

where X is the number of rGO layers, β is the line broadening at half the maximum intensity (FWHM), K is a dimensionless shape factor, λ is the X-ray beam wavelength, and θ is the Bragg angle.

It is also seen a sharp peak at 28.5°, which corresponds to Dy_2O_3 (222) plane. Diffraction peaks at 2θ = 20.3°, 28.5°, 33°, 35.7°, 39.6°, 43.1°, 47.7°, 56.7°, and 78° assigned to the (211), (222), (400), (411), (332), (433), (440), (611), and (622) crystalline planes, respectively clearly indicate the cubic phase of the Dy_2O_3, and the peaks fit very well with the JCPDS data card No. 78-0388 [31]. The other sharp peaks that appear at 2θ = 38.6°, 44.1°, 64.7°, and 77.75° attributed to the (111), (200), (222), and (311) planes, respectively, confirm the face-centred cubic (FCC) structure of the Pd NPs inside the rGO NSs with an average of 13 nm crystalline size, which is calculated by Scherrer (Equation (3)) and fits well to the JCPDS DATA CARD No.5-681 [22].

The surface morphology of the Dy_2O_3-Pd-PDA/rGO nanocomposite was characterised by SEM at different magnifications (Figure 4). SEM images illustrate the PDA functionalised rGO nanosheets with porous structure and some open edges, which confirms that the PDA/rGO nanosheets are entirely interconnected. Figure 4 also shows that PDA/rGO nanosheets are homogeneously decorated with Pd NPs and nanoclusters. Figure 4b,c shows that the PDA/rGO NSs structure has a crumpled and wrinkled form, and the Pd and Dy_2O_3 are dispersed on both sides of the PDA/rGO NSs. Moreover, these images confirm that Pd and Dy_2O_3 are fairly distributed on PDA/rGO NSs without substantial agglomerations.

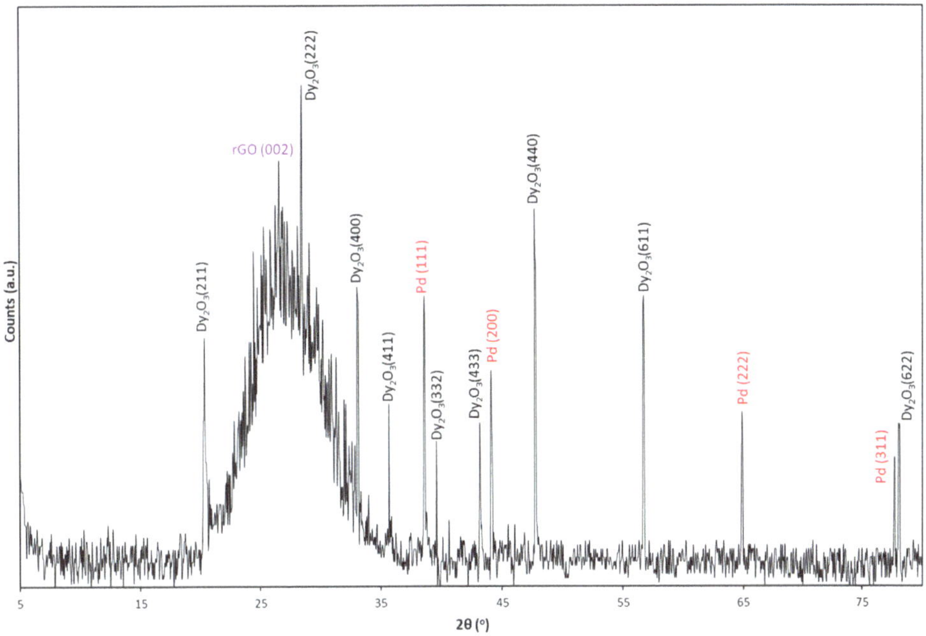

Figure 3. X-ray diffraction pattern of Dy_2O_3-Pd-PDA/rGO nanocomposite.

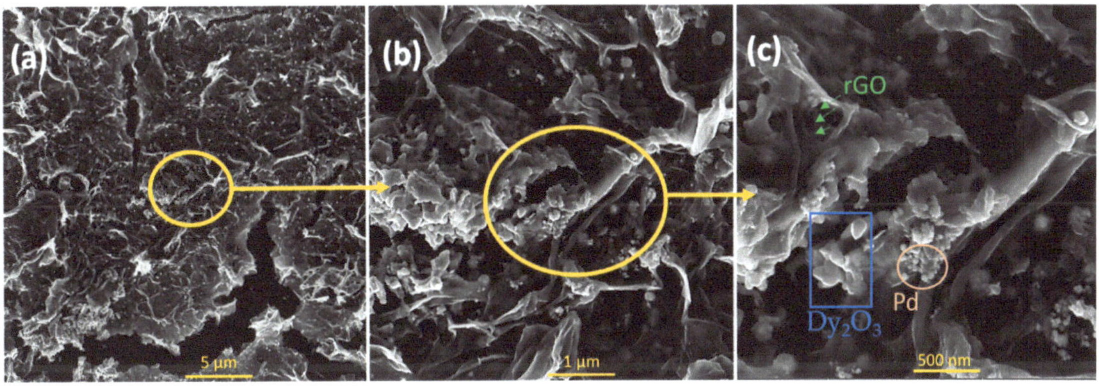

Figure 4. (**a**–**c**) SEM images of highly porous structure of Dy_2O_3-Pd-PDA/rGO nanocomposite with wrinkles and open edges at different magnifications. (**c**) shows the different elements of the nanocomposite.

Energy-dispersive X-ray spectroscopy (EDS) was also used to determine the elemental composition and surface coverage of each element on the Dy_2O_3-Pd-PDA/rGO nanocomposite (Figure 5). Figure 5b–e confirms the existence of carbon, oxygen, palladium, and dysprosium, respectively, which indicates the successful formation of Dy_2O_3-Pd-PDA/rGO nanocomposite. Figure 5f illustrates the elemental composition of the Dy_2O_3-Pd-PDA/rGO nanocomposite, confirming the Weight% and Atom% of each element.

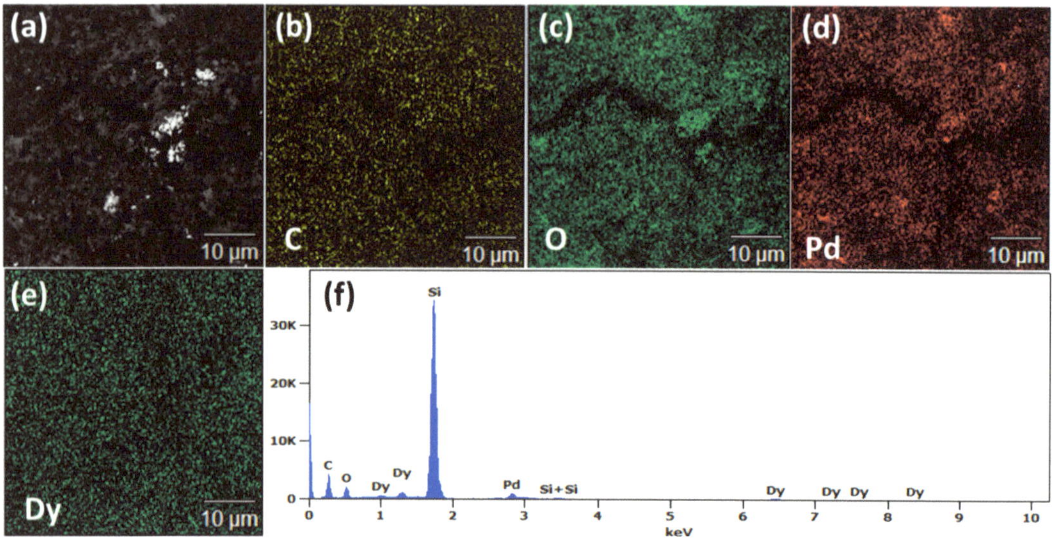

Figure 5. EDS image (**a**) and mapping of (**b**) carbon, (**c**) oxygen, (**d**) palladium, and (**e**) dysprosium; and (**f**) elemental measurement of the Dy_2O_3-Pd-PDA/rGO nanocomposite.

The Dy_2O_3-Pd-PDA/rGO nanocomposite was further analysed by HRTEM to evaluate its internal morphology and crystal structural features (Figure 6). HRTM images of Dy_2O_3-Pd-PDA/rGO nanocomposite in different magnifications indicate thin and transparent film-like morphology of the PDA-rGO NSs, which confirms the formation of a few layers of highly reduced graphene oxide decorated with Pd NPs and Dy_2O_3. Figure 6a,b illustrates the morphology of each element and distribution of the 2D structured Dy_2O_3 and Pd NPs into the rGO NSs (consistent with the SEM images). Pd NPs appear to have a nearly spherical shape, which is well dispersed onto the PDA/rGO NSs surface. It also indicates that Dy_2O_3 is distributed into the PDA/rGO NSs with some agglomerations. Figure 6c show the lattice fingers of the Pd and Dy_2O_3 planes. The interplanar spacings of 0.144 nm, 0.232 nm, and 0.205 nm are ascribed to cubic Pd (222), Pd (111), and Pd (200) lattice planes, respectively. In addition, Figure 6c indicates the polycrystalline cubic structured Dy_2O_3 (400), (440), and (222) planes with characteristic spacings of 0.271 nm, 0.235 nm, and 0.313 nm, respectively. According to Bragg's law (Equation (2)), XRD analysis of the Dy_2O_3-Pd-PDA/rGO nanocomposite supports the crystal structural features found by the HRTEM investigation. As shown in Figure 6a,b, PDA/rGO NSs are relatively thin layers that help support the Pd and Dy_2O_3 nanoclusters to form the Dy_2O_3-Pd-PDA/rGO nanocomposite.

Raman measurements at RT were performed using a 514 nm laser to characterise Dy_2O_3-Pd-PDA/rGO nanocomposite, Pd-PDA/rGO, and Dy_2O_3 and determine the structural change and formation of any defects inside the synthesised materials (Figure 7). The Raman spectrum of the 2D Dy_2O_3 indicates a sharp dominant peak at ~372 cm^{-1} assigned to F_g mode, confirming its cubic phase structure as also confirmed by XRD analysis, which is consistent with the TEM results. In comparison, this peak is shifted to a lower wavelength (~368 nm) in the Dy_2O_3-Pd-PDA/rGO nanocomposite Raman spectrum attributed to the formation of structural defects and oxygen vacancies on the nanocomposite surface due to strong interaction between the Dy_2O_3 and Pd-PDA/rGO NSs, facilitating the charge transfer from the Dy_2O_3 to rGO NSs [32]. A number of low-intensity Raman bands were also observed at F_g modes, including the 462 cm^{-1} and 585 cm^{-1}, which well describes the c-type structure of the Dy_2O_3. Comparing the 2D bands of the Dy_2O_3-Pd-PDA/rGO nanocomposite (2745 cm^{-1} and 2937 cm^{-1}) and Pd-PDA/rGO (2721 cm^{-1} and 2927 cm^{-1}) reveals that there is a general shifting to higher wavenumbers in the nanocomposite as the layer thickness increased [33]. In addition, the broadened 2D bands in the nanocomposite represent the exfoliation of the rGO NSs [32]. The 2D bands of the pure Dy_2O_3 are found to be at 2853 cm^{-1} and 2921 cm^{-1}. The Raman spectra of the Pd-PDA/rGO in Figure 7 indicate two characteristic peaks at ~1357 cm^{-1} and 1602 cm^{-1}, corresponding to the D and G bands, respectively. While the D and G bands on the Dy_2O_3-Pd-PDA/rGO Raman spectrum are slightly shifted to lower wavelengths (~1351 cm^{-1} and 1597 cm^{-1}) due to the formation of heterojunction between the Dy_2O_3 and Pd-PDA/rGO and electron transfer from Dy_2O_3 to the Pd-PDA/rGO [34]. In general, the D band is associated with the density of defects in the sp2 hybridised material, and the G band starts by active E_{2g} phonon mode of the symmetric structure in the graphite [35,36]. The presence of these two D and G bands confirms that the structure of the Pd-PDA/rGO NSs had been maintained after loading the 2D Dy_2O_3 [37]. The intensity ratio (I_D/I_G) of the D band to G band is associated with the density of defects that arises in Pd-PDA/rGO NSs due to doping with the 2D Dy_2O_3 in the formation of Dy_2O_3-Pd-PDA/rGO nanocomposite. The I_D/I_G for Pd-PDA/rGO NSs was found to be 0.99, while it increased to 1.04 for the Dy_2O_3-Pd-PDA/rGO nanocomposite due to the higher defect density in the nanocomposite and more oxygen vacancy on its surface [38]. This result confirms that the Dy_2O_3 is successfully incorporated into the Pd-PDA/rGO NSs, forming the heterojunction Dy_2O_3-Pd-PDA/rGO nanocomposite.

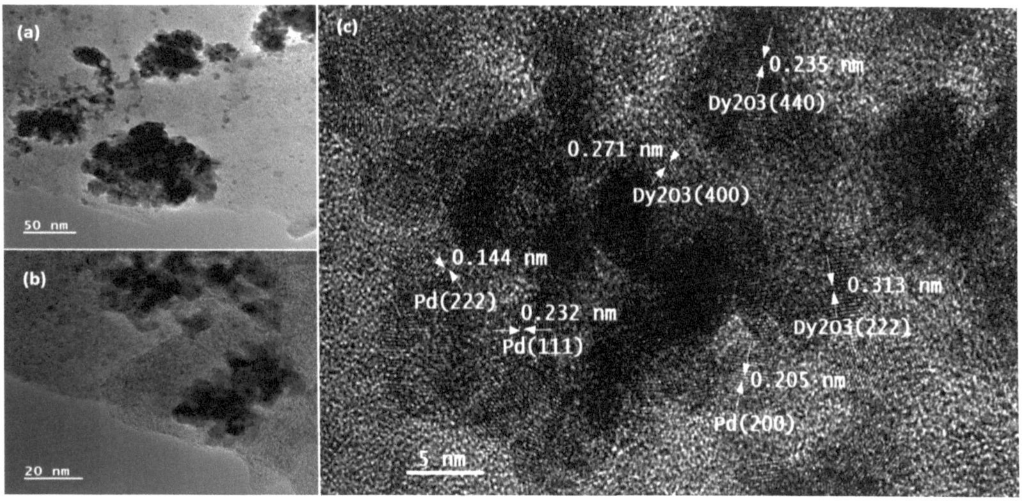

Figure 6. (**a**–**c**) HRTEM images of Dy_2O_3-Pd-PDA/rGO nanocomposite in different magnifications. (**c**) shows interplanar spacings of Pd NPs and Dy_2O_3.

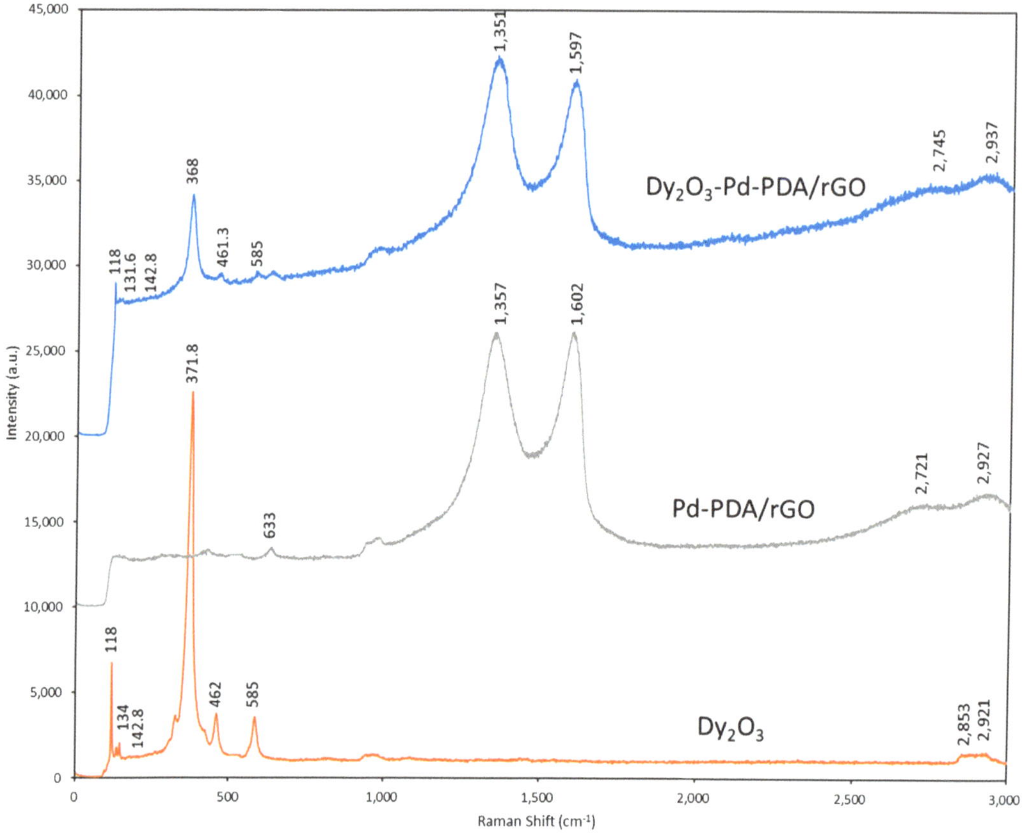

Figure 7. Raman analysis of Dy_2O_3-Pd-PDA/rGO nanocomposite, Pd-PDA/rGO, and Dy_2O_3.

3.2. Hydrogen Sensing Characterisation

The hydrogen gas sensing performance of the Dy_2O_3-Pd-PDA/rGO nanocomposite-based sensor was examined under different working conditions. The following sections report the effects of operating temperature, UV illumination, gas concentration, and humidity on the sensing parameters.

3.2.1. Effect of Working Temperature

Optimising the operating temperature is one of the critical factors in gas sensing, which directly relates to the experiment's power consumption and cost. The fabricated Dy_2O_3-Pd-PDA/rGO based conductometric sensor was examined towards 400 ppm and 600 ppm H_2 at different working temperatures under UV (365 nm) illumination and 0% relative humidity (%RH). The operating temperature was changed from RT (30 °C) to 200 °C, and a change in resistivity of the sensor was monitored throughout the experiment. It was observed that the sensor's response (resistivity) increased by increasing the operating temperature from RT (30 °C) up to 150 °C, due to the acceleration of the adsorption/desorption kinetics between the H_2 molecules and the Dy_2O_3-Pd-PDA/rGO composite sensing layer [39,40]. Further increasing the operating temperature to 200 °C led to a decrease in the sensing response. This result can be attributed to the increase in the desorption rate of H_2 molecules compared to the adsorption rate while raising the temperature further than the optimum temperature, which is called the Langmuir effect [40]. Therefore, the optimum working temperature was found to be 150 °C. The 150 °C temperature corresponds to the 86 meV potential, which is among typical overpotential values in electrochemical hydrogen evolution reaction (HER) of water splitting [41]. Figure 8a shows the typical p-type semiconducting sensing behaviour of the sensor as the conductivity decreases towards H_2 (reducing gas). The increase in the resistivity of the p-type sensor upon exposure to the reducing gas is due to the charge transfer from the target gas molecules to the sensing layer [1,41]. As an electron donor gas, H_2 transfers electrons to the valent band of the sensing layer, reducing the hole concentration on the surface, resulting in an increase in resistivity. Figure 8b illustrates the response of the sensor towards 400 and 600 ppm H_2 at different temperatures.

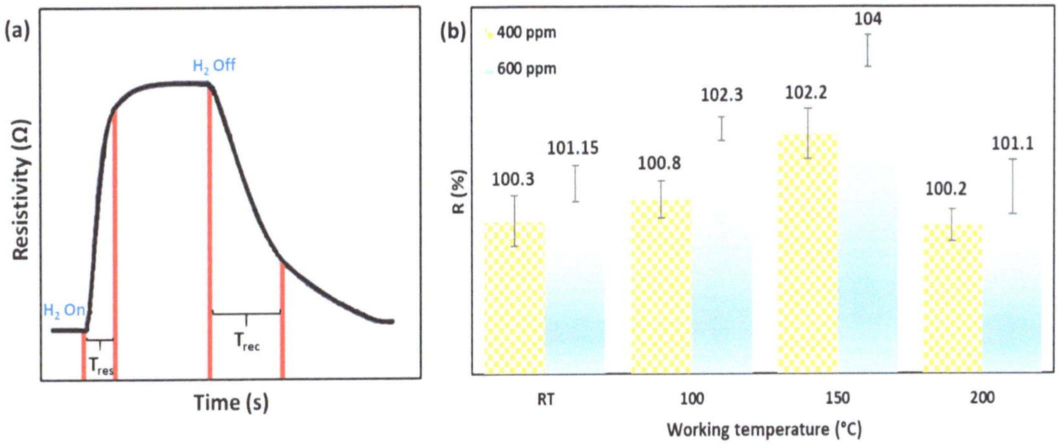

Figure 8. (**a**) P-type semiconducting sensing behaviour towards H_2 as a reducing gas, T_{res}: response time, and T_{rec}: recovery time; and (**b**) Response of the Dy_2O_3-Pd-PDA/rGO sensor towards 400 and 600 ppm hydrogen at different working temperatures.

3.2.2. Effect of UV Illumination

The effect of UV illumination on the H_2 sensing performance of the Dy_2O_3-Pd-PDA/rGO sensor was studied at 150 °C and 0 %RH as a function of hydrogen concentration. Figure 9a,b illustrate the dynamic response of the sensor towards H_2 with different concentrations (200–600 ppm) at 150 °C in the dark and under UV (365 nm) illumination, respectively. As explained in the previous section, the sensor shows a p-type semiconducting sensing behaviour as the conductivity decreases upon exposure to H_2 molecules [42]. The resistivity shifts happen based on the adsorption/desorption of the gas molecules onto the sensing layer and are correlated to the target gas concentrations [43]. This correlation could be due to the rise in gas molecules numbers interacting with the sensing layer surface. Therefore, the adsorption and desorption co-occur until the balanced state is reached [44].

As seen in Figure 9a, a baseline drift occurs without using UV illumination, indicating that the sensor cannot be fully recovered to its original baseline, showing partly reversible gas/material interactions. However, Figure 9b demonstrates fully reversible gas/material interactions under UV illumination, revealing that the sensor is fully recoverable [5,45,46]. This result can be attributed to the change of charge carrier density on the surface of the sensing layer by stimulating the electrons from the valence band of the Dy_2O_3-Pd-PDA/rGO nanocomposite while using UV radiation [47]. The desorption of H_2 molecules is accelerated by these photogenerated electrons from the pre-adsorbed ambient oxygen species [48]. In addition, the UV radiation weakens the bonding forces between the sensing layer (Dy_2O_3-Pd-PDA/rGO nanocomposite) and H_2 molecules during the physisorption, resulting in photoexcitation and reversible adsorption/desorption reactions [45,48,49]. The presence of 4f electron improves the photocatalytic activity and adsorption capacity of the Dy_2O_3 while exposed to UV. When the sensor is in contact with H_2 as a reducing agent, the electrons transfer from hydrogen into the Dy_{4f} orbitals and thus to oxygen, creating superoxide anion radicals [50]. Therefore, the formation of highly reactive oxidant species including °OH, HO_2° and $O_2^{\circ -}$ on the Dy_2O_3-Pd-PDA/rGO heterojunction surface results in shaping the electron-hole pairs where the excited electrons from the valence band move to the conduction band and consequently leaving holes in the valence band [51]. These photoproduced electron-hole pairs can transfer to the surface, inducing the electron acceptor (O_2) to active oxygen (°O_2^-) reduction process [52]. In addition, the holes oxidize H_2 (the electron donor) to produce hydroxyl radicals and/or H_2O molecules [52]. The active °O_2^- and OH° accelerate the oxidisation of the adsorbed H_2 molecules, enhancing the sensing performance [52]. Response magnitude of the sensor at 150 °C and 0 %RH was calculated for each experiment towards H_2 with different concentrations (200–600 ppm) in the dark and under UV radiation and presented in Table 1.

Table 1. Hydrogen sensing response of Dy_2O_3-Pd-PDA/rGO conductometric sensor in the dark and under UV illumination at 150 °C and 0 %RH towards H_2 with different concentrations.

		H_2 Concentration (ppm)				
		200	**300**	**400**	**500**	**600**
Response (%)	**In the dark**	100.3	100.68	100.72	100.75	100.75
	Under UV	100.7	101.3	102.2	103.1	104

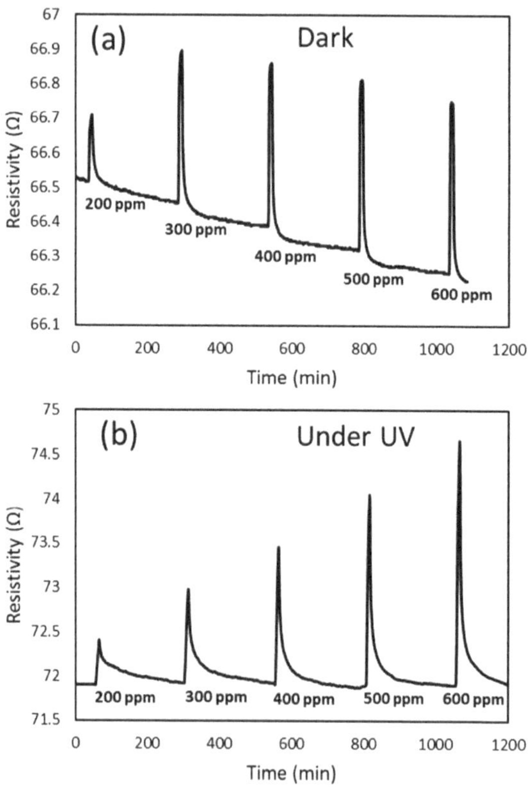

Figure 9. Dynamic response of Dy_2O_3-Pd-PDA/rGO sensor at 150 °C and 0 %RH as a function of hydrogen concentration, (**a**) without and (**b**) with UV illumination.

Table 1 confirms an improvement in the response of the photo-induced sensor under UV illumination because of the decrease in the adsorption energy barrier between the gas molecules and sensing film [53]. In addition, the sensor's response in the dark is nearly constant by rising the H_2 concentration from 400 to 600 ppm, which can be attributed to the competition between recombination of the electron-hole pairs and the re-adsorption of the H_2 molecules [54,55]. The H_2 sensing mechanism of the Dy_2O_3-Pd-PDA/rGO nanocomposite-based sensor is complex and cannot be simplified based on different physical and chemical adsorption/desorption interactions, including van der Waals forces, π-π interactions, formation of various chemical bonds, oxygen species on the surface, and charge transfer between the heterojunction nanocomposite and H_2 molecules [56].

3.2.3. Effect of Relative Humidity

Some environmental elements can influence the gas sensing performance, including humidity [57]. In a humid environment, water molecules could adsorb onto the sensor as well, thus affecting the physisorption and chemisorption processes of hydrogen [48,58]. In this work, we investigated the effect of humidity (from 0 %RH to 25 %RH) on the Dy_2O_3-Pd-PDA/rGO sensor's performance at the optimum temperature of 150 °C under different H_2 concentrations and UV radiation.

The sensing parameters, including the response magnitude and response and recovery times towards H_2 with different concentrations and various %RH, was calculated for each experiment and presented in Tables 2 and 3. The results reveal that the sensor's responses to H_2 are enhanced by increasing the H_2 concentration from 200 ppm to 600 ppm. Table 2 also indicates that the sensor's responses are improved by elevating the %RH up to 15% and then is reduced by further increasing to 25%. In addition, it was seen that the electrical properties of the sensor remained almost constant by further raising the %RH above 25% up to 47.5%. The sensitivity improvement of up to 15 %RH can be attributed to the role of water molecules in humid environments, participating in physical and chemical adsorption/desorption reactions [59]. Similar to H_2, water molecules could also act as electron donors, so when the relative humidity elevates to 15%, the resistance of the p-type sensor rises, improving the sensor's response due to the increased number of adsorbed water molecules [59,60]. In addition, the polar hydroxyl and carbonyl functional groups present on the sensing material's surface can interact with water, causing a resistance change as well [61]. Therefore, the formation of the hydroxyl active sites on the surface of the Dy_2O_3-Pd-PDA/rGO due to the interaction between the adsorbed H_2 molecules and the present oxygen species on the sensing layer surface leads to improvement in the sensor's response [62]. By further increasing the relative humidity, the adsorbed water molecules concentration increases on the sensor's surface, breaking the sensing material's sublattice and symmetry (especially the graphene oxide) and continuously widening its bandgap, consequently changing the electrical properties of the sensor, and decreasing the sensor's response [63].

Table 2. Sensing parameters of the sensor towards 200–600 ppm H_2 at 150 °C under UV illumination at different humidity level.

H_2 Conc (ppm)	Response (%)				
	0 %RH	10 %RH	15 %RH	20 %RH	25 %RH
200	100.7	100.9	101.9	101.3	101.9
300	101.3	101.4	102.9	101.5	102.1
400	102.2	102.5	104.1	101.6	102.3
500	103.1	103.9	105.1	101.6	102.3
600	104	104.15	107.1	101.6	101.9

Table 3. Sensing parameters towards H_2 with 2000–6000 ppm concentrations at 150 °C under UV illumination.

H_2 Conc (ppm)	Response (%)			Response Time (s)			Recovery Time (s)		
	0 %RH	15 %RH	20 %RH	0 %RH	15 %RH	20 %RH	0 %RH	15 %RH	20 %RH
2000	104.3	114.12	113.5	60	600	720	180	240	180
3000	107.4	122.1	130.4	180	420	600	180	240	180
4000	111.6	131.9	142.6	180	420	360	180	240	120
5000	115.8	140.4	145.2	150	360	180	210	180	120
6000	121	144.1	144	150	300	180	240	180	120

In addition, excess water molecules on the sensing surface could compete with H_2 molecules to be adsorbed on the sensor's surface. Consequently, due to the larger molecule size of the water, it is challenging for them to compete with H_2 molecules to enter the sensing material's surface, resulting in a decrease in the sensing response [57].

The humidity effect was also investigated when the sensor was exposed to high H_2 concentrations from 2000 ppm to 6000 ppm. As shown in Table 3, the sensor demonstrates a maximum response of 121% at 0 %RH in the presence of 6000 ppm H_2 at 150 °C. By increasing the H_2 concentration, the number of hydrogen molecules interacting with the sensing layer rises, enhancing the adsorption rate on the surface until it gets saturated.

Table 3 also reveals that the sensor's responses enhance by raising the humidity to 20%. The sensor's response upon exposure to 6000 ppm H_2 at 20 %RH almost doubled. The response is calculated to be 144%, comparable with the sensor's response (144.1%) at 15 %RH under the same operational conditions. However, by increasing the H_2 concentration up to 5000 ppm at 20 %RH, the sensor's response slightly enhanced to 145.2% while it dropped a little by further growing the H_2 concentration to 6000 ppm. This can be attributed to the competition between the H_2 and water molecules to be adsorbed on the sensing layer's surface [46,64].

Another explanation can be seen in Figure 10, which shows the dynamic response of the sensor towards H_2 at 0 %RH, 15 %RH, and 20 %RH and 150 °C. Comparing these results reveals that a slight baseline drift occurs when the sensor is operating at 20 %RH, which means the sensor does not recover fully, decreasing the sensor's response while exposed to 6000 ppm H_2.

In addition, Figure 10a shows that the sensor's baseline (i.e., sensor's resistivity) at 0 %RH is higher than what it is at 15 %RH and 20 %RH. This result could be attributed to increasing the charge carriers on the surface of a sensing material by elevating the %RH due to the adsorbed ionised water molecules, enhancing the sensor's conductivity [61,64]. Moreover, the presence of hydroxyl active sites (as electron donors) on the sensing layer in the humid environment enhances the electric charge density by forming the hydronium cations from the ionised water molecules (Equation (4)), leading to an increase in the conductivity [48,58].

$$2H_2O \rightleftarrows OH^- + H_3O^{+}. \quad (4)$$

Figure 10 also compares the response and recovery times while increasing the humidity level and H_2 concentrations. A fast response and recovery, 60 s and 180 s, respectively, were measured for 2000 ppm H_2 at 150 °C and 0 %RH. It can be seen that both response and recovery times decrease by rising the H_2 concentration at a higher humidity level than 0 %RH due to an increase in the surface coverage rate accelerating the adsorption/desorption of the H_2 molecules [65]. In addition, the high permeability of the Dy_2O_3-Pd-PDA/rGO to water molecules and the hoping proton mechanism play essential roles in the sensing performance [49,66]. When the concentration of the adsorbed water molecules increases on the surface (by increasing the %RH from 15% to 20%), the arisen protons (H^+), from hydroxyl functional groups of the rGO, bond with the water molecules. This leads to forming the hydronium ions, which enhance the electric charge density [39,63]. Thus, by increasing the concentration of the H_2 molecules, the adsorption/desorption interactions become quicker due to the increased gas diffusion and conductive routes with high mobility [67,68]. Consequently, by comparing the response and recovery times at 15% and 20 %RH, we observe quicker response and recovery by increasing the H_2 concentration at 20 %RH. Moreover, by raising the humidity to 20%, the sensor's recovery became faster than it was at 0 % RH due to the physical and chemical adsorption/desorption rate of the H_2 molecules.

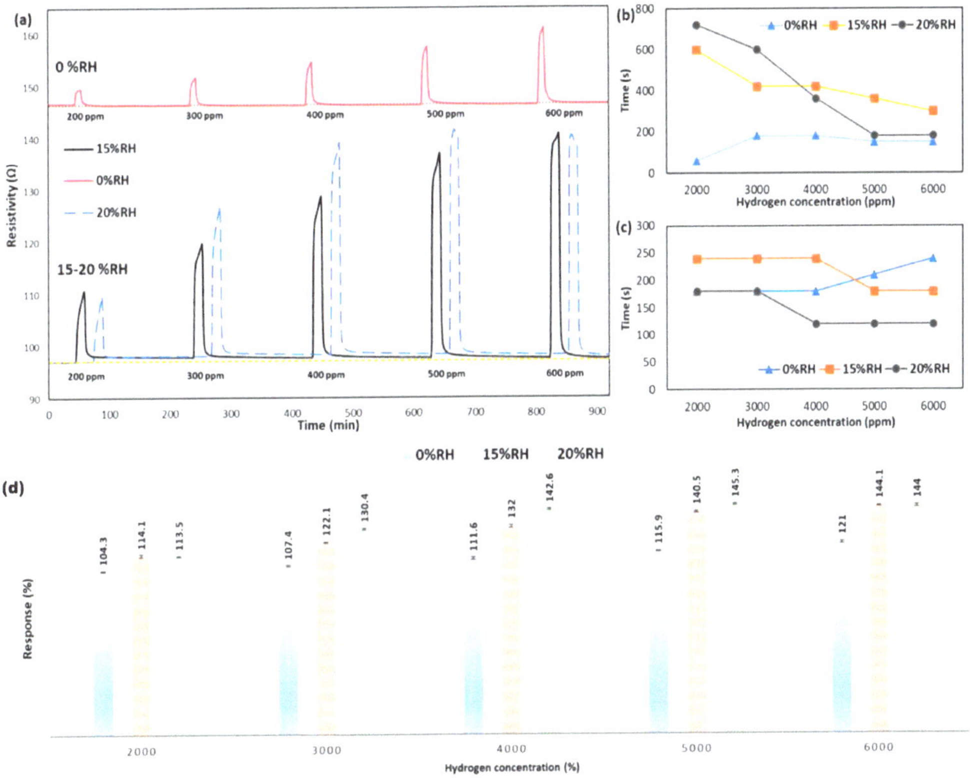

Figure 10. (**a**) Dynamic response of Dy_2O_3-Pd-PDA/rGO sensor towards H_2 with different concentrations (2000–6000 ppm) at 150 °C, and relative humidity levels (0 %RH, 15 %RH, and 20 %RH) under UV radiation; (**b**) response time and (**c**) recovery time of the sensor at 0 %RH, 15 %RH, and 20 %RH as a function of H_2 concentrations. (**d**) Response magnitude of the sensor towards H_2 with different concentrations at 0 %RH, 15 %RH, and 20 %RH.

A fast recovery (120 s) was measured for 4000 to 6000 ppm H_2 at 150 °C and 20 %RH. In a humid environment, it was observed that the response and recovery were quicker by increasing the H_2 concentration based on an increase in the surface coverage rate facilitating the adsorption/desorption of the H_2 molecules. In addition, the sensor saturates faster at higher concentrations. Moreover, the functional groups on the Dy_2O_3-Pd-PDA/rGO, including the hydroxyl and carbonyl groups, can interact with water, changing the adsorption/desorption rate and participating in the physisorption of the H_2 molecules on the Dy_2O_3-Pd-PDA/rGO surface, based on Van der Waals forces, and also forming the covalent bonds via the chemisorption reactions [65]. The water molecules usually behave as electron donors, inducing the ionisation of the OH and COOH functional groups in the sensing layer to generate a concentration gradient of protons [13]. This gradient accelerates the diffusion of the protons to the sensing material, delivering a voltage and current in the external circuit that leads to a quicker response [69]. Figure 10d shows the response percentage of the sensor at 0 %RH, 15 %RH, and 20 %RH and 150 °C. The highest response is calculated to be 145.2% when the sensor is exposed to 5000 ppm H_2 at 20 %RH.

3.2.4. Sensor Selectivity and Long-Term Stability

The gas selectivity of the Dy_2O_3-Pd-PDA/rGO sensor was investigated at the optimum temperature of 150 °C and 0 %RH. Figure 11a shows the response of the sensor to 50 ppm H_2 gas as compared with different interfering gases (all at 50 ppm concentrations), including methane (CH_4), ammonia (NH_3), and acetone (C_3H_6O). The sensor indicated a remarkable response of 100.5% towards hydrogen while showing a negligible response (2.1%) towards acetone and inadequate responses of 5.1% and 3.2% towards CH_4 and NH_3, respectively. As mentioned before, the sensing mechanism of the p-type Dy_2O_3-Pd-PDA/rGO nanocomposite is mainly based on the chemical adsorption/desorption interactions of the oxygen active sites with the H_2 molecules and the charge transfer from the H_2 molecules to the sensing layer [24,70]. The oxygen in the air interacts with the free electrons on the sensing layer surface, forming O^{2-}, O_2^-, and O^- ions. These reactive oxygen species interact with the H_2 molecules as an electron donor gas, enhancing the electron charge transfer to the sensing layer, reducing the hole concentration, and increasing resistivity [24,41,42]. The Dy_2O_3-Pd-PDA/rGO sensor was examined repeatedly over eight months towards 6000 ppm H_2 at 150 °C and 0 %RH and indicated long term stability with a negligible degradation (Figure 11b).

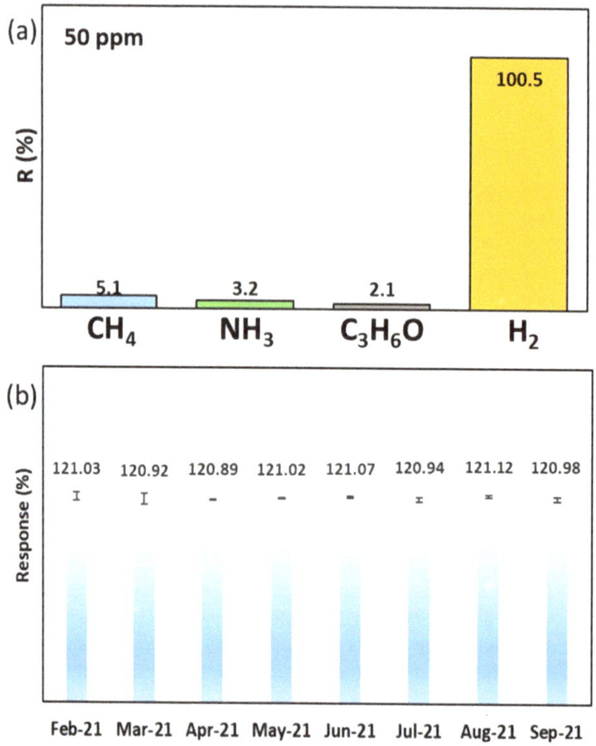

Figure 11. (**a**) Gas selectivity of the Dy_2O_3-Pd-PDA/rGO sensor towards 50 ppm of different gases, (**b**) long term stability of the sensor's response towards 6000 ppm H_2 at 150 °C and 0 %RH over eight months.

4. Discussion and Conclusions

The heterojunction Dy_2O_3-Pd-PDA/rGO nanocomposite was synthesised by an environmentally friendly and straightforward chemical procedure [22]. The surface characteristics of the sensor revealed a uniform distribution of 2D Dy_2O_3 and Pd NPs onto the PDA/rGO nanosheets. The SEM images illustrated a multi-layered porous structure of the Dy_2O_3-Pd-PDA/rGO nanocomposite that considerably impacted the H_2 gas sensing performance. Raman analysis showed no significant defects in Pd-PDA/rGO NSs while doping with 2D Dy_2O_3. The hydrogen gas sensing performance of the fabricated conductometric sensor was extensively studied under different operational conditions, including H_2 concentration (50–6000 ppm), temperature (up to 200 °C), UV radiation, and relative humidity (up to 25 %RH). The sensor showed a negligible response to hydrogen without UV illumination. In contrast, the response was enhanced significantly under UV illumination due to stimulated electrons from the valence band of the sensing material [47]. These photogenerated electrons accelerate the adsorption and desorption reactions of H_2 molecules with the ambient oxygen species or water molecules [47].

The experimental results confirmed a p-type sensing behaviour with a maximum response of 121% towards 6000 ppm H_2 at 0 %RH and an optimum working temperature of 150 °C. This result could be attributed to increasing the concentration of the H_2 molecules, facilitating their interactions with the sensing material. Thus, enhancing the adsorption/desorption rate on the surface until it saturates and a steady state is reached. The sensor was capable of detecting the H_2 concentration as low as 50 ppm (R = 100.5%).

Gas sensing performance is correlated with the adsorption of atmospheric H_2O molecules as a significant source of interference [9,18]. Based on the literature, humidity is a complex challenge faced by most resistive H_2 sensors [9,71,72]. Therefore, developing a sensor capable of detecting H_2 in a humid environment can be a critical solution for humidity-related issues in the real environment [73]. Hence, we studied the effect of humidity on the Dy_2O_3-Pd-PDA/rGO nanocomposite sensing performance and its behaviour on the adsorption/desorption of H_2 molecules as one of the critical factors of developing a commercial sensor for humid environments [74]. It was observed that by increasing the %RH up to 20%, the sensor's performance (i.e., response and recovery times) drastically improved so that its response towards 5000 ppm H_2 at 150 °C was changed from 115.8% at 0 %RH to 145.2% at 20 %RH.

In addition, a maximum response of 144% was obtained upon exposure to 6000 ppm H_2 at 20 %RH and 150 °C, which is much higher than its response in a dry ambient (121%) at similar operational conditions. This result could be due to rising the charge carriers on the sensing layer surface by enriching the %RH because of the adsorbed ionised water molecules, in addition to the presence of the electron donor hydroxyl active sites on the sensing layer in the humid environment. By raising the %RH (from 15% to 20%), the sensor's response and recovery times also improved from 360 s to 180 s and 180 s to 120 s, respectively. Table 4 summarises a comparison of H_2 sensing performance between selected graphene-based nanomaterials, which are functionalised by noble metals and/or metal oxides with the current work.

In this work, reducing the GO by adding the DA, as an environmentally friendly chemical, highly improved its reduction process and functionalising the PDA/rGO with Pd NPs and 2D Dy_2O_3 significantly enhanced the H_2 sensing parameters [19]. In addition, the developed Dy_2O_3-Pd-PDA/rGO sensor was highly selective to hydrogen and illustrated high effectiveness in the humid environment.

Table 4. H_2 sensing properties of different graphene-based materials functionalised by noble metals and/or metal oxides.

Material.	Synthesis Method	Hybrid Material	Temp. (°C)	RH (%)	H_2 Conc. (ppm)	Response Time (s)	Recovery Time (s)	Response (%)
Graphene [75]	Thermal CVD	Pd NPs-SiO_2	RT	-	500	213	600	4.1
rGO [76]	Freeze drying	NiO	50	-	10,000	28	142	0.64
rGO [8]	Wet chemistry	Pd NPs	100	10	5000	170	1440	18.2
rGO [77]	Hummers' method, Hydrothermal	Pd-Pt	25	-	8000	300	600	0.52
rGO [78]	Hummers' method, Hydrothermal	Pt-SnO_2	RT	-	5000	3	2	3
PDA/rGO (this work)	Wet chemistry	Pd NPs-2D Dy_2O_3	150	20	5000	180	120	145.2

According to the literature, the hybrid materials' sensing performance can be enhanced by surface engineering due to forming the new physical and electrical properties, including the changes in thickness and several reactive sites for gas molecules interactions [79,80]. Overall, these enhancements could be based on improved electrical properties and charge transfer of the hybrid sensing material [81]. However, the hybrid materials gas sensing mechanisms seem to be complex, so to achieve predictable sensing improvements, many variants need to be investigated, such as the impact of different fabrication techniques, functionalisation, and structural parameters such as the porosity, thickness, surface to volume ratio and crystallinity of the materials. The mentioned parameters play critical roles in the sensing material and gas interactions, consequently changing the sensing performance [82]. However, these effects on the chemical and electrical properties of the sensors and thus enhancement in gas sensing properties may not be directly related.

In this work, the presence of oxygen species (O_2^-, O^-, and O^{2-}) on the sensing material surface under UV illumination played a critical role in the molecular adsorption of the fabricated p-type gas sensors, forming the photo-induced electron-hole pairs and an electron depleted region [15,83]. When the Dy_2O_3-Pd-PDA/rGO heterojunction nanocomposite is exposed to hydrogen as a reducing gas, the reduction reaction between the oxygen species and H_2 molecules accrue on the Dy_2O_3 surface, creating the Dy^{3+} due to the oxygen vacancy. This results in releasing the electron to the sensing layer's surface, affecting the depletion layer and resistance of the sensor [24]. At the same time, hybridising the Dy_2O_3 with Pd-PDA/rGO formed a p-n junction nanocomposite, creating Dy-O bondings at the interface between Dy_2O_3 and rGO, which positively affected the H_2 sensing performance of the final nanocomposite at a low operating temperature [84,85]. Another benefit of hybridisation was the formation of Pd-O-Dy bonds as active sites in the interface between the Dy_2O_3 and Pd NPs, facilitating the chemisorption of the H_2 molecules with the lattice oxygen, which also creates a synergic effect on the durability of the material due to anchoring the Pd NPs with the active sites' support [85].

To conclude, the results confirmed the regulated electronic structure of the Dy_2O_3-Pd-PDA/rGO heterojunction nanocomposite at the interface region, resulting in reduced energy barrier, enhanced kinetics, and more balanced hydrogen adsorption/desorption rate [86]. Moreover, the experimentally determined performance of the Dy_2O_3-Pd-PDA/rGO heterojunction nanocomposite revealed to be reliable, sensitive, have a rapid response, and recovery sensing materials, which has benefited from the use of environmentally friendly and simple synthesis and fabrication techniques as well as improvements in the combinations of Pd NPs and 2D Dy_2O_3 that comprise the composite sensor.

Author Contributions: H.H.: Conceptualisation, methodology, investigation, software, sensor devices fabrications, material characterisations, gas sensing data collection, data analysis, data interpretation and validation, data visualisation, writing—original and final draft preparation. A.Y.: material synthesis, review and editing the draft. S.J.: supervision, review and editing the draft. M.S.: main supervision, review and editing the draft, funding acquisition, resources. All authors have read and agreed to the published version of the manuscript.

Funding: This research was funded by the Australian Renewable Energy Agency (ARENA) as part of ARENA's Research and Development Program—Renewable Hydrogen for Export (Contract No. 2018/RND012). This research was supported by using the Nectar Research Cloud, a collaborative Australian research platform supported by the NCRIS-funded Australian Research Data Commons (ARDC).

Data Availability Statement: Not applicable.

Acknowledgments: This research was funded by the Australian Renewable Energy Agency (ARENA) as part of ARENA's Research and Development Program—Renewable Hydrogen for Export (Contract No. 2018/RND012). The views expressed herein are not necessarily the views of the Australian Government, and the Australian Government does not accept responsibility for any information or advice contained herein. We would like to acknowledge Christopher Harrison and Andrew Moore for their support in developing the gas sensing system at Swinburne. This research was supported by using the Nectar Research Cloud, a collaborative Australian research platform supported by the NCRIS-funded Australian Research Data Commons (ARDC). The authors would like to thank Jamie Riches for assistance in obtaining TEM images at the Central Analytical Research Facility at QUT.

Conflicts of Interest: The authors declare no conflict of interest.

References

1. Hashtroudi, H.; Atkin, P.; Mackinnon, I.D.R.; Shafiei, M. Low-operating temperature resistive nanostructured hydrogen sensors. *Int. J. Hydrogen Energy* **2019**, *44*, 26646–26664. [CrossRef]
2. Buttner, W.J.; Post, M.B.; Burgess, R.; Rivkin, C. An overview of hydrogen safety sensors and requirements. *Int. J. Hydrogen Energy* **2011**, *36*, 2462–2470. [CrossRef]
3. Hübert, T.; Boon-Brett, L.; Black, G.; Banach, U. Hydrogen sensors—A review. *Sens. Actuators B Chem.* **2011**, *157*, 329–352. [CrossRef]
4. Li, B.L.; Wang, J.; Zou, H.L.; Garaj, S.; Lim, C.T.; Xie, J.; Li, N.B.; Leong, D. Low-Dimensional Transition Metal Dichalcogenide Nanostructures Based Sensors. *Adv. Funct. Mater.* **2016**, *26*, 7034–7056. [CrossRef]
5. Ilnicka, A.; Lukaszewicz, J.P. Graphene-Based Hydrogen Gas Sensors: A Review. *Processes* **2020**, *8*, 633. [CrossRef]
6. Pham, T.K.N.; Brown, J.J. Hydrogen Sensors Using 2-Dimensional Materials: A Review. *ChemistrySelect* **2020**, *5*, 7277–7297. [CrossRef]
7. Weng, Q.; Li, G.; Feng, X.; Nielsch, K.; Golberg, D.; Schmidt, O.G. Electronic and Optical Properties of 2D Materials Constructed from Light Atoms. *Adv. Mater.* **2018**, *30*, e1801600. [CrossRef]
8. Hashtroudi, H.; Kumar, R.; Savu, R.; Moshkalev, S.; Kawamura, G.; Matsuda, A.; Shafiei, M. Hydrogen gas sensing properties of microwave-assisted 2D Hybrid Pd/rGO: Effect of temperature, humidity and UV illumination. *Int. J. Hydrogen Energy* **2021**, *46*, 7653–7665. [CrossRef]
9. Joshi, N.; Hayasaka, T.; Liu, Y.; Liu, H.; Oliveira, O.N., Jr.; Lin, L.J.M.A. A review on chemiresistive room temperature gas sensors based on metal oxide nanostructures, graphene and 2D transition metal dichalcogenides. *Microchim. Acta* **2018**, *185*, 213. [CrossRef] [PubMed]
10. Fang, W.; Yang, Y.; Yu, H.; Dong, X.; Wang, R.; Wang, T.; Wang, J.; Liu, Z.; Zhao, B.; Wang, X. An In_2O_3 nanorod-decorated reduced graphene oxide composite as a high-response NOx gas sensor at room temperature. *New J. Chem.* **2017**, *41*, 7517–7523. [CrossRef]
11. Varghese, S.S.; Varghese, S.H.; Swaminathan, S.; Singh, K.K.; Mittal, V. Two-Dimensional Materials for Sensing: Graphene and Beyond. *Electronics* **2015**, *4*, 651–687. [CrossRef]
12. Shi, J.; Cheng, Z.; Gao, L.; Zhang, Y.; Xu, J.; Zhao, H. Facile synthesis of reduced graphene oxide/hexagonal WO3 nanosheets composites with enhanced H_2S sensing properties. *Sens. Actuators B Chem.* **2016**, *230*, 736–745. [CrossRef]
13. Luo, Y.; Zhang, C.; Zheng, B.; Geng, X.; Debliquy, M. Hydrogen sensors based on noble metal doped metal-oxide semiconductor: A review. *Int. J. Hydrogen Energy* **2017**, *42*, 20386–20397. [CrossRef]
14. Anand, K.; Singh, O.; Singh, M.P.; Kaur, J.; Singh, R.C. Hydrogen sensor based on graphene/ZnO nanocomposite. *Sens. Actuators B Chem.* **2014**, *195*, 409–415. [CrossRef]
15. Boudiba, A.; Roussel, P.; Zhang, C.; Olivier, M.-G.; Snyders, R.; Debliquy, M. Sensing mechanism of hydrogen sensors based on palladium-loaded tungsten oxide (Pd–WO3). *Sens. Actuators B Chem.* **2013**, *187*, 84–93. [CrossRef]
16. Novoselov, K. Nobel Lecture: Graphene: Materials in the Flatland. *Rev. Mod. Phys.* **2011**, *83*, 837–849. [CrossRef]
17. Armgarth, M.; Söderberg, D.; Lundström, I. Palladium and platinum gate metal-oxide-semiconductor capacitors in hydrogen and oxygen mixtures. *Appl. Phys. Lett.* **1982**, *41*, 654–655. [CrossRef]
18. Bindra, P.; Hazra, A. Capacitive gas and vapor sensors using nanomaterials. *J. Mater. Sci. Mater. Electron.* **2018**, *29*, 6129–6148. [CrossRef]
19. Johnson, J.L.; Behnam, A.; Pearton, S.; Ural, A. Hydrogen Sensing Using Pd-Functionalized Multi-Layer Graphene Nanoribbon Networks. *Adv. Mater.* **2010**, *22*, 4877–4880. [CrossRef]

20. Silva, C.; Simon, F.; Friedel, P.; Pötschke, P.; Zimmerer, C. Elucidating the Chemistry behind the Reduction of Graphene Oxide Using a Green Approach with Polydopamine. *Nanomaterials* **2019**, *9*, 902. [CrossRef]
21. Xu, L.Q.; Yang, W.J.; Neoh, K.-G.; Kang, E.-T.; Fu, G.D. Dopamine-Induced Reduction and Functionalization of Graphene Oxide Nanosheets. *Macromolecules* **2010**, *43*, 8336–8339. [CrossRef]
22. Fu, L.; Lai, G.; Zhu, D.; Jia, B.; Malherbe, F.; Yu, A. Advanced Catalytic and Electrocatalytic Performances of Polydopamine-Functionalized Reduced Graphene Oxide-Palladium Nanocomposites. *ChemCatChem* **2016**, *8*, 2975–2980. [CrossRef]
23. Su, Y.-D.; Shih, W.-C.; Lee, J.Y.-M. The characterization of retention properties of metal–ferroelectric ($PbZr_{0.53}Ti_{0.47}$ $47O_3$)–insulator (Dy_2O_3, Y_2O_3)–semiconductor devices. *Microelectron. Reliab.* **2007**, *47*, 619–622. [CrossRef]
24. Dong, X.; Cheng, X.; Zhang, X.; Sui, L.; Xu, Y.; Gao, S.; Zhao, H.; Huo, L. A novel coral-shaped Dy_2O_3 gas sensor for high sensitivity NH3 detection at room temperature. *Sens. Actuators B Chem.* **2017**, *255*, 1308–1315. [CrossRef]
25. Pan, T.-M.; Lin, C.-W. Structural and Sensing Characteristics of Dy_2O_3 and Dy_2TiO_5 Electrolyte−Insulator−Semiconductor pH Sensors. *J. Phys. Chem. C* **2010**, *114*, 17914–17919. [CrossRef]
26. Nur-E-Alam, M.; Vasiliev, M.; Kotov, V.; Alameh, K. Recent Developments in Magneto-optic Garnet-type Thin-film Materials Synthesis. *Procedia Eng.* **2014**, *76*, 61–73. [CrossRef]
27. Vuong, Q.L.; Van Doorslaer, S.; Bridot, J.-L.; Argante, C.; Alejandro, G.; Hermann, R.; Disch, S.; Mattea, C.; Stapf, S.; Gossuin, Y. Paramagnetic nanoparticles as potential MRI contrast agents: Characterization, NMR relaxation, simulations and theory. *Magn. Reson. Mater. Phys. Biol. Med.* **2012**, *25*, 467–478. [CrossRef]
28. He, B.; Zhao, L.; Wang, W.; Chen, F.; Xia, C. Electro-catalytic activity of Dy_2O_3 as a solid oxide fuel cell anode material. *Electrochem. Commun.* **2011**, *13*, 194–196. [CrossRef]
29. Andrade, E.N.D.C.; Lonsdale, K.Y. *William Henry Bragg, 1862–1942*; Royal Society Publishing: London, UK, 1943.
30. Klung, H.; Alexander, L. *X-ray Diffraction Procedures*; John Wiley & Sons: New York, NY, USA, 1962; Volume 1, p. 974.
31. Sharma, N.D.; Singh, J.; Dogra, S.; Varandani, D.; Poswal, H.K.; Sharma, S.M.; Bandyopadhyay, A.K. Pressure-induced anomalous phase transformation in nano-crystalline dysprosium sesquioxide. *J. Raman Spectrosc.* **2011**, *42*, 438–444. [CrossRef]
32. Verma, R.; Samdarshi, S.K. In Situ Decorated Optimized CeO_2 on Reduced Graphene Oxide with Enhanced Adsorptivity and Visible Light Photocatalytic Stability and Reusability. *J. Phys. Chem. C* **2016**, *120*, 22281–22290. [CrossRef]
33. Anema, J.R.; Li, J.-F.; Yang, Z.-L.; Ren, B.; Tian, Z.-Q. Shell-Isolated Nanoparticle-Enhanced Raman Spectroscopy: Expanding the Versatility of Surface-Enhanced Raman Scattering. *Annu. Rev. Anal. Chem.* **2011**, *4*, 129–150. [CrossRef]
34. Liu, Y.; Wang, F.; Liu, Y.; Wang, X.; Xu, Y.; Zhang, R. Charge transfer at carbon nanotube–graphene van der Waals heterojunctions. *Nanoscale* **2016**, *8*, 12883–12886. [CrossRef]
35. Zhang, B.; Cheng, M.; Liu, G.; Gao, Y.; Zhao, L.; Li, S.; Wang, Y.; Liu, F.; Liang, X.; Zhang, T.; et al. Room temperature NO_2 gas sensor based on porous Co_3O_4 slices/reduced graphene oxide hybrid. *Sens. Actuators B Chem.* **2018**, *263*, 387–399. [CrossRef]
36. Shen, Y.; Lu, S.; Xu, W.; Lv, A.; Wang, Z.; Wang, H.; Liu, G.; Zhang, Y. Fabrication of Composite Material with Pd Nanoparticles and Graphene on Nickel Foam for Its Excellent Electrocatalytic Performance. *Electrocatalysis* **2020**, *11*, 522–535. [CrossRef]
37. Eifert, B.; Becker, M.; Reindl, C.T.; Giar, M.; Zheng, L.; Polity, A.; He, Y.; Heiliger, C.; Klar, P.J. Raman studies of the intermediate tin-oxide phase. *Phys. Rev. Mater.* **2017**, *1*, 014602. [CrossRef]
38. Dresselhaus, M.S.; Jorio, A.; Filho, A.G.S.; Saito, R. Defect characterization in graphene and carbon nanotubes using Raman spectroscopy. *Philos. Trans. R. Soc. London. Ser. A Math. Phys. Eng. Sci.* **2010**, *368*, 5355–5377. [CrossRef] [PubMed]
39. Hashtroudi, H.; Mackinnon, I.D.R.; Shafiei, M. Emerging 2D hybrid nanomaterials: Towards enhanced sensitive and selective conductometric gas sensors at room temperature. *J. Mater. Chem. C* **2020**, *8*, 13108–13126. [CrossRef]
40. Kapoor, A.; Ritter, J.A.; Yang, R.T. An extended Langmuir model for adsorption of gas mixtures on heterogeneous surfaces. *Langmuir* **1990**, *6*, 660–664. [CrossRef]
41. Voiry, D.; Yang, J.; Chhowalla, M. Recent Strategies for Improving the Catalytic Activity of 2D TMD Nanosheets Toward the Hydrogen Evolution Reaction. *Adv. Mater.* **2016**, *28*, 6197–6206. [CrossRef] [PubMed]
42. Sun, Y.-F.; Liu, S.-B.; Meng, F.-L.; Liu, J.-Y.; Jin, Z.; Kong, L.-T.; Liu, J.-H. Metal Oxide Nanostructures and Their Gas Sensing Properties: A Review. *Sensors* **2012**, *12*, 2610–2631. [CrossRef]
43. Rajkumar, K.; Kumar, R.R. Gas sensors based on two-dimensional materials and its mechanisms. In *Fundamentals and Sensing Applications of 2D Materials*; Elsevier: Amsterdam, The Netherlands, 2019; pp. 205–258.
44. Du, H.; Xie, G.; Su, Y.; Tai, H.; Du, X.; Yu, H.; Zhang, Q. A New Model and Its Application for the Dynamic Response of RGO Resistive Gas Sensor. *Sensors* **2019**, *19*, 889. [CrossRef] [PubMed]
45. Rout, C. *Fundamentals and Sensing Applications of 2D Materials*; Woodhead Publishing: Duxford, UK, 2019.
46. Potje-Kamloth, K. Semiconductor Junction Gas Sensors. *Chem. Rev.* **2008**, *108*, 367–399. [CrossRef] [PubMed]
47. Kumar, R.; Liu, X.; Zhang, J.; Kumar, M. Room-Temperature Gas Sensors Under Photoactivation: From Metal Oxides to 2D Materials. *Nano-Micro Lett.* **2020**, *12*, 164. [CrossRef] [PubMed]
48. Sharma, R.; Rawal, D. *The Physics of Semiconductor Devices: Proceedings of IWPSD 2017*; Springer: Berlin/Heidelberg, Germany, 2018; Volume 215.
49. Meng, Z.; Stolz, R.M.; Mendecki, L.; Mirica, K.A. Electrically-Transduced Chemical Sensors Based on Two-Dimensional Nanomaterials. *Chem. Rev.* **2019**, *119*, 478–598. [CrossRef] [PubMed]

50. Bellardita, M.; Fiorenza, R.; Palmisano, L.; Sciré, S. Photocatalytic and photo-thermo-catalytic applications of cerium oxide-based material. In *Cerium Oxide (CeO_2): Synthesis, Properties and Applications*; a Volume in Metal Oxides Series; Elsevier: Amsterdam, The Netherlands, 2020; ISBN 978-0-12-815661-2. [CrossRef]
51. Chen, X.; Shen, S.; Guo, L.; Mao, S.S. Semiconductor-based Photocatalytic Hydrogen Generation. *Chem. Rev.* **2010**, *110*, 6503–6570. [CrossRef]
52. Chen, X.; Mao, S.S. Titanium dioxide nanomaterials: Synthesis, properties, modifications, and applications. *Chem. Rev.* **2007**, *107*, 2891–2959. [CrossRef]
53. Muckley, E.S.; Nelson, A.J.; Jacobs, C.B.; Ivanov, I.N. Effect of UV irradiation on adsorption/desorption of oxygen and water on carbon nanotubes. *Org. Photonic Mater. Devices* **2016**, *9745*, 97451. [CrossRef]
54. Sun, P.; Zhu, M.; Wang, K.; Zhong, M.; Wei, J.; Wu, D.; Cheng, Y.; Zhu, H. Photoinduced molecular desorption from graphene films. *Appl. Phys. Lett.* **2012**, *101*, 53107. [CrossRef]
55. Azpiroz, J.M.; Ronca, E.; De Angelis, F. Photoinduced Energy Shift in Quantum-Dot-Sensitized TiO_2: A First-Principles Analysis. *J. Phys. Chem. Lett.* **2015**, *6*, 1423–1429. [CrossRef]
56. Chauhan, P.S.; Bhattacharya, S. Hydrogen gas sensing methods, materials, and approach to achieve parts per billion level detection: A review. *Int. J. Hydrogen Energy* **2019**, *44*, 26076–26099. [CrossRef]
57. Moon, J.; Cheng, Y.; Daemen, L.L.; Li, M.; Polo-Garzon, F.; Ramirez-Cuesta, A.J.; Wu, Z. Discriminating the Role of Surface Hydride and Hydroxyl for Acetylene Semihydrogenation over Ceria through In Situ Neutron and Infrared Spectroscopy. *ACS Catal.* **2020**, *10*, 5278–5287. [CrossRef]
58. Chen, Y.; Pei, Y.; Jiang, Z.; Shi, Z.; Xu, J.; Wu, D.; Xu, T.; Tian, Y.; Wang, X.; Li, X. Humidity sensing properties of the hydrothermally synthesized WS2-modified SnO_2 hybrid nanocomposite. *Appl. Surf. Sci.* **2018**, *447*, 325–330. [CrossRef]
59. Yavari, F.; Koratkar, N. Graphene-Based Chemical Sensors. *J. Phys. Chem. Lett.* **2012**, *3*, 1746–1753. [CrossRef]
60. Dimiev, A.M.; Alemany, L.B.; Tour, J.M. Graphene Oxide. Origin of Acidity, Its Instability in Water, and a New Dynamic Structural Model. *ACS Nano* **2012**, *7*, 576–588. [CrossRef]
61. Yan, W.; Worsley, M.A.; Pham, T.; Zettl, A.; Carraro, C.; Maboudian, R. Effects of ambient humidity and temperature on the NO2 sensing characteristics of WS2/graphene aerogel. *Appl. Surf. Sci.* **2018**, *450*, 372–379. [CrossRef]
62. Inpaeng, S.; Muangrat, W.; Tedsree, K.; Pfeiler, W.; Chodjarusawad, T.; Issro, C. Effective hydrogen gas sensor based on palladium nanoparticles dispersed on graphene sheets by spin coating technique. *Mater. Sci.* **2020**, *38*, 305–311. [CrossRef]
63. Yao, Y.; Chen, X.; Zhu, J.; Zeng, B.; Wu, Z.; Li, X. The effect of ambient humidity on the electrical properties of graphene oxide films. *Nanoscale Res. Lett.* **2012**, *7*, 363. [CrossRef]
64. Piloto, C.; Shafiei, M.; Khan, H.; Gupta, B.; Tesfamichael, T.; Motta, N. Sensing performance of reduced graphene oxide-Fe doped WO_3 hybrids to NO_2 and humidity at room temperature. *Appl. Surf. Sci.* **2018**, *434*, 126–133. [CrossRef]
65. Buscema, M.; Island, J.O.; Groenendijk, D.J.; Blanter, S.I.; Steele, G.A.; van der Zant, H.S.J.; Castellanos-Gomez, A. Photocurrent generation with two-dimensional van der Waals semiconductors. *Chem. Soc. Rev.* **2015**, *44*, 3691–3718. [CrossRef]
66. Nair, R.R.; Wu, H.A.; Jayaram, P.N.; Grigorieva, I.V.; Geim, A.K. Unimpeded Permeation of Water Through Helium-Leak-Tight Graphene-Based Membranes. *Science* **2012**, *335*, 442–444. [CrossRef]
67. Toda, K.; Furue, R.; Hayami, S. Recent progress in applications of graphene oxide for gas sensing: A review. *Anal. Chim. Acta* **2015**, *878*, 43–53. [CrossRef] [PubMed]
68. Esfandiar, A.; Ghasemi, S.; Zad, A.I.; Akhavan, O.; Gholami, M. The decoration of TiO_2/reduced graphene oxide by Pd and Pt nanoparticles for hydrogen gas sensing. *Int. J. Hydrogen Energy* **2012**, *37*, 15423–15432. [CrossRef]
69. Borini, S.; White, R.; Wei, D.; Astley, M.; Haque, S.; Spigone, E.; Harris, N.; Kivioja, J.; Ryhänen, T. Ultrafast Graphene Oxide Humidity Sensors. *ACS Nano* **2013**, *7*, 11166–11173. [CrossRef] [PubMed]
70. Seiyama, T.; Kato, A.; Fujiishi, K.; Nagatani, M. A New Detector for Gaseous Components Using Semiconductive Thin Films. *Anal. Chem.* **1962**, *34*, 1502–1503. [CrossRef]
71. Hassan, K.; Chung, G.-S. Catalytically activated quantum-size Pt/Pd bimetallic core–shell nanoparticles decorated on ZnO nanorod clusters for accelerated hydrogen gas detection. *Sens. Actuators B Chem.* **2017**, *239*, 824–833. [CrossRef]
72. Zhang, J.; Liu, X.; Neri, G.; Pinna, N. Nanostructured Materials for Room-Temperature Gas Sensors. *Adv. Mater.* **2015**, *28*, 795–831. [CrossRef]
73. Chatterjee, S.G.; Chatterjee, S.; Ray, A.K.; Chakraborty, A.K. Graphene–metal oxide nanohybrids for toxic gas sensor: A review. *Sens. Actuators B Chem.* **2015**, *221*, 1170–1181. [CrossRef]
74. Tripathi, K.; Kim, T.; Losic, D.; Tung, T.T. Recent advances in engineered graphene and composites for detection of volatile organic compounds (VOCs) and non-invasive diseases diagnosis. *Carbon* **2016**, *110*, 97–129. [CrossRef]
75. Wu, W.; Liu, Z.; Jauregui, L.A.; Yu, Q.; Pillai, R.; Cao, H.; Bao, J.; Chen, Y.P.; Pei, S.-S. Wafer-scale synthesis of graphene by chemical vapor deposition and its application in hydrogen sensing. *Sens. Actuators B Chem.* **2010**, *150*, 296–300. [CrossRef]
76. Ren, H.; Gu, C.; Joo, S.W.; Zhao, J.; Sun, Y.; Huang, J. Effective hydrogen gas sensor based on NiO@rGO nanocomposite. *Sens. Actuators B Chem.* **2018**, *266*, 506–513. [CrossRef]
77. Peng, Y.; Ye, J.; Zheng, L.; Zou, K. The hydrogen sensing properties of Pt–Pd/reduced graphene oxide based sensor under different operating conditions. *RSC Adv.* **2016**, *6*, 24880–24888. [CrossRef]

78. Russo, P.; Donato, N.; Leonardi, S.G.; Baek, S.; Conte, D.E.; Neri, G.; Pinna, N. Room-Temperature Hydrogen Sensing with Heteronanostructures Based on Reduced Graphene Oxide and Tin Oxide. *Angew. Chem. Int. Ed.* **2012**, *51*, 11053–11057. [CrossRef] [PubMed]
79. Chen, A.; Liu, R.; Peng, X.; Chen, Q.; Wu, J. 2D Hybrid Nanomaterials for Selective Detection of NO_2 and SO_2 Using "Light On and Off" Strategy. *ACS Appl. Mater. Interfaces* **2017**, *9*, 37191–37200. [CrossRef]
80. Hoa, L.T.; Tien, H.N.; Luan, V.H.; Chung, J.S.; Hur, S.H. Fabrication of a novel 2D-graphene/2D-NiO nanosheet-based hybrid nanostructure and its use in highly sensitive NO_2 sensors. *Sens. Actuators B Chem.* **2013**, *185*, 701–705. [CrossRef]
81. Achary, L.S.K.; Kumar, A.; Barik, B.; Nayak, P.S.; Tripathy, N.; Kar, J.P.; Dash, P. Reduced graphene oxide-$CuFe_2O_4$ nanocomposite: A highly sensitive room temperature NH_3 gas sensor. *Sens. Actuators B Chem.* **2018**, *272*, 100–109. [CrossRef]
82. Yamazaki, T.; Okumura, H.; Jin, C.-J.; Nakayama, A.; Kikuta, T.; Nakatani, N. Effect of density and thickness on H_2-gas sensing property of sputtered SnO_2 films. *Vacuum* **2005**, *77*, 237–243. [CrossRef]
83. Yuan, K.-P.; Zhu, L.-Y.; Yang, J.-H.; Hang, C.-Z.; Tao, J.-J.; Ma, H.-P.; Jiang, A.-Q.; Zhang, D.W.; Lu, H.-L. Precise preparation of WO_3@SnO_2 core shell nanosheets for efficient NH3 gas sensing. *J. Colloid Interface Sci.* **2020**, *568*, 81–88. [CrossRef]
84. Liu, H.-H.; Wang, Y.; Jia, A.-P.; Wang, S.; Luo, M.-F.; Lu, J.-Q. Oxygen vacancy promoted CO oxidation over Pt/CeO_2 catalysts: A reaction at Pt–CeO_2 interface. *Appl. Surf. Sci.* **2014**, *314*, 725–734. [CrossRef]
85. Hinokuma, S.; Fujii, H.; Okamoto, M.; Ikeue, K.; Machida, M. Metallic Pd Nanoparticles Formed by Pd−O−Ce Interaction: A Reason for Sintering-Induced Activation for CO Oxidation. *Chem. Mater.* **2010**, *22*, 6183–6190. [CrossRef]
86. Cao, Q.; Hao, S.; Wu, Y.; Pei, K.; You, W.; Che, R. Interfacial charge redistribution in interconnected network of Ni_2P–Co_2P boosting electrocatalytic hydrogen evolution in both acidic and alkaline conditions. *Chem. Eng. J.* **2021**, *424*, 130444. [CrossRef]

chemosensors

MDPI

Communication

Potential Assessment of Dehydration during High-Intensity Training Using a Capacitance Sensor for Oral Mucosal Moisture: Evaluation of Elite Athletes in a Field-Based Survey

Gen Tanabe [1,2], Tetsuya Hasunuma [3,4], Yuto Inai [3,5], Yasuo Takeuchi [6], Hiroaki Kobayashi [6,7], Kairi Hayashi [1], Shintaro Shimizu [1], Nana S Kamiya [1,3], Hiroshi Churei [1], Yuka I Sumita [2], Katsuhiko Suzuki [8], Naoki Moriya [3,9] and Toshiaki Ueno [1,*]

1 Department of Sports Medicine/Dentistry, Graduate School of Medical and Dental Sciences, Tokyo Medical and Dental University, Tokyo 113-8510, Japan; gen.spmd@tmd.ac.jp (G.T.); k.hayashi.spmd@tmd.ac.jp (K.H.); sshispmd@tmd.ac.jp (S.S.); nshspmd@tmd.ac.jp (N.S.K.); chu.spmd@tmd.ac.jp (H.C.)
2 Department of Maxillofacial Prosthetics, Graduate School of Medical and Dental Sciences, Tokyo Medical and Dental University, Tokyo 113-8510, Japan; yuka.mfp@tmd.ac.jp
3 Public Interest Incorporated Association, Japan Triathlon Union, Tokyo 160-0013, Japan; hsnmtetsuya@gmail.com (T.H.); tri0410inai@gmail.com (Y.I.); moriya@jtu.or.jp (N.M.)
4 Faculty of Human Development and Culture, Fukushima University, Fukushima 960-1296, Japan
5 Graduate School of Sport Sciences, Waseda University, Saitama 359-1192, Japan
6 Department of Periodontology, Graduate School of Medical and Dental Sciences, Tokyo Medical and Dental University, Tokyo 113-8510, Japan; takeuchi.peri@tmd.ac.jp (Y.T.); h-kobayashi.peri@tmd.ac.jp (H.K.)
7 Oral Health Center, Sumitomo Corporation, Tokyo 100-8601, Japan
8 Faculty of Sport Science, Waseda University, Saitama 359-1192, Japan; katsu.suzu@waseda.jp
9 Bunka Gakuen University, Tokyo 151-8523, Japan
* Correspondence: t.ueno.spmd@tmd.ac.jp; Tel.: +81-3-5803-5867

Citation: Tanabe, G.; Hasunuma, T.; Inai, Y.; Takeuchi, Y.; Kobayashi, H.; Hayashi, K.; Shimizu, S.; Kamiya, N.S; Churei, H.; Sumita, Y.I.; et al. Potential Assessment of Dehydration during High-Intensity Training Using a Capacitance Sensor for Oral Mucosal Moisture: Evaluation of Elite Athletes in a Field-Based Survey. *Chemosensors* **2021**, *9*, 196. https://doi.org/10.3390/chemosensors9080196

Academic Editor: Pietro Salvo

Received: 23 June 2021
Accepted: 24 July 2021
Published: 28 July 2021

Abstract: Background: The aim of this clinical study was to reveal the relationship between body dehydration and oral mucosa moisture measured by the use of a capacitance sensor for oral epithelial moisture. Methods: The following clinical parameters were recorded from each one of 19 athletes in a one-week period of high-intensity exercise at the U-23 Triathlon Training Camp in summer and winter; body weight, urine specific gravity, oral mucosa moisture, subjective oral thirst, and subjective throat thirst (within 30 min after waking and before breakfast at 7:00 a.m. on Day2 and Day6). Results: There were no significant differences in the mean values of body weight, urine specific gravity, oral mucosa moisture, oral thirst, and throat thirst between Day2 and Day6 in both measurements in summer and winter. The oral mucosa moisture had a moderate negative correlation with urine specific gravity ($p < 0.05$, $r = -0.45$). Conclusions: This study suggests that oral mucosal moisture determined using an oral moisture-checking device could be a potential index for assessing dehydration during sports activities.

Keywords: dehydration; capacitance sensor; oral mucosal moisture; urine specific gravity

1. Introduction

Body temperature elevations due to physical activity elicit responses of increased skin blood flow and increased sweat secretion [1]. Sweat evaporation provides a primary function of heat loss in the body, but both water and electrolytes are also lost. Water and electrolyte imbalances (dehydration and hyponatremia) can develop and adversely impact individuals' exercise performance and health [2].

Dehydration of >2% body weight (BW) can degrade aerobic exercise performance [3] and cognitive/mental performance [4]. Dehydration of over 3% BW degrades muscular strength [5]. Furthermore, dehydration impairs exercise performance and contributes to serious heat exhaustion [6] and exertional heat stroke [7], rhabdomyolysis [8], and exercise-associated hyponatremia [9]. Dehydration can lead to death in severe cases; thus, dehydration assessment is important.

Table 1 shows an assessment of several dehydration biomarkers and subjective indicators. Total body weight and plasma osmolality are very valid and precise measures of body hydration but are of low practicality for use by most persons [2]. In physical activity, the biomarkers should be practical for use by individuals, coaches, and so on. Measures of urine biomarkers such as urine specific gravity and osmolality provide valuable insight [10], and urine color and volume are often used as subjective indicators. These urine indices are used in sports because of their practicality and especially urine specific gravity has been reported to be an invasive, inexpensive, simple, fast, and accurate indicator of hydration status before exercise [11]. Urine specific gravity is often used in clinicopathology to assess kidney function. The kidney minimizes water loss by concentrating urine. Therefore, urine specific gravity of <1.020 g/mL is indicative of dehydration.

Table 1. Assessment of dehydration biomarkers.

	Measure	Euhydration Cutoff
Blood	Plasma osmolality	<290 mOsmol
Urine	Urine specific gravity Urine osmolality Urine color	<1.020 g/mL <700 mOsmol
Weight	Body weight Total body water	<1% <2%
Oral	Thirst Mouth dryness	

Oral characteristics such as wrinkles and dryness of the tongue and subjective feelings of thirst in the mouth and/or throat have also been used to assess dehydration. Recent studies have reported the utility of salivary viscosity [12], saliva osmotic pressure [13], capillary refill time [14], and oral moisture [15] as diagnostic indicators.

All these indices have limitations such as the time, cost, and labor required for collection during exercise, the fact that they may be invasive and require technical expertise, and the fact that there is a time lag between the current symptoms and the indices, so there is a need for a simpler evaluation method that can provide immediate feedback.

Recently, in the field of medicine, evaluation of dehydration severity using a capacitance sensor for oral epithelial moisture has been attempted for dehydrated patients brought to emergency rooms, and it was reported that oral moisture measurement is useful for initial evaluation of dehydration [15]. Dry mouth is one of the most common clinical findings in dehydrated patients and is often associated with dehydration. McGee et al. reported that 85% of dehydrated patients experience dry mouth [16]. In recent years, the measurement of oral moisture has begun to be applied in not only the evaluation of oral dryness among elderly populations [17] but also a variety of situations in clinical practice, such as maxillofacial prosthesis patients [18,19], oropharyngeal cancer [20], xerostomia due to side effects of medication [21], and so on.

Mucus® (Life Co., Saitama, Japan; Figure 1), a capacitance sensor for oral epithelial moisture, easily measures oral moisture within two seconds [22]. This device is the first to measure oral moisture, and the principle of this device is that the epithelial moisture is measured as capacitance. The dielectric constant of water is much higher than that of other substances; therefore, the percentage of water in the substance can be checked by measuring the dielectric constant of the substance. The moisture of the substance is measured by calculating the capacitance from the dielectric constant of the substance, and the principle is the same as that of a skin moisture sensor. The indicated value (%) of an oral moisture meter is based on the gravimetric moisture content of a standard sample of a special protein membrane for medical use and is expressed as a percentage of moisture content. The definition of moisture content by the gravimetric method is the following: moisture content = B/(A + B) × 100% (A: weight of dry protein membrane, B: weight of water), where 100% represents the state when the sample is entirely water. However, the measurement range of the moisture meter is about 15 to 65%. The reliability of the data was previously demonstrated by comparison with the dry weight method [22]. Fukushima et al. reported that the sensitivity and specificity values are close to 80%. The oral moisture values range from 0 to 99.9, and values of ≥29.6, 28.0–29.5, and ≤27.9 are defined as normal, borderline dry mouth, and dry mouth, respectively [23]. Takano et al. reported that using an oral moisture-checking device has sufficient intra- and inter-investigator reliabilities [24]. As such, Mucus® has received manufacturing and marketing approval as a body composition analyzer (approval number: 22200BZX00640000) by the Pharmaceutical and Medical Devices Agency of Japan.

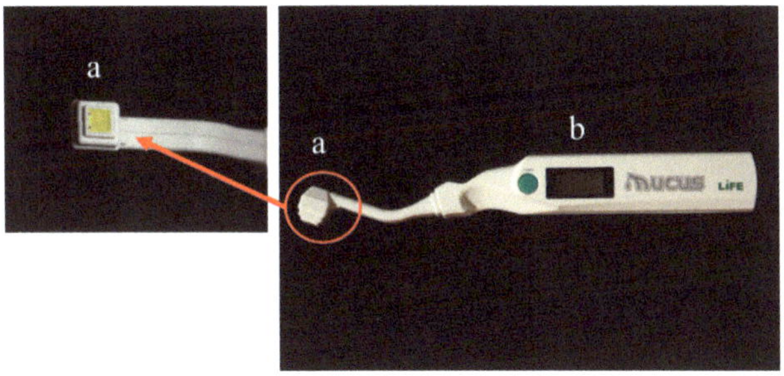

Figure 1. Mucus®; Life Co., Ltd., Saitama, Japan, (**a**) is a capacitance sensor; (**b**) is the score display.

We hypothesized that oral moisture measures by the use of such a device may provide a rapid and precise indication of dehydration in sports. The aim of this clinical study conducted in a one-week period of high-intensity exercise at the U-23 Triathlon Training Camp during summer and winter 2019 was to reveal the relationship between body dehydration and the oral mucosa moisture measured by the use of a capacitance sensor for oral epithelial moisture.

2. Materials and Methods

2.1. Study Subjects

This study was approved by the research ethics committee of Tokyo Medical and Dental University (approval number: D2019-031). Before the start of this study, all the study participants received verbal explanations regarding data collection and the protection of privacy and personal information, and they each signed an informed consent form.

Data were collected at the U-23 Triathlon Training Camp during summer and winter 2019. Samples of 6 subjects in the summer camp and 13 subjects in the winter camp were recruited from the participants of the U-23 Triathlon Training Camp; those diagnosed with oral mucosal abnormalities of the tongue and/or Sjögren's syndrome or xerostomia were not invited to participate in this study.

2.2. Data Collection

The camp lasted for seven days, of which the mornings of the second and sixth days were set as measurement days. Within 30 min after waking and before breakfast at 7:00 a.m., data regarding body weight, urine specific gravity, oral mucosa moisture, oral thirst, and throat thirst were collected from each subject. Body weight was measured and recorded on a scale (YAMAZEN HCF-40 Life Co., Ltd., Saitama, Japan). Urine specific gravity was measured by collecting urine early in the morning and measuring it with a urine specific gravity meter (PEN urine specific gravity meter Atago Co., Ltd., Tokyo, Japan). The extent of oral mucosa moisture was measured using a capacitance sensor for oral epithelial moisture (Mucus®; Life Co., Ltd., Saitama, Japan). It was measured on the center of the lingual mucosa approximately 10 mm from the tip of the tongue; the device was manually applied at a pressure of approximately 200 g by a single measurer (Figure 2). A dedicated disposable polyethylene cover was applied to the sensor for each subject. The measurements were done three times, and the median values were recorded. Subjective scores of oral and throat thirst were recorded on an 11-point scale from 0 to 10 (10: strong sense of thirst). Furthermore, the weather and maximum temperatures during the camp were recorded, and the session rating of perceived exertion (s-RPE) was recorded as the exercise intensity during the camp. For the measurement of the s-RPE, the modified Borg CR-10 scale was used to quantify session intensity [25]. Following the final training session each day, each participant individually recorded the subjective exercise intensity for that day.

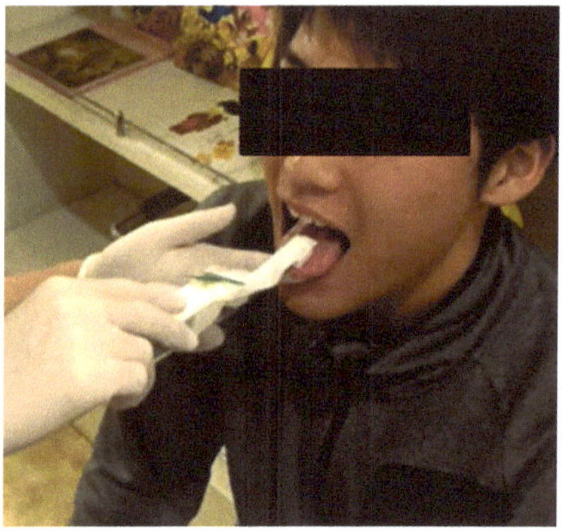

Figure 2. An example of measurement of oral mucosal moisture by using an oral moisture checking device (Mucus®; Life Co., Ltd., Saitama, Japan).

2.3. Data Analysis

Data regarding body weight, urine specific gravity, oral mucosa moisture, oral thirst, and throat thirst were statistically analyzed using Student's paired *t*-test to compare the second day with the sixth day. The correlations of the oral mucosa moisture and urine specific gravity results were examined via Spearman's correlation coefficient, and the regression line was calculated. The statistical analyses were performed using statistical software JMP14 (SAS Institute Inc., Cary, NC, USA) at a 5% significance level.

3. Results

The study participants' demographic data (sex, age, body weight) are summarized in Table 2. The weather and maximum temperature during the camp is shown in Table 3. The summer camp had mostly sunny and hot days over 30 °C, while the winter camp had many rainy, cloudy, and cold days. Transitions in body weight, urine specific gravity, oral mucosa moisture, oral thirst, and throat thirst in the summer and winter camps are shown in Figure 3. There were no significant differences in the mean values of body weight, urine specific gravity, oral mucosa moisture, oral thirst, and throat thirst between Day2 and Day6 in both measurements in summer and winter. In the summer, the urine specific gravity was above 1.020 g/mL, which is considered in sports to indicate dehydration, and the oral mucosa moisture measured by the Mucus® device was below 28.0, which is diagnosed as xerostomia. In the summer, all values tended to worsen at the end of the camp, but in the winter, they tended to improve because summer is the heat and humidity environment [26].

Table 2. Subject data, mean ± SD.

Number of subjects	19 (Male 10; Female 9)
Mean age	20.8 ± 0.8
Weight (kg)	58.1 ± 8.7

Table 3. The weather and maximum temperatures during the summer/winter camps.

Summer	Day1	Day2	Day3	Day4	Day5	Day6	Day7
Weather	Sunny	Sunny	Sunny	Sunny	Sunny	Sunny	Sunny
Maximum temperature (°C)	30.1	30.3	30.5	33.5	32.6	31.1	32.0
s-RPE [1]	6.6 ± 2.7 *	34.0 ± 10.5	25.5 ± 7.4	13.3 ± 4.5	33.0 ± 10.5	39.4 ± 8.8	10.3 ± 2.9
Winter	**Day1**	**Day2**	**Day3**	**Day4**	**Day5**	**Day6**	**Day7**
Weather	Rain	Rain	Sunny	Cloudy	Rain	Rain	Sunny
Maximum temperature (°C)	13.2	10.2	17.1	15.8	16.5	16.4	12.7
s-RPE	2.3 ± 3.60	9.9 ± 3.7	32.8 ± 12.3	17.6 ± 4.2	24.5 ± 11.3	20.7 ± 6.6	24.6 ± 6.7

[1]: s-RPE is the session rating of perceived exertion. *: Average ± S.D.

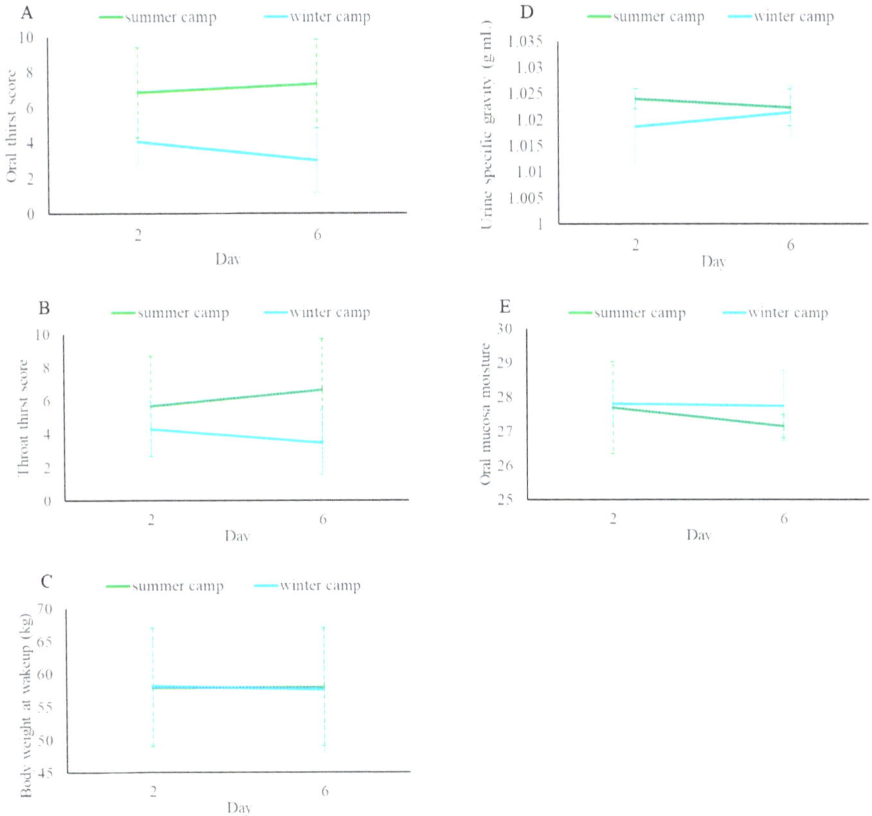

Figure 3. Transitions in oral thirst (**A**), throat thirst (**B**), body weight at wakeup (**C**), urine specific gravity (**D**), and oral mucosa moisture (**E**) during the summer and winter camps. Urine specific gravity above 1.020 g/mL indicates dehydration in sports. Oral moisture values range from 0 to 99.9, and values of ≥29.6, 28.0–29.5, and ≤27.9 are defined as normal, borderline dry mouth, and dry mouth, respectively. Error bars reflect standard deviation.

Meanwhile, the Spearman's correlation coefficient illustrated that the oral mucosa moisture had a moderate negative correlation with urine specific gravity ($p < 0.05$, $r = -0.45$), and the regression equation showed that the value of oral mucosa moisture predicting urine specific gravity of 1.020 g/mL, which is an index of dehydration, was ≈27.5 (Figure 4). It was analyzed the multivariate correlations among body weight, urine specific gravity, oral mucosa moisture, subjective oral thirst, and subjective throat thirst, but results correlating with oral mucosal moisture were not observed, except for urine specific gravity.

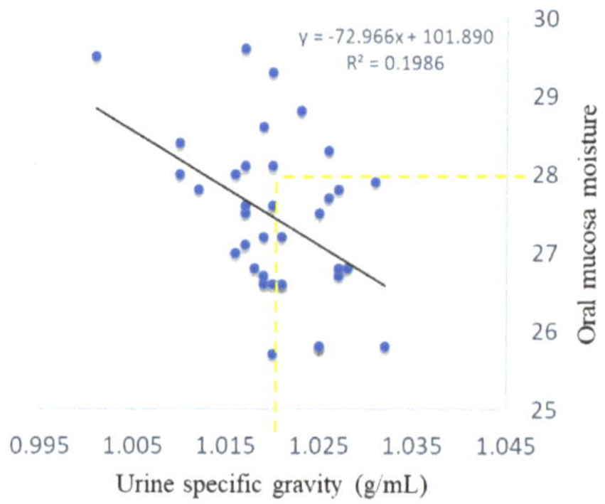

Figure 4. A scatter plot of the oral mucosa moisture and urine specific gravity for both the summer and winter camps. The oral mucosa moisture had a moderate negative correlation with urine specific gravity ($p < 0.05$, $r = -0.45$). The regression equation showed that the value of oral mucosa moisture predicting urine specific gravity of 1.020 g/mL, which is an index of dehydration, was ≈27.5.

4. Discussion

The findings of this study suggest that oral mucosa moisture measured using a capacitance sensor for oral epithelial moisture is a simple index for dehydration assessment during sports activities. Although no subjects were diagnosed with heat stroke in either camp, the urine specific gravity of the subjects ranged from 1.001 to 1.032 g/mL, which means that the subjects included both normal and possibly dehydrated conditions. These results suggest that the subjects in this study were appropriate subjects for the evaluation of dehydration during high-intensity training periods.

For these subjects, a moderate negative correlation of the oral mucosa moisture and urine specific gravity is shown in Figure 4. The results agree with the finding reported by Fukushima et al. that there is a negative correlation between the severity of dehydration and the degree of oral mucosal moisture ($p < 0.05$, $r = -0.69$) [9]. The index of mouth and throat thirst which has been used as an indicator of dehydration is subjective for the athletes themselves. The fact that oral mucosal wetness is a numerical and objective data, and its value is correlated with urine specific gravity is expected to have a great impact on future dehydration indicators. Although the correlation between the oral mucosa moisture and urine specific gravity was an important finding, the following points should be noted. As shown in the scatter plot, there were nine cases in which the urine specific gravity was less than 1.020, but the oral wetness was plotted lower than 28.0. In this case, the urine specific gravity may have failed to assess the state of dehydration, or the oral mucosal wetness may have overestimated, but this cannot be determined from this study design alone. Furthermore, there were two cases where the urine specific gravity was 1.020 or higher, but the oral wetness was higher than 28.0. These results suggest that the oral mucosa moisture measured using a capacitance sensor for oral epithelial moisture as a simple index for dehydration assessment on its own may not be sufficient, and combining multiple assessment methods may be necessary. Simplified approach to assessing daily dehydration is using a Venn diagram decision tool [27]. Weight loss (>1%), dark-colored urine (>5 arbitrary units, a.u), and thirst (conscious desire for water) are used as markers of inadequate fluid intake. In addition, it may be useful to continue to take daily timed measurements within the same athlete to capture individual data and compare it to the data in the field dehydration index.

The regression equation in Figure 4 shows that the value of oral mucosa moisture predicting urine specific gravity of 1.020 g/mL, which is an index of dehydration, was ≈27.5. Since this value is less than 28.0, which is diagnosed as xerostomia, the cutoff line for dehydration cannot be determined in this study design, but it is expected to be in a reasonable range. Epidemiological studies to examine the cutoff line for dehydration determined via oral mucosa moisture need to be done in the future. The Mucus® device is intended to measure oral mucosa moisture in elderly patients with xerostomia. Therefore, it may be possible to modify this measurement for dehydration by changing the dielectric ratio.

Although there are multiple measures of dehydration, it is necessary to consider the validated targets, in other words, acute dehydration index or chronic dehydration index. TBW and plasma osmolarity are indicators of acute and chronic dehydration, while urine specific gravity and urine osmolarity are indicators of chronic dehydration [2]. The oral mucosa moisture was previously compared with acute indications [15,28]. Suzuki et al. examined changes in oral mucosal moisture during a 60-min bike ride in the laboratory at temperatures of 15, 23, and 30 °C. The results showed a gradual downward trend in oral mucosal moisture at all temperatures, with a smaller value recorded at 30 °C throughout the 60 min [28]. In this study, a correlation was confirmed with urine specific gravity, which is an indicator of chronic dehydration. This suggests that it may be possible to evaluate not only acute dehydration in emergency situations but also chronic dehydration that can be applied in daily condition management.

The capacitance sensor for oral epithelial moisture, Mucus®, employed in this study currently has no restrictions on who can use it, and it can be easily employed by anyone. In addition, it is not large and can be carried easily. The measurement time is only a few seconds, which is less of a burden for the athlete and the measurer. However, when we performed measurements with it in this study, we found that it was somewhat technically sensitive in terms of the pressure and angle of the contact surface, and in some cases, practice was necessary before use. The design makes it difficult to measure by oneself. There is room for improvement in the design to make it more universal, so that it can be used by anyone, anywhere, at any time.

The design of this study was a field survey, and environmental factors and levels of exercise load that influence dehydration were recorded but not controlled. Comparing their oral mucosa moisture measurement with other gold standard methods such as salivary flow rate is necessary. The relationship with other indicators of dehydration, such as plasma osmolality, also needs to be evaluated.

5. Conclusions

This study indicated that oral mucosal moisture determined using an oral moisture-checking device could be a simple valuable index for assessing dehydration during sports activities.

Author Contributions: Conceptualization, G.T. and T.U.; methodology, G.T.; validation, Y.T. and H.K.; formal analysis, G.T.; investigation, G.T., T.H., Y.I., Y.T., H.K., K.H., S.S., N.S.K. and H.C.; data curation, G.T.; writing—original draft preparation, G.T.; writing—review and editing, Y.I.S. and T.U.; supervision, T.H., K.S., N.M. and T.U.; project administration, H.C.; funding acquisition, G.T. All authors have read and agreed to the published version of the manuscript.

Funding: This research was funded by JSPS KAKENHI, Grant Number 20 K18824, and by YAMAHA Motor Foundation for Sports, Grant Number 2983.

Institutional Review Board Statement: The study was conducted according to the Declaration of Helsinki and approved by the Institutional Review Board (or Ethics Committee) of the research ethics committee of Tokyo Medical and Dental University (approval number: D2019-031).

Informed Consent Statement: Informed consent was obtained from all subjects involved in the study.

Data Availability Statement: The data presented in this study are available on request from the corresponding author. The data are not publicly available due to elite athlete.

Conflicts of Interest: The authors declare no conflict of interest. The funders had no role in the design of the study; in the collection, analyses, or interpretation of data; in the writing of the manuscript; or in the decision to publish the results.

References

1. Sawka, M.N.; Young, A.J. Physiological Systems and Their Responses to Conditions of Heat and Cold. In *ACSM's Advanced Exercise Physiology*; Tipton, C.M., Sawka, M.N., Tate, C.A., Terjung, R.L., Baltimore, M.D., Eds.; Williams & Wilkins: Philadelphia, PA, USA, 2005; pp. 535–563.
2. Sawka, M.N.; Burke, L.M.; Eichner, E.R.; Maughan, R.J.; Montain, S.J.; Stachenfeld, N.S. American College of Sports Medicine position stand: Exercise and fluid replacement. *Med. Sci. Sports Exerc.* **2007**, *39*, 377–390. [CrossRef] [PubMed]
3. Cheuvront, S.N.; Carter, R., 3rd; Sawka, M.N. Fluid balance and endurance exercise performance. *Curr. Sports Med. Rep.* **2003**, *2*, 202–208. [CrossRef]
4. Hancock, P.A.; Vasmatzidis, I. Effects of heat stress on cognitive performance: The current state of knowledge. *Int. J. Hyperth.* **2003**, *19*, 355–372. [CrossRef] [PubMed]
5. Greiwe, J.S.; Staffey, K.S.; Melrose, D.R.; Narve, M.D.; Knowlton, R.G. Effects of dehydration on isometric muscular strength and endurance. *Med. Sci. Sports Exerc.* **1998**, *30*, 284–288. [CrossRef] [PubMed]
6. Sawka, M.N.; Young, A.J.; Latzka, W.A.; Neufer, P.D.; Quigley, M.D.; Pandolf, K.B. Human tolerance to heat strain during exercise: Influence of hydration. *J. Appl. Physiol.* **1992**, *73*, 368–375. [CrossRef]
7. Carter, R., 3rd; Cheuvront, S.N.; Williams, J.O.; Kolka, M.A.; Stephenson, L.A.; Sawka, M.N.; Amoroso, P.J. Epidemiology of hospitalizations and deaths from heat illness in soldiers. *Med. Sci. Sports Exerc.* **2005**, *37*, 1338–1344. [CrossRef]
8. Sayers, S.P.; Clarkson, P.M. Exercise-induced rhabdomyolysis. *Curr. Sports Med. Rep.* **2002**, *1*, 59–60. [CrossRef]
9. Dancaster, C.P.; Whereat, S.J. Fluid and electrolyte balance during the comrades marathon. *S. Afr. Med. J.* **1971**, *45*, 147–150.
10. Armstrong, L.E.; Maresh, C.M.; Castellani, J.W.; Bergeron, M.F.; Kenefick, R.W.; LaGasse, K.E.; Riebe, D. Urinary indices of hydration status. *Int. J. Sport Nutr.* **1994**, *4*, 265–279. [CrossRef]
11. Armstrong, L.E. Hydration assessment techniques. *Nutr. Rev.* **2005**, *63* (Suppl. 1), S40–S54. [CrossRef]
12. Yoshihara, A.; Hirotomi, T.; Takano, N.; Kondo, T.; Hanada, N.; Miyazaki, H. Serum markers of chronic dehydration are associated with saliva spinability. *J. Oral Rehabil.* **2007**, *34*, 733–738. [CrossRef] [PubMed]
13. Fortes, M.B.; Owen, J.A.; Raymond-Barker, P.; Bishop, C.; Elghenzai, S.; Oliver, S.J.; Walsh, N.P. Is this elderly patient dehydrated? Diagnostic accuracy of hydration assessment using physical signs, urine, and saliva markers. *J. Am. Med. Dir. Assoc.* **2015**, *16*, 221–228. [CrossRef] [PubMed]
14. Shimizu, M.; Kinoshita, K.; Hattori, K.; Ota, Y.; Kanai, T.; Kobayashi, H.; Tokuda, Y. Physical signs of dehydration in the elderly. *Intern. Med.* **2012**, *51*, 1207–1210. [CrossRef]
15. Fukushima, Y.; Sano, Y.; Isozaki, Y.; Endo, M.; Tomoda, T.; Kitamura, T.; Sato, T.; Kamijo, Y.; Haga, Y.; Yoda, T. A pilot clinical evaluation of oral mucosal dryness in dehydrated patients using a moisture-checking device. *Clin. Exp. Dent. Res.* **2019**, *5*, 116–120. [CrossRef]
16. McGee, S.; Abernethy, W.B., 3rd; Simel, D.L. The rational clinical examination. Is this patient hypovolemic? *JAMA* **1999**, *281*, 1022–1029. [CrossRef]
17. Saito, M.; Ono, Y.; Kitamura, N.; Yamaguchi, M.; Saito, C. A study of moisture content in the oral mucosa in the elderly, Part II. Investigating moisture content in the oral mucosa in the elderly population residing at a nursing home. *Ann. Jpn. Gerodont. Soc.* **2008**, *23*, 97–105.
18. Murakami, M.; Nishi, Y.; Kamashita, Y.; Nagaoka, E. Comparison of a saliva wetness tester and a moisture-checking device in patients with maxillary obturator prostheses. *Gerodontology* **2014**, *31*, 83–88. [CrossRef]
19. Kanazaki, A.; Hu, J.; Otomaru, T.; Sumita, Y.I. Evaluation of xerostomia and effect of maxillofacial prosthesis use on mouth dryness in head and neck tumor patients. *Int. J. Maxillofac. Prosthet.* **2021**, *4*, 9–17. [CrossRef]
20. Nishii, M.; Akashi, M.; Kakei, Y.; Hasegawa, T.; Minamikawa, T.; Furudoi, S.; Shibuya, Y.; Takahashi, M.; Otsuki, N.; Nibu, K.; et al. Sequential changes in oral dryness evaluated by a moisture-checking device in patients with oropharyngeal cancer during chemoradiotherapy: A pilot study. *Oral Health Dent. Manag.* **2014**, *13*, 507–511.
21. Takahashi, F.; Takahashi, M.; Toya, S.; Morita, O. Relationship between medicine and stimulated saliva and oral moisture. *Nihon Hotetsu Shika Gakkai Zasshi* **2008**, *52*, 537–542. [CrossRef] [PubMed]
22. Takahashi, F.; Koji, T.; Morita, O. The usefulness of an oral moisture checking device (Moisture Checker for Mucus). *Nihon Hotetsu Shika Gakkai Zasshi* **2005**, *49*, 283–289. [CrossRef] [PubMed]
23. Fukushima, Y.; Yoda, T.; Araki, R.; Sakai, T.; Toya, S.; Ito, K.; Sato, T. Evaluation of oral wetness using an improved moisture-checking device for the diagnosis of dry mouth. *Oral Sci. Int.* **2017**, *14*, 33–36. [CrossRef]
24. Takano, T.; Kugimiya, Y.; Morita, K.; Tazawa, S.; Ueda, T.; Sakurai, K. Intra- and inter-investigator reliabilities of oral moisture measured using an oral moisture-checking device. *J. Oral Rehabil.* **2020**, *47*, 480–484. [CrossRef]

25. Foster, C.; Florhaug, J.A.; Franklin, J.; Gottschall, L.; Hrovatin, L.A.; Parker, S.; Doleshal, P.; Dodge, C. A new approach to monitoring exercise training. *J. Strength Cond. Res.* **2001**, *15*, 109–115.
26. Yoshida, G.; Iwama, A.; Sakata, T.M.; Nakamura, N.; Eiraku, H.; Higashionna, A.; Akamine, T.; Yoshitake, Y. Fluid Balance during Training for Top University Canoe Sprint Athletes: A Preliminary Study Investigating the Differences in Controlling Energy Expenditure by Season. *Jpn. J. Nutr. Diet.* **2021**, *79*, 103–111. [CrossRef]
27. Cheuvront, S.N.; Kenefick, R.W. Am I Drinking Enough? Yes, No, and Maybe. *J. Am. Coll. Nutr.* **2016**, *35*, 185–192. [CrossRef] [PubMed]
28. Suzuki, E.; Kashimura, O. Relationship between physiological responses and thirst sensation at exercise. *Jpn. J. Biometeor.* **2015**, *52*, 165–174. [CrossRef]

chemosensors

MDPI

Article

Near-Infrared-Emitting *Meso*-Substituted Heptamethine Cyanine Dyes: From the Synthesis and Photophysics to Their Use in Bioimaging

Louise Kommers Reimann [1], Daniela de Souza Fortes [1], Fabiano da Silveira Santos [1], Henrique de Castro Silva Junior [1], Ana Moira Morás [2], Dinara Jaqueline Moura [2], Rodrigo da Costa Duarte [1] and Fabiano Severo Rodembusch [1,*]

[1] Grupo de Pesquisa em Fotoquímica Orgânica Aplicada, Instituto de Química, Departamento de Química Orgânica, Universidade Federal do Rio Grande do Sul (UFRGS), Porto Alegre 91501-970, Brazil

[2] Laboratório de Genética Toxicológica, Universidade Federal de Ciências da Saúde de Porto Alegre (UFCSPA), Porto Alegre 90240-511, Brazil

* Correspondence: fabiano.rodembusch@ufrgs.br

Abstract: Heptamethine cyanine dyes were synthesized in good yields by the reaction between quaternary indoles and a pentamethinic salt, under mild reaction conditions minimizing photooxidation. These compounds were used as precursors to prepare *meso*-substituted derivatives. The cyanine dye precursors presented UV-Vis absorption, related to fully allowed electronic transitions and fluorescence emission in the NIR region, without any evidence of aggregation in both ground and excited states. The substitution at the *meso* position showed a fundamental role in their photophysics, with the main absorption in the green-orange region related to the monomeric species. Moreover, the excited state photophysics presented emission profiles dependent on the excitation wavelengths, complicating the correlation of spectroscopy and structure. Density Functional Theory and OO-SCS-MP2 calculations under different solvation conditions revealed the heavy impact of conjugation effects on ground and excited states' geometries and electronic configurations of these compounds. Finally, the observed photophysical features of the *meso*-substituted heptamethine cyanine dyes were successfully used to explore their application as fluorescent probes in biological media, allowing stable staining in live and fixed cells.

Keywords: bioimaging; heptamethine cyanine; fluorescence; charge transfer; theoretical calculations

Citation: Reimann, L.K.; Fortes, D.d.S.; Santos, F.d.S.; Silva Junior, H.d.C.; Morás, A.M.; Moura, D.J.; Duarte, R.d.C.; Rodembusch, F.S. Near-Infrared-Emitting *Meso*-Substituted Heptamethine Cyanine Dyes: From the Synthesis and Photophysics to Their Use in Bioimaging. *Chemosensors* **2023**, *11*, 47. https://doi.org/10.3390/chemosensors11010047

Academic Editor: Pietro Salvo

Received: 13 December 2022
Revised: 31 December 2022
Accepted: 2 January 2023
Published: 5 January 2023

1. Introduction

Cyanines are a subclass of heterocyclic compounds ascribed as polymethine dyes presenting, in general, two heterocyclic moieties in their chemical structure where the bridges between them are π-conjugated systems. These compounds can be differentiated by the number of sp^2 carbons in the molecule central portion and the nature of the heterocyclic groups present in the peripheral molecular structure [1–4]. These structural changes allow large absorption intensities and spectral emissions ranging from ultraviolet to near-infrared regions [5–8]. Moreover, heptamethine cyanine dyes may be substituted at the *meso* position by reaction with nucleophilic compounds, such as amines, thiols, and aromatic or aliphatic alcohols [9–12]. In this regard, molecules presenting halogen atoms at the *meso* position provide a reactive site for this substitution in most cases, through a unimolecular radical nucleophilic substitution ($S_{RN}1$) mechanism, a substitution reaction in which a certain substituent on an aromatic compound is replaced by a nucleophile through an intermediary free radical species [13]. Additionally, cyanine dyes can be reactive to the environment, showing potential biomedical [14] and sensing applications [15,16], whereas the solvatochromic ones have been emerging mainly for analytical purposes [17]. Based on their electronic properties [18], they have been used as polarity sensors in several studies related to halochromism [19], to study micro-heterogeneity in solution [20] and solvent mixtures [21]. In addition, heptamethine cyanine dyes have attracted attention for utilization in both imaging and therapy in cancer cells, mainly due to their low toxicity, accumulation capability, low autofluorescence, and superior signal-to-noise ratios [22–24]. A recent study showed that cyanine dyes efficiently labeled cancer cells and spontaneous tumors in transgenic mice, indicating their potential application to detect cancer metastasis and cancer cells in blood with a high degree of sensitivity [25].

Herein, we present the synthesis and photophysical characterization of new pyrrolidine-substituted heptamethine dyes, where the presence of the non-bonding electrons may be useful for different conjugation pathways where ICT may be present, and its preliminary investigation as fluorescent probes in biological media for cell staining.

2. Materials and Methods

2.1. Materials

2,3,3-Trimethylindolenine (CAS 1640-39-7), 1-iodomethane (CAS 74-88-4), 1-iodobutane (CAS 542-69-8), phosphorus(V) oxychloride (CAS 10025-87-3), 4-ethyl cyclohexanone (CAS 5441-51-0), triethylamine (CAS 121-44-8), hydrochloric acid (37%) (CAS 7647-01-0), aniline (CAS 62-53-3), and pyrrolidine (CAS 123-75-1) were purchased from Sigma-Aldrich (San Luis, MI, USA), and used as received. The syntheses were monitored by thin-layer chromatography using silica gel 60GF254 Merck (Darmstadt, Germany). The solvents acetonitrile (CAS 75-05-8), ethyl acetate (CAS 141-78-6), ethanol (CAS 67-63-0), toluene (CAS 108-88-3), dichloromethane (CAS 75-09-2), methanol (CAS 67-56-1), and hexane (CAS 110-54-3) were purchased from Honeywell (Charlotte, NC, USA), and used as received or purified according to the literature.

2.2. Characterization

Infrared absorption spectra were obtained on a Shimadzu-IR PRESTIGE-21 spectrometer on a KBr disc or operating on attenuated reflectance (ATR) mode, with a resolution of 4.0 $cm^{-1}$and 20 scans in the 4000–750 cm^{-1} range. Melting points were determined on a Fisatom 430D apparatus and are uncorrected. ^{1}H and ^{13}C NMR spectra were achieved on a Varian Inova operating at 300 or 400 MHz for ^{1}H and 75 or 100 MHz for ^{13}C, respectively, using deuterated solvents (DMSO-d_6 or $CDCl_3$). The chemical shifts (δ) are recorded in ppm and coupling constants (J) in Hz. ^{1}H NMR data have their data expressed as multiplicity (s, singlet; ls, large singlet; d, doublet; t, triplet; q, quintuplet; m, multiplet), coupling constant, and relative number of hydrogens. High-resolution mass spectra (HRMS-ESI) were recorded with electrospray ionization in the positive mode using a Bruker Impact II. The operating conditions were 4.5 kV for capillary voltage, and 33V and 2.5 V for sample and extraction cone voltage, respectively. Nitrogen was used as desolvation gas (N_2) to a temperature of 200 °C. Sodium formate was used for calibration. The original spectra from the spectroscopic characterization can be found as (Supplementary material Figures S1–S20). Spectroscopic grade solvents (Aldrich) were used for absorption in the Vis-NIR and fluorescence emission spectroscopies. All experiments were carried out at ambient room temperature (25 °C) using solutions of concentrations between 10^{-5} and 10^{-6} M. The measurements were performed using 10 × 10 mm path-length quartz cuvettes. Absorption spectroscopic measurements in the Vis-NIR region were obtained on a Shimadzu UV-2450 spectrometer. Fluorescence emission measurements were obtained on a Shimadzu RF-5301PC spectrofluorometer. The relative fluorescence quantum yields (Φ_{FL}) of the *meso*-substituted cyanines **8a-b** were obtained using cresyl violet in MeOH (Φ_{FL} = 0.54) as a standard [26].

2.3. Synthesis

2.3.1. Indoles **3a-b**

To the reaction flask containing acetonitrile (10 mL), previously saturated for 15 min with nitrogen (N_2), the 2,3,3-trimethylindolenine **1** (1.9 mmol, 0.302 g) and the respective

1-iodoalkane (**2a** or **2b**) (9.9 mmol, 1.4 g for **2a** or 1.8 g for **2b**) were added under stirring. The reaction system was heated under reflux and nitrogen atmosphere for 6 h. After this period, the mixture was poured into 50 mL of ethyl acetate under stirring and heated for 30 min, after which time the supernatant was removed. This process was repeated 3 times. The obtained solids were filtered and dried at room temperature (25 °C). Indole **3a**: Pale yellow solid. Melting point: 240–242 °C. Yield: 72% (411 mg). ^{1}H NMR (DMSO-d_6, 300 MHz): δ (ppm) 7.91 (m, 1H); 7.82 (m, 1H); 7.60 (m, 2H); 3.97 (s, 3H); 2.78 (s, 3H); 1.52 (s, 6H). ^{13}C NMR (DMSO-d_6, 75 MHz): δ (ppm) 196.4; 142.5; 142.0; 129.7; 129.2; 123.7; 115.6; 54.4; 35.3; 22.1; 14.9. Indole **3b**: Pale yellow solid. Melting point: (217–219 °C) [27]. Yield: 70% (456 mg). ^{1}H NMR ($CDCl_3$, 400 MHz): δ (ppm) 7.64 (m, 1H); 7.61 (m, 3H); 4.67 (t, 2H, J = 7.5 Hz); 3.13 (s, 3H); 1.96 (m, 2H); 1.67 (s, 6H); 1.53 (m, 2H); 1.01 (t, 3H, J = 6.0 Hz). ^{13}C APT NMR (DMSO-d_6, 75 MHz): δ (ppm) 195.5; 141.5; 140.7; 130.0; 129.4; 123.4; 115.3; 54.6; 49.7; 29.8; 23.1; 20.0; 17.0; 13.6.

2.3.2. Pentamethine Salt **5** [28]

In an ice bath, phosphorus(V) oxychloride (22 mmol, 2.05 mL) was added dropwise in dimethylformamide (3 mL) and kept under stirring for 30 min. After this period, 4-ethyl cyclohexanone **4** (9.6 mmol, 1.35 mL) was added, and the temperature was increased up to reflux for 2 h. Then, the reaction was brought to room temperature (25 °C), an aniline/ethanol solution (6 mL) (1:1 v/v) was added, and the reaction was kept under stirring at room temperature for an additional 1 h. After this time, the reaction crude was poured into ice-cold H_2O and 20 mL of hydrochloric acid (37%) (10:1 v/v). The reaction was allowed to stand at room temperature for 12 h. Then, the solid was filtered, washed with H_2O, and recrystallized with toluene. The obtained solid was filtered, dried, and used for the next step without further purification. Pentamethine salt **5**: Violet solid. Melting point: (219–222 °C). Yield: 58% (2.15 g). FTIR (KBr, cm^{-1}): 3608, 3521, 3442, 3302, 2953, 2871, 1607, 1565, 1472. ^{1}H NMR (DMSO-d_6, 300 MHz): δ (ppm) 11.37 (ls, 2H); 8.51 (s, 2H); 7.60 (m, 4H); 7.44 (m, 4H); 7.25 (m, 2H); 3.07 (m, 2H); 2.18 (m, 2H); 1.65 (m, 1H); 1.47 (m, 2H, J = 7.2 Hz); 1.05 (t, 3H, J = 7.5 Hz).

2.3.3. Heptamethine Cyanine Dyes **6a-b**

In a round bottom amber flask containing acetonitrile (20 mL), previously dried and saturated for 15 min with nitrogen (N_2), the previously prepared pentamethine salt **5** (0.5 mmol, 0.193 g) was added under stirring and heated up to reflux temperature. In a second flask, acetonitrile (5 mL), triethylamine (1.5 mmol, 0.210 mL), and the respective indole (**3a** or **3b**) (1.5 mmol, 0.261 g for **3a** or 0.282 g for **3b**) were added and allowed to stir at room temperature (25 °C) for 5 min. After this time, the second flask solution was poured into the first one and then the reaction was allowed to stir at reflux temperature under a nitrogen atmosphere for a further 6 h. The reaction was checked for completion by TLC (dichloromethane/methanol 9:1, v/v). Finally, acetonitrile was evaporated on a rotatory evaporator and 2 mL of methanol was added to the reaction crude. Then, the reaction crude was poured into a solution of 100 mL of ethyl acetate/hexane (1:1 v/v) and placed in an ultrasonic bath for 30 min. After 12 h of solid decantation, the solid was filtered. This process was repeated 4 times. The final solids (**6a-b**) were filtered and dried. Heptamethine cyanine dye **6a**: Green solid. Melting point: (222–223 °C). Yield: 62% (190 mg). FTIR (KBr, cm^{-1}): 3603, 3521, 3326, 2962, 1651, 1551, 1365. ^{1}H NMR ($CDCl_3$, 300 MHz): δ (ppm) 8.34 (d, 2H, J = 12.0 Hz); 7.38 (m, 4H); 7.22 (m, 4H); 6.15 (d, 2H, J = 12.0 Hz); 3.73 (s, 6H); 2.88 (m, 2H); 2.24 (m, 2H); 1.71 (s, 12H); 1.57 (m, 2H); 1.05 (t, 3H, J = 7.5 Hz). ^{13}C NMR ($CDCl_3$, 75 MHz): δ (ppm) 172.9; 150.7; 144.5; 142.7; 140.8; 128.8; 127.0; 125.3; 122.1; 110.9; 101.3; 49.2; 33.6; 32.6; 28.2; 28.0; 11.8. Heptamethine cyanine dye **6b**: Green solid. Melting point: 227–228 °C. Yield: 60% (210 mg). FTIR (KBr, cm^{-1}): 3517, 3330, 2957, 2862, 1665, 1556, 1514, 1428. ^{1}H NMR ($CDCl_3$, 400 MHz): δ (ppm) 8.36 (d, 2H, J = 16.0 Hz); 7.41 (m, 4H); 7.29 (m, 2H); 7.21 (m, 2H); 6.20 (d, 2H, J = 12.0 Hz); 4.21 (t, 2H, J = 8.0 Hz); 2.90 (m, 2H); 2.27 (m, 2H); 1.74 (s, 12H); 1.68 (m, 1H); 1.61 (m, 2H); 1.51 (q, 2H; J = 8.0 Hz); 1.04 (m, 9H).

^{13}C NMR ($CDCl_3$, 100 MHz): δ (ppm) 172.4; 150.4; 144.4; 142.1; 141.0; 128.8; 126.5; 125.4; 122.3; 111.0; 101.1; 49.4; 44.8; 33.6; 32.0; 29.4; 28.1; 20.3; 13.8; 11.7.

2.3.4. *Meso*-Substituted Heptamethine Cyanine Dyes **8a-b**

In a round bottom amber flask containing acetonitrile (2 mL), previously dried and saturated for 15 min with nitrogen (N_2), the respective cyanine **6a** (7.80×10^{-2} mmol, 0.05 g) or **6b** (7.80×10^{-2} mmol, 0.05 g) was added under stirring. To this solution, pyrrolidine **7** (24.4 mmol, 1.5 mL) was added. The reaction was allowed to stir at reflux temperature under a nitrogen atmosphere for a further 30 min. The reaction was followed by thin-layer chromatography (TLC) using dichloromethane/ethyl acetate/methanol (45:45:10 $v/v/v$) as eluent. After completion of the reaction, the mixture was evaporated to dryness. The resulting crude residue was purified by column chromatography over silica gel and using the same mixture used for TLC as eluent to afford the derivatives **8a-b**. *Meso*-substituted heptamethine cyanine dye **8a**: violet solid. Melting point: 95–98 °C. Yield: 78% (39 mg). FTIR (KBr, cm^{-1}): 3387, 2955, 2917, 2854, 1538, 1305, 1187, 794. ^{1}H NMR (DMSO-d_6, 300 MHz): δ (ppm) 7.39 (d, 2H, J = 6.0 Hz); 7.25 (t, 2H, J = 9.0 Hz); 7.16 (d, 2H, J = 12.0 Hz); 7.03 (d, 2H, J = 6.0 Hz); 6.97 (t, 2H, J = 9.0 Hz); 5.44 (d, 2H, J = 6.0 Hz); 4.04 (m, 4H); 3.33 (s, 6H); 2.80 (d, 2H, J = 15.0 Hz); 2.17 (m); 2.0 (m, 4H); 1.57 (s, 6H); 1.54 (s, 6H); 1.46 (m, 2H); 1.02 (t, 3H, J = 9.0 Hz). ^{13}C NMR (DMSO-d_6, 75 MHz): δ (ppm) 173.5; 164.3; 144.4; 139.5; 133.9; 128.4; 122.3; 121.6; 121,1; 108.3; 92.2; 55.5; 46.6; 34.2; 29.6; 28.6; 28.2; 24.2; 12.2. HRMS (ESI-qTOF) m/z: $[M + H]^+$ calculated for $C_{38}H_{48}N_3$: 546.3841; found 546.3843. *Meso*-substituted heptamethine cyanine dye **8b**: violet solid. Melting point: 114–115 °C. Yield: 75% (38 mg). FTIR (KBr, cm^{-1}): 3401, 2952, 2918, 2861, 1541, 1187, 1102, 739. ^{1}H NMR (DMSO-d_6, 400 MHz): δ (ppm) 7.36 (d, 2H, J = 8.0 Hz), 7.21 (m), 7.13 (d, 2H, J = 12.0 Hz), 6.99 (d, 2H, J = 8.0 Hz), 6.94 (t, 2H, J = 4.0 Hz), 5.46 (d, 2H, J = 12.0 Hz), 4.02 (m, 4H), 3.82 (m), 2.75 (d, 2H, J = 16.0 Hz), 2.14 (m), 1.98 (m, 4H), 1.62 (m, 5H), 1.54 (s, 6H), 1.50 (s, 6H), 1.42 (m, 2H), 1.35 (m, 4H), 0.97 (t, 3H, J = 8.0 Hz), 0.92 (t, 6H, J = 8.0 Hz). ^{13}C NMR (DMSO-d_6, 100 MHz): δ (ppm) 172.9; 162.9; 143.4; 139.1; 133.5; 127.9; 121.9; 121.1; 120.4; 107.9; 91.7; 55.1; 46.2; 33.7; 33.5; 29.2; 27.9; 27.6; 23.7; 19.6; 13.7; 11.6. HRMS (ESI-qTOF) m/z: $[M+H]^+$ calculated for $C_{44}H_{60}N_3$: 630.4771; found 630.4782.

2.4. *Theoretical Calculations*

All DFT and ab initio calculations were performed using the ORCA v5.0.2 quantum chemistry package [29]. Initial geometries for heptamethine cyanine dyes **6a** and **8a** were obtained after conformational sampling using the CREST software [30]. This procedure was, in turn, based on energies obtained by the accurate semi-empirical method GFN2-xTB [31]. The lowest conformers were subsequently optimized to the ground (GS) and first excited states (ES) using Density Functional Theory via the B97-3c scheme [32], which is a revamped version of Becke's B97 GGA functional [33], with a slightly modified version of the triple-ζ Def2-TZVP basis set [34] and including the D3 dispersion correction [35] and a Short-Range Basis (SRB) correction [36]. We used an improved version of the Conductor-like Continuum Polarization Model called LR-CPCM to simulate solvent relaxation during geometric optimizations to the ground and excited states [37]. A better description of charge transfers and energy gaps was achieved on a higher OO-SCS-MP2/Def2-TZVP computational level as properties dependent on virtual orbitals are not always correctly described by the DFT method [38]. All calculations were set to tight convergence criteria of 1.0E-08 a.u with isosurfaces using $\rho = 0.05$ e/Å^3 and 3D molecular representations using the CPK color scheme.

2.5. *Cellular Stain*

The glioblastoma multiform T89G cells were acquired by American Type Culture Collection® (ATCC). The cells were grown in Dulbecco's modified Eagle's medium (Gibco™ DMEM, Life Technology, Grand Island, NY, USA) supplemented with 10% heat-inactivated fetal bovine serum (Gibco™ FBS, Life Technology, Grand Island, NY, USA), 100 units·$mL^{-1}$of penicillin, and 100 g·mL^{-1}of streptomycin at 37 °C in a humidified atmosphere with 5% CO_2. To investigate the fluorescence stain capacity of the studied heptamethine cyanine dyes, cells grown on 96-well plates were fixed in paraformaldehyde (4%), washed with PBS, and permeabilized in Triton X-100 (0.01%). The stock solutions of the cyanine dyes were solubilized in ethanol at 5 μM. All solutions were vortexed for 1 min and sonicated for 30 min. The cells were then washed with PBS and incubated with the cyanine dyes chosen as the model, in stock solutions concentrations (30 min, 37 °C). Nuclei were contrasted with Hoechst (4 μg·mL^{-1} for 10 min). Finally, the cells were imaged in INCell analyzer 2200 (GE Healthcare Life Sciences, Piscataway, NJ, USA) after 30 min and 24 h of staining. To investigate the fluorescence stain capacity in non-fixed samples, cells grown on 96-well plates were washed with PBS and incubated with the studied cyanine dyes in stock solution concentrations (30 min, 37 °C). Nuclei were contrasted with Hoechst (4 μg·mL^{-1} for 10 min). Finally, the cells were imaged in an INCell analyzer 2200 (GE Healthcare Life Sciences, Piscataway, NJ, USA). To verify if there was a loss of confluence after treatment with cyanine dye **8a**, T98G cells were grown in a 96-well plate and incubated with increasing concentrations of the compound (0.005–5μM) and, after 30 min, photos were acquired with an increase of 5x in an inverted optical microscope (TCM400, Labo America, Fremont, CA, USA).

3. Results and Discussion

3.1. Synthesis

The heptamethine cyanine dyes **6a-b** were prepared as presented in Scheme 1. The quaternary indoles **3a-b** were synthesized in previously dried, and N_2-saturated acetonitrile using an excess of the respective alkyl halides **2a-b** [39]. It is worth mentioning that in this reaction, the protection from light and the N_2 atmosphere played an important role in this reaction [40]. The heptamethine cyanine dyes **6a-b** were synthesized by the reaction between the quaternary indoles **3a-b** and the pentamethine salt **5 [41–43]**. The reaction was performed protected from light in amber round bottom flasks to minimize their photooxidation [43]. HPLC-grade solvents were previously saturated with N_2 for at least 20 min, and the reaction took place under an N_2 atmosphere during the heating step. These procedures allowed for obtaining the reaction crude presenting only the characteristics of the desired products, suggesting the importance of reducing the photooxidation process during the reaction. It is important to note that the order of substrate addition played a fundamental role in cyanine synthesis. Using the following sequence: pentamethine salt **5**, indole **3a-b**, and triethylamine, the reaction medium, which is predominantly red due to the high molar absorptivity of the pentamethine salt, turns yellow, and the conversion of reagents is affected after 24 h of reaction. To prove that the yellow color arises from the deprotonated compound **5**, triethylamine was added dropwise into an acetonitrile solution of **5**. The observed behavior was similar to that expected, where the reaction media became yellow. This result indicates that the pentamethine salt presents higher reactivity in the protonated specie as the literature reports reaction yields between 32 and 45% using pyridine as the solvent [44]. Finally, the *meso*-substituted heptamethine cyanine dyes **8a-b** were prepared as also presented in Scheme 1. In this reaction, an $S_{RN}1$ mechanism is believed to be present in the presence of aprotic polar solvents, such as dimethylformamide, dimethylsulfoxide, and acetonitrile [13].

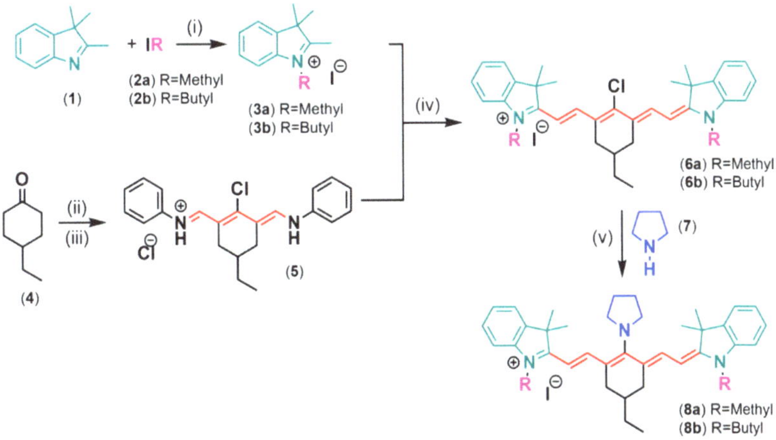

Scheme 1. Synthesis of heptamethine cyanine dyes **6a-b** and the respective *meso*-substituted derivatives **8a-b**: (i) acetonitrile, N_2, reflux, 6 h; (ii) $POCl_3$/dimethylformamide, reflux 2 h; (iii) aniline/ethanol, 1h, H_2O/HCl, r.t. 15 h; (iv) acetonitrile, triethylamine, reflux, 6 h; (v) acetonitrile, N_2, reflux, 30 min.

3.2. Photophysical Characterization

The photophysical study in solution was performed using organic solvents with different dielectric constants (ethyl acetate, dichloromethane, methanol, and acetonitrile). The relevant data from the steady-state ground and excited states characterization are summarized in Table 1. Figure 1 presents the UV-Vis absorption and fluorescence emission spectra of the heptamethine cyanine dye **6a**, which was chosen as a representative structure of the synthesized cyanine precursor. It is worth mentioning that the cyanine dye **6b** presented absorption and emission curves with similar shapes and absorption maxima locations (Figure S21).

Table 1. Photophysical data of heptamethine cyanine dyes **6a-b** in solution, where λ_{abs} and λ_{em} are the absorption and emission maxima (nm), respectively; ε is the molar extinction coefficient ($M^{-1}\cdot cm^{-1}$); f_e is the calculated oscillator strength; k_e^0 is the calculated radiative rate constant ($\times 10^8$ s^{-1}); τ^0 is the calculated pure radiative lifetime (ns); $\Delta\lambda_{ST}$ is the Stokes shift (nm/cm^{-1}); E_g is the optical bandgap (eV).

Cyanine	Solvent	λ_{abs}	ε	f_e	k_e^0k	τ^0	E_g	λ_{em}	$\Delta\lambda_{ST}$
6a	Ethyl acetate	778	197,000	0.303	0.50	1.998	1.49	785	7/115
	Dichloromethane	789	288,000	0.345	0.55	1.806	1.49	790	1/16
	Methanol	774	205,000	0.289	0.48	2.075	1.48	782	8/132
	Acetonitrile	775	225,000	0.329	0.55	1.826	1.50	787	12/197
6b	Ethyl acetate	783	189,000	0.249	0.41	2.465	1.48	784	1/16
	Dichloromethane	789	206,000	0.217	0.35	2.864	1.48	793	4/64
	Methanol	780	365,000	0.445	0.73	1.367	1.48	784	4/65
	Acetonitrile	780	284,000	0.330	0.54	1.845	1.49	792	12/194

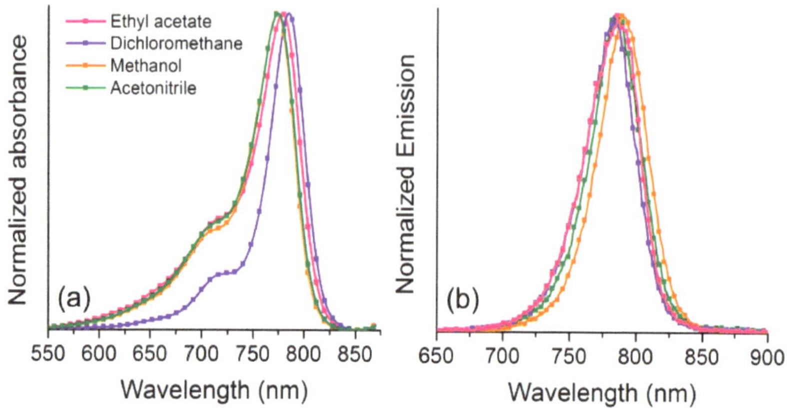

Figure 1. (**a**) UV-Vis absorption and (**b**) steady-state fluorescence emission spectra in solution of different organic solvents (~10^{-6} M) of the heptamethine cyanine dye **6a**.

The heptamethine cyanine dye 6a presents absorption maxima in the NIR region, between 774 and 789 nm, depending on the solvent polarity with no clear tendency on the environment polarity. It could be observed that the cyanine dyes presented an absorption maxima variation ($\Delta\lambda_{abs}$) of 14 nm/229 cm^{-1} (**6a**) and 9 nm/146 cm^{-1} nm (**6b**), which can be related to cyanine-type electronic transitions, as already observed in similar compounds [9]. In this evaluation, methanol was excluded due to its specific interactions with the compounds. In addition, the UV-Vis spectra plotted in epsilon value showed similar intensities (Figure S22), which exclude a possible ion-pairing effect that may occur in the less polar environment, especially considering an iodine counterion [45]. The photophysical investigation in the ground state also allowed for obtaining the experimental extinction coefficient (ε) and, from the Strickler–Berg relations (Equations (1) and (2)), the theoretical rate constant for emission (k_e^0k) and the respective oscillator strengths (f_e) [46]:

$$f_e \approx 4.3 \times 10^{-9} \int \varepsilon d\overline{v} \tag{1}$$

$$k_e^0 \approx 2.88 \times 10^{-9} \overline{v}_2^0 \int \varepsilon d\overline{v} \tag{2}$$

For an electronic transition, the respective oscillator strength can be obtained from Equation (1), which relates the area under the absorption curve from a plot of the molar absorptivity coefficient ε ($M^{-1}\cdot cm^{-1}$) against wavenumber $\overline{v}$ (cm^{-1}). The same integral, applying Equation (2), allows the obtention of the theoretical rate constant for emission (k_e^0k) (Equation (2), where the definition of $\overline{v}_0$ν is the wavenumber (energy in 1/λ units) of the absorption band maximum). In addition, from the k_e^0k, the pure radiative lifetime τ^0 can be obtained, defined as 1/k_e^0k [47]. The high molar absorptivity coefficient values (ε~10^5 $cm^{-1}\cdot M^{-1}$), as well as the respective calculated radiative rate constants (k_e^0k), indicate, for all studied compounds, spin- and symmetry-allowed electronic transitions, which could be related to $^1\pi$–π^* transitions (Table 1). It could also be observed in methanol that these cyanines presented an increase in the molar absorptivity coefficient value, increasing the alkyl chain of the indolic ring, as already reported in the literature [41]. The observed electronic transitions are associated with a small optical bandgap between the HOMO and LUMO orbitals (~1.5 eV) as already reported for different chromophores absorbing in the NIR region [48,49]. Moreover, such transitions are related to non-aggregated species, so-called monomeric ones [50]. In this sense, it is worth mentioning that the studied heptamethine cyanine dyes presented shoulder-like blue-shifted absorption, located around 720 nm, which is usually reported in the literature as an electronic transition related to the formation of H-type aggregates [51,52]. To clarify this topic, fluorescence emission spectra of these compounds were acquired at different concentrations and excitation wavelengths, including 720 nm. In addition, their respective excitation spectra were also obtained (Figures S23–S27). These results indicate that the observed blue-shifted bands are probably related to vibronic structure. In addition, an almost constant radiative lifetime τ^0 suggests that after the radiation absorption, the heptamethine cyanine dyes populate the same excited state. Figure 1b depicts the fluorescence emission spectra, obtained by exciting the compounds at the absorption maxima (Table 1). In general, these compounds present a relatively sharp emission between 700 and 850 nm, with a small Stokes shift (up to 12 nm/197 cm^{-1}), which is characteristic for this class of molecules, showing almost absent differences between the electronic structure of the ground and excited states.

The photophysical study in solution was also performed for the *meso*-substituted heptamethine cyanine dyes **8a-b**. The relevant data from the steady-state ground and excited states characterization are summarized in Table 2. The fluorescence emission spectra were obtained by exciting the compounds at the absorption located between 559 and 611 nm (Table 2). Figure 2 shows the absorption of cyanine dye **8a** (Figure 2a) and **8b** (Figure 2b), as well as a concentration study of compound **8b** in acetonitrile.

Table 2. Photophysical data of heptamethine cyanine dyes **8a-b** in solution, where λ_{abs} is the absorption maximum (nm), ε is the molar extinction coefficient ($M^{-1}\cdot cm^{-1}$), f_e is the calculated oscillator strength, k_e^0k is the calculated radiative rate constant ($\times 10^8$ s^{-1}), and τ^0 is the calculated pure radiative lifetime (ns).

Cyanine	Solvent	λ_{abs}	ε	f_e	k_e^0k	τ^0
8a	Dichloromethane	602	47,400	0.578	1.60	0.627
	Ethanol	578	41,900	0.622	1.86	0.537
	Methanol	572	42,900	0.578	1.77	0.566
	Acetonitrile	559	36,000	0.525	1.68	0.595
8b	Dichloromethane	611	65,300	0.728	1.95	0.513
	Ethanol	587	55,600	0.753	2.19	0.458
	Methanol	581	43,000	0.618	1.83	0.546
	Acetonitrile	570	46,000	0.626	1.93	0.519

It can be observed that both cyanines present absorption in the visible region, with any significant dependence on the alkyl chain. In addition, the main absorption bands shift to lower wavelengths (559–611 nm) if compared to the cyanine precursors **6a-b** (~780 nm), as well as present a more significant solvatochromism (43 nm for **8a** and 41 nm for **8b**). Although it does not present a clear tendency, this latter seems to present a negative solvatochromic effect, indicating a higher dipole moment in the ground state. All these features indicate a loss of cyanine character, allowed by the substitution at the *meso* position of the cyanine [10]. In this case, the change of the chlorine atom by pyrrolidine seems to increase the electron-donating ability of the central substituent, allowing a progressive localization of the positive charge on the central carbon atom of the polymethine chain. Thus, the charge delocalization on the π-conjugated structure is reduced, and the cyanine character is lost. Instead, a bis-dipole structure is reached, described as a central cationic acceptor bearing two electron-donating moieties [53]. In addition, by applying the Strickler–Berg relations, the main electronic transitions could be related to fully allowed transitions, related to π–π^* (Table 2). It is worth mentioning that these derivatives presented additional absorption bands located around 450 nm, as already observed in the literature for similar compounds [11]. As already discussed for precursors **6a-b**, it is believed that these bands arise from the monomeric species, being related to vibronic structures of the cyanines and not from H-aggregates, as usually reported in the literature [54]. Finally, an additional red-shifted absorption band located around 700 nm was already observed [11], which is absent in the cyanine precursors. The correlation between J-aggregates and this absorption band located at longer wavelengths was discarded as it is well-known that these aggregates present quite particular features. The absorption maxima shift to longer wavelengths (~100 nm) with a dramatic sharpening in width (10–20 nm) [55], and a strong increase in the molar absorbance in comparison with isolated monomers [56]. These aggregates are concentration-dependent, a behavior that was not observed in the studied cyanine dye **8b** in acetonitrile (Figure 2c). For the concentration study, similar results were found in dichloromethane, 1,4-dioxane, and ethanol (Figures S28 and S29). Additional experiments were also performed as already discussed in the literature to induce the aggregation in these compounds [57], without any clear indication of its formation (Figures S31 and S32), which corroborates with the affirmation that all observed absorption bands are related to their monomeric species. Although the excited state investigation concerning heptamethinecyanines is less studied if compared to the UV-Vis data, probably due to its particular photophysics, some studies can be found and are related to the presence of monomeric species in solution or even aggregates [58]. In this way, aiming to bring some light to this discussion and based on the ground state results obtained in this investigation, the steady-state fluorescence emission was performed with cyanine dyes **8a-b**. In this sense, PL spectra were obtained at different excitation wavelengths (Figure 3a,b). The relevant data from this investigation are summarized in Table 3. It is worth mentioning that the emission spectra of cyanine dye **8a** presented the same emission profile (Figures S33 and S34). It can be observed that cyanine dye **8b** after excitation between 581 and 611 nm presents in all studied solvents a dual fluorescence emission located around 650 and 750 nm. The main emission in dichloromethane and polar protic media is located at higher energies, and in acetonitrile, the emission band is located at longer wavelengths (Figure 3a). However, under excitation between 708 and 715 nm, the main fluorescence emission can be observed located around 725 nm (Figure 3b). Based on the ground and excited state photophysics, the observed emission profiles seem not to be related to aggregation in solution [58]. In addition, relatively large Stokes shift values (3329–4868 cm^{-1}) could be calculated for the synthesized compounds under excitation between 581 and 611 nm, as shown in Table 3. Usually, this parameter is polarity dependent, as observed in this study, where more polar solvents presented higher Stokes shift values. This behavior indicates that different electronic structures are present between the ground and the excited states, as reported in the literature for similar compounds [18]. Thus, it is believed that the observed dual fluorescence emission with a large Stokes shift may be evidence of intramolecular charge transfer (ICT) between electron donor and acceptor groups present in the molecular structure [18]. In this way, as observed in Figure 3a, the emission profiles can be related to both locally excited (blue-shifted, Stokes shift ~1500 cm^{-1}) and charge transfer (red-shifted, Stokes shift ~4000 cm^{-1}) species. The fluorescence quantum yield values were measured from emission curves obtained using the absorption maxima as excitation wavelengths, and the values are in agreement with the literature [11].

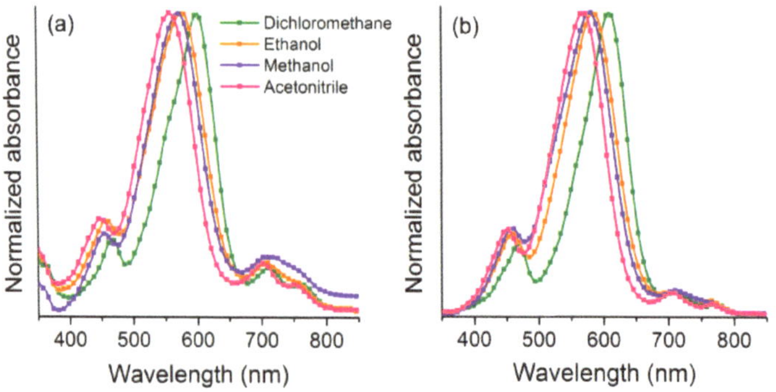

Figure 2. *Cont.*

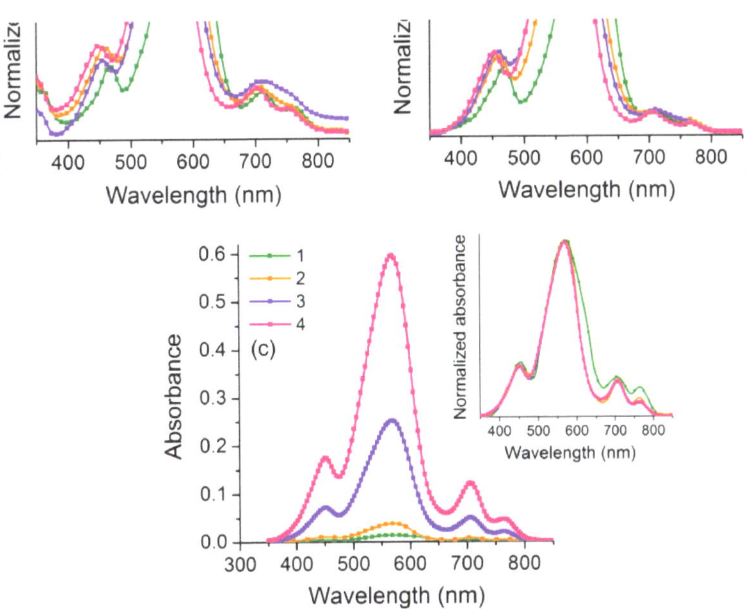

Figure 2. Normalized UV-Vis absorption spectra of heptamethine cyanine dyes (**a**) **8a** and (**b**) **8b** in a solution of different organic solvents (~10^{-6} M) and (**c**) cyanine dye **8b** in acetonitrile at different concentrations (1: 5.0×10^{-7} M, 2: 1.0×10^{-6} M, 3: 5.0×10^{-6} M and 4: 5.0×10^{-5} M). The inset shows the respective normalized UV-Vis spectra. (DCM = dichloromethane).

Figure 2. UV-Vis absorption spectra of heptamethine cyanine dyes (**a**) **8a** and (**b**) **8b** in a solution of different organic solvents (~10^{-6} M) and (**c**) cyanine dye **8b** in acetonitrile at different concentrations (1: 1.0×10^{-7} M, 2: 1.0×10^{-6} M, 3: 5.0×10^{-6} M, and 4: 5.0×10^{-5} M). The inset shows the respective normalized UV-Vis spectra. (DCM = dichloromethane).

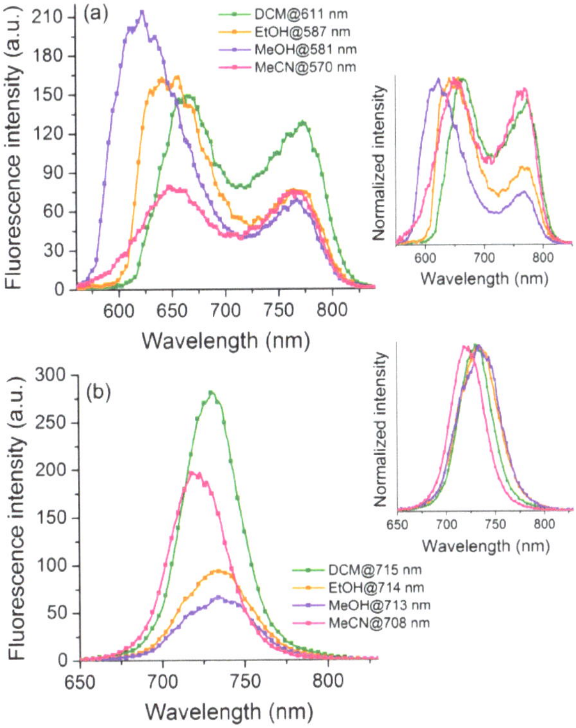

Figure 3. Steady-state fluorescence emission spectra of heptamethine cyanine dye **8b** in a solution of different organic solvents (~10^{-6} M) at excitation wavelength between (**a**) 581 and 611 nm (Table 2) and (**b**) 708 and 715 nm (Table S2). The inset shows the respective normalized spectra. (DCM = dichloromethane, MeCN = acetonitrile and MeOH = methanol).

Figure 3. Steady-state fluorescence emission spectra of heptamethine cyanine dye **8b** of different organic solvents (~10^{-6} M) at excitation wavelength between (**a**) 581 and 6 2) and (**b**) 708 and 715 nm (Table S2). The inset shows the respective normalized spe dichloromethane, MeCN = acetonitrile and MeOH = methanol).

On the other hand, due to the single fluorescence emission with a small (~500 cm^{-1}), the observed emission in Figure 3b seems to arise from a loc specie. To better understand the observed photophysics, the difference ir moments between the excited and ground states was obtained for cyanine

Table 3. Excited-state photophysical data of *meso*-substituted heptamethine cyanine dyes **8a-b** in solution, where λ_{em} is the emission maximum (nm) at different excitation wavelengths (λ_{exc} in nm), $\Delta\lambda_{ST}$ is the Stokes shift (cm^{-1}), and Φ_{FL} is the fluorescence quantum yield.

Cyanine	Solvent	λ_{exc} [(a)]	LE		ICT		λ_{exc} [(b)]	LE		ICT		Φ_{FL}
			λ_{em}	$\Delta\lambda_{ST}$	λ_{em}	$\Delta\lambda_{ST}$		λ_{em}	$\Delta\lambda_{ST}$	λ_{em}	$\Delta\lambda_{ST}$	
8a	Dichloromethane	602	666	1596	774	3691	712	731	365	-	-	0.25
	Ethanol	578	647	1845	769	4297	708	735	519	-	-	0.16
	Methanol	572	624	1457	768	4462	703	736	638	-	-	0.32
	Acetonitrile	559	652	2552	768	4868	712	721	175	-	-	0.17
8b	Dichloromethane	611	660	1215	767	3329	715	727	231	-	-	0.27
	Ethanol	587	654	1745	764	3947	714	729	288	-	-	0.25
	Methanol	581	647	1756	763	4106	713	732	364	-	-	0.32
	Acetonitrile	570	641	1943	763	4438	708	718	197	-	-	0.25

[(a)] obtained from the absorption spectra (maxima). [(b)] obtained from the absorption spectra (redshifted bands).

On the other hand, due to the single fluorescence emission with a small Stokes shift (~500 cm^{-1}), the observed emission in Figure 3b seems to arise from a locally excited specie. To better understand the observed photophysics, the difference in the dipole moments between the excited and ground states was obtained for cyanine dye **8b** by applying the simplified Lippert–Mataga correlation presented in Equation (3) [59], where h is Planck's constant, c is the speed of light, a is the Onsager cavity radius, and μ_g and μ_e are the dipole moments of the solute in the ground and excited states, respectively. In this plot, a linear relation of the absorbance or fluorescence maxima versus the solvent polarity function can be related to the internal charge transfer character [60,61]. The relevant data from this investigation are summarized in Table S1. Macroscopically, Equation (3) relates solvatochromic shifts from the Stokes shift versus the orientation polarization function (Δf). This latter is given by Equation (4) [47], where ε and n are the dielectric constant and the refractive index, respectively, for a mixture of solvents (Equations (5) and (6)) [62]. In this mixture, f_A and f_B are the volumetric fractions of the two solvents.

$$\Delta\overline{v}_{st} = \frac{2(\mu_e - \mu_g)^2}{hca^3}\Delta f + \Delta\overline{v}_0 \quad (3)$$

$$\Delta f = \frac{(\varepsilon - 1)}{(2\varepsilon + 1)} - \frac{(n^2 - 1)}{(2n^2 + 1)} \quad (4)$$

$$\varepsilon_{mix} = f_A \cdot \varepsilon_A + f_B \cdot \varepsilon_B \quad (5)$$

$$n_{mix}^2 = f_A n_A^2 + f_B n_B^2 \quad (6)$$

Figure 4 presents the Lippert–Mataga relationship between the absorption or emission maxima, as well as the Stokes shift. It is worth mentioning that this investigation was focused on the locally excited and intramolecular charge transfer emissions (see Figure 3a). The linear relationship between the absorption maxima with increasing solvent polarity (Figure 4a) and the upward curvature indicates that cyanine dye **8b** seems to be significantly destabilized in a more polar environment [63]. Taking the two emission bands into account (Figure 4b), a similar feature can be observed, once again with a positive slope with the solvent polarity function Δf. These results indicate that despite quaternary nitrogen in the indolic moiety, this compound seems to present low polarity in both ground and excited states. This latter can be related to a very effective delocalization of the positive charge into the cyanine skeleton. Surprisingly, the emission bands also seem to be equally affected by the solvent polarity (Figure 4c). Excitation spectra obtained from the fluorescence emission presented in Figure 3 indicate a quite similar absorption profile to those observed in the UV-Vis spectra (Figures S35–S40). Moreover, using 650 nm and 750 nm as the observation effective delocalization of the positive charge into the cyanine skeleton. Surprisi[…] emission bands also seem to be equally affected by the solvent polarity (Fi[…] Excitation spectra obtained from the fluorescence emission presented in Figure 3[…] a quite similar absorption profile to those observed in the UV-Vis spectra […] S35–S40). Moreover, using 650 nm and 750 nm as the observation wavelengths, it can be observed that the higher intensities are located around 600 nm, as expected. In contrast, the excitation spectra from Figure 4b present a magnification of the signal located around 700 nm, if compared to the UV-Vis (Figure 2b). This behavior was also expected, indicating that the main fluorescence emission observed at 725 nm is really due to the species with absorption around 700 nm, which is different from the species absorbing between 581 and 611 nm.

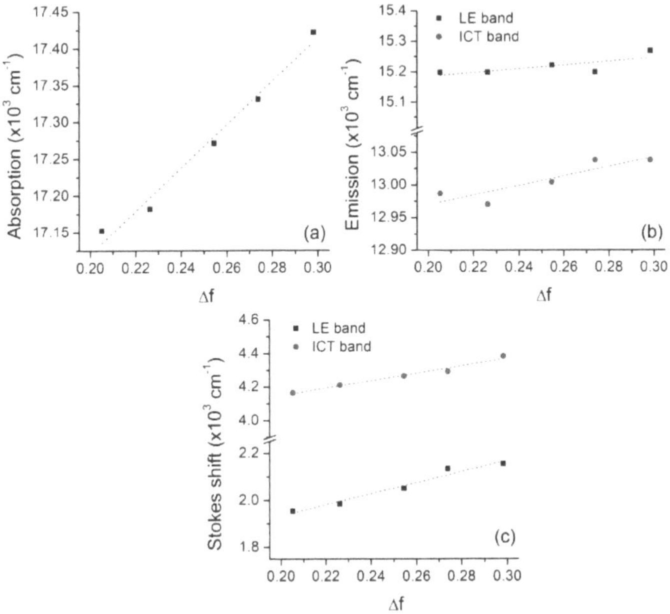

Figure 4. Solvent effect on the spectral position of the (**a**) absorption, (**b**) fluorescence emission maxima at LE and ICT bands, and respective (**c**) Stokes shift for cyanine dye **8b** in different acetonitrile/1,4-dioxane mixtures.

3.3. Theoretical Calculations

Cyanine dyes **6a** and **8a**, chosen as models for the theoretical investigation, have a large chemical space composed, respectively, of 201 and 33 unique conformers. Structure **6a** has the lowest conformer appearing at 38.6% of the Boltzmann population separated from the second lowest conformer (8.0% of the Boltzmann population) by 0.010 kcal·mol^{-1}. Similarly, the Boltzmann distributions indicate that the most stable conformer for cyanine **8a** corresponds to 36.13% of the population, separated from the second lowest conformer (1.9% of the Boltzmann population) by 0.047 kcal·mol^{-1} (Figure 5).

The selected lower conformers underwent geometric optimizations to the ground and first excited states using Density Functional Theory on the B97-3c scheme. All optimizations accounted for relaxed solvation effects and the ground states are confirmed to be true energy minima by the absence of imaginary vibrational modes. The final ground state geometries showed small distortions due to solvent effects. Cyanine dye **6a** showed small distortions after optimizations with acetonitrile, dichloromethane, and methanol with an average RMSD of 0.007Å. Similarly, cyanine **8a** after being optimized under acetonitrile, dichloromethane, ethyl acetate, and methanol differs by an average RMSD of only 0.23 Å from each other. As expected, differences between ground and first excited state geometries are small (Figure S41), as electronic absorption occurs extremely fast [62]. As can be seen in Figures S42 and S43, the change in the solvation environment also has just a small effect on the HOMO-LUMO gap described by Density Functional Theory, leading to variations in the order of milihartrees. An additional investigation based on the CHarges from

Electrostatic Potentials using a Grid-based method (CHELPG) [64] showed, also, that the solvation environment itself has little to no impact on atomic charge distribution. A standard deviation (σ) of 0.001 charge units for **6a** and 0.003 charge units for cyanine dye **8a** was obtained after comparing the implicit solvation models by Density Functional Theory, as can be seen in data available in Tables S2 and S3. After confirming that solvation effects have only small impacts on geometric and electronic configurations, our analysis of ground and excited state geometries was based on a higher computational level based on Orbital-Optimized, Spin-Component Scaled Second Order Many-Body Perturbation Theory (OO-SCS-MP2) to avoid restrictions imposed by the Koopmans theorem on Density Functional Theory [65,66]. The HOMO-LUMO gap observed for cyanine dye **8a** on the ground state is 6.57 eV with a slight reduction in the excited state configuration to a gap of 6.42 eV. Charge transfer seems to be extremely relevant during the GS-ES transition due to an observed increase in dipole moment measured by the relaxed MP2 electronic density, going from 2.93 to 3.14 Debye, a 7% increase with noticeable dipole vector repositioning (Figure 6). The same trend is observed for cyanine precursor **6a,** but with a remarkable increase of 27.32%, from 1.83 to 2.33 Debye; the orientation of the total dipole moment vector is, also, drastically changed, as can be seen in Figure 6, indicating a relevant charge transfer effect between GS and ES.

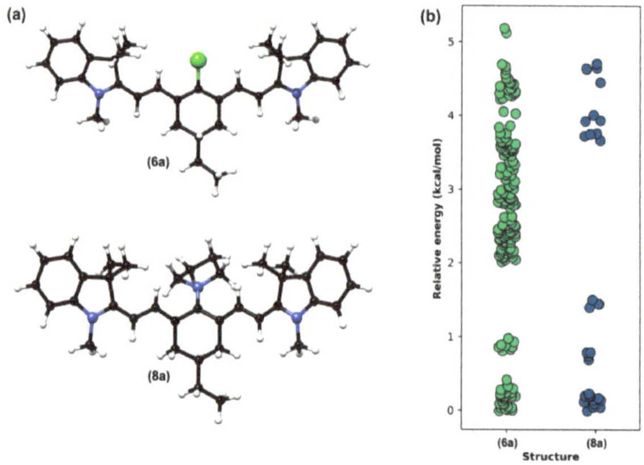

Figure 5. (**a**) Lowest and more stable conformers for cyanine dyes **6a** and **8a**, and (**b**) relative energy distribution of every unique conformer on its respective chemical space.

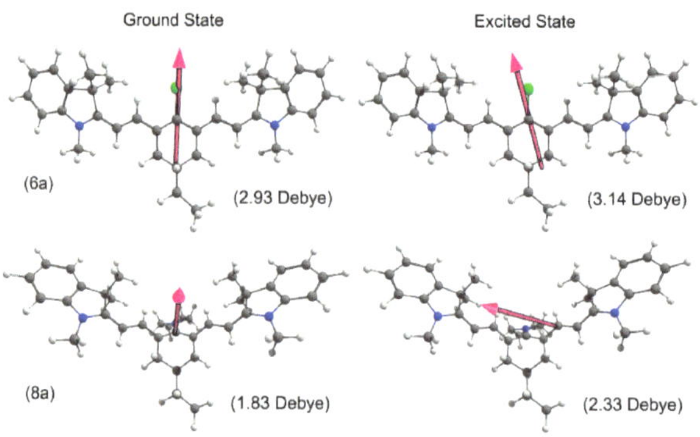

Figure 6. Change in dipole moment between ground and excited state of cyanine dyes **6a** (**top**) and **8a** (**bottom**).

Our analysis using the Molecular Electrostatic Potential Surface (MEPS) reveals that both molecules have a higher concentration of negative charge (represented in red) toward the center (Figure S44), while the indoline "wings" remain, overall, positively charged. Upon transition to the first excited state, the *meso*-substituted heptamethine cyanine dye **8a** migrates a considerable amount of negative charge to the center-left side of the molecule, in the opposite direction of the ethyl "tail". Charge transfer visualization using MEPS is not as straightforward for cyanine dye **6a,** because the dipole moment change is also due to a greater geometric change on the excited state and not just due to an almost vertical ground-to-excited transition as expected on cyanine **8a**. It is still possible to discern a slight increase in negative charge toward the upper center of the molecule as the wings become more positive. The Configuration Interaction Singles (CIS) calculated for both cyanines corroborates that only the immediate frontier orbitals are substantially involved in the first electronic transition with an 89% HOMO-LUMO character for cyanine **8a** and 86% HOMO-LUMO character for cyanine **6a**. The Electronic Density Difference (EDD) [67] was used as a tool to describe the dynamic vertical charge transfer between the frontier orbitals by exploring the hole–particle formalism (Figure 7a). It could also be observed that, to some extent, conjugation plays an important role in the ground state of the molecules and their first excited state (Figure 7b). Applying the Pipek–Mezey localization method [68,69], it was possible to quantify the bonds with high delocalization character. Cyanine dye **6a** accounts for a significantly higher delocalized orbital composition, presenting 10 bonds with a small electronic population. On the other hand, cyanine dye **8a** presents only 6 delocalized orbitals, probably due to not being as planar as dye **6a** and reducing the conjugation effects.

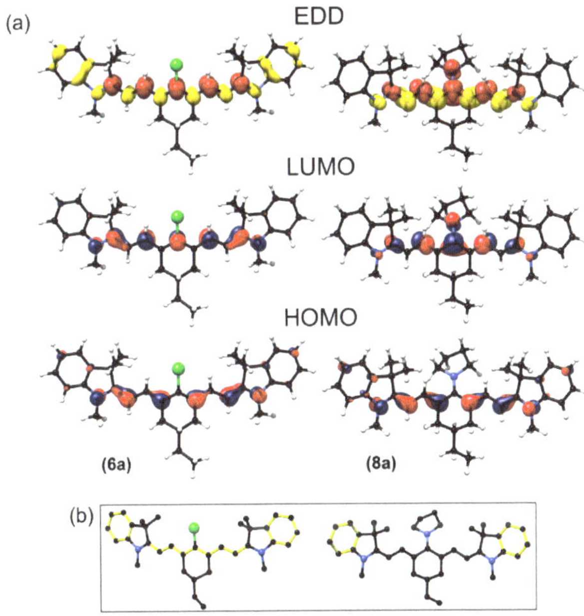

Figure 7. (**a**) Electronic Density Difference (EDD) with isosurface = 0.0016 for cyanine dyes **6a** and **8a**. The red color indicates a vertical charge increase and the yellow indicates avertical charge decrease for the GS-ES transition. HOMO and LUMO orbitals have isosurface = 0.05 with the (+) phase color coded as red and the (−) phase as blue. (**b**) Bonds with strong delocalized character (yellow) calculated by the Pipek–Mezey method. Hydrogens were omitted for clarity.

Figure 7. (**a**) Electronic Density Difference (EDD) with isosurface = 0.0016 for cyanine dyes **6a** and **8a**. The red color indicates a vertical charge increase and the yellow indicates avertical charge decrease for the GS-ES transition. HOMO and LUMO orbitals have isosurface = 0.05 with the (+) phase color coded as red and the (−) phase as blue. (**b**) Bonds with strong delocalized character (yellow) calculated by the Pipek–Mezey method. Hydrogens were omitted for clarity.

3.4. Bioimaging

3.4. Bioimaging

It was observed that all studied *meso*-substituted cyanines kept their optical properties in the dark at room temperature for over 4 weeks, presenting UV-Vis and PL spectra

It was observed that all studied *meso*-substituted cyanines kept their optical properties in the dark at room temperature for over 4 weeks, presenting UV-Vis and PL spectra with similar intensity and shape. It was reported that traditional cyanine dyes display poor photostability [16,70,71]; however, it was observed that the incorporation of a rigid cyclohexenyl ring in the polymethine chain leads to increased photostability [72]

Chemosensors **2023**, *11*, x FOR PEER REVIEW 16 of 21

with similar intensity and shape. It was reported that traditional cyanine dyes display poor photostability [16,70,71]; however, it was observed that the incorporation of a rigid cyclohexenyl ring in the polymethine chain leads to increased photostability [72]. In addition, it was also reported that an improved photostability on the cyanine scaffold takes place upon the incorporation of electron-withdrawing groups [73]. Based on these results, and the studies regarding very close structures with their respective photostability studies [71,73], it was decided to investigate their ability in cell staining without a previous photostability experiment, as usually presented in the literature [42,73]. In this sense, their ability to stain cellular compartments in fixed and live cells was explored (Figures 8 and 9). As shown in Figure 8, cyanine dye **8a**, used as a model, showed strong staining of cellular cytoplasm with long-lasting staining, remaining 24 h after the incubation in fixed cells. Similarly, the cyanine could also stain the cellular cytoplasm of live cells (Figure 9). It was also verified whether this cyanine could compromise cell adhesion (Figure 10), where there is no impairment of cell adhesion. These results demonstrated that cyanine dye **8a** could be used as a fluorescent probe in biotechnology and, more specifically, in protocols of cytochemistry for cytoplasm labeling. Further investigations are in progress for specific staining of selected organelles.

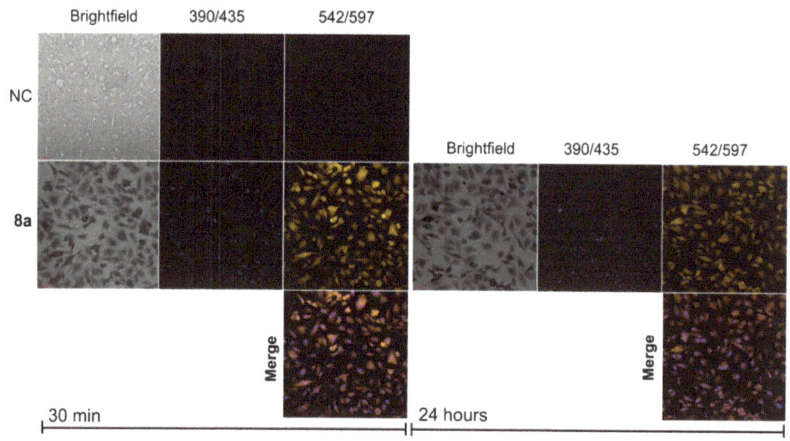

Figure 8. Fluorescent staining of fixed (paraformaldehyde 4%) cells using the *meso*-substituted cyanine **8a** (5 μM). The first column shows the brightfield, the central column shows the fluorescence of cells incubated with the nuclear dye Hoescht, and the third column shows cyanine **8a**. The result after 30 min of incubation is shown on the left, and that after 24 h of incubation is shown on the right. The first line represents cells without dye (negative control) and the third line represents the merge of Hoescht and cyanine **8a**. Scale bar (red) = 60 μm.

Figure 8. Fluorescent staining of fixed (paraformaldehyde 4%) cells using the *meso*-substituted cyanine **8a** (5 μM). The first column shows the brightfield, the central column shows the fluorescence of cells incubated with the nuclear dye Hoescht, and the third column shows cyanine **8a**. The result after 30 min of incubation is shown on the left, and that after 24 h of incubation is shown on the right. The first line represents cells without dye (negative control) and the third line represents the merge of Hoescht and cyanine **8a**. Scale bar (red) = 60 μm.

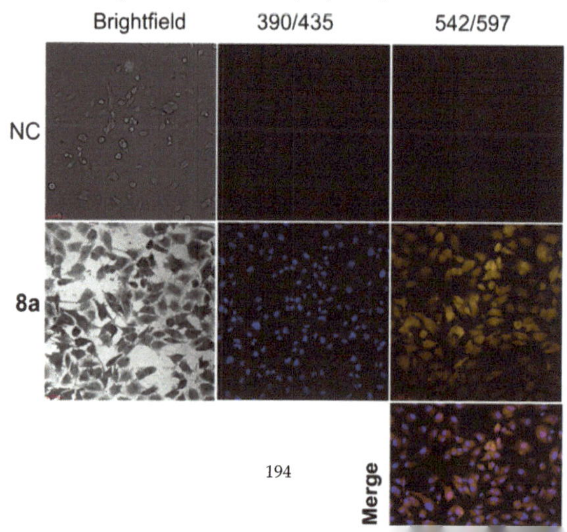

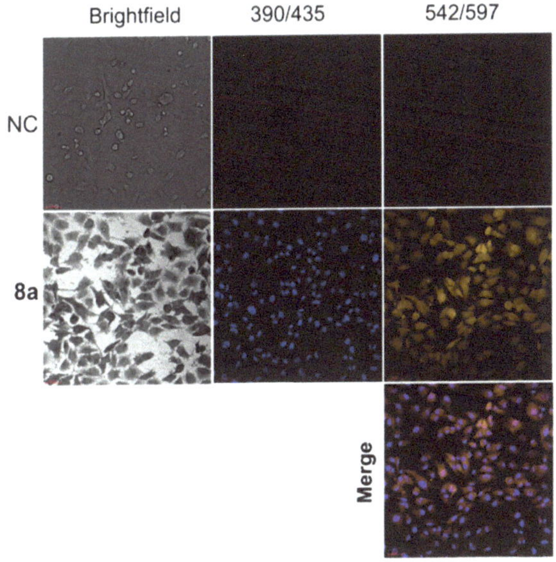

Figure 9. Fluorescent staining of live cells using the *meso*-substituted cyanine **8a** (5 μM). The images were obtained 30 min after incubation. NC = control. The first column shows the brightfield, the central column showsthe fluorescence of cells incubated with the nuclear dye Hoescht, and the third column shows cyanine 8a. The first line represents cells without dye (negative control), the second line represents cells exposed to cyanine **8a,** and the third line represents the merge of Hoescht and cyanine **8a**. Scale bar (red) = 60 μm.

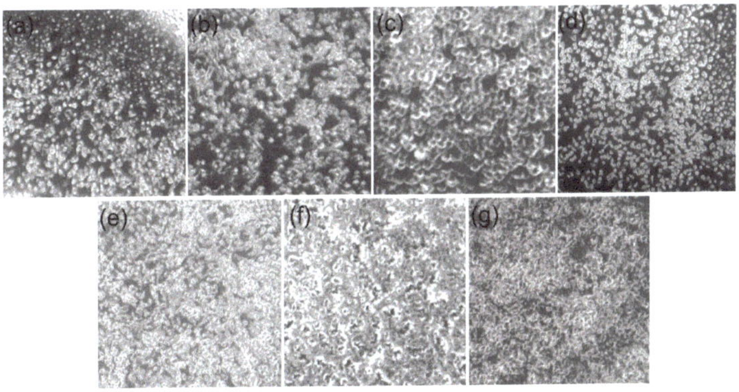

Figure 10. Optical images of cellular confluence after treatment of increasing doses of cyanine dye**8a**: (**a**) negative control, (**b**) 0.005 μM, (**c**) 0.01 μM, (**d**) 0.05 μM, (**e**) 0.1 μM, (**f**) 0.5 μM, and (**g**) 5.0 μM after 30 min of incubation.

4. Conclusions

Chlorine heptamethine cyanine dyes were obtained in good yields using a methodology between quaternary indoles and a pentamethine salt, minimizing photooxidation. These cyanines presented a simple and behaved photophysical behavior with no evidence of aggregation in both ground and excited states. In a second step, they were used as molecular scaffolds to synthesize *meso*-substituted derivatives by reaction with pyrrolidine via the $S_{RN}1$ mechanism with good yields. In this case, the substitution at the meso position showed a fundamental role in their photophysics, presenting intricate behavior but still related to monomeric species. DFT calculations showed that solvation effects, despite being present, are not responsible for relevant geometric or electronic changes on the studied cyanine dyes. Charge transfers between the ground and first excited states are responsible for an increase in negative charge to the center of the molecule, while the external rings become more positive. It is also clear that conjugation effects are much more active on the *meso*-substituted heptamethine cyanine dye due to its greater planarity in comparison to its chlorine heptamethine cyanine dye precursor. Finally, the observed photophysical features of the *meso*-substituted heptamethine cyanines were successfully used to explore their application as fluorescent probes in biological media, showing potential for cytoplasm labeling.

Supplementary Materials: The Supplementary Materials are available online at https://www.mdpi.com/article/10.3390/chemosensors11010047/s1. Figures S1–S20 (original spectra from the spectroscopic characterization), Figures S21–S40 and Table S1 (additional photophysical data), and Figures S41–S44, and Tables S2 and S3 (additional theoretical data).

Author Contributions: Conceptualization, F.S.R., D.J.M., H.d.C.S.J. and R.d.C.D.; methodology, R.d.C.D., H.d.C.S.J., D.J.M. and F.d.S.S.; validation, F.S.R. and D.J.M.; investigation, D.d.S.F., A.M.M., L.K.R., F.d.S.S., H.d.C.S.J. and R.d.C.D.; resources, F.S.R. and D.J.M.; writing—original draft preparation, F.S.R., D.J.M. and H.d.C.S.J.; writing—review and editing, R.d.C.D., H.d.C.S.J., D.J.M. and F.S.R.; visualization, F.S.R.; supervision, F.S.R. and D.J.M.; project administration, F.S.R.; funding acquisition, F.S.R. and D.J.M. All authors have read and agreed to the published version of the manuscript.

Funding: This research was funded by Fundação de Amparo à Pesquisa do Estado do Rio Grande do Sul (FAPERGS), grant number 17/2551-0000968-1; National Council for Scientific and Technological Development (CNPq), grant number 409855/2018-9 and 305954/2019-9 and Coordenação de Aperfeiçoamento de Pessoal de Nível Superior (CAPES), Finance Code 001.

Institutional Review Board Statement: Not applicable.

Informed Consent Statement: Not applicable.

Data Availability Statement: Not applicable.

Acknowledgments: All theoretical calculations were performed using the Lobo Carneiro supercomputer from Núcleo Avançado de Computação de Alto Desempenho (NACAD), under Project ID a20006e.

Conflicts of Interest: The authors declare no conflict of interest.

References

1. Reichardt, C. Chiral polymethine dyes: A remarkable but forgotten conjugated π system. *J. Phys. Org. Chem.* **1995**, *8*, 761–773. [CrossRef]
2. Panigrahi, M.; Dash, S.; Patel, S.; Mishra, B.K. Syntheses of cyanines: A review. *Tetrahedron* **2012**, *68*, 781–805. [CrossRef]
3. Bricks, J.L.; Kachkovskii, A.D.; Slominskii, Y.L.; Gerasov, A.O.; Popov, S.V. Molecular design of near infrared polymethine dyes: A review. *Dyes Pigm.* **2015**, *121*, 238–255. [CrossRef]
4. Shindy, H.A. Fundamentals in the chemistry of cyanine dyes: A review. *Dyes Pigm.* **2017**, *145*, 505–513. [CrossRef]
5. Bouit, P.A.; Aronica, C.; Toupet, L.; Le Guennic, B.; Andraud, C.; Maury, O. Continuous symmetry breaking induced by ion pairing effect in heptamethine cyanine dyes: Beyond the cyanine limit. *J. Am. Chem. Soc.* **2010**, *132*, 4328–4335. [CrossRef]
6. Strehmel, B.; Schmitz, C.; Kütahya, C.; Pang, Y.; Drewitz, A.; Mustroph, H. Photophysics and photochemistry of NIR absorbers derived from cyanines: Key to new technologies based on chemistry. *Beilstein J. Org. Chem.* **2020**, *16*, 415–444. [CrossRef]
7. Štacková, L.; Muchová, E.; Russo, M.; Slavíček, P.; Štacko, P.; Klán, P. Deciphering the structure-property relations in substituted heptamethine cyanines. *J. Org. Chem.* **2020**, *85*, 9776–9790. [CrossRef]
8. Usama, S.M.; Thompson, T.; Burgess, K. Productive manipulation of cyanine dye π-networks. *Angew. Chem. Int. Ed.* **2019**, *8*, 8974–8976. [CrossRef]
9. Exner, R.M.; Cortezon-Tamarit, F.; Pascu, S.I. Explorations into the effect of meso-substituents in tricarbocyanine dyes: A path to diverse biomolecular probes and materials. *Angew. Chem. Int. Ed.* **2020**, *60*, 6230–6241. [CrossRef]
10. Pascal, S.; Haefele, A.; Monnereau, C.; Charaf-Eddin, A.; Jacquemin, D.; Le Guennic, B.; Andraud, C.; Maury, O. Expanding the polymethine paradigm: Evidence for the contribution of a bis-dipolar electronic structure. *J. Phys. Chem. A* **2014**, *118*, 4038–4047. [CrossRef]
11. Zhang, J.; Moemeni, M.; Yang, C.; Liang, F.; Peng, W.T.; Levine, B.G.; Lunt, R.R.; Borhan, B. General strategy for tuning the Stokes shifts of near infrared cyanine dyes. *J. Mater. Chem. C* **2020**, *8*, 16769–16773. [CrossRef]

12. Sissa, C.; Painelli, A.; Terenziani, F.; Trotta, M.; Ragni, R. About the origin of the large Stokes shift in aminoalkyl substituted heptamethine cyanine dyes. *Phys. Chem. Chem. Phys.* **2020**, *22*, 129–135. [CrossRef]
13. Gragg, J.L. Synthesis of Near-Infrared Heptamethine Cyanine Dyes. Ph.D. Thesis, Georgia State University, Atlanta, GA, USA, 2010.
14. Gorka, A.P.; Nani, R.R.; Schnermann, M.J. Cyanine polyene reactivity: Scope and biomedical applications. *Org. Biomol. Chem.* **2015**, *13*, 7584–7598. [CrossRef] [PubMed]
15. Dimer, L.M.Z.; Machado, V.G. Chromogenic and fluorogenic chemosensors for detection of anionic analites. *Química Nova* **2008**, *31*, 2134–2146. [CrossRef]
16. Sun, W.; Guo, S.; Hu, C.; Fan, J.; Peng, X. Recent Development of chemosensors based on cyanine platforms. *Chem. Rev.* **2016**, *116*, 7768–7817. [CrossRef] [PubMed]
17. Reichardt, C. Pyridinium-N-phenolate betaine dyes as empirical indicators of solvent polarity: Some new findings. *Pure Appl. Chem.* **2008**, *80*, 1415–1432. [CrossRef]
18. Peng, X.; Song, F.; Lu, E.; Wang, Y.; Zhou, W.; Fan, J.; Gao, Y. Heptamethine cyanine dyes with a large Stokes shift and strong fluorescence: A paradigm for excited-state intramolecular charge transfer. *J. Am. Chem. Soc.* **2005**, *127*, 4170–4171. [CrossRef]
19. Machado, V.G.; Nascimento, M.G.; Rezende, M.C. The halochromism of the 1-methyl-8-oxyquinolinium dye. *Spectrosc. Lett.* **1998**, *31*, 359–367. [CrossRef]
20. Tada, E.B.; Novaki, L.P.; El Seoud, O.A. Solvatochromism in cationic micellar solutions: Effects of the molecular structures of the solvatochromic probe and the surfactant headgroup. *Langmuir* **2001**, *17*, 652–658. [CrossRef]
21. Keum, S.R.; Roh, S.J.; Ahn, S.M.; Lim, S.S.; Kim, S.H.; Koh, K. Solvatochromic behavior of non-activated indolinobenzospiropyran 6-carboxylates in aqueous binary solvent mixtures. Part II. *Dyes Pigm.* **2007**, *74*, 343–347. [CrossRef]
22. Medeiros, N.G.; Braga, C.A.; Câmara, V.C.; Duarte, R.C.; Rodembusch, F.S. Near-Infrared fluorophores based on heptamethine cyanine dyes: From their synthesis and photophysical properties to recent optical sensing and bioimaging applications. *Asian J. Org. Chem.* **2022**, *11*, 118–149. [CrossRef]
23. Thomas, R.G.; Jeong, Y.Y. NIRF heptamethine cyanine dye nanocomplexes for multi modal theranosis of tumors. *Chonnam Med. J.* **2017**, *53*, 83–94. [CrossRef]
24. Cosco, E.D.; Spearman, A.L.; Ramakrishnan, S.; Lingg, J.G.; Saccomano, M.; Pengshung, M.; Arús, B.A.; Wong, K.C.; Glasl, S.; Ntziachristos, V.; et al. Shortwave infrared polymethine fluorophores matched to excitation lasers enable non-invasive, multicolour in vivo imaging in real time. *Nat. Chem.* **2020**, *12*, 1123–1130. [CrossRef]
25. Yang, X.; Shi, C.; Tong, R.; Qian, W.; Zhau, H.E.; Wang, R.; Zhu, G.; Cheng, J.; Yang, V.W.; Cheng, T.; et al. Near IR heptamethine cyanine dye-mediated cancer imaging. *Clin. Cancer Res.* **2010**, *16*, 2833–2844. [CrossRef]
26. Myochin, T.; Kiyose, K.; Hanaoka, K.; Kojima, H.; Terai, T.; Nagano, T. Rational design of ratiometric near-infrared fluorescent pH probes with various pKa values, based on aminocyanine. *J. Am. Chem. Soc.* **2011**, *133*, 3401–3409. [CrossRef]
27. Duarte, R.C.; Reimann, L.K.; Rodembusch, F.S.; Duarte, L.G.T.A. 1-Butyl-2,3,3-trimethylindol-1-ium iodide. *IUCrData* **2018**, *3*, x181130. [CrossRef]
28. Menéndez, G.O.; Pichel, M.E.; Spagnuolo, C.C.; Jares-Erijman, E.A. NIR fluorescent biotinylated cyanine dye: Optical properties and combination with quantum dots as a potential sensing device. *Photochem. Photobiol. Sci.* **2013**, *12*, 236–240. [CrossRef]
29. Neese, F.; Wennmohs, F.; Becker, U.; Riplinger, C. The ORCA quantum chemistry program package. *J. Chem. Phys.* **2020**, *152*, 224108. [CrossRef]
30. Pracht, P.; Bohle, F.; Grimme, S. Automated exploration of the low-energy chemical space with fast quantum chemical methods. *Phys. Chem. Chem. Phys.* **2020**, *22*, 7169–7192. [CrossRef]
31. Bannwarth, C.; Ehlert, S.; Grimme, S. GFN2-XTB—An accurate and broadly parametrized self-consistent tight-binding quantum chemical method with multipole electrostatics and density-dependent dispersion contributions. *J. Chem. Theory Comput.* **2019**, *15*, 1652–1671. [CrossRef]
32. Brandenburg, J.G.; Bannwarth, C.; Hansen, A.; Grimme, S. B97-3c: A revised low-cost variant of the B97-D density functional method. *J. Chem. Phys.* **2018**, *148*, 064104. [CrossRef] [PubMed]
33. Becke, A.D. Density-functional thermochemistry. V. Systematic optimization of exchange-correlation functionals. *J. Chem. Phys.* **1997**, *107*, 8554–8560. [CrossRef]
34. Weigend, F.; Ahlrichs, R. Balanced basis sets of split valence, triple zeta valence and quadruple zeta valence quality for H to Rn: Design and assessment of accuracy. *Phys. Chem. Chem. Phys.* **2005**, *7*, 3297–3305. [CrossRef] [PubMed]
35. Grimme, S.; Antony, J.; Ehrlich, S.; Krieg, H. A consistent and accurate ab initio parametrization of density functional dispersion correction (DFT-D) for the 94 elements H-Pu. *J. Chem. Phys.* **2010**, *132*, 154104. [CrossRef] [PubMed]
36. Sure, R.; Grimme, S. Corrected small basis set Hartree-Fock method for large systems. *J. Comput. Chem.* **2013**, *34*, 1672–1685. [CrossRef]
37. Cammi, R.; Mennucci, B.; Tomasi, J. Fast evaluation of geometries and properties of excited molecules in solution: A Tamm-Dancoff model with application to 4-dimethylaminobenzonitrile. *J. Phys. Chem. A* **2000**, *104*, 5631–5637. [CrossRef]
38. Neese, F.; Schwabe, T.; Kossmann, S.; Schirmer, B.; Grimme, S. Assessment of orbital-optimized, spin-component scaled second-order many-body perturbation theory for thermochemistry and kinetics. *J. Chem. Theory Comput.* **2009**, *5*, 3060–3073. [CrossRef]

39. Dost, T.L.; Gressel, M.T.; Henary, M. Synthesis and optical properties of pentamethine cyanine dyes with carboxylic acid moieties. *Anal. Chem. Insights* **2017**, *12*, 1–6. [CrossRef]
40. Shi, Q.Q.; Sun, R.; Ge, F.J.; Xu, Q.F.; Li, N.J.; Lu, J.M. A comparative study of symmetrical and unsymmetrical trimethine cyanine dyes bearing benzoxazolyl and benzothiazolyl groups. *Dyes Pigm.* **2012**, *93*, 1506–1511. [CrossRef]
41. Chapman, G.; Henary, M.; Patonay, G. The effect of varying short-chain alkyl substitution on the molar absorptivity and quantum yield of cyanine dyes. *Anal. Chem. Insights* **2011**, *6*, 29–36. [CrossRef]
42. Shershov, V.E.; Spitsyn, M.A.; Kuznetsova, V.E.; Timofeev, E.N.; Ivashkina, O.A.; Abramov, I.S.; Nasedkina, T.V.; Zasedatelev, A.S.; Chudinov, A.V. Near-infrared heptamethine cyanine dyes. Synthesis, spectroscopic characterization, thermal properties and photostability. *Dyes Pigm.* **2013**, *97*, 353–360. [CrossRef]
43. Owens, E.A.; Bruschi, N.; Tawney, J.G.; Henary, M. A microwave-assisted and environmentally benign approach to the synthesis of near-infrared fluorescent pentamethine cyanine dyes. *Dyes Pigm.* **2015**, *113*, 27–37. [CrossRef]
44. Salon, J.; Wolinska, E.; Raszkiewicz, A.; Patonay, G.; Strekowski, L. Synthesis of benz[e]indolium heptamethine cyanines containing C-substituents at the central portion of the heptamethine moiety. *J. Heterocycl. Chem.* **2005**, *42*, 959–961. [CrossRef]
45. Pascal, S.; Chi, S.H.; Perry, J.W.; Andraud, C.; Maury, O. Impact of ion-pairing effects on linear and nonlinear photophysical properties of polymethine dyes. *ChemPhysChem* **2020**, *21*, 2536–2542. [CrossRef]
46. Strickler, S.J.; Berg, R.A. Relationship between absorption intensity and fluorescence lifetime of molecules. *J. Chem. Phys.* **1962**, *37*, 814–822. [CrossRef]
47. Turro, N.J.; Scaiano, J.C.; Ramamurthy, V. *Principles of Molecular Photochemistry: An Introduction*, 1st ed.; University Science Book: Sausalito, CA, USA, 2008.
48. Yao, Y.; Liang, Y.; Shrotriya, V.; Xiao, S.; Yu, L.; Yang, Y. Plastic near-infrared photodetectors utilizing low band gap polymer. *Adv. Mater.* **2007**, *19*, 3979–3983. [CrossRef]
49. Qian, G.; Zhong, Z.; Luo, M.; Yu, D.; Zhang, Z.; Wang, Z.Y.; Ma, D. Simple and efficient near-infrared organic chromophores for light-emitting diodes with single electroluminescent emission above 1000 nm. *Adv. Mater.* **2009**, *21*, 111–116. [CrossRef]
50. Shapiro, B.I. Molecular assemblies of polymethine dyes. *Russ. Chem. Rev.* **2006**, *75*, 433–456. [CrossRef]
51. Lee, H.; Mason, J.C.; Achilefu, S. Synthesis and spectral properties of near-infrared aminophenyl-, hydroxyphenyl-, and phenyl-substituted heptamethinecyanines. *J. Org. Chem.* **2008**, *3*, 723–725. [CrossRef]
52. Pydzińska, K.; Ziółek, M. Solar cells sensitized with near-infrared absorbing dye: Problems with sunlight conversion efficiency revealed in ultrafast laser spectroscopy studies. *Dyes Pigm.* **2015**, *122*, 272–279. [CrossRef]
53. Arjona-Esteban, A.; Stolte, M.; Würthner, F. Conformational switching of π-conjugated junctions from merocyanine to cyanine states by solvent polarity. *Angew. Chem. Int. Ed.* **2016**, *55*, 2470–2473. [CrossRef]
54. Egorov, V.V. Nature of the optical band shapes in polymethine dyes and H-aggregates: Dozy chaos and excitons. Comparison with dimers, H*- and J-aggregates. *R. Soc. Open Sci.* **2017**, *4*, 160550. [CrossRef]
55. Bricks, J.L.; Slominskii, Y.L.; Panas, I.D.; Demchenko, A.P. Fluorescent J-aggregates of cyanine dyes: Basic research and applications review. *Methods Appl. Fluoresc.* **2018**, *6*, 012001. [CrossRef]
56. Eisfeld, A.; Briggs, J. The shape of the J-band of pseudoisocyanine. *Chem. Phys. Lett.* **2007**, *446*, 354–358. [CrossRef]
57. Würthner, F.; Kaiser, T.E.; Saha-Möller, C.R. J-aggregates: From serendipitous discovery to supramolecular engineering of functional dye materials. *Angew. Chem. Int. Ed.* **2011**, *50*, 3376–3410. [CrossRef]
58. Rösch, U.; Yao, S.; Wortmann, R.; Würthner, F. Fluorescent H-aggregates of merocyanine dyes. *Angew. Chem. Int. Ed.* **2006**, *45*, 7026–7030. [CrossRef]
59. El-Daly, S.A.; Asiri, A.M.; Alamry, K.A. Experimental determination of ground and excited state dipole moments of *N*,*N*-bis (2,5-di-tert-butylphenyl)-3,4:9,10-perylenebis (dicarboximide) (DBPI) a photostable laser dye. *J. Fluoresc.* **2014**, *24*, 1307–1311. [CrossRef]
60. Lippert, E.; Lüder, W.; Moll, F.; Nägele, W.; Boos, H.; Prigge, H.; Seibold-Blankenstein, I. Umwandlung von elektronenanregungsenergie. *Angew. Chem.* **1961**, *73*, 695–706. [CrossRef]
61. Dahiya, P.; Choudhury, S.D.; Maity, D.K.; Mukherjee, T.; Pal, H. Solvent polarity induced structural changes in 2,6-diamino-9,10-anthraquinone dye. *Spectrochim. Acta A* **2008**, *69*, 134–141. [CrossRef]
62. Lakowicz, J.R. *Principles of Fluorescence Spectroscopy*, 3rd ed.; Springer: New York, NY, USA, 2006.
63. Gündüz, M.G.; da Silva, C.B.; Zanotto, G.M.; Toldo, J.M.; Şimşek, R.; Şafak, C.; Gonçalves, P.F.B.; Rodembusch, F.S. Theoretical and experimental study of the ground and excited states of 1,4-dihydropyridine based hexahydroquinoline derivatives achieved by microwave irradiation. *New J. Chem.* **2017**, *41*, 11686–11694. [CrossRef]
64. Breneman, C.M.; Wiberg, K.B. Determining atom-centered monopoles from molecular electrostatic potentials. The need for high sampling density in formamide conformational analysis. *J. Comput. Chem.* **1990**, *11*, 361–373. [CrossRef]
65. Stowasser, R.; Hoffmann, R. What do the Kohn-Sham orbitals and eigenvalues mean? *J. Am. Chem. Soc.* **1999**, *121*, 3414–3420. [CrossRef]
66. Tsuneda, T.; Song, J.W.; Suzuki, S.; Hirao, K. On Koopmans' theorem in density functional theory. *J. Chem. Phys.* **2010**, *133*, 174101. [CrossRef]
67. Harrison, J.F. On the role of the electron density difference in the interpretation of molecular properties. *J. Chem. Phys.* **2003**, *119*, 8763–8764. [CrossRef]

68. Pipek, J.; Mezey, P.G. A fast intrinsic localization procedure applicable for ab initio and semiempirical linear combination of atomic orbital wave functions. *J. Chem. Phys.* **1989**, *90*, 4916–4926. [CrossRef]
69. HØyvik, I.M.; Jansik, B.; JØrgensen, P. Pipek-Mezey localization of occupied and virtual orbitals. *J. Comput. Chem.* **2013**, *34*, 1456–1462. [CrossRef]
70. Shi, C.; Wu, J.B.; Pan, D. Review on near-infrared heptamethine cyanine dyes as theranostic agents for tumor imaging, targeting, and photodynamic therapy. *J. Biomed. Opt.* **2016**, *21*, 050901. [CrossRef]
71. Usama, S.M.; Thavornpradit, S.; Burgess, K. Optimized heptamethinecyanines for photodynamic therapy. *ACS Appl. BioMater.* **2018**, *1*, 1195–1205. [CrossRef]
72. Lee, H.; Mason, J.C.; Achilefu, S. Heptamethine cyanine dyes with a robust C-C bond at the central position of the chromophore. *J. Org. Chem.* **2006**, *71*, 7862–7865. [CrossRef]
73. Samanta, A.; Vendrell, M.; Das, R.; Chang, Y.T. Development of photostable near-infrared cyanine dyes. *Chem. Commun.* **2010**, *46*, 7406–7408. [CrossRef]

chemosensors

Article

Development of Polydiacetylene-Based Testosterone Detection as a Model Sensing Platform for Water-Insoluble Hormone Analytes

Jaewon Jung [1], Sung-Min An [1], Eun-Kyung Lim [2,3], Seung-Chul Kim [4], Beum-Soo An [1] and Sungbaek Seo [1,*]

1 Department of Biomaterials Science (BK21 FOUR Program), Science/Life and Industry Convergence Research Institute, Pusan National University, Miryang 50463, Korea; jenifer5192@pusan.ac.kr (J.J.); ied234@naver.com (S.-M.A.); anbs@pusan.ac.kr (B.-S.A.)
2 BioNanotechnology Research Center, KRIBB, 125 Gwahak-Ro, Yuseong-Gu, Daejeon 34141, Korea; eklim1112@kribb.re.kr
3 Department of Nanobiotechnology, KRIBB School of Biotechnology, UST, 217 Gajeong-Ro, Yuseong-Gu, Daejeon 34113, Korea
4 Department of Obstetrics and Gynecology, College of Medicine, Pusan National University, Yangsan, Gyeongnam 50612, Korea; ksch0127@naver.com
* Correspondence: sbseo81@pusan.ac.kr

Abstract: We have developed a polydiacetylene (PDA)-based sensing platform to detect testosterone (T) as a potential biomarker of preterm birth. The insolubility of the steroid hormone in water, where PDA assemblies are dispersed, poses a major issue, since they can hardly interact with each other. To overcome this challenge, acetonitrile was used as a suitable solvent. In addition, to minimize false signals of PDA assemblies caused by the solvent, a mixture of acetonitrile and distilled water was selected. To prove a concept of PDA-based sensing platform for targeting T hormone, we conjugated anti-T antibodies to surface of PDA assemblies to induce selective binding between T and anti-T antibodies. The fluorescence sensory signaling of the PDA-anti-T antibody conjugate was selectively generated for T, over 3.4 times higher sensitivity of the signaling compared to that from other sex steroid hormones studied (β-estradiol and progesterone).

Keywords: polydiacetylene; preterm birth; steroid hormone detection

Citation: Jung, J.; An, S.-M.; Lim, E.-K.; Kim, S.-C.; An, B.-S.; Seo, S. Development of Polydiacetylene-Based Testosterone Detection as a Model Sensing Platform for Water-Insoluble Hormone Analytes. *Chemosensors* **2021**, *9*, 176. https://doi.org/10.3390/chemosensors9070176

Academic Editor: Pietro Salvo

Received: 24 May 2021
Accepted: 8 July 2021
Published: 12 July 2021

1. Introduction

The low birth rate of developed countries is a growing concern, mainly because it will cause national productivity to plummet in the near future. Preterm birth—early delivery of a baby before 37 weeks of pregnancy—is considered a reason for low birth rates [1,2]. Globally, approximately 15 million infants are born preterm every year, and this number is increasing [3]. Preterm birth not only causes health risks, such as deformities or infant death, but also raises the country's economic burden of newborn care.

Placental alpha macroglobulin-1 (PAMG-1) levels, fetal fibronectin levels, and ultrasound have been generally used to diagnose preterm birth [4,5]. PAMG-1 is found in amniotic fluid during pregnancy, and its concentration is 1000 times higher than that in normal vaginal discharge or maternal blood [6,7]. PAMG-1 can be detected using a lateral flow immunoassay, which are typically tests strips containing monoclonal anti-PAMG-1 antibodies with gold nanoparticles as detection label that can be visualized in the presence of PAMG-1. Fetal fibronectin is a glycoprotein of the basal decidual membrane, located near the amniotic fluid and the space between the placental tissue and placenta. It is released by mechanical or inflammation-mediated damage to the membrane or placenta prior to birth. The fetal fibronectin in cervical or posterior vaginal fornix can be detected by enzyme-linked immunosorbent assay, including monoclonal antibodies [8]. However, the detection of PAMG-1 or fetal fibronectin is usually available at the earliest at 7–10 days before birth delivery. Although obstetric ultrasound can be used to predict the risk of preterm delivery, it requires expensive equipment and trained operators. Therefore, developing low-priced and user-friendly diagnostic technology for the early prediction of preterm birth is essential.

Polydiacetylene (PDA) is a remarkable sensing material due to its dual-mode optical transitions, which produce easily detectable colorimetric and fluorogenic signaling responses. The optical transition of PDA responds to external stimuli, such as biomolecules [9–11] and chemical analytes [12–14]. Assemblies of PDA, such as PDA liposomes in aqueous phase, react or bind with target analytes, which generates an optical transition by distorting the conjugated backbone of PDA. The liposomes consisting of amphiphilic diacetylene or/and lipid molecule are readily formulated in aqueous and the surface of the liposomes could be immobilized with various biomolecules such as peptide, enzyme and antibody [15–20]. These characteristics of simple formulation and easy-to-surface modulation enable for the PDA liposomes to be used as easily accessible and economical sensing platform.

Based on previous studies, including our research [21–23], in patients showing symptoms of preeclampsia, the level of sex steroid hormones, such as estradiol, is 5-fold lower than that in normal pregnant women [21]. Also, there is a study that very low birth weight (less than 1500 g) preterm infants could be influenced by prenatal exposure to high levels of testosterone (T) [24]. Thus, sex steroid hormones could be potential biomarkers of preterm birth, and their early detection could be vital to prevent premature deliveries. As a detection of T in clinical practice, mass spectrometry-based technique is commonly used. The analytical methods such as gas chromatography–mass spectrometry and liquid chromatography–mass spectrometry provide an accurate quantification but require time-consuming and expensive instruments [25]. To our knowledge, the development of PDA-based sensors for hormone detection has rarely been explored [26,27]. Cho et al. studied the selective detection of progesterone using phospholipid-incorporated PDA assembly [26]. This method was inspired by the interactions between steroids and phospholipids of the cellular membrane. In addition, Jung et al. developed a glutathione substrate-tagged PDA assembly for the detection of glutathione S-transferase enzyme-based human growth hormone [27]. However, the main challenge in developing PDA-based hormone detection is that PDA assemblies in aqueous solution cannot efficiently meet and bind with water-insoluble hormones.

Herein, we investigated a suitable co-solvent medium for solubilizing the hydrophobic hormones and dispersing them in water simultaneously while minimizing optical false signaling by the medium. Sex steroid hormones, namely T, progesterone, and β-estradiol, were screened and incubated to generate sensory signals with PDA assembly of albumin or anti-T antibody conjugates via colorimetric and fluorogenic transitions.

2. Experimental Details

2.1. Materials

All solvents were purchased from Daejung Chemicals (Seoul, Korea). 10,12-Pentacosadiynoic acid (PCDA) was purchased from Alfa Aesar (Waltham, MA, USA). 1-(3-Dimethylaminopropyl)-3-ethylcarbodiimide hydrochloride (EDC), N-hydroxysuccinimide (NHS), and T were purchased from Tokyo Chemical Industry (Tokyo, Japan). Progesterone and Pierce^{TM} anti-T antibodies (T Ab, Product # MIT0103) were purchased from Thermo Fisher Scientific (Waltham, MA, USA). β-estradiol and albumin from human serum were purchased from Sigma Aldrich (Darmstadt, Germany). Phosphate buffered saline (1 × PBS) was purchased from Biosesang (Seoul, Korea).

2.2. Preparation of PDA Assemblies

To prepare PDA liposomes consisting of PCDA monomers, PCDA (3.75 mg) was dissolved in acetone (300 μL) and then injected into distilled water (DI water, 20 mL). The suspension was sonicated using a bath sonicator (P0000VUV, Kodo, Seoul, Korea) for 1 min and kept at 4 °C overnight.

We modified a protocol by Kim et al. that enables conjugating albumin or T Ab to PDA [28]. 50 mM EDC in DI water (1 mL) was added to an aqueous solution of 0.5 mM PDA (1 mL), after which 50 mM NHS in DI water (1 mL) was added. The mixture was stirred at room temperature for 2 h. The solution was centrifuged (Centurion Scientific, West Sussex, UK) at 15,000 rpm (1411× *g*) for 15 min to remove residual EDC/NHS. The supernatant of the centrifuged solutions was removed and re-dispersed in 1 × PBS (1 mL). Later, albumin or T Ab were added to the dispersion at a concentration of 0.1 mg/mL, and were stirred at room temperature for 2 h. To deactivate residual NHS, 1 mM ethylenediamine in 1 × PBS (1 mL) was added to the dispersion and stirred at room temperature for 2 h. The solution was centrifuged at 15,000 rpm (1411× *g*) for 15 min to remove unreacted residues. After washing three times with 1 × PBS by centrifugation, it was re-dispersed in 1 × PBS (1 mL) with sonication. The PDA assemblies were polymerized for 5 min with a UV lamp (254 nm, 1 mW·cm^{-2}, Vilber, Marne-la-Vallée, France).

2.3. *Characterization of PDA Assemblies and Steroid Hormones*

The morphology of the PDA assemblies was observed using an ultra-high resolution low-voltage-scanning electron microscope (JSM-7900F, JEOL, Tokyo, Japan) at an accelerating voltage of 5.0 kV. The specimens were coated with platinum with an 8 nm thickness. A Zetasizer (Zetasizer Nano ZS90, Malvern Panalytical, Worcestershire, UK) was used to measure the alteration in zeta potential and size of the PDA assemblies during the conjugation of albumin or T Ab to PDA. The chemical bonding or interaction of the PDA assemblies was monitored using Fourier transform infrared (FT-IR) spectroscopy (Vertex 80 v, Bruker Optics Co., Leipzig, Germany). To check the solubility of steroid hormones in the solvents, 3 mg/mL of hormone (T, progesterone, and β-estradiol) was added to candidate solvents (dimethyl sulfoxide [DMSO], acetonitrile [ACN], ethanol [EtOH], and methanol [MeOH]), and the transmittance of the solutions was measured using UV-Vis spectroscopy (Libra S70, Cambridge, UK)

2.4. *Hormone Detection Tests Using Colorimetric Response (CR) and Fluorescence Measurement*

To measure the *CR* of the PDA assemblies by adding medium (mixture of ACN:DI water, EtOH:DI water, MeOH:DI water) or hormone (T, progesterone, and β-estradiol) solution (0.1, 0.2, 0.5, 1, 2, and 3 mg/mL), the solution of PDA assemblies (120 μL) and the medium/the hormone solution (80 μL) were incubated for 1 min, and the absorption spectra were measured using UV-Vis spectroscopy.

The *CR* was calculated as follows:

$$CR(\%) = \frac{PB_{before} - PB_{after}}{PB_{before}} \times 100\,,\ PB = \frac{A_{blue}}{A_{blue} + A_{red}} \tag{1}$$

where A_{blue} is the absorbance intensity at 650 nm, and A_{red} is the absorbance intensity at 550 nm. PB_{before} and PB_{after} are the values before and after incubating with medium or hormone samples, respectively.

To measure the fluorescence (FL) intensity of the PDA assemblies, a solution of these assemblies (120 μL) was incubated with hormone solution (80 μL) for 1 min, and the emission spectra were measured using a fluorescence reader (iD5 Multi-Mode Microplate Reader, Molecular Devices, San Jose, CA, USA). The measurement was set to excitation at 485 nm, and the emissive intensity was recorded at 630 nm.

The data of FL intensity at each concentration of analyte hormone were expressed as mean value with error bar (n = 3). The sensitivity of sensing platform for each hormone was calculated from value of slope in the linear fitting. We analyzed limit of detection (*LOD*) was calculated as follows:

$$LOD = Standard\ Error\ (SE) \times \sqrt{N} \times 3.3 \div Slope \tag{2}$$

where N is the number of data elements (herein $N = 5$, five data point of hormone concentrations: 0.2, 0.5, 1, 2, 3 mg/mL), *Slope* is the value of linear fitting, and standard error (*SE*) is a standard deviation of the regression line (red line in the graphs) calculated using the OriginPro 8 software (Northampton, MA, USA).

2.5. T Hormone Detection Tests in Human Serum (1%)

To prevent the interfering sensory signals of PDA assemblies from human serum proteins in human serum, it was diluted with Tris-HCl buffer (20 mM, pH 7.0) and filtered using centrifugal filter (MWCO of 100–150 kDa) or centrifuging at 1500 rpm for 20 min. Then T hormone was spiked into 50% v/v of the filtered human serum (1%) and ACN, making various concentrations (0.1, 0.2, 0.5, 1, 2 and 3 mg/mL).

3. Results and Discussion

In mammalian blood plasma, water-insoluble sex steroid hormones (Figure 1a) exist as water-soluble complexes via binding to serum albumin or sex hormone-binding globulin [29,30]. Accordingly, these complexes transport and maintain the affinity of hormones in the aqueous phase. Few studies have been conducted to develop PDA-based sensors that target steroid hormones because water-insoluble hormones hardly meet and bind to PDA assemblies in aqueous media. Inspired by the hormone-albumin and hormone-globulin complexes in aqueous plasma, we utilized albumin or hormone-specific binding antibodies that could form a mediator of intermediate solubility (between water-insoluble hormones and water-dispersible PDA assemblies).

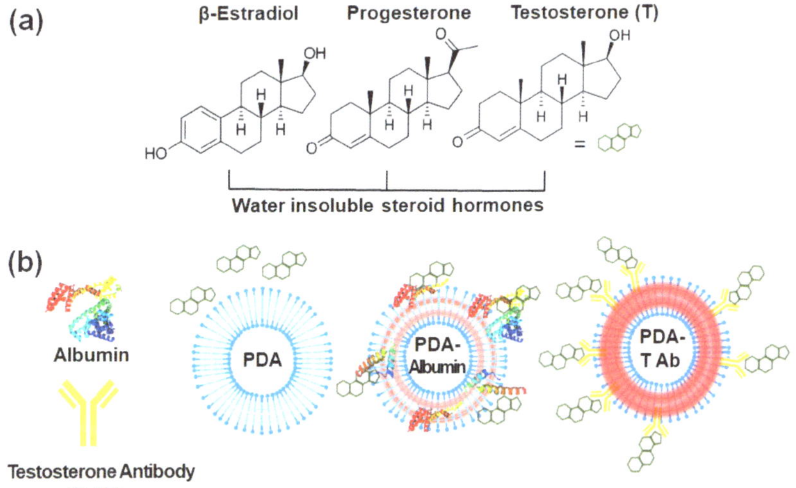

Figure 1. (**a**) Chemical structure of three sex steroid hormones (β-estradiol, progesterone, testosterone (T)) used as preterm biomarkers. (**b**) Schematic scenario of interaction of T on three types of polydiacetylene assemblies; liposome consisting of 10,12-pentacosadiynoic acid monomers (PDA), albumin-conjugated PDA (PDA-Albumin), and anti-T antibody-conjugated PDA (PDA-T Ab), when incubated with T. Depending on the interaction of T and PDA assemblies, colorimetric and fluorogenic sensory signal (red color) generated was drawn in the middle of the PDA assemblies.

Sex steroid hormones such as β-estradiol, progesterone, and T are thought to be pregnancy/preterm birth-related biomarkers [21–23]. Based on the current study that preterm infants could be influenced by T level [24], the PDA sensory platform was designed to target T for preterm birth prediction. We developed three types of PDA-based sensing platforms to compare sensory signals in the presence of T: (1) PDA liposomes consisting of PCDA monomers (PDA), (2) albumin conjugated to PDA (PDA-Albumin), and (3) anti-T

Sex steroid hormones such as β-estradiol, progesterone, and T are thought to be pregnancy/preterm birth-related biomarkers [21–23]. Based on the current study that preterm infants could be influenced by T level [24], the PDA sensory platform was designed to target T for preterm birth prediction. We developed three types of PDA-based sensing platforms to compare sensory signals in the presence of T: 1) PDA liposomes consisting of PCDA monomers (PDA), 2) albumin conjugated to PDA (PDA-Albumin), and 3) anti-T antibody conjugated to PDA (PDA-T Ab) (Figure 1b). We assumed that three types of PDA assemblies could have different numbers of binding and binding affinities of T—the highest binding available onto the PDA-T Ab by pairing interactions between T and T Ab.

The morphology of PDA assemblies was observed using SEM (Figure 2a). While PDA resulted in ~5 μm particle-shaped fragments, PDA-Albumin and PDA-T Ab presented an angular shape of 10–100 μm. Similar aggregated or angular structures of polydiacetylene-antibody conjugates have been reported in the literature [28,31]. The changes in surface charge and size distribution of PDA assemblies during conjugation to albumin or T Ab were measured by dynamic light scattering measurements (Figure 2b,c and Figure S1). PDA has a zeta potential of −46.9 ± 1.4 mV. During EDC/NHS activation of carboxylic acid on PDA, the zeta potential became less negatively charged, resulting in a value of −17.5 ± 0.3 mV (Figure S1). After conjugation with albumin and T Ab, the zeta potential values were increased to −5.8 ± 0.1 and −8.6 ± 0.5 mV, respectively. We hypothesize that positively charged albumin and T Ab were conjugated to the negatively charged PDA, inducing a change in the zeta potential after conjugation [32]. Based on the light scattering measurement, PDA has a size of 130 ± 10 nm, and after conjugation with albumin and T Ab, the size of PDA-Albumin and PDA-T Ab increased to 4400 ± 400 and 3800 ± 500 nm, respectively. Since PDA-Albumin and PDA-T Ab became less negatively charged than PDA, less repulsion between the assemblies would make being aggregated or bigger sized. In Figure S2, the transmittance of PDA-Albumin and PDA-T Ab at 1637 cm^{-1} (-COOH) recorded by FT-IR spectroscopy was relatively lower compared to that of PDA, implying that the carboxylic acid of PDA was consumed by the chemical reaction with EDC/NHS.

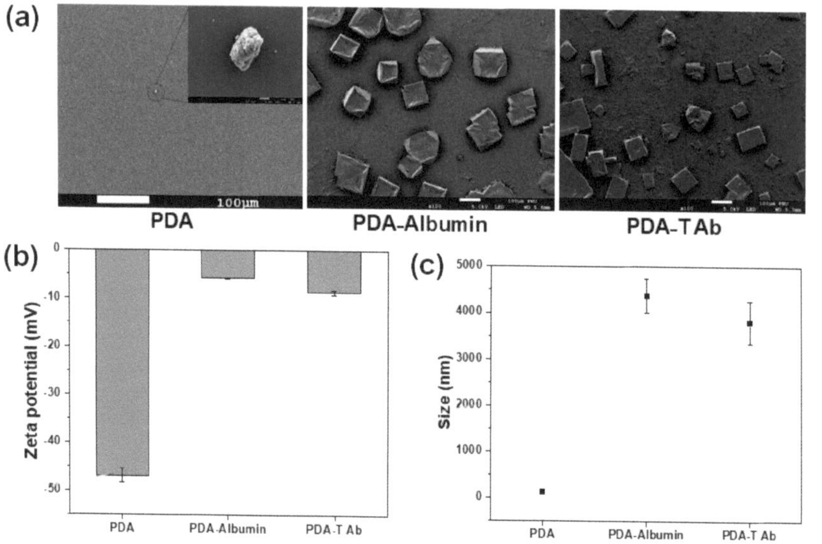

Figure 2. (**a**) SEM image of PDA, PDA-Albumin and PDA-T Ab. Scale bars represent 100 μm. The scale bar of the inset (magnified image) represents 1 μm. (**b**) Zeta potential values and (**c**) size of PDA, PDA-Albumin, and PDA-T Ab.

Since the solubility of steroid hormones in aqueous solution is low, we first aimed to find a suitable solvent medium that would both solubilize the hormones and enable them to encounter and react with PDA assemblies in aqueous solution. It is also important that the solvent minimize false signals from the PDA assemblies when added, since solvents, except for water, usually tend to generate a colorimetric transition [33,34]. Relatively polar solvent candidates, DMSO, ACN, EtOH, and MeOH, were screened for their ability to solubilize hormones by measuring the transmittance at 400 nm using a UV-Vis spectrophotometer

(Figure 3a). DMSO showed the lowest transmittance, which translates as the poorest solubility, and was thus excluded from the solvent candidates. We then assessed the *CR* of the PDA assemblies when introducing the candidate solvents (Figure 3b). Pure solvents (100% ACN, EtOH, and MeOH) induced a certain degree of *CR* (ACN: 15.4%, EtOH: 15.2%, and MeOH: 15.6%, calculated from absorption intensity at 650 nm and 550 nm) with pale violet colorimetric change (see inset image of Figure 3b). Therefore, the solvent ratio was reduced to 50% (v/v) (mixed with DI water), which decreased the false signals (ACN: 6.7%, EtOH: 7.7%, and MeOH: 4.1%). ACN was selected as the most suitable solvent because it has higher solubility than MeOH and causes fewer false signals than EtOH.

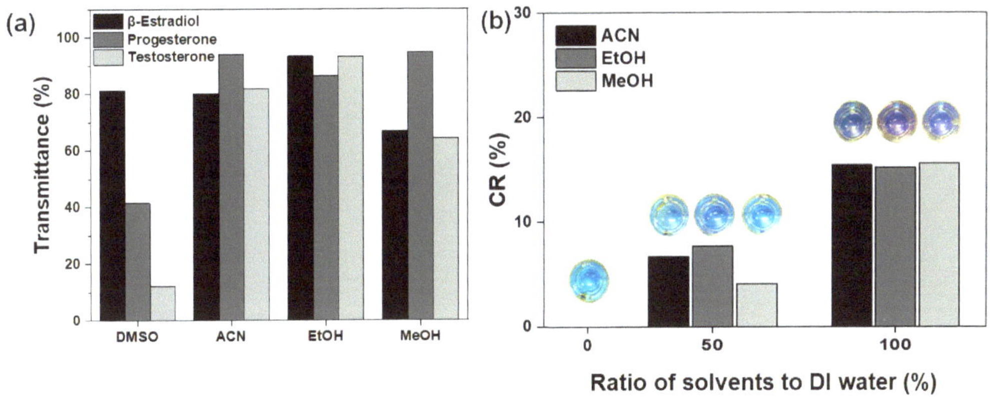

Figure 3. (**a**) Solubility test of sex steroid hormones in solvent medium (Dimethyl sulfoxide (DMSO), acetonitrile (ACN), ethanol (EtOH), methanol (MeOH)). (**b**) Colorimetric response (*CR*) of PDA when adding candidate solvent to the medium. Inset: photograph of the PDA solution after adding solvent to the medium.

The *CR* values of PDA-based sensors are commonly used as sensory signals. In all three types of PDA assemblies (PDA, PDA-Albumin, PDA-T Ab), hormone concentration-dependent *CR* values were observed with a trend of gradual increase (Figure 4a–c), indicating non-specific optical signals from the hormones. In particular, the *CR* of PDA-T Ab for T hormone was not significantly different from that of PDA or PDA-Albumin (Figure 4c), even though PDA-T Ab was designed to selectively react with or bind to T. The *CR* values were calculated by comparing the absorption intensity at 550 nm (wavelength of red absorption) and 650 nm (wavelength of blue absorption). In this study, estimating the correct *CR* values proved difficult since the overall spectral intensity was interfered by the absorption of water-insoluble hormone (Figure 4d–f). A possible explanation is that the instability and insolubility of hormones in aqueous solution inhibited the absorption-based colorimetric sensory signals, and no significant difference was observed in any of the three types of PDA assemblies as a result.

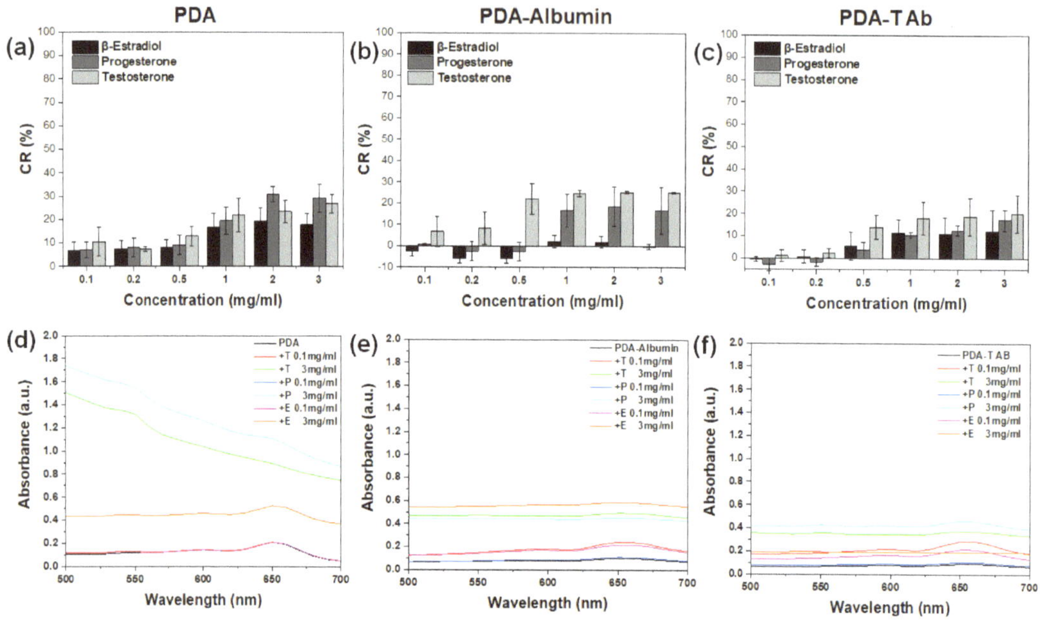

Figure 4. *CR* of (**a**) PDA, (**b**) PDA-Albumin and (**c**) PDA-T Ab after incubating with sex steroid hormones (β-estradiol, progesterone and T). UV-Vis absorption spectra of (**d**) PDA, (**e**) PDA-Albumin and (**f**) PDA-T Ab when incubating with sex steroid hormones at low (0.1 mg/mL) and high (3 mg/mL) concentrations.

To avoid the interruption of signals from the absorbance of the hormones, the FL response of the PDA assemblies was measured using a fluorescence reader (λ_{ex} = 485 nm, λ_{em} = 630 nm). The FL of polydiacetylene is not generally affected by the optical properties of other materials in the same environment, and PDA shows higher sensitivity in fluorogenic mode than in colorimetric transition [35]. As shown in Figure 5a, none of the three hormones (β-estradiol, progesterone, and testosterone) produced FL when incubated with PDA. As shown in Figure 5b, the presence of T increased the FL of PDA-Albumin in a concentration-dependent manner, starting from 0.5 mg/mL of T. When incubated with progesterone, the FL of PDA-Albumin only increased from a concentration 2 mg/mL. In contrast, no FL changes were found in PDA-Albumin when incubated with β-estradiol. The sensitivity of the PDA-Albumin for T and progesterone was 630,000 a.u./mg·mL^{-1} and 130,000 a.u./mg·mL^{-1}, respectively, when considering a range of 0.2–3 mg/mL of either hormone (Figure 5e). In general, progesterone and T have considerable, but non-specific affinity towards albumin in blood plasma [36]; for example, 53–55% of T binds to serum albumin and is transported as a complex in human blood [37]. Therefore, introducing progesterone or T could generate a certain of FL signals originated from distorting conjugated backbone of PDA by binding of progesterone or T onto albumin at the surface of PDA-Albumin.

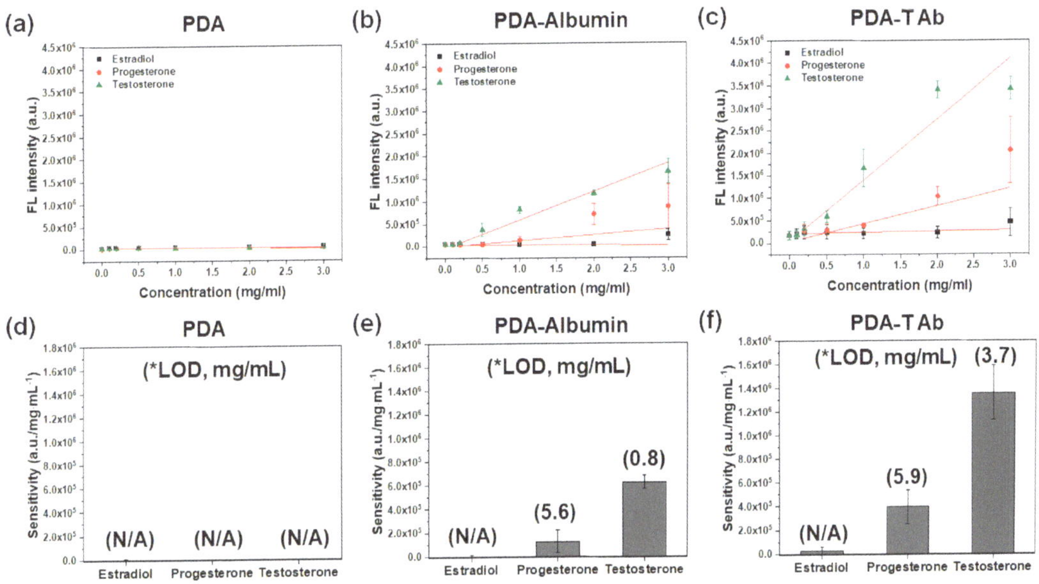

Figure 5. Fluorescence (FL) intensity of (**a**) PDA, (**b**) PDA-Albumin and (**c**) PDA-T Ab after incubation with sex steroid hormones (β-estradiol, progesterone and T). Sensitivity of sensing platforms based on (**d**) PDA, (**e**) PDA-Albumin and (**f**) PDA-T Ab for each sex steroid hormone. * *LOD*: limit of detection.

In order to prove a concept of PDA-based sensing platform for targeting T, FL signals of PDA-T Ab by introducing T was compared with those of PDA-T Ab in presence of β-estradiol and progesterone. As shown in Figure 5c, the PDA-T Ab did not generate FL after incubation with β-estradiol. The sensitivity of PDA-T Ab for T and progesterone was 1,400,000 a.u./mg·mL^{-1}, sensitivity for progesterone: 400,000 a.u./mg·mL^{-1}, in the range of 0.2–3 mg/mL of either hormone (Figure 5f). These results suggest that PDA-T Ab showed more discernable and selective sensory signaling than PDA-Albumin. In both PDA-Albumin and PDA-T Ab, *LOD* values for T and progesterone was similar, respectively. In summary, PDA-T Ab displayed a more selective fluorescent sensory signaling in the presence of T compared to PDA and PDA-Albumin.

To demonstrate T detection using the PDA-T Ab in any biologically complex matrix, we conducted spike tests with filtered human serum. In the spike test, the T hormone concentration-dependent sensory signaling was observed (Figure S3) as a similar trend in Figure 5f, however, the sensitivity was reduced comparing with that without biologically complex matrix. We supposed that non-filtered human serum proteins and small molecules of hormones could interrupt the sensitivity of the sensory signaling.

4. Conclusions

T Ab was conjugated to PDA to selectively detect T as a biomarker of preterm birth. The PDA-based sensory signals were evaluated by comparing with those of other two sex steroid hormones (progesterone and β-estradiol). To overcome the issue of the poor hormone solubility in aqueous solution, we adjusted a co-solvent medium of ACN and DI water (50% v/v) that enables solubilizing the hydrophobic steroid hormones and minimizing false signals from the medium. Although an accurate colorimetric response of PDA assemblies could not be determined due to the interfering absorption of the hormones, we demonstrated that fluorescence sensory signaling of PDA-T Ab was dose-dependent on T concentration and was selective for T hormone. This finding would give an insight for designing PDA-based sensors to detect broad spectrum of water-insoluble target analytes.

Supplementary Materials: The following are available online at https://www.mdpi.com/article/10.3390/chemosensors9070176/s1, Figure S1: Zeta potential values and size of PDA during conjugating albumin to PDA, Figure S2: FT-IR spectra of PDA, PDA-Albumin and PDA-T Ab, Figure S3: Fluorescence intensity of PDA-T Ab after incubation with testosterone hormones in filtered human serum.

Author Contributions: Conceptualization, S.-C.K., B.-S.A. and S.S.; validation, J.J. and S.-M.A.; formal analysis, investigation, data curation, J.J.; writing—original draft preparation, J.J. and S.S.; writing—review and editing, E.-K.L., S.-C.K., B.-S.A. and S.S.; supervision, project administration, S.S. All authors have read and agreed to the published version of the manuscript.

Funding: This work was supported by a National Research Foundation of Korea (NRF) grant funded by the Korean government (MSIP: Ministry of Science, ICT & Future Planning) (NRF-2018R1D1A1B07050070), a grant of the Korea Health Technology R&D Project through the Korea Health Industry Development Institute (KHIDI), funded by the Ministry of Health & Welfare, Republic of Korea (grant number: HI19C0185), the Korea Research Institute of Bioscience and Biotechnology (KRIBB) Research Initiative Program (1711134038) and the BK21 FOUR project (F21YY8109033) through the NRF funded by the Ministry of Education, Korea.

Conflicts of Interest: There are no conflicts of interest to declare.

References

1. Liu, L.; Oza, S.; Hogan, D.; Chu, Y.; Perin, J.; Zhu, J.; Lawn, J.E.; Cousens, S.; Mathers, C.; Black, R.E. Global, regional, and national causes of under-5 mortality in 2000–15: An updated systematic analysis with implications for the Sustainable Development Goals. *Lancet* **2016**, *388*, 3027–3035. [CrossRef]
2. Blencowe, H.; Cousens, S.; Oestergaard, M.Z.; Chou, D.; Moller, A.-B.; Narwal, R.; Adler, A.; Vera Garcia, C.; Rohde, S.; Say, L.; et al. National, regional, and worldwide estimates of preterm birth rates in the year 2010 with time trends since 1990 for selected countries: A systematic analysis and implications. *Lancet* **2012**, *379*, 2162–2172. [CrossRef]
3. Vogel, J.P.; Chawanpaiboon, S.; Watananirun, K.; Lumbiganon, P.; Petzold, M.; Moller, A.-B.; Thinkhamrop, J.; Laopaiboon, M.; Seuc, A.H.; Hogan, D.; et al. Global, regional and national levels and trends of preterm birth rates for 1990 to 2014: Protocol for development of World Health Organization estimates. *Reprod. Health* **2016**, *13*, 76. [CrossRef]
4. Peaceman, A.M.; Andrews, W.W.; Thorp, J.M.; Cliver, S.P.; Lukes, A.; Iams, J.D.; Coultrip, L.; Eriksen, N.; Holbrook, R.H.; Elliott, J.; et al. Fetal fibronectin as a predictor of preterm birth in patients with symptoms: A multicenter trial. *Am. J. Obstet. Gynecol.* **1997**, *177*, 13–18. [CrossRef]
5. Dziadosz, M.; Bennett, T.-A.; Dolin, C.; West Honart, A.; Pham, A.; Lee, S.S.; Pivo, S.; Roman, A.S. Uterocervical angle: A novel ultrasound screening tool to predict spontaneous preterm birth. *Am. J. Obstet. Gynecol.* **2016**, *215*, 376.e1–376.e7. [CrossRef] [PubMed]
6. Çekmez, Y.; Kıran, G.; Haberal, E.T.; Dizdar, M. Use of cervicovaginal PAMG-1 protein as a predictor of delivery within seven days in pregnancies at risk of premature birth. *BMC Pregnancy Childbirth* **2017**, *17*, 246. [CrossRef]
7. Lee, S.E.; Park, J.S.; Norwitz, E.R.; Kim, K.W.; Park, H.S.; Jun, J.K. Measurement of Placental Alpha-Microglobulin-1 in Cervicovaginal Discharge to Diagnose Rupture of Membranes. *Obstet. Gynecol.* **2007**, *109*, 634–640. [CrossRef]
8. Honest, H. Accuracy of cervicovaginal fetal fibronectin test in predicting risk of spontaneous preterm birth: Systematic review. *BMJ* **2002**, *325*, 301. [CrossRef] [PubMed]
9. Qian, X.; Städler, B. Recent Developments in Polydiacetylene-Based Sensors. *Chem. Mater.* **2019**, *31*, 1196–1222. [CrossRef]
10. Kim, D.; Cao, Y.; Mariappan, D.; Bono, M.S.; Hart, A.J.; Marelli, B. A Microneedle Technology for Sampling and Sensing Bacteria in the Food Supply Chain. *Adv. Funct. Mater.* **2021**, *31*, 2005370. [CrossRef]
11. Son, S.U.; Seo, S.B.; Jang, S.; Choi, J.; Lim, J.; Lee, D.K.; Kim, H.; Seo, S.; Kang, T.; Jung, J.; et al. Naked-eye detection of pandemic influenza a (pH1N1) virus by polydiacetylene (PDA)-based paper sensor as a point-of-care diagnostic platform. *Sens. Actuators B Chem.* **2019**, *291*, 257–265. [CrossRef]
12. Yoon, B.; Lee, S.; Kim, J.-M. Recent conceptual and technological advances in polydiacetylene-based supramolecular chemosensors. *Chem. Soc. Rev.* **2009**, *38*, 1958. [CrossRef] [PubMed]
13. Nguyen, L.H.; Oveissi, F.; Chandrawati, R.; Dehghani, F.; Naficy, S. Naked-Eye Detection of Ethylene Using Thiol-Functionalized Polydiacetylene-Based Flexible Sensors. *ACS Sens.* **2020**, *5*, 1921–1928. [CrossRef]
14. Kang, D.H.; Kim, K.; Son, Y.; Chang, P.-S.; Kim, J.; Jung, H.-S. Design of a simple paper-based colorimetric biosensor using polydiacetylene liposomes for neomycin detection. *Analyst* **2018**, *143*, 4623–4629. [CrossRef]
15. Lee, S.; Kim, J.-Y.; Chen, X.; Yoon, J. Recent progress in stimuli-induced polydiacetylenes for sensing temperature, chemical and biological targets. *Chem. Commun.* **2016**, *52*, 9178–9196. [CrossRef]
16. Oh, J.; Jeon, I.; Kim, D.; You, Y.; Baek, D.; Kang, S.J.; Lee, J. Highly Stable Upconverting Nanocrystal–Polydiacetylenes Nanoplates for Orthogonal Dual Signaling-Based Detection of Cyanide. *ACS Appl. Mater. Interfaces* **2020**, *12*, 4934–4943. [CrossRef]

17. Cai, G.; Yu, Z.; Tong, P.; Tang, D. Ti_3C_2 MXene quantum dot-encapsulated liposomes for photothermal immunoassays using a portable near-infrared imaging camera on a smartphone. *Nanoscale* **2019**, *11*, 15659–15667. [CrossRef] [PubMed]
18. Lin, Y.; Zhou, Q.; Zeng, Y.; Tang, D. Liposome-coated mesoporous silica nanoparticles loaded with L-cysteine for photoelectrochemical immunoassay of aflatoxin B1. *Microchim. Acta* **2018**, *185*, 311. [CrossRef]
19. Ren, R.; Cai, G.; Yu, Z.; Tang, D. Glucose-loaded liposomes for amplified colorimetric immunoassay of streptomycin based on enzyme-induced iron(II) chelation reaction with phenanthroline. *Sens. Actuators B Chem.* **2018**, *265*, 174–181. [CrossRef]
20. Lin, Y.; Zhou, Q.; Tang, D. Dopamine-Loaded Liposomes for in-Situ Amplified Photoelectrochemical Immunoassay of AFB 1 to Enhance Photocurrent of Mn^{2+}-Doped $Zn_3(OH)_2V_2O_7$ Nanobelts. *Anal. Chem.* **2017**, *89*, 11803–11810. [CrossRef] [PubMed]
21. Shin, Y.; Jeong, J.; Park, M.; Lee, J.; An, S.; Cho, W.; Kim, S.; An, B.; Lee, K. Regulation of steroid hormones in the placenta and serum of women with preeclampsia. *Mol. Med. Rep.* **2018**, *17*, 2681–2688. [CrossRef] [PubMed]
22. Goldenberg, R.L.; Goepfert, A.R.; Ramsey, P.S. Biochemical markers for the prediction of preterm birth. *Am. J. Obstet. Gynecol.* **2005**, *192*, S36–S46. [CrossRef] [PubMed]
23. Norwitz, E.R.; Caughey, A.B. Progesterone supplementation and the prevention of preterm birth. *Rev. Obstet. Gynecol.* **2011**, *4*, 60–72. [CrossRef] [PubMed]
24. Cho, J.; Holditch-Davis, D. Effects of Perinatal Testosterone on Infant Health, Mother–Infant Interactions, and Infant Development. *Biol. Res. Nurs.* **2014**, *16*, 228–236. [CrossRef]
25. Taieb, J.; Mathian, B.; Millot, F.; Patricot, M.-C.; Mathieu, E.; Queyrel, N.; Lacroix, I.; Somma-Delpero, C.; Boudou, P. Testosterone Measured by 10 Immunoassays and by Isotope-Dilution Gas Chromatography–Mass Spectrometry in Sera from 116 Men, Women, and Children. *Clin. Chem.* **2003**, *49*, 1381–1395. [CrossRef]
26. Cho, E.; Hu, Y.; Choi, Y.; Jung, S. A dimyristoyl phosphatidylcholine/polydiacetylene biomimetic assembly for the selective screening of progesterone. *J. Ind. Eng. Chem.* **2018**, *63*, 288–295. [CrossRef]
27. Jung, Y.K.; Park, H.G. Colorimetric polydiacetylene (PDA) liposome-based assay for rapid and simple detection of GST-fusion protein. *Sens. Actuators B Chem.* **2019**, *278*, 190–195. [CrossRef]
28. Kim, C.; Lee, K. Polydiacetylene (PDA) Liposome-Based Immunosensor for the Detection of Exosomes. *Biomacromolecules* **2019**, *20*, 3392–3398. [CrossRef]
29. Södergard, R.; Bäckström, T.; Shanbhag, V.; Carstensen, H. Calculation of free and bound fractions of testosterone and estradiol-17β to human plasma proteins at body temperature. *J. Steroid Biochem.* **1982**, *16*, 801–810. [CrossRef]
30. Baker, M.E. Albumin's role in steroid hormone action and the origins of vertebrates: Is albumin an essential protein? *FEBS Lett.* **1998**, *439*, 9–12. [CrossRef]
31. Jeong, J.-P.; Cho, E.; Lee, S.-C.; Kim, T.; Song, B.; Lee, I.-S.; Jung, S. Detection of Foot-and-Mouth Disease Virus Using a Polydiacetylene Immunosensor on Solid-Liquid Phase. *Macromol. Mater. Eng.* **2018**, *303*, 1700640. [CrossRef]
32. Bae, S.J.; Choi, H.; Choi, J.S. Synthesis of Polymerizable Amphiphiles with Basic Oligopeptides for Gene Delivery Application. *Polym. Korea* **2013**, *37*, 94–99. [CrossRef]
33. Lee, J.; Chang, H.T.; An, H.; Ahn, S.; Shim, J.; Kim, J.-M. A protective layer approach to solvatochromic sensors. *Nat. Commun.* **2013**, *4*, 2461. [CrossRef]
34. Pumtang, S.; Siripornnoppakhun, W.; Sukwattanasinitt, M.; Ajavakom, A. Solvent colorimetric paper-based polydiacetylene sensors from diacetylene lipids. *J. Colloid Interface Sci.* **2011**, *364*, 366–372. [CrossRef] [PubMed]
35. Cui, C.; Kim, S.; Ahn, D.J.; Joo, J.; Lee, G.S.; Park, D.H.; Kim, B.-H. Unusual enhancement of fluorescence and Raman scattering of core-shell nanostructure of polydiacetylene and Ag nanoparticle. *Synth. Met.* **2018**, *236*, 19–23. [CrossRef]
36. André, C.; Jacquot, Y.; Truong, T.; Thomassin, M.; Robert, J.; Guillaume, Y. Analysis of the progesterone displacement of its human serum albumin binding site by β-estradiol using biochromatographic approaches: Effect of two salt modifiers. *J. Chromatogr. B* **2003**, *796*, 267–281. [CrossRef]
37. Czub, M.P.; Venkataramany, B.S.; Majorek, K.A.; Handing, K.B.; Porebski, P.J.; Beeram, S.R.; Suh, K.; Woolfork, A.G.; Hage, D.S.; Shabalin, I.G.; et al. Testosterone meets albumin-the molecular mechanism of sex hormone transport by serum albumins. *Chem. Sci.* **2019**, *10*, 1607–1618. [CrossRef] [PubMed]

chemosensors

MDPI

Article

A Novel Colorimetric Biosensor for the Detection of Catalase-Positive *Staphylococcus aureus* Based on an Onion-like Carbon Nanozyme

Yining Fan [1,2], Guanyue Gao [1,2,*] and Jinfang Zhi [1,2]

[1] Key Laboratory of Photochemical Conversion and Optoelectronic Materials, Technical Institute of Physics and Chemistry, Chinese Academy of Sciences, Beijing 100190, China; fanyining20@mails.ucas.ac.cn (Y.F.); zhi-mail@mail.ipc.ac.cn (J.Z.)
[2] University of Chinese Academy of Sciences, Beijing 100049, China
* Correspondence: gaoguanyue@mail.ipc.ac.cn

Abstract: *Staphylococcus aureus* is one of the leading causes of skin and soft tissue infections, and it is even life-threatening if it enters the bloodstream, lung or heart. In the present work, we proposed a novel colorimetric biosensor for the detection of *S. aureus* through hydrogen peroxide consumption. An onion-like carbon nanozyme with high peroxidase-like activity was prepared, which competed with the endogenous catalase of *S. aureus* in consuming hydrogen peroxide. This reaction was further characterized by the colorimetric reaction of 3,3′,5,5′-tetramethylbenzidine. The results showed that our approach allowed for the simple and rapid determination of *S. aureus*, with a linear range of 2×10^4 to 2×10^7 CFU/mL. Moreover, our method displayed good selectivity, with *Bacillus subtilis* and *Escherichia coli* showing negligible responses at the concentration of 2×10^5 CFU/mL. The application of the as-prepared biosensor to analyze *S. aureus* in real water samples yielded recovery rates ranging from 95% to 112%, with relative standard deviations less than 7%. The method demonstrated good accuracy and specificity, which offers a novel approach for the simple and selective detection of *S. aureus*.

Keywords: *Staphylococcus aureus*; colorimetric biosensor; endogenous catalase; onion-like carbon; peroxidase-like nanozyme

Citation: Fan, Y.; Gao, G.; Zhi, J. A Novel Colorimetric Biosensor for the Detection of Catalase-Positive *Staphylococcus aureus* Based on an Onion-like Carbon Nanozyme. *Chemosensors* **2024**, *12*, 176. https://doi.org/10.3390/chemosensors12090176

Academic Editors: Pietro Salvo and Lorena Tedeschi

Received: 28 June 2024
Revised: 27 August 2024
Accepted: 30 August 2024
Published: 2 September 2024

1. Introduction

Bacterial infections are among the leading causes of illness and death, which presents a growing risk to public health. *Staphylococcus aureus* is among the most prevalent pathogens, causing a spectrum of infections including severe inflammation, skin ulceration, and suppuration [1–3], as well as more serious conditions such as endocarditis [4], pneumonia [5], and meningitis [6]. Despite the fact that regulations have been enacted to restrict *S. aureus* contamination, persistent risks remain in environments such as medical facilities [7]. Moreover, the presence of food-borne pathogens in meat and meat products poses a significant risk of infection to consumers, with livestock-associated workers at heightened risk of colonization and potential transmission [8,9]. Therefore, there is an urgent need to develop rapid and efficient techniques capable of detecting *S. aureus* and promptly identifying potential outbreaks.

The plate-counting approach, which has been the standard method for *S. aureus* detection, is a conventional culture-dependent microbiological method that is used to observe and enumerate viable microorganisms. In this method, bacterial concentration is reported as colony-forming units (CFUs), which is presumed to originate from a certain microorganism, and the bacterial colonies can also reflect specific characteristic features, such as color, size and shape. However, it usually requires several days to achieve trustworthy results, which no longer meets the demands for rapid pathogenic detection [10]. Consequently, considerable efforts have been directed towards the development of rapid detection methods. Polymerase chain reaction (PCR) and fluorescence-linked immunosorbent assay (FLISA) have significantly reduced the detection time to a few hours [11,12]. Biosensors are capable of converting signals arising from interactions between biological components and analytes into detectable physicochemical signals, which have been extensively used in applications ranging from medical treatment to environmental monitoring. Among them, colorimetric biosensors have emerged as a promising platform for rapid bacterial recognition [13,14]. These biosensors enable the quantitative measurement of specific analytes through chromogenic or discoloration reactions, such as nanozyme-mediated colorimetric reactions or the localized surface plasmon resonance of noble metal nanoparticles. Their merits of cost-effectiveness, straightforward interpretation of results, and rapid response render them well-suited for the on-site detection of *S. aureus* [15–17].

S. aureus is one of the representative Gram-positive bacteria, with most strains being catalase (CAT)-positive and capable of catalyzing the decomposition of hydrogen peroxide (H_2O_2) into O_2 and water [18,19]. Catalase production represents a defense mechanism, allowing bacteria to better resist intracellular and extracellular killing by H_2O_2. The catalase-negative strain is observed to possess much lower pathogenic potential than the catalase-positive strain, which may contribute to the predominance of the catalase-positive strain in bacterial infections. Utilizing the endogenous catalase of *S. aureus*, Guarín et al. proposed an electrochemical sensor to monitor H_2O_2 consumption by *S. aureus*. Using a screen-printed gold electrode modified with cysteine and peroxidase (POD), this method demonstrated a sensitive response to H_2O_2, enabling the detection of *S. aureus* ranging from 3×10^2 to 3×10^8 CFU/mL, with a detection limit of 10^2 CFU/mL [20]. Majumdar et al. applied an amperometric approach to detect H_2O_2 consumption for the quantification of *S. aureus* at a Pt microelectrode. This work realized the quantification of *S. aureus* in the range from 10 CFU/mL to 10^6 CFU/mL within 10 min [21]. Abdelhamid et al. reported a fluorescent biosensor based on chitosan modified CdS quantum dots, in which *S. aureus* showed accelerated decomposition of H_2O_2, resulting in a significant quenching of fluorescence emission, thereby offering high selectivity for CAT-positive *S. aureus* over *Bacillus subtilis* and *Escherichia coli* [22]. In comparison to other assays for the detection of *S. aureus*, these three biosensors offer the benefits of short detection periods and broad detection ranges. Also, the sensor preparation process is simplified as no additional recognition units are required.

Nanozymes, defined as nanomaterials with enzyme-like catalytical activity, have attracted increasing attention in biosensing applications over the past few years [23–25]. These nanozyme-based enzymatic reactions can generate or amplify signals for precise target detection, among which POD-like and oxidase-like nanozymes are commonly employed in biosensor construction. POD is defined as a class of oxidoreductases that catalyze the oxidation of a substrate by hydrogen peroxide or an organic peroxide [26]. Nanozymes that exhibit mimic peroxidase activity are known as POD-like nanozymes [27]. In addition to catalytic properties, nanomaterials offer enhanced electrical conductivity and compatibility with biorecognition elements, thereby improving biosensor performance. Onion-like carbon (OLC) is a carbonaceous nanomaterial with stacked layers of sp^2 graphene sheets and exhibits high POD catalytic ability [28–34]. Its highly symmetric structure and large volume-to-surface ratio make it promising for applications in energy storage and pollutant degradation [35,36]. In the field of biosensing, OLC can be utilized as a modification material to increase the conductivity and biocompatibility of the electrodes [37–40]. OLC can serve as an effective linker between biomolecules and nanomaterials, facilitating electron transfer and signal amplification. Therefore, in combination with the intrinsic POD-like property of OLC, it is promising in the construction of biosensors with high sensitivity and specificity.

Herein, we proposed a novel approach for *S. aureus* detection utilizing its endogenous CAT activity. The accelerated consumption of H_2O_2 catalyzed by CAT was characterized using a colorimetric reaction involving 3,3′,5,5′-tetramethylbenzidine (TMB) in the presence of OLC. The analytical performance and anti-interference ability of the as-prepared colorimetric biosensor were examined. Additionally, the recovery experiments were conducted using real water samples to assess the feasibility of the present method in practical scenarios. The present work provided a simple and rapid approach for the detection of CAT-positive bacteria, which can be used in pathogen monitoring and environmental early warning.

2. Materials and Methods

2.1. Reagents and Materials

Staphylococcus aureus (ATCC 6538), *Escherichia coli* (ATCC 25922) and *Bacillus subtilis* (CGMCC 1.1086) were provided by the Testing Center for Antimicrobial Materials, Technical Institute of Physics and Chemistry, Chinese Academy of Sciences. Additionally, 3,3′,5,5′-tetramethylbenzidine (TMB), acetic acid (>99.7%) and dimethyl sulfoxide (DMSO) were provided by Macklin Co., Ltd. (Shanghai, China). Sodium acetate and 2,2′-azinobis(3-ethylbenzothiazoline-6-sulfonic acid ammonium salt) (ABTS) were provided by InnoChem Co., Ltd. (Shanghai, China). Hydrogen peroxide (H_2O_2) was purchased from Aladdin Co., Ltd. (Shanghai, China). Lysostaphin was provided by Sangon Co., Ltd. (Shanghai, China). Nutrient broth and nutrient agar were purchased from Haibo Co., Ltd. (Qingdao, China). Nanodiamonds (NDs) were purchased from Carbodeon (Finland). Phosphate-buffered saline (PBS, 0.1 M, pH = 7) and Tris-HCl buffer (0.05 M, pH = 7.4, sterile) were provided by Regan Co., Ltd. (Beijing, China). All chemicals were of reagent grade and used as received. Aqueous solutions were prepared with Milli-Q water (Millipore Merck, Burlington, MA, USA, R > 18.2 MΩ·cm). Real water samples were collected from the tap-water pipe.

2.2. Preparation of Onion-like Carbon Nanozyme

OLC was obtained by annealing NDs in an N_2 atmosphere at 800 °C for 1 h. The morphology of as-prepared OLC was characterized by high-resolution transmission electron microscopy (TEM, JEOL JEM-2100F, Tokyo, Japan). The phase structure of OLC was obtained using an X-ray diffractometer (Bruker D8 focus, Karlsruhe, Germany). The composition of OLC was characterized using a Raman spectrometer (Renishaw InVia-Qontor, New Mills, UK) with an excitation light at 532 nm.

2.3. Detection of the POD-like Catalytic Activity of Onion-like Carbon Nanozyme

Chromogenic reactions based on TMB-H_2O_2 and ABTS-H_2O_2 double-substrate systems were used to detect the POD-like properties of the OLC nanozyme. Experiments on the TMB-H_2O_2 reaction were carried out with 250 μg/mL of OLC, 1 mM of H_2O_2 and 0.5 mM of TMB in 0.2 M NaAc-HAc buffer (pH = 4). The mixed solution was incubated at 37 °C for 10 min. Experiments on the ABTS-H_2O_2 reaction were performed in 0.2 M NaAc-HAc buffer (pH = 4) containing 250 μg/mL of OLC, 1 mM of H_2O_2 and 1 mM of ABTS. The mixed solution was placed at 37 °C for 10 min.

2.4. Catalytic Kinetic Determination of OLC's POD-like Activity

The catalytic kinetic determination of OLC was carried out using TMB and H_2O_2 as substrates. The catalytic reaction was carried out in 0.2 M NaAc-HAc buffer (pH = 4). The concentration of OLC was 100 μg/mL. The concentration of TMB was 1 mM in the kinetic assay of H_2O_2 and the concentration of H_2O_2 was 1 mM in the kinetic assay of TMB. After reacting at room temperature for 10 min, the reaction mixture was filtered by a needle filter (0.22 μm × 13 mm), and the absorbance of the supernatant at 652 nm was measured on an ultraviolet-visible spectrophotometer (SECOMAM UVIKONXL). The Michaelis–Menten equation was used to calculate the Michaelis constant (K_M) and maximum reaction rate (V_{max}).

2.5. Bacterial Culture

The bacteria were cultured in LB broth that contained 10 g/L peptone, 3 g/L beef extract and 5 g/L NaCl. The broth was sterilized by autoclaving at 121 °C for 15 min. After cooling to room temperature, the bacteria were inoculated into the sterilized culture medium and cultured aerobically in a shaker at 37 °C. The culture time for *S. aureus*, *E. coli* and *B. subtilis* were 24 h, 16 h and 24 h, respectively. Subsequently, the microbial cells were harvested by centrifugation at 6000 rpm for 5 min and washed twice with PBS. The harvested bacteria were re-dispersed in Tris-HCl buffer, and the OD_{600} value of the bacterial solution was adjusted to 1.5 using an ultraviolet-visible spectrophotometer (SECOMAM UVIKONXL).

The concentration of bacteria was further determined by the plate-counting method according to the following procedure: First, the agar liquid medium that contained 10 g/L of peptone, 5 g/L of sodium chloride, 3 g/L of beef extract and 15 g/L of agar was well mixed and sterilized at 121 °C. After cooling to 46 °C, the liquid agar medium was poured onto the culture dishes and allowed to solidify at room temperature. Then, 100 μL of bacteria solution was added to the agar plate and spread evenly using a cell spreader. The plate was then flipped over and incubated for 24 h at 37 °C. After incubating, a colony counter was used to read the total number of bacterial colonies.

2.6. Colorimetric Detection of *S. aureus*

The detection principle of the colorimetric biosensor is illustrated in Scheme 1. To enhance detection specificity, lysostaphin was employed, as it selectively hydrolyzed the cell wall of Staphylococcus species [41,42]. Following incubation with lysostaphin, CAT was released from *S. aureus* and catalyzed the decomposition of H_2O_2. With the increase in *S. aureus* concentration, CAT concentration in the bacterial lysate increased, leading to the accelerated consumption of H_2O_2. As a result, less H_2O_2 was involved in the subsequent TMB oxidation, resulting in a decrease in the final absorbance of TMB oxide and an increase in the inhibition rate.

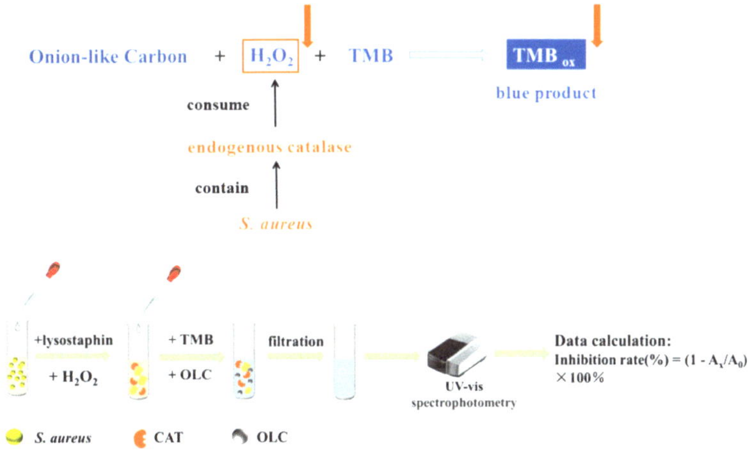

Scheme 1. Schematic representation of the detection process.

The detailed detection procedure was as follows: lysostaphin (13 μg/mL) was mixed with *S. aureus* solution to lyse *S. aureus* and release endogenous CAT. At the same time, H_2O_2 (0.5 mM) was added and consumed under the catalysis of CAT. The reaction was carried out at 37 °C for 20 min. After incubation, TMB (1 mM), OLC (250 μg/mL) and NaAc-HAc buffer (0.2 M, pH = 4) were added to the above solution. The reaction mixture was placed at room temperature for 10 min. Then, the solution was filtered with a needle

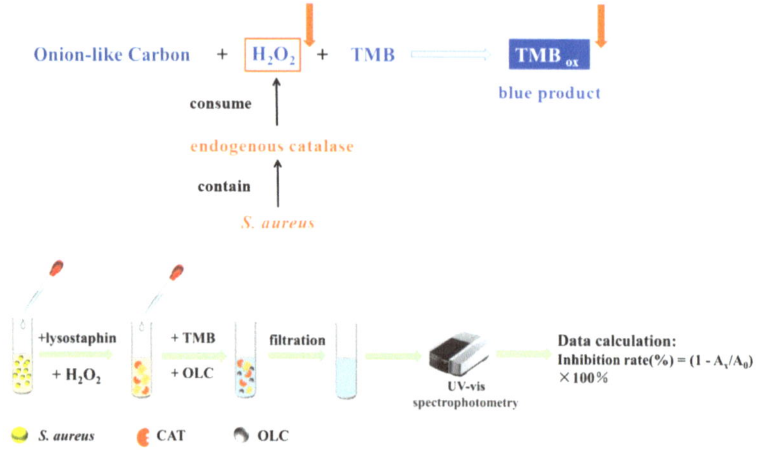

Scheme 1. Schematic representation of the detection process.

Then, the absorbance ...k group, the *S. aureus* ...conditions remained ...al reference and was ...erial concentration:

$$(1)$$

inhibition rate eliminated the influence of subjective errors and fluctuations in nanozyme activity on detection accuracy, which improved the precision and reliability of the results.

3. Results and Discussion

3.1. Structural and Catalytic Activity Characterization of OLC Nanozyme

OLC was obtained by annealing NDs at high temperature under an inert atmosphere. As shown in Figure 1a, the TEM image showed the presence of (111) planes of sp^3-hybridized cubic diamond and (002) planes of sp^2-hybridized graphitized carbon, with lattice spacing of 0.206 and 0.34 nm, respectively. OLC exhibited a quasi-spherical shape with a diameter of 5–10 nm. The X-ray diffraction pattern showed broad peaks of graphitic carbon at 27° and diffraction peaks of diamond at 44° and 75°, suggesting the presence of both sp^2 and sp^3 carbon in OLC (Figure 1b). We also observed peaks in the Raman spectra at 1320 cm^{-1} and 1535 cm^{-1}, which belong to the D and G bands of graphene (Figure 1c). The above characterization confirmed the partial conversion from sp^3 to sp^2 carbon in pristine NDs, suggesting the successful synthesis of OLC. The abovementioned structural characterization in the present work is also consistent with our previous works, which can verify the formation of OLC [35,36].

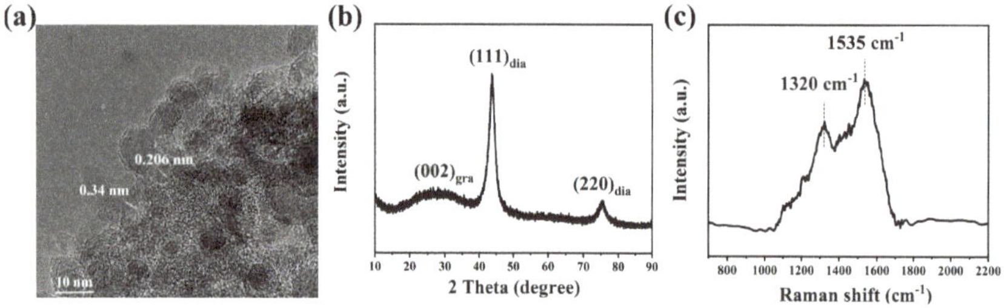

Figure 1. Structural and compositional characterizations of OLC: (**a**) TEM image; (**b**) XRD pattern; (**c**) Raman spectra.

The POD-like property of OLC was validated by the chromogenic reactions with TMB/ABTS and H_2O_2. Under the catalysis of OLC, the •OH was generated from the decomposition of H_2O_2, which subsequently oxidized TMB/ABTS into charge-transfer complex TMB_{ox} (blue)/$ABTS^{•+}$ (green) (Equations (2) and (3)). As shown in Figure 2, in the presence of H_2O_2 and OLC, TMB and ABTS underwent oxidation to produce blue and green products, respectively. However, the experimental groups in the absence of OLC showed almost no color change, indicating that OLC acted as an effective catalyst for the oxidation of TMB and ABTS in the presence of H_2O_2. Considering that TMB exhibits greater sensitivity in biosensing when compared to other chromogenic substrates, including ABTS [43,44], TMB was chosen as the substrate for subsequent detection.

$$TMB + H_2O_2 \overset{OLC}{\rightarrow} TMB_{OX} + H_2O \quad (2)$$

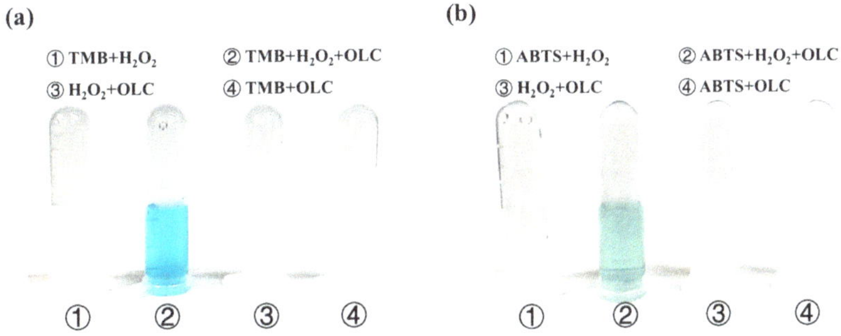

Figure 2. (**a**) The digital photos of color changes in different groups: (**a**) TMB substrate; (**b**) ABTS substrate, respectively.

$$ABTS + H_2O_2 \overset{OLC}{\rightarrow} ABTS^{\bullet +} + H_2O \tag{3}$$

The POD-mimicking activity of the OLC nanozyme was further characterized according to the Michaelis–Menten equation (Equation (4)) and the Lineweaver–Burk equation (Equation (5)), which reflect the relationship between reaction rate and substrate concentration:

$$v = \frac{V_{max}[S]}{K_M + [S]} \tag{4}$$

$$\frac{1}{v} = \frac{K_M}{V_{max}}\frac{1}{[S]} + \frac{1}{V_{max}} \tag{5}$$

$$v = \frac{\Delta c}{t} \tag{6}$$

$$A = \varepsilon L c \tag{7}$$

where $[S]$ is the concentration of the substrate TMB or H_2O_2, v is the reaction rate, and V_{max} refers to the maximum reaction rate when the enzyme is saturated by the substrate. The reaction rate v is determined by Equation (6), where Δc is the concentration change of the chromogenic substance during the reaction, and t is the reaction time. The concentration of the chromogenic substance can be calculated from the absorbance value according to Lambert–Beer law (Equation (7)). A refers to the absorbance value measured by the UV-vis spectrophotometer. ε is the molar absorption coefficient. The ε value for TMB is 39,000 M^{-1} cm^{-1}. c is the concentration of the chromogenic substance, and L is the optical path length. By calculating the Δc values based on the changes of absorbance intensity, the reaction rates of TMB can be determined. K_M is the Michaelis constant, representing the substrate concentration at which the reaction rate reaches half of V_{max}. A smaller K_M indicates a higher affinity between the enzyme and the substrate. Figure 3 shows the Michaelis–Menten and Lineweaver–Burk curves of OLC. V_{max} and K_M were determined from the fitted straight line between 1/v and 1/[S]. The intercept of the line refers to $1/V_{max}$. Therefore, V_{max} can be obtained from the intercept. The slope refers to K_M/V_{max}. The value of K_M can be calculated from the slope of the fitted line and V_{max}. The K_M and V_{max} of both substrates were calculated and are listed in Table 1. OLC exhibited a higher V_{max} when compared to other carbon nanomaterials with POD-like properties [45–47]. Furthermore, a smaller K_M value was observed for OLC compared with graphdiyne oxide and $C_{60}[C(COOH)_2]_2$, indicating that OLC had a high affinity to H_2O_2 and exhibited high POD-like catalytic activity.

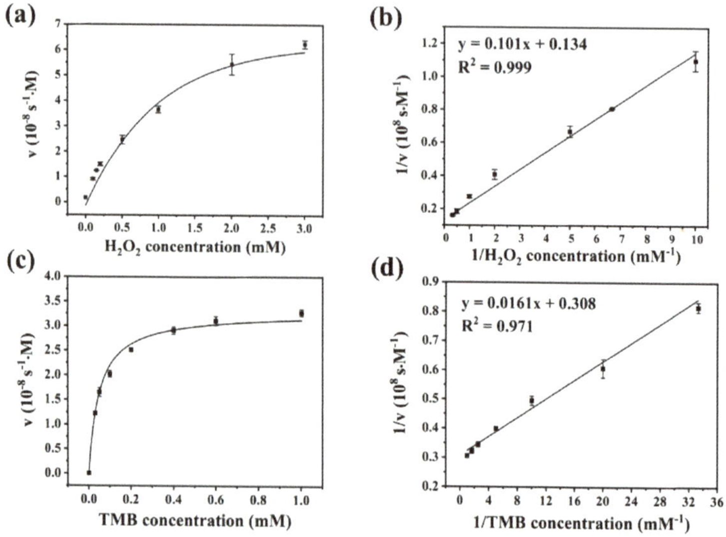

Figure 3. Michaelis–Menten and Lineweaver–Burk curves of OLC: (**a**,**b**) H_2O_2 substrate; c(TMB) = 1 mM, c(OLC) = 100 μg/mL; (**c**,**d**) TMB substrate, c(H_2O_2) = 1 mM, c(OLC) = 100 μg/mL.

Table 1. Comparisons of kinetic parameters between OLC and other carbon nanozymes.

Nanozyme	K_M (mM)		V_{max} (10^{-8} M s^{-1})		Reference
	H_2O_2	TMB	H_2O_2	TMB	
graphdiyne oxide	2.59	0.62	1.77	1.92	[45]
$C_{60}[C(COOH)_2]_2$	24.58	0.23	0.40	0.35	[46]
GO-COOH	3.99	0.02	3.85	3.45	[47]
OLC	0.77	0.05	7.48	3.25	This work

3.2. Feasibility Verification of the As-Prepared Biosensor for S. aureus Detection

CAT is a biological enzyme capable of catalyzing the decomposition of H_2O_2 into water and oxygen [48,49], thereby playing a crucial role in protecting cells from oxidative damage induced by reactive oxygen species [50,51]. Firstly, the presence of CAT in *S. aureus* (ATCC 6538) cells was confirmed by incubating 10^8 CFU/mL *S. aureus* with 5 mM H_2O_2 at 37 °C for 30 min. Clear bubbles were observed in the solution (Figure 4a), corresponding to the generation of O_2 under the catalysis of bacterial endogenous CAT.

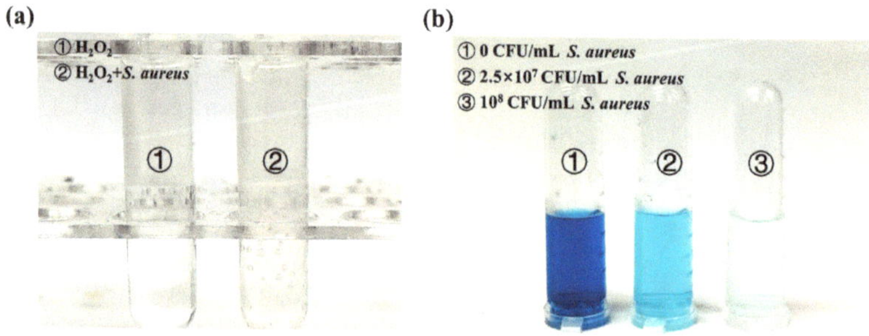

Figure 4. (**a**) The digital photos of the bubbles after mixing *S. aureus* with H_2O_2; (**b**) The digital photos of color changes caused by different concentrations of *S. aureus*.

Furthermore, to verify the feasibility of the as-prepared biosensor for *S. aureus* detection, the colorimetric responses of *S. aureus* solutions at different concentrations were recorded using TMB as the chromogenic agent. The bacteria were first incubated with 5 mM H_2O_2 for 30 min. Then, the bacteria were removed by filtration to avoid interfering with the subsequent absorbance detection. After filtration, OLC and TMB were added, and the colorimetric changes were observed after 5 min. In the presence of H_2O_2 and nanozymes, TMB easily loses one electron and produces a charge-transfer complex that displays a blue color. When TMB is exposed to an acidic condition or the amount of TMB is insufficient, it can then undergo further oxidation to produce a yellow diimine compound [52]. In present work, the TMB amount we used was sufficient, and the blue-colored reaction product TMB_{ox} was observed. As shown in Figure 4b, the intensity of the blue color decreased with increasing concentrations of *S. aureus*, owing to the higher consumption of H_2O_2 by *S. aureus* and less H_2O_2 being involved in subsequent TMB oxidation. In summary, the method proposed in the present work is feasible for *S. aureus* detection according to the colorimetric change of TMB.

To demonstrate the superior performance of the OLC nanozyme in our detection system, control experiments were performed using nanodiamonds and Fe_3O_4 nanoparticles as representative carbonaceous and non-carbonaceous nanomaterials, respectively. The control experiments were conducted at a *S. aureus* concentration of 10^5 CFU/mL. The detection procedures and parameters were consistent with those described in Section 2.6. The results are presented in Table 2. Under the same concentration of nanozymes (250 µg/mL), the highest change in TMB absorbance intensity was observed for OLC, indicating that OLC possessed the strongest POD-like properties. Additionally, OLC also demonstrated the highest inhibition rate, reflecting its enhanced sensitivity for *S. aureus* detection.

Table 2. Results of different materials in *S. aureus* detection.

Materials	OLC	Nanodiamonds	Fe_3O_4 Nanoparticles
Absorbance of experimental group	1.18	0.13	0.45
Absorbance of blank group	1.45	0.14	0.53
Inhibition rates	18.6%	7.1%	15.1%

3.3. *Optimization of Detection Conditions*

3.3.1. Optimization of the Incubation Time of Lysostaphin and *S. aureus* (t_1)

To enhance the detection efficacy of the as-prepared biosensor, experimental conditions were systematically optimized. Firstly, considering the potential interference posed by the bacterial cell wall on the interaction between CAT within the *S. aureus* cells and H_2O_2 in the solution, lysostaphin was employed to facilitate bacterial lysis and ensure the release and adequate reaction between endogenous CAT and H_2O_2. Evaluation revealed that at a concentration of 10^6 CFU/mL *S. aureus*, lysostaphin treatment resulted in an inhibition rate of 49.11%, significantly higher than the 19.84% inhibition rate observed without lysostaphin, which proved the sensitivity enhancement facilitated by lysostaphin.

In the detection process, *S. aureus* was first incubated with lysostaphin to proceed bacterial lysis and release endogenous CAT, after which H_2O_2 was added. Considering a high concentration of H_2O_2 may hinder the lytic activity of lysostaphin, the incubation time of lysostaphin was first optimized by using 2×10^5 CFU/mL *S. aureus* as the target analyte (Figure 5). The increase in t_1 from 0 to 15 min did not yield significant changes in the inhibition rate, suggesting a rapid and complete lysis of *S. aureus* in a short period of time. Given that the inhibition rate was not sensitive to t_1, lysostaphin and H_2O_2 were introduced to *S. aureus* simultaneously to simplify the detection process in the subsequent experiments.

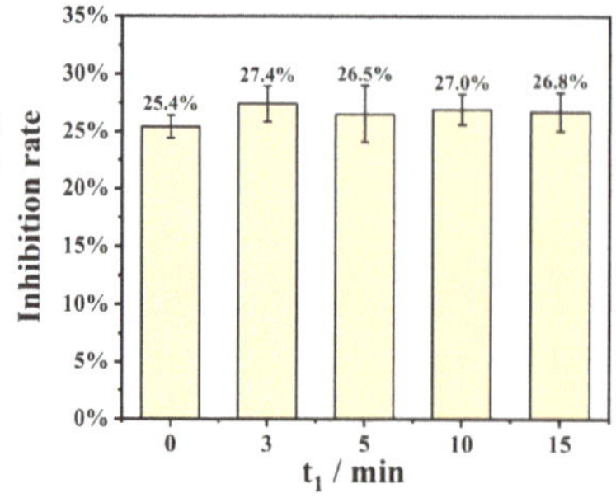

Figure 5. Effect of lysostaphin incubation time on detection sensitivity, c(*S. aureus*) = 2 × 10^5 CFU/mL, c(H_2O_2) = 0.5 mM, t_2 = 20 min.

3.3.2. Optimization of H_2O_2 Concentration and H_2O_2 Reaction Time (t_2)

Considering that the determination of *S. aureus* concentration relies on the enzymatic consumption of H_2O_2 by endogenous CAT, the initial concentration of H_2O_2 is essential in achieving optimal detection sensitivity. As depicted in Figure 6a, the inhibition rate exhibited an upward trend as the H_2O_2 concentration varied from 0.1 to 0.5 mM. However, the response of the biosensor to *S. aureus* exhibited a diminishing trend as the H_2O_2 concentration increased from 0.5 to 1 mM, with the maximum detection sensitivity observed at 0.5 mM. Consequently, 0.5 mM was selected as the optimal concentration.

Moreover, the influence of H_2O_2 reaction time was also investigated. As shown in Figure 6b, the inhibition rate increased with the prolonged reaction time ranging from 5 to 20 min and reached a plateau after 20 min, indicating that 20 min was sufficient for a complete catalytic reaction. Therefore, 20 min was selected as the optimal time for subsequent experiments.

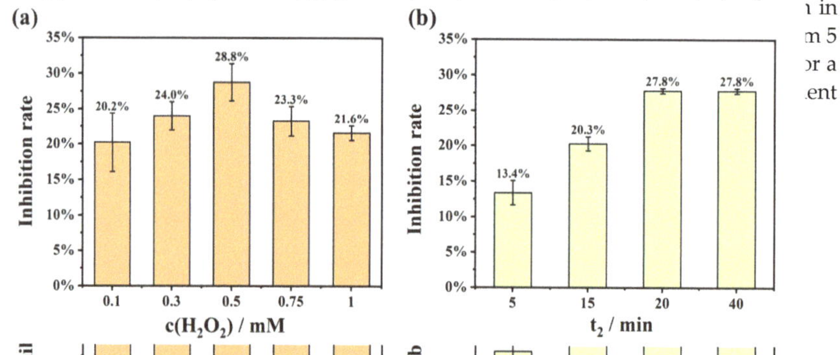

Figure 6. (**a**) Effect of H_2O_2 concentration on inhibition rate, c(*S. aureus*) = 2 × 10^5 CFU/mL, t_1 = 0 min, t_2 = 20 min; (**b**) Effect of H_2O_2 reaction time on inhibition rate, c(*S. aureus*) = 2 × 10^5 CFU/mL, c(H_2O_2) = 0.5 mM, t_1 = 0 min.

3.4. Application of the As-Prepared Colorimetric Biosensor for S. aureus Detection

Under optimized conditions, the application of the as-prepared colorimetric biosensor for *S. aureus* detection was assessed. As depicted in Figure 7a, the final absorbance intensity decreased as the concentration of *S. aureus* was elevated. The relationship between the inhibition rate and the logarithm of bacterial concentration was fitted and is depicted in Figure 7b, revealing a strong linear correlation within the concentration range of 2 × 10^4 to 2 × 10^7 CFU/mL (R^2 = 0.983). Since the sensitivity of our method relies on the measurement accuracy of the UV-vis spectrophotometer, we was reached when the absorbance intensity difference between the experimental and blank groups was less than 0.02. The limit of detection (LOD) was calculated as 3 S_b/m, where m represents the slope at low bacterial concentrations and S_b is the standard deviation of the blank group. The LOD value of our method was 9.2 × 10^3 CFU/mL.

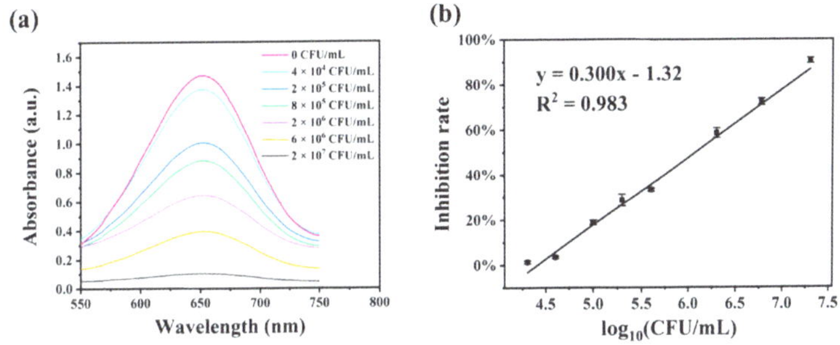

Figure 7. (**a**) UV-visible absorption spectrum at varying concentrations of *S. aureus*; (**b**) Linear fitting curve of inhibition rate–logarithm of bacterial concentration.

3.5. Assessment of Anti-Interference Ability

Two other types of pathogenic bacteria, Gram-positive *B. subtilis* and Gram-negative *E. coli*, were selected to evaluate the anti-interference capability of the present method. The detection of other pathogenic bacteria followed the same procedure. The colorimetric results showed an inhibition rate of 28.76% for *S. aureus* at the concentration of 2 × 10^5 CFU/mL, while the *E. coli* and *B. subtilis* under the same concentration exhibited nearly no response (Figure 8). The good selectivity of the present method is attributed to two key factors. Firstly, the strain of *S. aureus* is CAT-positive, so it exhibits specific CAT catalytic activity that is different to other bacteria strains [22]. Secondly, the *S. aureus*-specific lysostaphin can selectively lyse the cell wall of *S. aureus* without affecting other bacteria. Together with the specific enzymatic activity of target *S. aureus* and the selective cell processing method, the as-prepared biosensor exhibited good anti-interference ability against other pathogenic bacteria.

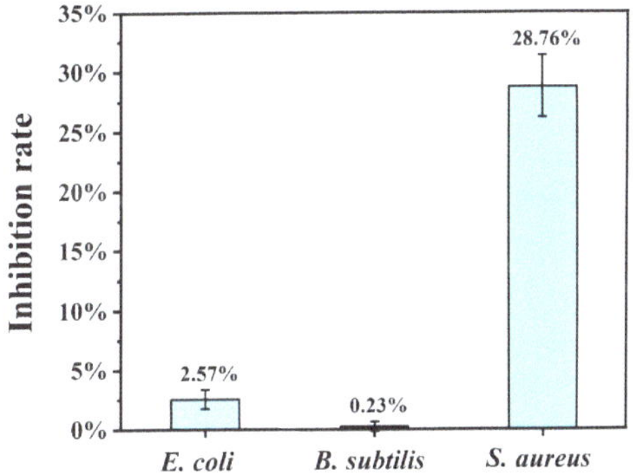

Figure 8. Comparison of the detection results of 2 × 10^5 CFU/mL *E. coli*, *B. subtilis* and *S. aureus*.

3.6. Detection of *S. aureus* in Real Water Samples

3.6.1. Single Bacterial Test

Recovery experiments in real water samples were conducted to confirm the feasibility of the present method for *S. aureus* detection. Three samples with varying *S. aureus* concentrations were added to the collected real water sample, and the colorimetric assay was performed. The inhibition rates of real water samples spiked with different concentrations of *S. aureus* were calculated, and the detected bacterial concentration was converted according to the fitting relationship in Figure 7b. Real sample results for *S. aureus* are shown in Table 3. The recovery rates between the measured concentration and the spiked *S. aureus* concentration demonstrated the good accuracy of the present method in real water detection.

Table 3. Results of *S. aureus* detection in real water samples.

Spiked Concentration of *S. aureus* (CFU/mL)	Inhibition Rate	Found Concentration (CFU/mL)	Recovery Rate	RSD
5×10^4	9.14%	5.16×10^4	103.20%	2.71%
2×10^5	27.16%	2.06×10^5	103.00%	6.45%
2×10^7	88.17%	2.24×10^7	112.00%	3.94%

The practical anti-interference performance of our method was also evaluated by detecting the real water samples spiked with *B. subtilis* or *E. coli* at a concentration of 2×10^5 CFU/mL, respectively. As listed in Table 4, negligible responses were observed for samples containing *B. subtilis* and *E. coli*. The results proved the effectiveness of our approach in detecting CAT-positive bacteria such as *S. aureus*, thereby offering a promising solution for rapid pathogen detection.

Table 4. Results of 2×10^5 CFU/mL *E. coli* or *B. subtilis* detection in real water samples.

	Inhibition Rate	RSD
E. coli	3.74%	0.97%
B. subtilis	1.25%	0.55%

3.6.2. Mixed Bacterial Test

Natural water bodies often contain a wide range of bacteria, which potentially complicates the detection of target *S. aureus* due to interferences among different species. To assess the robustness of our assay under such conditions, we added the mixtures of *S. aureus*, *E. coli* and *B. subtilis* to the real water samples with each at a concentration of 2×10^5 CFU/mL, and the found concentrations of *S. aureus* were determined. The found *S. aureus* concentrations in the mixed bacterial samples, as well as recovery rates and RSDs, are listed in Table 5. The results indicated that the presence of *E. coli* and *B. subtilis* had a minor effect on the detection of *S. aureus*, with recovery rates ranging from 95.5% to 101.5%. The above experiments demonstrated the capability of the present method in effectively mitigating interferences from co-existing bacteria. However, given that our method determines the *S. aureus* concentration by measuring the H_2O_2 consumption, it is plausible that substances that can react with H_2O_2 may interfere with the detection results. This could include sulfur ions and iodine ions, which may undergo redox reactions with H_2O_2. Additionally, industrial wastewater may also contain inorganic metal ions, such as Fe^{3+}, which can accelerate the decomposition of H_2O_2 and potentially interfere with detection accuracy.

Table 5. Results of the mixed bacterial test in real water samples.

Spiked Samples	Inhibition Rate	Found Concentration (CFU/mL)	Recovery Rate	RSD
S. aureus + *E. coli*	26.93%	2.03×10^5	101.50%	4.94%
S. aureus + *B. subtilis*	26.20%	1.92×10^5	96.00%	2.42%
S. aureus + *E. coli* + *B. subtilis*	26.14%	1.91×10^5	95.50%	4.00%

4. Conclusions

In this study, we proposed a colorimetric method for the detection of CAT-positive *S. aureus* based on its endogenous CAT activity. The POD-like property of an onion-like carbon nanozyme and a classical H_2O_2-TMB reaction were employed to monitor H_2O_2 consumption, which also served as the indicator for *S. aureus* concentration. In contrast to conventional biosensors for specific bacterial detection, our approach did not require the use of antibodies and aptamers, which greatly reduced the detection cost and manufacturing complexity. The method demonstrated a linear response to *S. aureus* from 2×10^4 to 2×10^7 CFU/mL with an LOD of 9.2×10^3 CFU/mL. Furthermore, it exhibited robust anti-interference capability, enabling us to effectively distinguish *S. aureus* from co-existing pathogenic bacteria in real water samples. The underlying mechanism of the present sensing method could be further expanded to the detection of other CAT-positive bacteria, providing a straightforward and specific approach for pathogen detection.

Author Contributions: Conceptualization, Y.F. and G.G.; methodology, Y.F. and G.G.; validation, Y.F.; formal analysis, Y.F.; investigation, Y.F.; resources, G.G. and J.Z.; writing—original draft preparation, Y.F.; writing—review and editing, G.G. and J.Z.; visualization, Y.F. and G.G.; supervision, G.G. and J.Z.; project administration, G.G. and J.Z.; funding acquisition, G.G. and J.Z. All authors have read and agreed to the published version of the manuscript.

Funding: This research was funded by the National Key Research and Development Program, grant number 2022YFB3304003; and the Presidential Foundation of Technical Institute of Physics and Chemistry, grant number E2A8F301.

Institutional Review Board Statement: Not applicable.

Informed Consent Statement: Not applicable.

Data Availability Statement: The data supporting the conclusions of this article will be made available by the authors on request.

Conflicts of Interest: The authors declare no conflicts of interest.

References

1. Tong, S.Y.C.; Davis, J.S.; Eichenberger, E.; Holland, T.L.; Fowler, V.G. *Staphylococcus aureus* Infections: Epidemiology, Pathophysiology, Clinical Manifestations, and Management. *Clin. Microbiol. Rev.* **2015**, *28*, 603–661. [CrossRef] [PubMed]
2. Knox, J.; Uhlemann, A.-C.; Lowy, F.D. *Staphylococcus aureus* infections: Transmission within households and the community. *Trends Microbiol.* **2015**, *23*, 437–444. [CrossRef] [PubMed]
3. Trilla, A.; Miro, J.M. Identifying high risk patients for *Staphylococcus aureus* infections: Skin and soft tissue infections. *J. Chemother.* **1995**, *7*, 37–43. [PubMed]
4. Nicolau, D.P.; Marangos, M.N.; Nightingale, C.H.; Quintiliani, R. Influence of aspirin on development and treatment of experimental *Staphylococcus aureus* endocarditis. *Antimicrob. Agents Chemother.* **1995**, *39*, 1748–1751. [CrossRef]
5. Robertson, L.; Caley, J.P.; Moore, J. Importance of *Staphylococcus aureus* in pneumonia in the 1957 epidemic of influenza A. *Lancet* **1958**, *272*, 233–236. [CrossRef]
6. Kim, J.H.; van der Horst, C.; Mulrow, C.D.; Corey, G.R. *Staphylococcus aureus* Meningitis: Review of 28 Cases. *Clin. Infect. Dis.* **1989**, *11*, 698–706. [CrossRef]
7. Solberg, C.O. Spread of *Staphylococcus aureus* in Hospitals: Causes and prevention. *Scand. J. Infect. Dis.* **2000**, *32*, 587–595. [CrossRef] [PubMed]
8. Strommenger, B.; Layer, F.; Werner, G. *Staphylococcus aureus* and Methicillin-Resistant *Staphylococcus aureus* in Workers in the Food Industry. In *Staphylococcus aureus*; Fetsch, A., Ed.; Academic Press: Cambridge, MA, USA, 2018; pp. 163–188.
9. Kadariya, J.; Smith, T.C.; Thapaliya, D. *Staphylococcus aureus* and Staphylococcal Food-Borne Disease: An Ongoing Challenge in Public Health. *BioMed Res. Int.* **2014**, *2014*, 827965. [CrossRef]
10. Zhao, X.; Wei, C.; Zhong, J.; Jin, S. Research advance in rapid detection of foodborne *Staphylococcus aureus*. *Biotechnol. Biotechnol. Equip.* **2016**, *30*, 827–833. [CrossRef]
11. Warren, D.K.; Liao, R.S.; Merz, L.R.; Eveland, M.; Dunne, W.M. Detection of Methicillin-Resistant *Staphylococcus aureus* Directly from Nasal Swab Specimens by a Real-Time PCR Assay. *J. Clin. Microbiol.* **2004**, *42*, 5578–5581. [CrossRef]
12. Kim, J.-K.; Yun, H.-Y.; Kim, J.-S.; Kim, W.; Lee, C.-S.; Kim, B.-G.; Jeong, H.-J. Development of fluorescence-linked immunosorbent assay for rapid detection of *Staphylococcus aureus*. *Appl. Microbiol. Biotechnol.* **2023**, *108*, 2. [CrossRef] [PubMed]
13. Nguyen, Q.H.; Kim, M.I. Nanomaterial-mediated paper-based biosensors for colorimetric pathogen detection. *TrAC Trends Anal. Chem.* **2020**, *132*, 116038. [CrossRef]

14. Zhang, R.; Belwal, T.; Li, L.; Lin, X.; Xu, Y.; Luo, Z. Nanomaterial-based biosensors for sensing key foodborne pathogens: Advances from recent decades. *Compr. Rev. Food Sci. Food Saf.* **2020**, *19*, 1465–1487. [CrossRef] [PubMed]
15. Yao, S.; Li, J.; Pang, B.; Wang, X.; Shi, Y.; Song, X.; Xu, K.; Wang, J.; Zhao, C. Colorimetric immunoassay for rapid detection of *Staphylococcus aureus* based on etching-enhanced peroxidase-like catalytic activity of gold nanoparticles. *Microchim. Acta* **2020**, *187*, 504. [CrossRef]
16. Zhou, W.; Li, Y.; Zhu, G.; Xu, X.; Hu, Q.; Yang, Z.; Gu, X. High-sensitivity and high-specificity colorimetric detection of viable *Staphylococcus aureus* in ready-to-eat foods using a temperate-bacteriophage-based system with peroxidase-like activity. *Sens. Actuators B Chem.* **2024**, *399*, 134810. [CrossRef]
17. Liu, Y.; Sun, M.; Qiao, W.; Cong, S.; Zhang, Y.; Wang, L.; Hu, Z.; Liu, F.; Wang, D.; Wang, P.; et al. Multicolor colorimetric visual detection of *Staphylococcus aureus* based on Fe_3O_4-Ag-MnO_2 composites nano-oxidative mimetic enzyme. *Anal. Chim. Acta* **2023**, *1239*, 340654. [CrossRef] [PubMed]
18. Bukharin, O.V.; Cherkasov, S.V.; Sgibnev, A.V.; Zabirova, T.M.; Ivanov, Y.B. Effects of microbial metabolites on catalase activity and growth of *Staphylococcus aureus* 6538 P. *Bull. Exp. Biol. Med.* **2000**, *130*, 679–681. [CrossRef] [PubMed]
19. Bukharin, O.V.; Sgibnev, A.V.; Cherkasov, S.V.; Ivanov, Y.B. The Effect of the Intra- and Extracellular Metabolites of Microorganisms Isolated from Various Ecotopes on the Catalase Activity of *Staphylococcus aureus* ATCC 6538 P. *Microbiology* **2002**, *71*, 183–186. [CrossRef]
20. Guarín, P.; Cristancho, J.; Castillo, J.J. Rapid electrochemical detection of *Staphylococcus aureus* based on screen-printed gold electrodes modified with cysteine and Guinea grass (*Panicum maximum*) peroxidase. *Rev. Acad. Colomb. Cienc. Exactas Físicas Nat.* **2020**, *44*, 835–844. [CrossRef]
21. Majumdar, T.; Chakraborty, R.; Raychaudhuri, U. Rapid Electrochemical Quantification of Food Borne Pathogen *Staphylococcus aureus* Based on Hydrogen Peroxide Degradation by Catalase. *J. Electrochem. Soc.* **2013**, *160*, G75–G78. [CrossRef]
22. Abdelhamid, H.N.; Wu, H.-F. Selective biosensing of *Staphylococcus aureus* using chitosan quantum dots. *Spectrochim. Acta Part A Mol. Biomol. Spectrosc.* **2018**, *188*, 50–56. [CrossRef] [PubMed]
23. Wang, W.; Gunasekaran, S. Nanozymes-based biosensors for food quality and safety. *TrAC Trends Anal. Chem.* **2020**, *126*, 115841. [CrossRef]
24. Baranwal, A.; Shukla, R.; Bansal, V. Nanozyme-enhanced paper-based biosensor technologies. *TrAC Trends Anal. Chem.* **2024**, *172*, 117573. [CrossRef]
25. Wang, X.; Dong, S.; Wei, H. Recent Advances on Nanozyme-based Electrochemical Biosensors. *Electroanalysis* **2022**, *35*, e202100684. [CrossRef]
26. Everse, J. Heme Proteins. In *Encyclopedia of Biological Chemistry*, 2nd ed.; Lennarz, W.J., Lane, M.D., Eds.; Academic Press: Cambridge, MA, USA, 2013; pp. 532–538.
27. Ye, Y.; Zou, J.; Wu, W.; Wang, Z.; Wen, S.; Liang, Z.; Liu, S.; Lin, Y.; Chen, X.; Luo, T.; et al. Advanced nanozymes possess peroxidase-like catalytic activities in biomedical and antibacterial fields: Review and progress. *Nanoscale* **2024**, *16*, 3324–3346. [CrossRef] [PubMed]
28. Lin, Y.; Sun, X.; Su, D.S.; Centi, G.; Perathoner, S. Catalysis by hybrid sp2/sp3 nanodiamonds and their role in the design of advanced nanocarbon materials. *Chem. Soc. Rev.* **2018**, *47*, 8438–8473. [CrossRef] [PubMed]
29. Zeiger, M.; Jäckel, N.; Mochalin, V.N.; Presser, V. Review: Carbon onions for electrochemical energy storage. *J. Mater. Chem. A* **2016**, *4*, 3172–3196. [CrossRef]
30. Dhand, V.; Yadav, M.; Kim, S.H.; Rhee, K.Y. A comprehensive review on the prospects of multi-functional carbon nano onions as an effective, high-performance energy storage material. *Carbon* **2021**, *175*, 534–575. [CrossRef]
31. Su, D.S.; Maksimova, N.; Delgado, J.J.; Keller, N.; Mestl, G.; Ledoux, M.J.; Schlögl, R. Nanocarbons in selective oxidative dehydrogenation reaction. *Catal. Today* **2005**, *102–103*, 110–114. [CrossRef]
32. Su, D.; Maksimova, N.I.; Mestl, G.; Kuznetsov, V.L.; Keller, V.; Schlögl, R.; Keller, N. Oxidative dehydrogenation of ethylbenzene to styrene over ultra-dispersed diamond and onion-like carbon. *Carbon* **2007**, *45*, 2145–2151. [CrossRef]
33. Szymański, G.S.; Wiśniewski, M.; Olejnik, P.; Koter, S.; Castro, E.; Echegoyen, L.; Terzyk, A.P.; Plonska-Brzezinska, M.E. Correlation between the catalytic and electrocatalytic properties of nitrogen-doped carbon nanoonions and the polarity of the carbon surface: Experimental and theoretical investigations. *Carbon* **2019**, *151*, 120–129. [CrossRef]
34. Butsyk, O.; Olejnik, P.; Romero, E.; Plonska-Brzezinska, M.E. Postsynthetic treatment of carbon nano-onions: Surface modification by heteroatoms to enhance their capacitive and electrocatalytic properties. *Carbon* **2019**, *147*, 90–104. [CrossRef]
35. Shi, Y.; Zhao, Q.; Guan, B.; Li, J.; Gao, G.; Zhi, J. Onion-like carbon with controllable carbon atom hybridization embedded in BiVO4 photoanode for enhanced photoelectrochemical water splitting. *Appl. Surf. Sci.* **2023**, *614*, 156120. [CrossRef]
36. Shi, Y.; Zhao, Q.; Li, J.; Gao, G.; Zhi, J. Onion-liked carbon-embedded graphitic carbon nitride for enhanced photocatalytic hydrogen evolution and dye degradation. *Appl. Catal. B Environ.* **2022**, *308*, 121216. [CrossRef]
37. Ghanbari, M.H.; Norouzi, Z. A new nanostructure consisting of nitrogen-doped carbon nanoonions for an electrochemical sensor to the determination of doxorubicin. *Microchem. J.* **2020**, *157*, 105098. [CrossRef]
38. Sok, V.; Fragoso, A. Amperometric biosensor for glyphosate based on the inhibition of tyrosinase conjugated to carbon nano-onions in a chitosan matrix on a screen-printed electrode. *Microchim. Acta* **2019**, *186*, 569. [CrossRef] [PubMed]
39. Kim, Y.; Kang, E. A graphitic nano-onion/molybdenum disulfide nanosheet composite as a platform for HPV-associated cancer-detecting DNA biosensors. *J. Nanobiotechnol.* **2023**, *21*, 187. [CrossRef] [PubMed]

40. Rizwan, M.; Hazmi, M.; Lim, S.A.; Ahmed, M.U. A highly sensitive electrochemical detection of human chorionic gonadotropin on a carbon nano-onions/gold nanoparticles/polyethylene glycol nanocomposite modified glassy carbon electrode. *J. Electroanal. Chem.* **2019**, *833*, 462–470. [CrossRef]
41. Yoon, J.H.; Han, Y.D.; Lee, P.C.; Yoon, H.C. Lysostaphin-mediated fragmentation of microbial peptidoglycans for label-free electrochemical impedance immunoanalysis of *Staphylococcus aureus*. *BioChip J.* **2013**, *7*, 344–352. [CrossRef]
42. Wang, M.; Fan, E.; Wu, Y.; Fu, Z. Daptomycin-modified magnetic beads integrated with lysostaphin for selective analysis of *Staphylococcus*. *J. Pharm. Biomed. Anal.* **2019**, *175*, 112785. [CrossRef]
43. Lee, H.Y.; Atlasevich, N.; Granzotto, C.; Schultz, J.; Loike, J.; Arslanoglu, J. Development and application of an ELISA method for the analysis of protein-based binding media of artworks. *Anal. Methods* **2015**, *7*, 187–196. [CrossRef]
44. Li, M.; Su, H.; Tu, Y.; Shang, Y.; Liu, Y.; Peng, C.; Liu, H. Development and Application of an Efficient Medium for Chromogenic Catalysis of Tetramethylbenzidine with Horseradish Peroxidase. *ACS Omega* **2019**, *4*, 5459–5470. [CrossRef]
45. Ma, W.; Xue, Y.; Guo, S.; Jiang, Y.; Wu, F.; Yu, P.; Mao, L. Graphdiyne oxide: A new carbon nanozyme. *Chem. Commun.* **2020**, *56*, 5115–5118. [CrossRef] [PubMed]
46. Li, R.; Zhen, M.; Guan, M.; Chen, D.; Zhang, G.; Ge, J.; Gong, P.; Wang, C.; Shu, C. A novel glucose colorimetric sensor based on intrinsic peroxidase-like activity of C60-carboxyfullerenes. *Biosens. Bioelectron.* **2013**, *47*, 502–507. [CrossRef] [PubMed]
47. Song, Y.; Qu, K.; Zhao, C.; Ren, J.; Qu, X. Graphene Oxide: Intrinsic Peroxidase Catalytic Activity and Its Application to Glucose Detection. *Adv. Mater.* **2010**, *22*, 2206–2210. [CrossRef]
48. Sepasi Tehrani, H.; Moosavi-Movahedi, A.A. Catalase and its mysteries. *Prog. Biophys. Mol. Biol.* **2018**, *140*, 5–12. [CrossRef] [PubMed]
49. Kaushal, J.; Mehandia, S.; Singh, G.; Raina, A.; Arya, S.K. Catalase enzyme: Application in bioremediation and food industry. *Biocatal. Agric. Biotechnol.* **2018**, *16*, 192–199. [CrossRef]
50. Galasso, M.; Gambino, S.; Romanelli, M.G.; Donadelli, M.; Scupoli, M.T. Browsing the oldest antioxidant enzyme: Catalase and its multiple regulation in cancer. *Free Radic. Biol. Med.* **2021**, *172*, 264–272. [CrossRef]
51. Kirkman, H.N.; Gaetani, G.F. Mammalian catalase: A venerable enzyme with new mysteries. *Trends Biochem. Sci.* **2007**, *32*, 44–50. [CrossRef]
52. Josephy, P.D.; Eling, T.; Mason, R.P. The horseradish peroxidase-catalyzed oxidation of 3,5,3′,5′-tetramethylbenzidine. Free radical and charge-transfer complex intermediates. *J. Biol. Chem.* **1982**, *257*, 3669–3675. [CrossRef]

chemosensors

MDPI

Article

Cost-Effective Real-Time Metabolic Profiling of Cancer Cell Lines for Plate-Based Assays

Wiktoria Blaszczak †, Zhengchu Tan † and Pawel Swietach *

Department of Physiology, Anatomy & Genetics, University of Oxford, Oxford OX1 3PT, UK; wiktoria.blaszczak@dpag.ox.ac.uk (W.B.); zhengchu.tan11@imperial.ac.uk (Z.T.)
* Correspondence: pawel.swietach@dpag.ox.ac.uk; Tel.: +44-18-6528-2515
† These authors contributed equally to this work.

Abstract: A fundamental phenotype of cancer cells is their metabolic profile, which is routinely described in terms of glycolytic and respiratory rates. Various devices and protocols have been designed to quantify glycolysis and respiration from the rates of acid production and oxygen utilization, respectively, but many of these approaches have limitations, including concerns about their cost-ineffectiveness, inadequate normalization procedures, or short probing time-frames. As a result, many methods for measuring metabolism are incompatible with cell culture conditions, particularly in the context of high-throughput applications. Here, we present a simple plate-based approach for real-time measurements of acid production and oxygen depletion under typical culture conditions that enable metabolic monitoring for extended periods of time. Using this approach, it is possible to calculate metabolic fluxes and, uniquely, describe the system at steady-state. By controlling the conditions with respect to pH buffering, O_2 diffusion, medium volume, and cell numbers, our workflow can accurately describe the metabolic phenotype of cells in terms of molar fluxes. This direct measure of glycolysis and respiration is conducive for between-runs and even between-laboratory comparisons. To illustrate the utility of this approach, we characterize the phenotype of pancreatic ductal adenocarcinoma cell lines and measure their response to a switch of metabolic substrate and the presence of metabolic inhibitors. In summary, the method can deliver a robust appraisal of metabolism in cell lines, with applications in drug screening and in quantitative studies of metabolic regulation.

Keywords: cell lines; screening; acidity; hypoxia; glycolysis; respiration; flux; buffers; PDAC

Citation: Blaszczak, W.; Tan, Z.; Swietach, P. Cost-Effective Real-Time Metabolic Profiling of Cancer Cell Lines for Plate-Based Assays. *Chemosensors* **2021**, *9*, 139. https://doi.org/10.3390/chemosensors9060139

Academic Editor: Pietro Salvo

Received: 18 May 2021
Accepted: 11 June 2021
Published: 13 June 2021

Publisher's Note: MDPI stays neutral with regard to jurisdictional claims in published maps and institutional affiliations.

1. Introduction

Cell growth and proliferation is ultimately powered by metabolic pathways that harness energy from substrates such as glucose [1,2]. Ever since the discovery that deregulated energetics is a hallmark of cancer [3], metabolic phenotyping has become a pillar of cancer research [3–5]. The two main energy-yielding pathways are glycolysis and respiration, and their rates can be gauged from the production of lactic acid and consumption of oxygen, respectively [6]. In tumors, glycolytic and respiratory fluxes can be very large, which explains their characteristic acidotic and hypoxic signatures [7–9]. Glycolytic and respiratory rates can be probed using dedicated platforms, such as Seahorse XF [10–12], or bespoke plates imprinted with pH and/or O_2 sensors for use with fluorescence or absorbance readers [13–18]. However, acid production and oxygen consumption are rarely measured under culture conditions over extended periods of time and then reported in terms of molar fluxes per cell. The Seahorse XF analyzer, for example, takes a snapshot of metabolic rates within a small but poorly controlled catchment of cells [19]. Data generated in this way are prone to measurement error, and as a result, comparisons between laboratories or even consecutive runs may not be advisable [20]. Measurements of pH or O_2 in bulk media require additional calculations to derive fluxes, but these are not always performed or reported [21,22]. Given that metabolism is a cellular process, it would be prudent to normalize these flux estimates to the number of cells [23,24]. A concern for many laboratories are the prohibitive costs associated with metabolic profiling, which can make continuous or high-throughput measurements impractical, despite a strong scientific justification [17,25,26]

Here, we describe a simple, fluorescence-based protocol for measuring pH and O_2 simultaneously and in real-time using standard culture plates. We demonstrate how the method can determine acid production (J_{acid}) and oxygen consumption (J_{oxygen}) fluxes per cell as a function of time, including their approach to the steady-state. The method uses commercially available chemical sensors that reduce the cost of fluorescence substances to under 5 cents per plate. We also present a workflow for calculating fluxes from pH and O_2 time courses, and illustrate the utility of our approach in metabolically phenotyping various cancer cell lines using pancreatic ductal adenocarcinoma (PDAC) cells as an example. With the aid of this accessible and simple approach, our intention is to standardize the process of metabolic phenotyping. In doing so, our overarching objective is to improve data quality, exchange, and reproducibility.

2. Materials and Methods

2.1. Plates and Plate Readers

Suitable plates for metabolic phenotyping are those designed for use with fluorescence readers. For illustrative purposes, this study used flat-bottom, black 96-well plates (655090, Greiner, Kremsmünster, Austria) that can be read using a Cytation 5 device (BioTek, Agilent, Winooski, VT, USA) [27]. Excitation was provided by a monochromator, and fluorescence emission was detected sequentially at five wavelengths, which were optimized for the dye combination used. Optimal settings on our system were excitation wavelengths of 400, 416, 450, 460, and 540 nm, and the corresponding emissions were 510, 510, 620, 510, and 580 nm. For improved signal-to-noise ratio, fluorescence can be acquired in time-resolved mode (e.g., delay: 0 μs; acquisition: 500 μs).

2.2. Fluorescent Dyes and Drugs

HPTS (8-Hydroxypyrene-1,3,6-trisulfonic acid trisodium salt) and RuBPY (tris(bipyridine)ruthenium(II) chloride) were purchased from Sigma-Aldrich (H1529 and 224758, St Louis, MO, USA) and dissolved in water at stocks of 4 and 100 mM, respectively. To maintain a consistent molar ratio of HPTS and RuBPY, stocks were mixed accordingly (1:1 v/v) and stored at −20 °C. CellTracker Orange CMRA (CTO) was obtained from Invitrogen (C34551) and kept in DMSO stock (sc-202581, Santa Cruz Biotechnology). Rotenone (R8875, Sigma-Aldrich) was dissolved in DMSO and stored at −20 °C.

2.3. Media

Powdered DMEM medium was obtained from Sigma-Aldrich (D5030). This formulation was chosen because it could be supplemented with buffers, as desired [28]. HEPES (H3375), MES (M3671), glucose (G7021), galactose (G5388), and NaCl (S7653) were obtained from Sigma-Aldrich. Low-buffer medium was prepared by dissolving Phenol Red-Free DMEM 5030 powder in deionized water, then adding 2 mM HEPES, 2 mM MES, 40 mM NaCl, 25 mM glucose, 10% FBS, 1% penicillin–streptomycin mixture, 1% GlutaMAX, and 1% sodium pyruvate. After the ingredients dissolved completely, the medium was tittered to pH 7.4, at 37 °C with 0.4 M NaOH and filtered (0.22 μm pore). High-buffer media were prepared similarly, with the exception that 20 mM HEPES was used instead of 2 mM HEPES plus 2 mM MES, and NaCl was reduced to 25 mM NaCl.

2.4. Cells

Pancreatic ductal adenocarcinoma (PDAC) cell lines AsPC1, MIA-PaCa-2 and PANC1 were a gift from Prof. Anna Trauzold (Christian-Albrecht University, Kiel, Germany), maintained in RPMI (Sigma-Aldrich, R0883) supplemented with 10% FBS, 1% penicillin–streptomycin mixture, 1% GlutaMAX (35050-038, Gibco, Waltham, MA, USA), 1% sodium pyruvate (11360-039, Gibco, Waltham, MA, USA), and used within passage 3–8 [29,30]. Adult mouse myocytes were isolated from Langendorff-perfused mice hearts using a previously published method [31,32]. Mice were killed humanely by an approved Schedule 1 method following ASPA regulations and University guidelines. Cells were kept in media for up to 20 h after isolation.

3. Results

3.1. Selecting Cost-Effective pH and Oxygen Probes for Assaying Glycolysis and Respiration

The criteria for appropriate pH- and O_2-sensitive dyes for a metabolic assay system are (i) low cost, (ii) non-toxicity to cells, (iii) impermeability to measure extracellular levels, (iv) temporal stability, (v) spectral compatibility, and (vi) ability to produce a calibrated signal. Furthermore, the ensemble spectral properties of the dye-pair must reserve a wavelength range for assessing cell numbers using fluorescent dyes, such as those of the CellTracker family. There are multiple pH-sensitive dyes available, including commonly used cSNARF derivatives [33,34], BCECF [35,36], and HPTS [37–39]. Of the three, HPTS has a superior value for money ($0.05 per µmole). Furthermore, HPTS is a ratiometric dye that can be calibrated, and its spectrum is shifted towards low wavelengths, giving spectral space for red-shifted O_2 sensors [40]. Conveniently, HPTS is membrane impermeable, chemically stable, and produces a very strong signal, which means that concentrations below 10 µM are usually sufficient. There are a number of O_2-sensing dyes, including those that are quenched by molecular oxygen [18,41–43]. A cost-effective dye that emits fluorescence in a range that has no overlap with HPTS is RuBPY.

The spectral characterization of HPTS and RuBPY was performed in a 96-well plate using a fluorescence plate reader. The degree of spectral overlap between HPTS and RuBPY fluorescence was measured in DMEM-based media buffered with 10 mM HEPES and 10 mM MES, and titrated to pH 6.0, 7.0, or 8.0 with 0.4 M NaOH and 0.5 N HCl. Excitation was delivered at wavelengths previously reported as suitable for HPTS in dual excitation mode (400 nm and 460 nm) and RuBPY in a single-excitation model (450 nm). The spectra, shown in Figure 1A, show a clear distinction between the dyes. Moreover, fluorescence emitted by HPTS, but not RuBPY, was pH-sensitive. To confirm that the choices of excitation wavelengths were optimal in the biological context of culture media, excitation spectra were probed at emission peaks of 510 nm for HPTS and 620 nm for RuBPY. The spectrum for HPTS was measured over a range of pH, from 6 to 8 (Figure 1B); the peaks producing maximal pH sensitivity for ratiometry were 400 nm and 460 nm. Excitation spectra intersected at 416 nm, which defines the pH-insensitive wavelength. This so-called isosbestic point is a useful reference point that relates to the concentration of HPTS. Unlike HPTS, RuBPY is not ratiometric, and its single excitation peak was 450 nm (Figure 1C). The lack of an oxygen-insensitive reference wavelength makes it difficult to calibrate RuBPY fluorescence into units of dissolved O_2. However, as the HPTS:RuBPY concentration ratio can be fixed by careful and consistent mixing, the HPTS isosbestic point could serve as a convenient reference point to O_2-sensitive RuBPY fluorescence. This is particularly useful for offsetting the effects of changes in volumes, such as evaporation, where both isosbestic HPTS and RuBPY would be affected equally.

In order to use the dyes in combination, it is first necessary to determine the concentrations that produce comparable peaks for best resolving power. HPTS and RuBPY were mixed in various molar ratios, and the emission spectra above 480 nm were recorded with 450 nm excitation—a wavelength that is sufficiently close to the excitation peaks of both dyes. A suitable combination of HPTS and RuBPY that produced a bimodal spectrum was determined empirically to be 2 µM and 50 µM, respectively. At this molar ratio, the spectral distinction between the dyes was >100 nm, and the optimal emission peaks for the pH- and O_2-sensitive channels were 510 nm and 620 nm, respectively (Figure 1D).

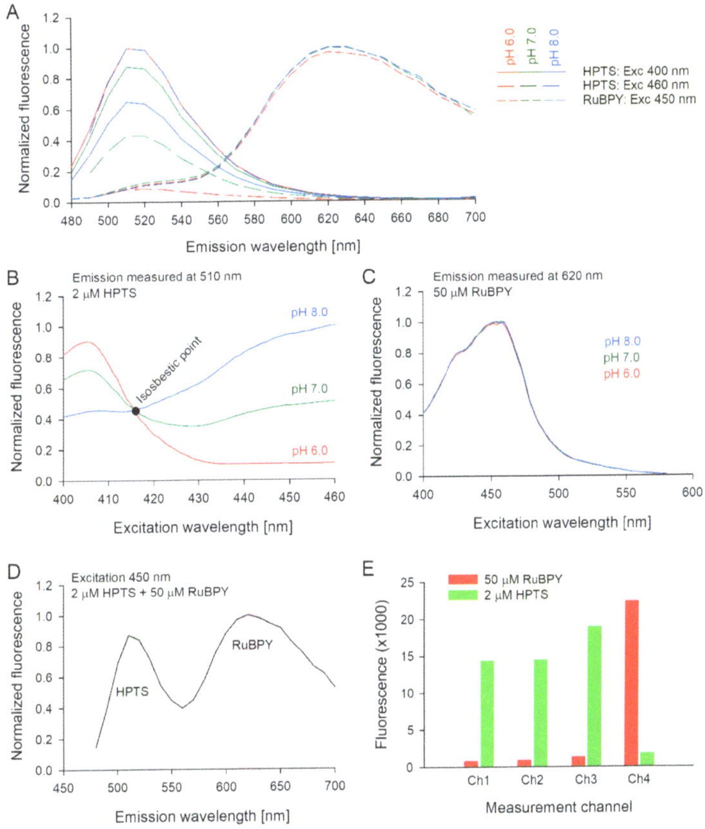

Figure 1. Spectral analyses of HPTS and RuBPY. (**A**) Emission spectra for HPTS and RuBPY measured separately in DMEM-based media buffered with 10 mM HEPES/10 mM MES and titrated to either pH 6.0, 7.0, or 8.0. (**B**) The excitation spectrum for HPTS fluorescence collected at its optimal emission (510 nm). Isosbestic point is the wavelength at which fluorescence is pH-insensitive. (**C**) The excitation spectrum for RuBPY fluorescence collected at its optimal emission (620 nm). Note the pH insensitivity. (**D**) Concentrations of HTPS (2 μM) and RuBPY (50 μM) determined empirically in DMEM-based media to produce comparable emission peaks when excited at 450 nm. (**E**) Test for fluorescence bleed-through between recording channels. Media contained either HTPS (2 μM) or RuBPY (50 μM) at pH 7.3, and measurement protocols were performed to obtain signals across all four channels sequentially (Ch1–Ch4). Ch1–Ch3 are nominally HPTS channels and found to detect a predominantly HPTS-emitted signal. Ch4 is nominally a RuBPY channel and detects a predominantly RuBPY-emitted signal.

Based on the optimization experiments performed, the most suitable measurements settings were determined to be:

- Ch1: Excitation 400 nm/emission 510 nm, optimized for HPTS fluorescence at low pH;
- Ch2: Excitation 460 nm/emission 510 nm, optimized for HPTS fluorescence at high pH;
- Ch3: Excitation 416 nm/emission 510 nm, optimized for pH-insensitive HPTS fluorescence as an O_2-insensitive reference to O_2-sensitive RuBPY;
- Ch4: Excitation 450 nm/emission 620 nm, optimized for RuBPY fluorescence, which is quenched by oxygen, i.e., inversely related to oxygen tension.

Ratios Ch2/Ch1 and Ch3/Ch4 increase with a rise in pH and oxygen, respectively. In order for these ratios to provide a truly selective measure of pH and O_2, the fluorescence order for these ratios to provide a truly selective measure of pH and O_2, the fluorescence emitted by one dye cannot bleed-through to another's wavelength range. To test for bleed-through at the optimized settings, measurements were performed in wells that contained one dye only, and recordings on all four fluorescence channels are presented in Figure 1E. These data show that RuBPY-emitted fluorescence detected at the three, nominally HPTS channels (Ch1-3) is small (<5%) compared to the fluorescence emitted from HPTS-containing wells. This indicates minimal bleed-through and does not warrant a correction. Similarly, HTPS-emitted fluorescence detected at the RuBPY channel (Ch4) was <7% of the signal measured in RuBPY-containing wells. Again, there was no requirement to implement any correction factor for this minimal bleed-through.

3.2. Calibration of Signals to Units of pH and Oxygen Tension

Having optimized the wavelengths for detecting a pH-sensitive ratio (R_{pH} = Ch2/Ch1) and an O_2-sensitive ratio (R_{O2} = Ch3/Ch4), calibration curves were obtained by carefully titrating culture media pH to a value over the range 5–8 and altering O_2 tension. Media were buffered with 10 mM HEPES and 10 mM MES, but no bicarbonate because calibrations were performed in the absence of CO_2. The calibration curves were measured in atmospheric O_2 (21%) and then in hypoxia (<1% O_2). In order to maintain oxygen tension during the measurement period, wells were covered with 150 μL of mineral oil, which serves as a diffusion barrier to minimize oxygen ingress during the brief recording period. The calibration curves are presented in Figure 2.

HPTS provides a highly pH-sensitive ratio across the physiological range. The calibration curves in normoxia and hypoxia were superimposable, indicating that HPTS is insensitive to oxygen and reports a pH-selective signal (Figure 2A). The best-fit equation is the Grinkiewicz type:

$$pH = pK_a - \log_{10} \frac{r_{max} - R_{pH}}{R_{pH} - r_{min}} \quad (1)$$

RuBPY fluorescence was pH-insensitive but responded to hypoxia, which is consistent with the quenching effect of oxygen (Figure 2B). The ratio of isosbestic HPTS to RuBPY fluorescence was deemed adequately O_2-sensitive, and its calibration curve can be described using a modification of the Stern-Volmer equation (Figure 2C):

$$O_2 = 21 \times \left(1 - \frac{1 - R_{O2}/R_{normoxia}}{1 - r_{anoxia}}\right) \quad (2)$$

Table 1. Best fit to calibration curves in Equations (1) and (2).

Calibration Variable	Best-Fit
pK_a	7.5383
r_{max}	3.9503
r_{min}	0.0603
r_{anoxia}	0.6945

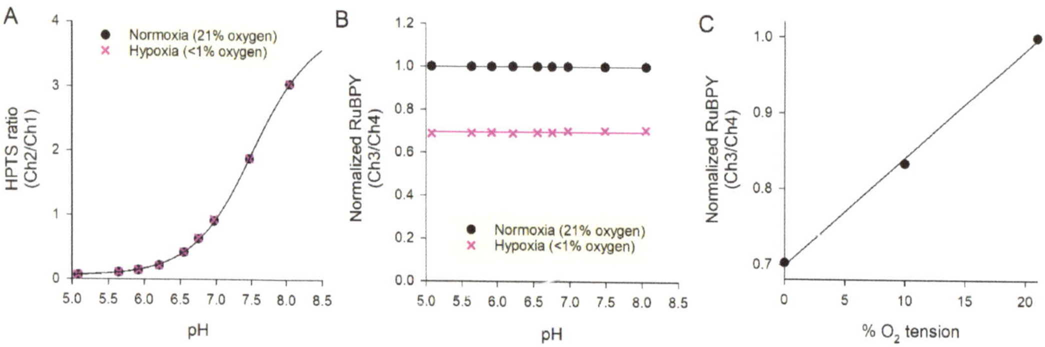

Figure 2. (**A**) The calibration curve for the HPTS ratio (Ch2/Ch1) in normoxic and hypoxic conditions. Best-fit to Equation (1) with constants listed in Table 1. (**B**) Hypoxia-sensitive ratio of isosbestic HPTS to RuBPY (as Ch3/Ch4). Hypoxia (<1% O_2) reduces the Ch3/Ch4 ratio but is pH-insensitive. (**C**) Three-point calibration for 0.1%, 10%, and 21% O_2.

HPTS provides a highly pH-sensitive ratio across the physiological range. The calibration curves in normoxia and hypoxia were superimposable, indicating that HPTS is insensitive to oxygen and reports a pH-selective signal (Figure 2A). The best-fit equation is the Grinkiewicz type:

$$pH = pK_a - \log_{10} \frac{r_{max} - R_{pH}}{R_{pH} - r_{min}} \quad (1)$$

RuBPY fluorescence was pH-insensitive but responded to hypoxia, which is consistent with the quenching effect of oxygen (Figure 2B). The ratio of isosbestic HPTS to RuBPY fluorescence was deemed adequately O_2-sensitive, and its calibration curve can be described using a modification of the Stern-Volmer equation (Figure 2C):

$$O_2 = 21 \times \left(1 - \frac{1 - R_{O2}/R_{normoxia}}{1 - r_{anoxia}}\right) \quad (2)$$

The value for $R_{normoxia}$ at 21% O_2, i.e., atmospheric, can be obtained in cell-free wells measured in parallel to wells that include cells. In contrast, r_{anoxia} must be obtained in separate experimental reads by placing plates in an O_2-free environment. In our system, the best-fit values are listed in Table 1. It is important that calibrations are determined for each laboratory individually to account for differences in equipment.

3.3. Converting pH and Oxygen Time Courses into Fluxes

The convertion of pH or O_2 time courses to molar fluxes must consider chemical buffering and whether the system being studied is open (i.e., can exchange freely with the atmosphere) or closed.

3.3.1. Buffering

Fluorescent dyes such as HPTS and RuBPY probe the level of free ligands, i.e., H^+ ions and O_2, respectively. It is, however, common for chemical systems to manifest buffering: pH buffering takes the form of H^+-binding weak acid/base pairs, whereas O_2 buffering can involve chelators such as hemoglobin. Fluxes of H^+ or O_2 caused by glycolysis and respiration should refer to the total concentration rather than the free pool. Thus, flux is best described as the product of buffering capacity and rate of change. Buffering can be quantified in terms of a buffer ratio (B), such as the number of molecules bound to buffer per free molecule. In most culture media, O_2 buffering is close to zero, and therefore the rate of change of measured O_2 gives an adequate estimate of flux. This can be used to describe oxygen flux (J_{O2}):

$$J_{O2} = \frac{d[O_2]^{total}}{dt} = \frac{d[O_2]^{free}}{dt} \times (1 + B) \approx \frac{d[O_2]^{free}}{dt} \quad (3)$$

H^+ ions, in contrast, can be heavily buffered; indeed, that is the rationale of including buffers such as HEPES in media formulations. By convention, pH buffering is described in terms of a buffering capacity (β) that may vary in a pH-dependent manner. Whilst it is appropriate to express acid flux in terms of H^+ ions, free H^+ ions are, by convention, reported as pH. Thus, the corresponding definition of buffering capacity is the concentration of acid added, divided by the pH change. Therefore, the acid flux (J_{H+}) is defined as follows:

$$J_{H+} = \frac{d[H^+]^{total}}{dt} = -\frac{dpH}{dt} \times \beta \quad (4)$$

Medium buffering capacity can be measured by adding known amounts of HCl or NaOH and measuring the resulting pH change. Figure 3A shows an exemplar recording of pH after sequential additions of 0.5 mM HCl to media from an alkaline starting pH and 0.5 mM NaOH to media from an acidic starting pH. For this experiment, media contained 2 mM HEPES and 2 mM MES to provide a low level of buffering. The calculated buffering capacity is plotted in Figure 3B.

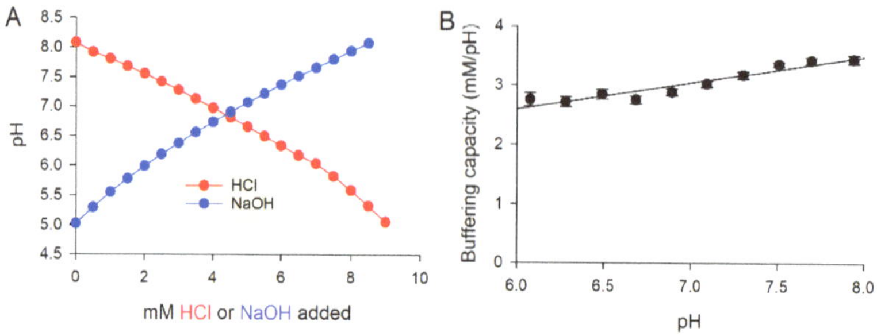

Figure 3. Measuring the pH buffering capacity. (**A**) Stepwise addition of 0.5 mM HCl or NaOH to media of undetermined buffering capacity. The inverse of the slope gives average buffering capacity for the entire pH range. (**B**) Buffering capacity expressed as the ratio of HCl (or NaOH) added to pH change, plotted against the midpoint of the pH change. The best fit line accounts for pH-sensitivity of buffering capacity. This function should be used whenever converting between pH changes and H^+ fluxes.

3.3.2. Open and Closed Systems

In a closed system, a change in O_2 or pH can be attributed to an internal process, such as respiration and glycolysis. Lactic acid is an example of a non-volatile acid, and media with glycolytic cells can be considered a closed system with respect to pH. Oxygen, in contrast, is gaseous, and any system that consumes (or produces) O_2 should be considered an open system in dynamic equilibrium with its immediate atmosphere. The exchange of O_2 between the well and its atmosphere introduces an additional flux that can be described by a permeability constant (P_{O2}) and gradient:

$$J_{O2}{}^{exchange} = P_{O2} \times \left([O_2]_{atm} - [O_2]_{well}\right) \quad (5)$$

Thus, O_2 dynamics in a well that (i) contains respiring cells and (ii) is open to the atmosphere should be described as follows:

$$J_{O2}{}^{measured} = J_{O2}{}^{respiration} + J_{O2}{}^{exchange} \quad (6)$$

Using Equation (6), it is possible to calculate the respiratory flux from O_2 measurements once passive O_2 exchange is factored in. In order to calculate the latter, an a priori estimate of permeability is required. This will depend on various parameters, among which is oxygen diffusivity in media and the height of the column of a medium. A column of medium can support relatively fast gas diffusion, which risks that any change in medium O_2 would be difficult to resolve. To impose a controlled restriction on gas diffusion, media in wells were covered with a layer of mineral oil (Sigma-Aldrich, M5904, St Louis, MO, USA). The ensemble permeability P_{O2} across media and oil can be estimated in cell-free wells that had been pre-equilibrated in hypoxia (e.g., 0.1% O_2). Returning plates to the normal atmosphere of the plate reader drives O_2 ingress, which was reported using RuBPY (Figure 4A). The time course of re-oxygenation gives a measure of time constant, inversely related to P_{O2} (Figure 4B). This experiment was performed for a range of oil volumes to vary the diffusive restriction (Figure 4C). These data provide an estimate of the time constant of O_2 diffusion in the media and the effect that oil has on restricting gas diffusion. For 150 μL of mineral oil, P_{O2} is 0.029 min^{-1}. Applying this to Equation (6) enables real-time measurements of respiratory fluxes.

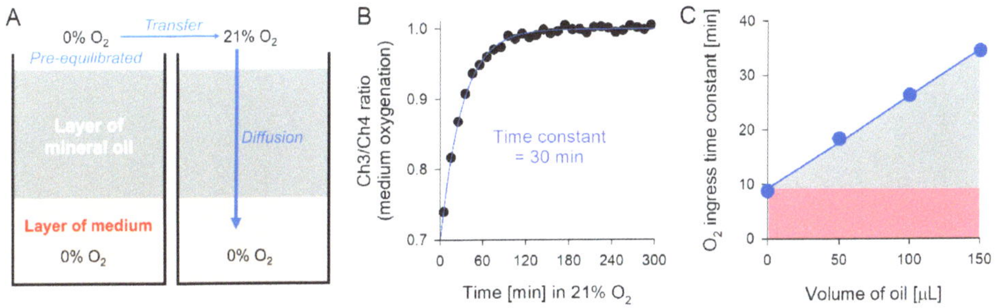

Figure 4. Characterizing the open system with respect to O_2 dynamics. (**A**) The protocol for measuring O_2 permeability across a layer of medium (typically 100 μL) and oil (varied between 0 and 150 μL). (**B**) Exemplar time course from well covered with 150 μL oil. Transferring a plate to an atmosphere of normal oxygen in the reader triggers re-oxygenation, which was monitored every 10 min. The best fit gives a time constant, which is inversely related to O_2 permeability. (**C**) Relationship between the time constant of O_2 ingress and the volume of oil added to wells. Red shading denotes time constant attributable to medium, and gray denotes that due to diffusion through oil.

3.4. Implementation of Assay to Metabolically Phenotype Cancer Cells

Proof-of-principle studies of glycolytic and respiratory fluxes were performed on pancreatic adenocarcinoma cell lines (AsPC1, MIA-PaCa-2, PANC1). Reported fluxes will depend on the density of cells; therefore, it was necessary to obtain a measure of cell count for normalization purposes.

3.4.1. Normalizing for Cell Number

The number of cells in a well can be estimated using fluorescent dyes loaded to cells; a higher signal would indicate higher density. The appropriate dye is required to be stable, membrane impermeant once loaded, spectrally non-overlapping with HTPS and RuBPY, as well as pH- and O_2-insensitive. Suitable candidates for cell counting are CellTracker dyes, and the optimal choice considering the spectral constraints is CellTracker Orange (CTO). The CTO emission spectrum is shown in Figure 5A. CTO fluorescence is resolvable from HPTS and RuBPY, as it is excited at a considerably higher wavelength (540 nm vs. <460 nm) (Figure 5B). As determined using serially-diluted human red blood cells, the CTO signal is linearly related to cell number (Figure 5C). A similar calibration curve can be obtained for adherent cells grown at the bottom of wells, as is shown for MIA-PaCa-2 and AsPC1 (Figure 5D). Note that the fluorescence per PDAC cell (0.455 F.U.) is higher than per red blood cell (0.19 F.U.) because the latter is smaller in size and contains hemoglobin, which absorbs some fluorescence.

The degree to which CTO is pH- and O_2-sensitive was determined experimentally by loading MIA PaCa-2 cells with the dye at 12.5 μM for 15 min, followed by superfusion maneuvers that change pH or O_2 levels. Cells were plated in LabTek slides that are suitable for superfusion. Superfusates contained 125 mM NaCl, 4.5 mM KCl, 11 mM glucose, 1 mM $CaCl_2$, 1 mM $MgCl_2$, and 20 mM HEPES, titrated to either pH 6.4 or 7.4. To make anoxic superfusates, solutions were bubbled in 100% N_2, and anoxia was stabilized chemically by the addition of 1 mM sodium dithionite, an oxygen scavenger. Superfusates were delivered to cells by a peristaltic pump and heated to 37 °C. Images were taken on a Leica confocal microscope using settings optimized previously for CTO. Fluorescence was first imaged under baseline conditions (normoxia of 21% O_2 and pH 7.4), and then after a change in pH to 6.4 (Figure 5E), allowing 10 min for equilibration, or change in O_2 tension to anoxic levels (Figure 5F), allowing 2 min for equilibration. Finally, superfusion was returned to baseline conditions. No effect of hypoxia or acidity was observed, indicating that CTO provides a stable fluorescence signal.

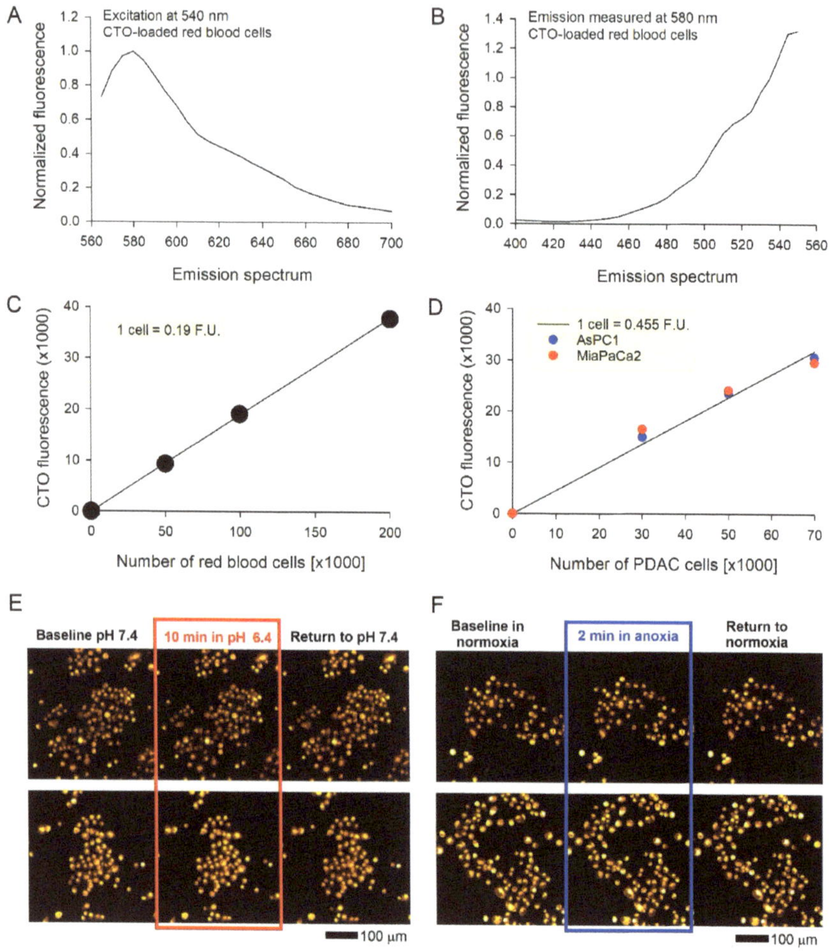

Figure 5. Using CellTracker Orange (CTO) to normalize for cell number. (**A**) The emission spectrum of CTO, excited at 540 nm. (**B**) The optimal CTO excitation is 540 nm, which ensures that the signal is resolvable from HPTS and RuBPY. (**C**) The correlation between CTO fluorescence and count of human red blood cells. (**D**) The correlating CTO signal with the number of plated AsPC1 and MIA-PaCa2 cells in wells of a 96-well plate. (**E**) Superfused MIA-PaCa2 cells imaged confocally. Fluorescence images taken during superfusion at pH 7.4 (baseline; 20 mM HEPES-buffered solution), after 10 min of superfusion at pH 6.4, and 10 min after returning to normal pH. Two exemplar experiments are shown. (**F**) Fluorescence images taken during superfusion in normoxic solution (baseline), after 2 min of superfusion in anoxic conditions (superfusate contained 1 mM dithionite, 100% N_2), and 2 min after returning to normoxia. Two exemplar experiments shown.

3.4.2. Real-Time Monitoring of Metabolic Fluxes

To monitor glycolytic and respiratory fluxes in real-time, AsPC1, MIA-PaCa2, and PANC1 cells were plated in 96-well plates at various cell densities (30k, 50k 70k) and left to adhere overnight. Next, the cells were loaded with 25 µM CTO for 15 min, after which media were replaced with 100 µL of a low-buffer formulation containing 2 µM HPTS and 50 µM RuBPY. Wells were immediately covered with a layer of mineral oil (150 µL), and the plate was inserted into a reader for recordings at regular intervals (e.g., 10 min). Some wells had no cells in order to measure background changes to pH during incubation and for obtaining reference values for normoxia ($R_{normoxia}$). Figure 6A shows a time course of

This information was used to calculate cumulative acid production (C_{H+}) and oxygen consumption (C_{O2}), where V is the volume of the well:

$$C_{H+} = V \times \left(\sum_{t=0}^{T} J_{H+}{}^{cell} - \sum_{t=0}^{T} J_{H+}{}^{cell-free} \right) = V \times \left(- \sum_{t=0}^{T} \left(\frac{dpH^{cell}}{dt} \times \beta \right) + \sum_{t=0}^{T} \left(\frac{dpH^{cell-free}}{dt} \times \beta \right) \right) \quad (7)$$

$$C_{O2} = V \times \sum_{t=0}^{T} \left(J_{O2}{}^{cell} - P_{O2} \times \left([O_2]_{atm} - [O_2]_{well}^{cell} \right) \right) = V \times \sum_{t=0}^{T} \left(\frac{d[O_2]^{cell}}{dt} - P_{O2} \times \left([O_2]_{atm} - [O_2]_{well}^{cell} \right) \right) \quad (8)$$

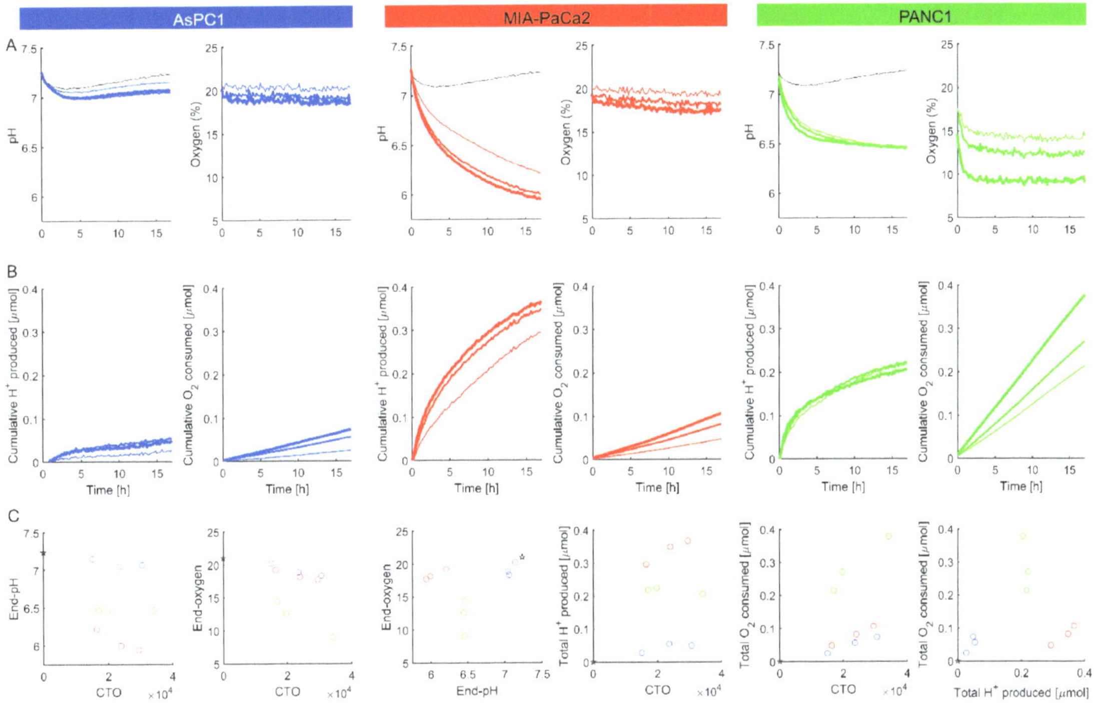

Figure 6. Metabolic profiling of PDAC cell lines. (**A**) Time courses of pH and O_2 for AsPC1 (blue), MIA-PaCa2 (red), and PANC1 (green) plated at 30, 50, or 70k cells per well. Thicker lines denote higher cell density. Blank time courses were obtained from cell-free wells. Media were prepared using DMEM 5030 and contained 2 mM HEPES and 2 mM MES, 25 mM glucose, supplemented with 10% FBS, 1% penicillin–streptomycin mixture. (**B**) Cumulative H^+ production and O_2 consumption, calculated using Equations (7) and (8). (**C**) Data from a and b replotted as a function of cell number, measured by CTO fluorescence. Plots with white background relate to end-point pH and O_2, and plots with gray background relate to cumulative H^+ production and O_2 consumption. The black star symbol denotes data from cell-free wells. Plots represent a mean of 3 biological repeats.

The time courses of C_{H+} and C_{O2} are shown in Figure 6B. The three PDAC cells presented strikingly different phenotypes: AsPC1 had low glycolytic and respiratory rates; MIA-PaCa2 had elevated glycolysis, whereas PANC1 had raised glycolysis and a very high respiratory rate. The relationship between cell number and end-point O_2 and pH, and cumulative O_2 consumption and H^+ production are shown in Figure 6C. The approach can therefore monitor glycolysis and respiration in real-time, investigate their interplay and the effect of cell density, delivering information that has hitherto been largely inaccessible to many researchers.

3.4.3. Measuring Metabolic Responses

The imaging system can be used to study the effect of changing respiratory substrate on glycolytic and respiratory fluxes. MIA-PaCa-2 and PANC1 cells were seeded at 70k per well and left to adhere overnight. Standard medium was then replaced with 100 µL of a high-buffering formulation that contained either 25 mM glucose or 25 mM galactose as a primary substrate, plus fluorescent dyes (2 µM HPTS and 50 µM RuBP). After covering each well with 150 µL of mineral oil, the plate was immediately transferred into the plate-reader for measurements. Consistent with the literature, galactose is less effective as a glycolytic substrate, which re-routes energetic flows towards respiration, as shown by the increase in O_2 consumption (Figure 7A).

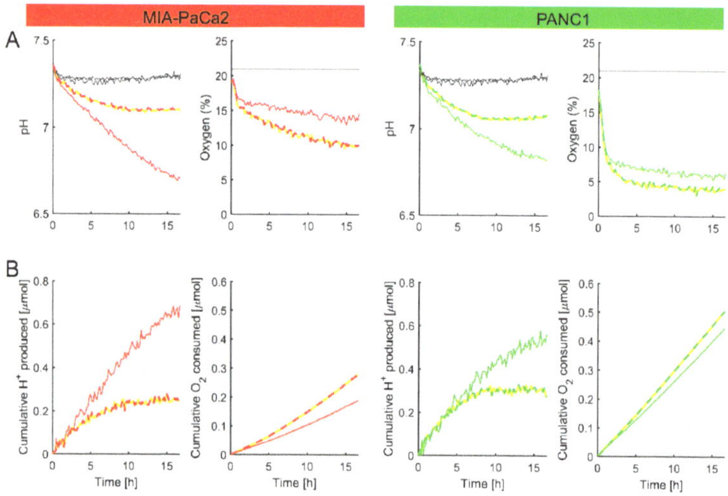

Figure 7. Using the method to study the response to a change in metabolic substrate. (**A**) Time courses of pH and O_2 measured for MIA-PaCa2 (red) and PANC1 (green) plated at 70k cells per well. Thick dashed lines denote media containing galactose (25 mM) in place of glucose. Media contained 20 mM HEPES, 25 mM glucose or galactose, 25 mM NaCl and was supplemented with 10% FBS, 1% penicillin–streptomycin mixture, 1% GlutaMAX, and 1% sodium pyruvate. (**B**) The cumulative H^+ production and O_2 consumption, calculated using Equations (7) and (8). Plots represent a mean of 3 biological repeats.

The method can also study responses to inhibitors, such as the effect of blocking mitochondrial respiration with electron transport chain inhibitors (e.g., rotenone) [44]. To measure the metabolic responses to mitochondrial inhibition in real-time, MIA-PaCa2 and PANC1 cells were plated at 70k per well and left to attach overnight. Next, the medium was replaced with 100 µL of the low-buffering formulation containing 2 µM HPTS and 50 µM RuBP, with or without 10 µM rotenone. Then, 150 µL of mineral oil was added to each well, and the plate was transferred into the plate-reader for measurements (Figure 8). As expected, pharmacological inhibition of the electron transport chain resulted in a shift from O_2 consumption towards a higher level of glycolysis, which verifies the assay's ability to report responses to metabolic inhibition.

3.5. Proof-of-Principle Measurements on Non-Adherent Cells

This plate-based assay can also measure O_2 and pH in suspensions of cells, such as non-adherent primary cells. As proof-of-principle, ventricular myocytes, freshly isolated from a mouse heart, were re-suspended in low-buffer media and aliquoted into wells of a plate for measurements according to the methods described for cancer cells. Myocytes have a high respiratory rate, as shown by the extent to which O_2 is depleted. A smaller

Chemosensors **2021**, *9*, 139

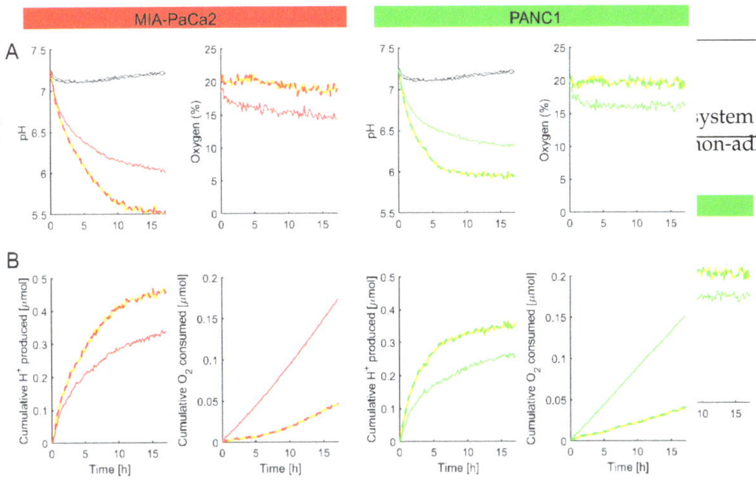

Figure 8. Using the method to study the effect of metabolic inhibitors. (**A**) Time courses of pH and O_2 for MIA-PaCa2 (red) and PANC1 (green) plated at 70k cells per well. The metabolic substrate was glucose throughout. Media contained 2 mM HEPES and 2 mM MES, 25 mM glucose, supplemented with 10% FBS, 1% penicillin–streptomycin mixture, 1% GlutaMAX, and 1% sodium pyruvate. Thick dashed lines indicate the effect of 10 μM rotenone. (**B**) The cumulative H^+ production and O_2 consumption, calculated using Equations (7) and (8). Plots represent a mean of 3 biological repeats.

3.5. Proof-of-Principle Measurements on Non-Adherent Cells

This plate-based assay can also measure O_2 and pH in suspensions of cells, such as non-adherent primary ventricular myocytes, freshly isolated from a mouse heart, which were resuspended in low-buffer media and aliquoted into wells of a plate for measurements according to the methods described for cancer cells. Myocytes have a high respiratory rate, as shown by the extent to which O_2 is depleted. A smaller but significant rate of glycolysis was observed (Figure 9). Thus, the system can be used to characterize the metabolism of non-cancer cells, including primary non-adherent cells.

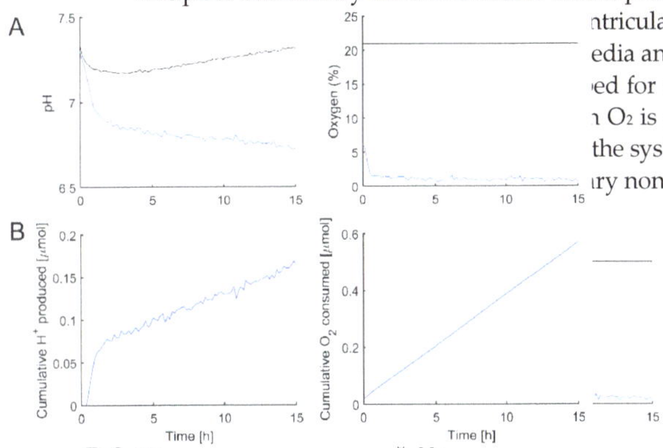

Figure 9. Using the method to study metabolism in primary myocytes freshly isolated from the mouse ventricle. (**A**) Time courses of pH and O_2. A suspension of myocytes in media was probed for O_2 and pH over a period of 15 h, shortly after enzymic isolation from a mouse heart. Media contained 2 mM HEPES and 2 mM MES, 25 mM glucose, supplemented with 10% FBS, 1% penicillin–streptomycin mixture, 1% GlutaMAX, and 1% sodium pyruvate. (**B**) The cumulative H^+ production and O_2 consumption, calculated using Equations (7) and (8). Myocytes have a high oxidative metabolism, causing the rapid depletion of O_2. Some degree of glycolytic metabolism was also noted. Mean of 3 technical repeats.

4. Discussion

This article presents a complete workflow for assaying the metabolic fluxes in cells. Most experiments were performed on cultured cancer cell lines, but the system can also be used to measure metabolic rate in primary non-adherent cells, as illustrated using ventricular myocytes. The key advantages of our method relate to its (i) low running costs and (ii) delivery of direct measurements of molar fluxes of H^+ and O_2 in real-time and under culture conditions over an extended period of time.

The protocol can be readily implemented using a suitable fluorescence plate reader, which is standard equipment in many laboratories. The running costs are very low thanks to the choice of sensors optimized for the assay. RuBPY is available for the list price of £33 for 250 mg, and the cost per plate, assuming a working concentration of 50 µM across 96 wells, is £0.05. HPTS is available at a list price of £70 for 1 g, and the equivalent cost per plate is £0.0068. Greiner plates are available at £5 per unit, which means an assay can be performed for £5.05 per plate. The cost will increase if normalization using CTO is required for specific experiments.

The protocol's workflow includes a step-by-step guide to calculating glycolytic and respiratory fluxes and normalizing these to the cell number. These calculations consider buffering for H^+ ions as well as O_2 ingress in the open system of plate wells. Molar H^+ and O_2 fluxes are the most direct readouts of the state of metabolism and we recommend that they are adopted as the reporting standard to facilitate data sharing, exchange, and comparison between studies. Fluxes normalized to cell number are less prone to measurement error than raw measurements such as pH or O_2. In principle, these values can be compared between runs and even between laboratories, as they are characteristic of cell lines under a given state and not subject to errors that arise from the lack of adequate normalization or flawed calculations.

The ability to follow metabolism in real-time and for extended periods (e.g., many hours) is a major advantage as it allows time for cells to implement more complete biological responses, e.g., to inhibitors or substrates. Critically, cells can be maintained in appropriate cell culture media as required, provided that these are characterized with respect to pH buffering and O_2 diffusivity. Importantly, longer-term follow-up can include the attainment of steady-state between pH, O_2, and metabolism, which is rarely reported in metabolic studies in vitro. Such a steady-state represents the balance of fluxes and may relate accurately to the in vivo scenario. Understanding the steady-state can explain the genesis of particular levels of pH and O_2 tension in tissues.

The ability to perform metabolic assays cheaply and on a large scale opens opportunities for high-throughput studies. Combining this with molar flux measurements and integrating the data over longer time periods can supply high-quality data for mechanistic studies on cancer and other disciplines of life science.

5. Appendix: Step-by-Step Protocol

(1) Dissolve HPTS and RuBPY in sterile, deionized water to obtain stocks of 4 and 100 mM, respectively. Mix both dyes in a 1:1 ratio, divide them into aliquots, and store them at −20 °C. Avoid multiple thaw-freeze cycles.

(2) Seed cells onto a black, fluorescence-compatible, flat-bottom 96-well plate at the desired density and leave to attach overnight. Higher densities will produce larger and more resolvable fluxes. When planning the plate, ensure that some wells are cell-free (blanks) to serve as reference points for pH and O_2. Recommendation: Add PBS to the outermost wells to help maintain humidity and prevent evaporation in the remainder of the plate.

(3) Thaw and vortex the dye mixture. Dissolve (1:1000 *v/v*) in Phenol Red-free medium of choice. Allow aliquots of 100 µL per well.

(4) Replace media that had bathed cells during the settling period with the dye-containing medium of the desired composition (e.g., pH, buffering, inhibitors etc). Whilst tilting the plate slightly, gently add 150 µL of mineral oil to each well to cover the medium

and introduce a controlled diffusion barrier to gas exchange. Volumes of medium and mineral oil should be optimized for the given oxygen consumption and acid production rate.

(5) Place the plate into the plate-reader, keeping the lid on. Start collecting the data immediately to capture the initial state.

(6) Perform calculations according to the equations described herein. Note: Media prepared for the assay must be characterized in separate experiments in terms of buffering capacity and oxygen diffusivity.

Author Contributions: Conceptualization, P.S.; methodology, W.B., Z.T. and P.S.; collecting data and analysis, W.B. and Z.T.; resources, P.S.; writing P.S., W.B. and Z.T. All authors have read and agreed to the published version of the manuscript.

Funding: This research was funded by the European Research Council, grant number SURVIVE, and the Marie Skłodowska-Curie Innovative Training Network (ITN) (grant Agreement number: 813834—*pHioniC*—H2020-MSCA-ITN-2018).

Institutional Review Board Statement: The study was conducted according to the guidelines of the Declaration of Helsinki, and approved by Oxford University ethical review board.

Informed Consent Statement: Not applicable.

Data Availability Statement: All data relevant to the work has been presented herein. Raw traces are available on request made to the corresponding author.

Acknowledgments: We thank Kyung Chan (KC) Park for isolating ventricular myocytes.

Conflicts of Interest: The authors declare no conflict of interest. The funders had no role in the design of the study; in the collection, analyses, or interpretation of data; in the writing of the manuscript, or in the decision to publish the results.

References

1. Birsoy, K.; Possemato, R.; Lorbeer, F.K.; Bayraktar, E.C.; Thiru, P.; Yucel, B.; Wang, T.; Chen, W.W.; Clish, C.B.; Sabatini, D.M. Metabolic determinants of cancer cell sensitivity to glucose limitation and biguanides. *Nature* **2014**, *508*, 108–112. [CrossRef]
2. Nowell, P.C. The clonal evolution of tumor cell populations. *Science* **1976**, *194*, 23–28. [CrossRef]
3. Hanahan, D.; Weinberg, R.A. Hallmarks of cancer: The next generation. *Cell* **2011**, *144*, 646–674. [CrossRef]
4. Li, Z.; Wang, Z.; Tang, Y.; Lu, X.; Chen, J.; Dong, Y.; Wu, B.; Wang, C.; Yang, L.; Guo, Z.; et al. Liquid biopsy-based single-cell metabolic phenotyping of lung cancer patients for informative diagnostics. *Nat. Commun.* **2019**, *10*, 3856. [CrossRef]
5. Ke, C.; Li, A.; Hou, Y.; Sun, M.; Yang, K.; Cheng, J.; Wang, J.; Ge, T.; Zhang, F.; Li, Q.; et al. Metabolic phenotyping for monitoring ovarian cancer patients. *Sci. Rep.* **2016**, *6*, 23334. [CrossRef]
6. Jain, M.; Nilsson, R.; Sharma, S.; Madhusudhan, N.; Kitami, T.; Souza, A.L.; Kafri, R.; Kirschner, M.W.; Clish, C.B.; Mootha, V.K. Metabolite profiling identifies a key role for glycine in rapid cancer cell proliferation. *Science* **2012**, *336*, 1040–1044. [CrossRef]
7. San-Millan, I.; Brooks, G.A. Reexamining cancer metabolism: Lactate production for carcinogenesis could be the purpose and explanation of the Warburg Effect. *Carcinogenesis* **2017**, *38*, 119–133. [CrossRef] [PubMed]
8. Mayer, A.; Vaupel, P. Hypoxia, lactate accumulation, and acidosis: Siblings or accomplices driving tumor progression and resistance to therapy? *Adv. Exp. Med. Biol.* **2013**, *789*, 203–209. [CrossRef]
9. Kato, Y.; Ozawa, S.; Miyamoto, C.; Maehata, Y.; Suzuki, A.; Maeda, T.; Baba, Y. Acidic extracellular microenvironment and cancer. *Cancer Cell Int.* **2013**, *13*, 89. [CrossRef]
10. Sun, C.; Liu, X.; Wang, B.; Wang, Z.; Liu, Y.; Di, C.; Si, J.; Li, H.; Wu, Q.; Xu, D.; et al. Endocytosis-mediated mitochondrial transplantation: Transferring normal human astrocytic mitochondria into glioma cells rescues aerobic respiration and enhances radiosensitivity. *Theranostics* **2019**, *9*, 3595–3607. [CrossRef]
11. O'Neill, S.; Porter, R.K.; McNamee, N.; Martinez, V.G.; O'Driscoll, L. 2-Deoxy-D-Glucose inhibits aggressive triple-negative breast cancer cells by targeting glycolysis and the cancer stem cell phenotype. *Sci. Rep.* **2019**, *9*, 3788. [CrossRef]
12. Guo, J.; Su, Y.; Zhang, M. Circ_0000140 restrains the proliferation, metastasis and glycolysis metabolism of oral squamous cell carcinoma through upregulating CDC73 via sponging miR-182-5p. *Cancer Cell Int.* **2020**, *20*, 407. [CrossRef]
13. John, G.T.; Goelling, D.; Klimant, I.; Schneider, H.; Heinzle, E. PH-sensing 96-well microtitre plates for the characterization of acid production by dairy starter cultures. *J. Dairy Res.* **2003**, *70*, 327–333. [CrossRef] [PubMed]
14. Heux, S.; Philippe, B.; Portais, J.C. High-throughput workflow for monitoring and mining bioprocess data and its application to inferring the physiological response of Escherichia coli to perturbations. *Appl. Environ. Microbiol.* **2011**, *77*, 7040–7049. [CrossRef]
15. Wu, S.; Wu, S.; Yi, Z.; Zeng, F.; Wu, W.; Qiao, Y.; Zhao, X.; Cheng, X.; Tian, Y. Hydrogel-Based Fluorescent Dual pH and Oxygen Sensors Loaded in 96-Well Plates for High-Throughput Cell Metabolism Studies. *Sensors* **2018**, *18*, 564. [CrossRef] [PubMed]

16. Kieninger, J.; Aravindalochanan, K.; Sandvik, J.A.; Pettersen, E.O.; Urban, G.A. Pericellular oxygen monitoring with integrated sensor chips for reproducible cell culture experiments. *Cell Prolif.* **2014**, *47*, 180–188. [CrossRef] [PubMed]
17. Steinegger, A.; Wolfbeis, O.S.; Borisov, S.M. Optical Sensing and Imaging of pH Values: Spectroscopies, Materials, and Applications. *Chem. Rev.* **2020**, *120*, 12357–12489. [CrossRef]
18. Wang, X.D.; Wolfbeis, O.S. Optical methods for sensing and imaging oxygen: Materials, spectroscopies and applications. *Chem. Soc. Rev.* **2014**, *43*, 3666–3761. [CrossRef]
19. Nonnenmacher, Y.; Palorini, R.; Hiller, K. Determining Compartment-Specific Metabolic Fluxes. *Methods Mol. Biol.* **2019**, *1862*, 137–149. [CrossRef]
20. Zhang, J.; Zhang, Q. Using Seahorse Machine to Measure OCR and ECAR in Cancer Cells. *Methods Mol. Biol.* **2019**, *1928*, 353–363. [CrossRef]
21. Hynes, J.; Swiss, R.L.; Will, Y. High-Throughput Analysis of Mitochondrial Oxygen Consumption. *Methods Mol. Biol.* **2018**, *1782*, 71–87. [CrossRef] [PubMed]
22. Naciri, M.; Kuystermans, D.; Al-Rubeai, M. Monitoring pH and dissolved oxygen in mammalian cell culture using optical sensors. *Cytotechnology* **2008**, *57*, 245–250. [CrossRef] [PubMed]
23. Arguello, R.J.; Combes, A.J.; Char, R.; Gigan, J.P.; Baaziz, A.I.; Bousiquot, E.; Camosseto, V.; Samad, B.; Tsui, J.; Yan, P.; et al. SCENITH: A Flow Cytometry-Based Method to Functionally Profile Energy Metabolism with Single-Cell Resolution. *Cell Metab.* **2020**, *32*, 1063–1075.e1067. [CrossRef] [PubMed]
24. Tanumihardja, E.; Slaats, R.H.; van der Meer, A.D.; Passier, R.; Olthuis, W.; van den Berg, A. Measuring Both pH and O2 with a Single On-Chip Sensor in Cultures of Human Pluripotent Stem Cell-Derived Cardiomyocytes to Track Induced Changes in Cellular Metabolism. *ACS Sens.* **2021**, *6*, 267–274. [CrossRef] [PubMed]
25. Diepart, C.; Verrax, J.; Calderon, P.B.; Feron, O.; Jordan, B.F.; Gallez, B. Comparison of methods for measuring oxygen consumption in tumor cells in vitro. *Anal. Biochem.* **2010**, *396*, 250–256. [CrossRef] [PubMed]
26. Ge, X.; Hanson, M.; Shen, H.; Kostov, Y.; Brorson, K.A.; Frey, D.D.; Moreira, A.R.; Rao, G. Validation of an optical sensor-based high-throughput bioreactor system for mammalian cell culture. *J. Biotechnol.* **2006**, *122*, 293–306. [CrossRef]
27. Alim Uysal, B.A.; Kotan, G.; Guneser, M.B.; Dincer, A.N.; Senturk, H.; Rafiqi, A.M. Investigation of the effect of different chelation solutions on penetration of resin-based and bioceramic sealers with a novel method. *Microsc. Res. Tech.* **2021**. [CrossRef]
28. Michl, J.; Park, K.C.; Swietach, P. Evidence-based guidelines for controlling pH in mammalian live-cell culture systems. *Commun. Biol.* **2019**, *2*, 144. [CrossRef]
29. Yao, J.; Czaplinska, D.; Ialchina, R.; Schnipper, J.; Liu, B.; Sandelin, A.; Pedersen, S.F. Cancer Cell Acid Adaptation Gene Expression Response Is Correlated to Tumor-Specific Tissue Expression Profiles and Patient Survival. *Cancers* **2020**, *12*, 2183. [CrossRef]
30. Forciniti, S.; Dalla Pozza, E.; Greco, M.R.; Amaral Carvalho, T.M.; Rolando, B.; Ambrosini, G.; Carmona-Carmona, C.A.; Pacchiana, R.; Di Molfetta, D.; Donadelli, M.; et al. Extracellular Matrix Composition Modulates the Responsiveness of Differentiated and Stem Pancreatic Cancer Cells to Lipophilic Derivate of Gemcitabine. *Int. J. Mol. Sci.* **2020**, *22*, 29. [CrossRef]
31. Ford, K.L.; Moorhouse, E.L.; Bortolozzi, M.; Richards, M.A.; Swietach, P.; Vaughan-Jones, R.D. Regional acidosis locally inhibits but remotely stimulates Ca2+ waves in ventricular myocytes. *Cardiovasc. Res.* **2017**, *113*, 984–995. [CrossRef]
32. Chung, Y.J.; Luo, A.; Park, K.C.; Loonat, A.A.; Lakhal-Littleton, S.; Robbins, P.A.; Swietach, P. Iron-deficiency anemia reduces cardiac contraction by downregulating RyR2 channels and suppressing SERCA pump activity. *JCI Insight* **2019**, *4*, e125618. [CrossRef]
33. Hulikova, A.; Harris, A.L.; Vaughan-Jones, R.D.; Swietach, P. Regulation of intracellular pH in cancer cell lines under normoxia and hypoxia. *J. Cell Physiol.* **2013**, *228*, 743–752. [CrossRef]
34. Jo, J.; Lee, C.H.; Kopelman, R.; Wang, X. In vivo quantitative imaging of tumor pH by nanosonophore assisted multispectral photoacoustic imaging. *Nat. Commun.* **2017**, *8*, 471. [CrossRef]
35. Contreras-Baeza, Y.; Sandoval, P.Y.; Alarcon, R.; Galaz, A.; Cortes-Molina, F.; Alegria, K.; Baeza-Lehnert, F.; Arce-Molina, R.; Guequen, A.; Flores, C.A.; et al. Monocarboxylate transporter 4 (MCT4) is a high affinity transporter capable of exporting lactate in high-lactate microenvironments. *J. Biol. Chem.* **2019**, *294*, 20135–20147. [CrossRef] [PubMed]
36. Bachmeier, C.J.; Trickler, W.J.; Miller, D.W. Drug efflux transport properties of 2′,7′-bis(2-carboxyethyl)-5(6)-carboxyfluorescein acetoxymethyl ester (BCECF-AM) and its fluorescent free acid, BCECF. *J. Pharm. Sci.* **2004**, *93*, 932–942. [CrossRef] [PubMed]
37. Lucien, F.; Harper, K.; Pelletier, P.P.; Volkov, L.; Dubois, C.M. Simultaneous pH measurement in endocytic and cytosolic compartments in living cells using confocal microscopy. *J. Vis. Exp.* **2014**, *86*, 51395. [CrossRef] [PubMed]
38. Hinton, A.; Sennoune, S.R.; Bond, S.; Fang, M.; Reuveni, M.; Sahagian, G.G.; Jay, D.; Martinez-Zaguilan, R.; Forgac, M. Function of a subunit isoforms of the V-ATPase in pH homeostasis and in vitro invasion of MDA-MB231 human breast cancer cells. *J. Biol. Chem.* **2009**, *284*, 16400–16408. [CrossRef]
39. Ray, A.; Koo Lee, Y.E.; Epstein, T.; Kim, G.; Kopelman, R. Two-photon nano-PEBBLE sensors: Subcellular pH measurements. *Analyst* **2011**, *136*, 3616–3622. [CrossRef]
40. Willoughby, D.; Schwiening, C.J. Electrically evoked dendritic pH transients in rat cerebellar Purkinje cells. *J. Physiol.* **2002**, *544*, 487–499. [CrossRef] [PubMed]
41. Lima, S.; Bonifacio, R.L.; Azzellini, G.C.; Coichev, N. Ruthenium(II) tris(bipyridyl) ion as a luminescent probe for oxygen uptake on the catalyzed oxidation of HSO3-. *Talanta* **2002**, *56*, 547–556. [CrossRef]

42. Zhou, C.; Zhao, W.X.; You, F.T.; Geng, Z.X.; Peng, H.S. Highly Stable and Luminescent Oxygen Nanosensor Based on Ruthenium-Containing Metallopolymer for Real-Time Imaging of Intracellular Oxygenation. *ACS Sens.* **2019**, *4*, 984–991. [CrossRef] [PubMed]
43. Jiang, Z.; Yu, X.; Zhai, S.; Hao, Y. Ratiometric Dissolved Oxygen Sensors Based on Ruthenium Complex Doped with Silver Nanoparticles. *Sensors* **2017**, *17*, 548. [CrossRef] [PubMed]
44. Scholler-Mann, A.; Matt, K.; Hochecker, B.; Bergemann, J. Ex vivo Assessment of Mitochondrial Function in Human Peripheral Blood Mononuclear Cells Using XF Analyzer. *Bio. Protoc.* **2021**, *11*, e3980. [CrossRef]

chemosensors

MDPI

Review

Recent Advances in Perylene Diimide-Based Active Materials in Electrical Mode Gas Sensing

Salman Ali [1], Akhil Gupta [1], Mahnaz Shafiei [2,*] and Steven J. Langford [1,*]

[1] Department of Chemistry and Biotechnology, Faculty of Science, Engineering and Technology, Swinburne University of Technology, Hawthorn, VIC 3122, Australia; salmanali@swin.edu.au (S.A.); akhilgupta@swin.edu.au (A.G.)

[2] School of Software and Electrical Engineering, Faculty of Science, Engineering and Technology, Swinburne University of Technology, Hawthorn, VIC 3122, Australia

* Correspondence: mshafiei@swin.edu.au (M.S.); sjlangford@swin.edu.au (S.J.L.)

Abstract: This review provides an update on advances in the area of electrical mode sensors using organic small molecule *n*-type semiconductors based on perylene. Among small organic molecules, perylene diimides (PDIs) are an important class of materials due to their outstanding thermal, chemical, electronic, and optical properties, all of which make them promising candidates for a wide range of organic electronic devices including sensors, organic solar cells, organic field-effect transistors, and organic light-emitting diodes. This is mainly due to their electron-withdrawing nature and significant charge transfer properties. Perylene-based sensors of this type show high sensing performance towards various analytes, particularly reducing gases like ammonia and hydrazine, but there are several issues that need to be addressed including the selectivity towards a specific gas, the effect of relative humidity, and operating temperature. In this review, we focus on the strategies and design principles applied to the gas-sensing performance of PDI-based devices, including resistive sensors, amperometric sensors, and operating at room temperature. The device properties and sensing mechanisms for different analytes, focusing on hydrazine and ammonia, are studied in detail, and some future research perspectives are discussed for this promising field. We hope the discussed results and examples inspire new forms of molecular engineering and begin to open opportunities for other rylene diimide classes to be applied as active materials.

Keywords: *n*-type organic molecules; perylene diimide; hydrazine; ammonia; sensors

Citation: Ali, S.; Gupta, A.; Shafiei, M.; Langford, S.J. Recent Advances in Perylene Diimide-Based Active Materials in Electrical Mode Gas Sensing. *Chemosensors* **2021**, 9, 30. https://doi.org/10.3390/chemosensors9020030

Academic Editor: Pietro Salvo

Received: 19 November 2020

Accepted: 26 January 2021

Published: 1 February 2021

1. Introduction

The increase in globalization, digitization, and the rapid production techniques of industry come with the need to be able to measure our environment, either local or more global, to ensure safe working and living environments for our communities to prosper. In particular, the monitoring and detection of toxic and harmful gases, air pollutants, and explosives are an essential need in industries such as pharmaceuticals, propulsion systems, fertilizers, renewable energy, and manufacturing, as well as in everyday life including the need to safely travel by air. Sensors are the primary devices used for the detection and monitoring of the toxic, explosive, and flammable analytes present in our surroundings, and a great deal of attention has been paid to the development of various types of gas sensors, including electrical, optical, acoustic, and fluorescent [1–9]. Major analytes include hydrazine (N_2H_4), ammonia (NH_3), hydrogen (H_2), hydrogen peroxide (H_2O_2), nitrogen dioxide (NO_2), methane (CH_4), and volatile organic compounds (VOCs), all of which cause adverse effects on human and environmental health [10–12]. The commercially available sensors are primarily based on inorganic materials [13], and, although market leading, they suffer from a number of disadvantages, including complex fabrication techniques, a lack of selectivity, the need for a high working temperature, and above all, high-costs [14,15]. Therefore, it is still important to develop new and suitable alternatives to fabricate gas sensors that are reliable, low-cost, stable at room temperature, and have a high degree of selectivity towards the desired analyte.

Over the past few years, organic semiconductors have gained some attention in gas-sensing applications, primarily due to their unique properties, including solution processability, low-cost fabrication, room temperature (RT) operation, ease of integration, and scalable and straightforward synthetic strategies [16]. Along with the development of organic field-effect transistors (OFETs), the design and synthesis of novel organic materials has provided a vast material library for electrical gas-sensing applications, e.g., pigments and dyes such as phthalocyanines [17–20], perylene diimide derivatives [21–24], and conjugated polymers [25–31]. There are two main types of organic semiconductors: the so-called *p*- and *n*-types. The *p*-type organic semiconductors are more widely applied in gas-sensing applications [32] compared to their *n*-type counterparts. This is primarily because, when *p*-type organic materials are exposed to oxidizing gases such as NO_2, an increase in conductivity occurs, whereas the reverse happens for reducing gases, such as NH_3 and H_2; thus, a positive response result is attained over a reduction in response, which could be interpreted in other ways, such as device failure or not providing a measurable linear response past a certain level. A variety of *p*-type organic materials, including pentacene, phthalocyanines, and thiophene, have become of interest, and they have all been used for gas-sensing applications [32]. The *n*-type materials continue to have limited utility for gas-sensing applications, mainly due to their instability in the air [32]. However, there are several *n*-type organic materials that have been designed and developed for gas-sensing applications [33] by incorporating a variety of electron-withdrawing groups, e.g., cyano (CN), fluoro (F), and chloro (Cl) within their structure. Due to the inherent electron-deficient nature of *n*-type materials, an increase in conductivity occurs when they are exposed to reducing gases, and the opposite happens for oxidizing gases [22,23,34,35]. Hence, the development of a library of organic semiconductors that demonstrate specific properties for specific analytes and gases, with specific responses to reducing or oxidizing gases, could be an important advance to the field. In this review, we elaborate on the chemistry of perylene diimides as an example of the *n*-type class that led to its use in sensor development.

Perylene diimide (PDI), as shown in Figure 1, is an *n*-type organic semiconductor discovered by Kardos in 1913 [36]. Initially used as a textile dye, this high-performing pigment comes in red, violet, and near-black shades [37]. PDIs possess a high thermal stability and photostability under ambient conditions [38], making them viable for self-assembly processing and applications in various electronic devices [39]. Several PDI molecules have been prepared into well-shaped supramolecular nanostructures; these achievements have also encouraged the area of the self-assembly of many other *n*-type organic semiconductor molecules [40]. The existence of electron-withdrawing imide fragments in PDIs permits them to be simply reduced chemically or electrochemically, forming radical anions, yet remaining stable to oxidation [41]. Their resistance to oxidation is the main reason why PDIs generally act as stable *n*-type organic materials [42]. Together with their additional properties, including facile structural functionalization, excellent light-absorption in the visible region, and near-unity fluorescence quantum yields in the molecular state, PDIs have been known as appealing candidates for preparing organic optoelectronic materials and devices, such as organic photovoltaics (OPVs), organic field effect transistors (OFETs), dye lasers, organic light-emitting diodes (OLEDs), and sensors [22,42–44]. The well-shaped 1D nanostructures of PDIs can be prepared via supramolecular self-assembly procedures [44]. This 1D self-assembly is mostly governed by the intrinsic π–π stacking interaction between the aromatic perylene planes, which support the 1D growth of assembly in cooperation with other intermolecular noncovalent interactions. The 1D nanostructures of PDIs display excellent optical and electronic properties along the 1D π–π stacking direction, which offers great potential for electronic device application like OPVs and sensors [44,45].

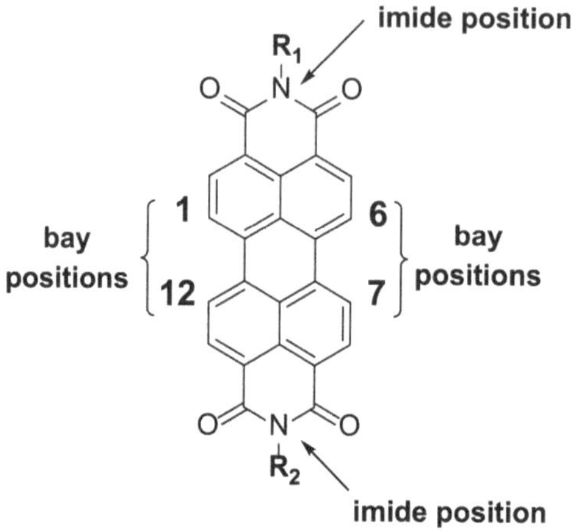

Figure 1. Structure of perylene diimide (PDI) core.

PDI and its derivatives have been studied as active materials in numerous organic electronic devices such as OPVs, OLEDs, and OFETs [36,46,47]. As is the focus of this review, PDI derivatives have also been widely used in gas-sensing applications, mainly due to their electron-deficient nature, desired self-assembled and crystal structures, and high chemical and thermal stabilities [36]. To improve the sensing performance of PDI-based sensors, research has focused on the molecular modification of PDI [21,22,48–50]. The modifications highlighted in this review within the PDI structure can be achieved through either substitution on nitrogen atoms or via incorporating aryl and/or alkyl substituents on the core and, in particular, the bay positions [41,51,52]. PDIs generated through imide nitrogen substitution usually possess similar optoelectronic properties because the electron density nodes in the lowest unoccupied molecular orbital (LUMO) and the highest occupied molecular orbital (HOMO) levels at imide positions reduce the coupling between the PDI unit and the imide substituents to a minimum [41]. In this respect, this is an excellent way of modification for processability or the introduction of connectors. PDIs obtained through substitution at the core are generally highly conjugated targets, and, depending upon the substitution, the optical band gap, crystal structure, absorption profile, and solubility of the target can be altered [53,54]. Moreover, such substitutions may help to minimize the π–π overlap between aggregating naphthalene subunits to improve the sensing performance of the resulting devices [43,55]. The structure modification through imide or core positions can affect the morphology and self-assembly of the PDIs, which can improve the sensing performance [22,56].

This review presents a summary of the reports on PDI-based electrical mode gas sensors that have been developed over the past decade. These sensors include resistive and amperometric devices that have been tested and evaluated towards reductive gases like hydrazine and ammonia at RT. We begin by describing PDI incorporation into electrical mode gas sensors, their configuration, and their sensing mechanisms. We then critically review reports on sensors incorporating PDIs for hydrazine, ammonia, and some other analytes. Finally, we describe the future outlook of PDIs and other rylene diimides as appropriate active components for electrical gas sensors.

2. Electrical Gas Sensors and Sensing Mechanism

Electrical gas sensors have been extensively studied, reported, and utilized, mainly because of their ease of processing and fabrication, portability, low-cost, and compatibility with various standard electronics [57]. An electrical gas sensor usually consists of two major components: the active material and the transducer. Generally, the mechanism of electrical gas sensors using organic materials is based on gas adsorption/desorption at the active material surface. A charge-transfer complex between the active layer and an analyte is formed, thus leading to a change in the charge carrier mobility of the sensing material and. hence, in the electrical response through conductivity, resistivity, and current [28,58,59]. This mechanism of adsorption may contain either chemical or weak interactions, i.e., chemisorption or physisorption, depending upon the chemical properties of both the sensing materials and the analyte [60]. To bring the sensor to its initial state, the target gas is removed from the testing chamber and air or nitrogen as a reference gas is introduced. The sensitivity or response of active materials mainly depends on two factors: (1) a high surface area, which can provide effective diffusion of the gas molecules, and (2) the chemical nature of the active materials, such as molecular packing, redox potential, and energy level, which are associated with the formation of charge carriers and efficient transport into the active materials [58]. As an *n*-type organic semiconductor, PDI can accept electrons from the electron donor gas, such as ammonia and hydrazine, via donor–acceptor complexation. Thus, the electrons can be efficiently transported through π-electron delocalization along the long-axis of perylene dyes and ultimately lead to a measurable increase in resistance or current [21,22,58].

Applying standard electronic devices such as resistors, capacitors, and field-effect transistors (FETs) allows the variations in the sensors active layer's physical properties to be recorded in the form of change in resistance or current (ΔR or ΔI), capacitance (ΔC), and voltage (ΔV), respectively, [61–63] as shown in Figure 2. Electrical gas sensors that record these gas–solid interactions mainly take the three device configurations of FET, capacitor, and chemiresistor (amperometric and resistive) [64–66].

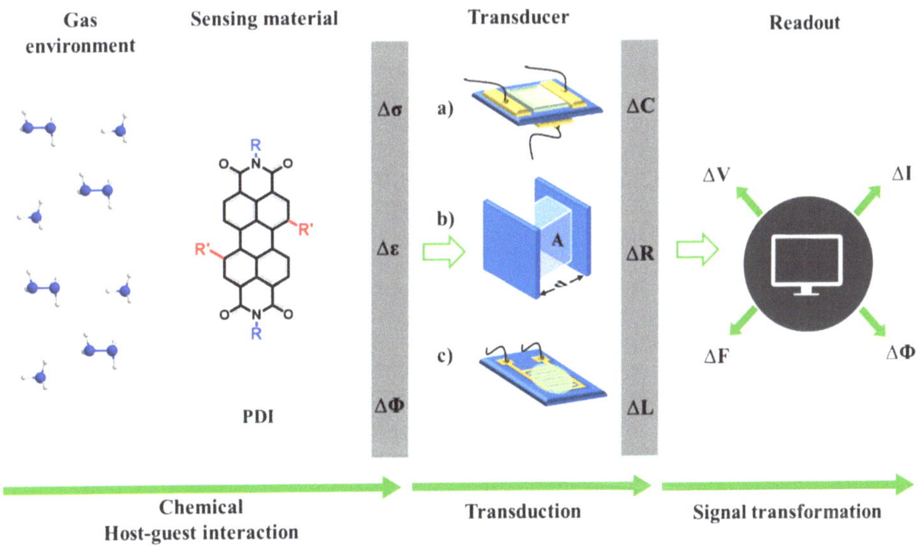

Figure 2. Schematic diagram illustrating the major components required to evaluate an electrical mode gas-sensing mechanism. The transducers are (**a**) FETs, (**b**) capacitors and (**c**) chemiresistors. Analytes interact with the sensing material changes some of its physical properties such as conductivity (s), work function (j), and permittivity (e). The transducer converts one of these physical quantities into the variation of its electric parameters such as capacitance, *C*, and resistance, *R*. Finally, the circuit to which the sensor is connected gives rise to the sensing electrical signal that can be in either current (*I*) or voltage (*V*), and each can be measured in frequency (*F*) and phase (Φ) [67].

In 1975, Lundstrom developed the core components of an FET-based gas sensor for the first time [68]. Generally, an FET contains two electrodes (the source and the drain), connected by a sensing layer as the channel, and a gate electrode usually located on the underside of the substrate (Figure 2a). When a source-drain voltage is applied, current flows through the channel, which is matched through the charge carriers by applying a gate voltage. This provides supplementary ways to control the current response in the sensing layer upon interacting with a target analyte. Outputs other than variations in channel current, such as the threshold voltage and sub-threshold swing, can also be used to reflect the sensing process [69]. Despite their complex structure and fabrication, these advantages make FETs more stable and reliable for gas sensing compared to chemiresistors. PDIs and other organic materials are being considered for FETs where flexibility, low processing temperatures, and print deposition are desired or additional chemical functionality to create the chemical selectivity needed for functioning gas sensors is required. As PDIs and other organic molecules have inferior mobilities than inorganic materials, they need higher drain and gate voltages to produce viable currents.

Capacitive-type sensors are mainly used for molecular detection including large entities such as DNA [70], biomolecules [71], smaller molecules related to humidity [72], and general gases [73]. In most cases, a capacitive-type sensor comprises a layer of active material sandwiched between two parallel electrodes (Figure 2b). The capacitance of the sensor is usually expressed as $C = \varepsilon_0\varepsilon_r A/d$, where ε_0 is the permittivity in vacuum, ε_r is the relative permittivity of the active material, A is the capacitor area, and d is the gap between the electrodes [74]. Gas or humidity adsorption on the active material alters ε_r and, as a result, changes the sensor capacitance. For the case of polymer-based active materials, gas-induced swelling and the subsequent alteration of d or A between the electrodes can also cause a change in the capacitance [75]. The capacitor set-up also permits the measurement of the change in material impedance, which is associated with both the resistive and capacitive variations in response to target analyte adsorption, typically in the range of zero to several megahertz [76]. PDIs and other organic molecules have been used in capacitive-type sensors for humidity and gas-sensing applications. PDI-based humidity sensors have only shown good sensitivity in high-humidity environments and are unable to show notable responses below 60% RH [72].

To date, the majority of PDI-based sensors have been developed as chemiresistive, i.e., resistive or amperometric, configurations, as shown in Figure 2c. A chemiresistor usually comprises interdigitated electrodes (IDEs) linked with a sensing material deposited onto an insulating substrate [77]. The resistance, i.e., current, of the sensing material changes upon exposure to the target analytes. Chemiresistive sensors comprise one of the most extensively used device structures for gas-sensing applications due to their simplicity, compatibility with the conventional direct current (DC) circuits, low-cost, predictable electrical properties, ease of high accuracy measurements [77], and suitability when speedy prototyping is essential. The simple design and operation of this type of device renders it appealing starting platforms for studying the chemical response of novel materials. Despite their simple setup, chemiresistors are restricted by their single type of output, which is easily influenced by environmental perturbations.

To analyze the general performance of a gas sensor against others, a number of parameters extracted from static and dynamic measurements are needed [78,79]. Critical parameters include sensitivity, selectivity, stability, hysteresis, response and recovery time, the limit of detection (LOD), and working temperature. Ideally, a gas sensor should possess a high sensitivity to the target gas with a linear response, quick response and recovery, high selectivity, low hysteresis, low LOD, and long-term stability. One of the disadvantages of organic materials is their perceived long-term instability when compared to inorganic materials.

Table 1. Comparison of sensing performance of different derivatives of PDIs at room temperature (RT). Compare the descriptors used in this table to Figure 10 for structural clarification.

Material Description (Structure #)	N-Imide Structural Variations	Core Structural Variations	Analyte	Electrical Mode of Sensor	Response/Recovery Time	LOD (ppm)	Response Calculation	Response	Ref
PDI-nanotubes **PDI-nanorods** **(1)**	Tertiary amine–iodide complex	NA	Hydrazine	Resistive	180 s/[a] 102 s/[a]	10	R/R_o	100 1000 at 10 ppm	[48]
PDI-single nanoribbon **PDI-nanofiber bundle** **(2)**	Sugar-based achiral alkyl chains	NA	Hydrazine	Resistive	300 s/[a] 400 s/[a]	10	R/R_o	10^4 100 at 10 ppm	[56]
PDI-OBAG **PDI-DBAG** **PDI-HBAG** **(3)**	Sugar-based achiral and n-alkyl chains	NA	Hydrazine	Amperometric	350 s/[a] 350 s/[a] 350 s/[a]	10	I/I_o	2.5×10^4 5.2×10^4 4×10^4 at 10 ppm	[80]
PDI-C$_{12}$ **PDI-ClC$_{12}$** **PDI-ClC$_4$F$_7$** **(4)**	Dodecyl and perfluorobutyl groups	H and Cl	Hydrazine	Resistive	160 s/[a] 600 s/[a] 600 s/[a]	8	R/R_o	100 10^4 10^4 at 10 ppm	[22]
PDI-d **(5)**	*n*-Ethylhexyl	Substituted acetylenes	Hydrazine	Resistive	4 s/4s	0.7	$\Delta R/R_o \times 100$	46.67% at 10 ppm	[81]
PDI-Bp$_2$C$_{10}$ **PDI-C$_{10}$** **PDI-Br$_2$C$_{10}$** **(6)**	*n*-Alkyl substituents with chiral center	H, Br, and aryloxy substituents	Hydrazine	Amperometric	450 s/150 s 300 s/350 s 700 s/200 s	0.75 0.75 1	I/I_o	10^4 10^3 10^2 at 10 ppm	[82]
PDI-DEY **PDI-DSPY** **PDI-DTPY** **PDI-DFPY** **(7)**	n-Alkyl substituents	Substituted acetylenes	Hydrazine	Amperometric	800 s/[a] 800 s/[a] 750 s/[a] 800 s/[a]	10	I/I_o	10^5 $\sim 10^4$ $\sim 10^4$ 10^4 at 10 ppm	[58]
PDI-ClC$_{10}$ **PDI-CNC$_{10}$** **(8)**	*n*-Alkyl substituents with chiral center	CN and Cl	Hydrazine	Amperometric	350 s/350 s 350 s/350 s	1 2	I/I_o	10^4 $\sim 10^3$ at 10 ppm	[34]
PDI-pyrimidine **(9)**	Fluorinated alkyl substituents	Phenyl and pyrimidine	Hydrazine	Resistive	23.2 s/35.4 s	0.1	$\Delta R/R_o \times 100$	90.72% at 100 ppm	[50]

Table 1. *Cont.*

Material Description (Structure #)	N-Imide Structural Variations	Core Structural Variations	Analyte	Electrical Mode of Sensor	Response/Recovery Time	LOD (ppm)	Response Calculation	Response	Ref
PDI-PH **PDI-TPH** **PDI-OPH** **(10)**	Cyclohexane	Aryloxy substituents	Hydrazine	Amperometric	116 s/19 s 163 s/52 s 170 s/58 s	0.6 0.5 0.5	I/I_o	130 520 1800 at 5 ppm	[83]
PDI-OT **PDI-ROT** **PDI-SOT** **(11)**	*n*-Alkyl chains	Substituted acetylenes	Hydrazine	Amperometric	400 s/[a] 350 s/[a] 360 s/[a]	10	I/I_o	1.2×10^4 2.25×10^4 1.7×10^4 at 10 ppm	[84]
PDI-PY **PDI-PI** **PDI-HE** **(12)**	*n*-Alkyl chains	N-atom containing alkyl and cycloalkyls	Hydrazine	Amperometric	40 s/10 s 50 s/10 s 7 s/7 s	0.46 0.24 0.17	I/I_o	2.5×10^4 1.3×10^5 1.5×10^5 at 10 ppm	[85]
PDI-DMB **PDI-TMSA** **PDI-TESA** **PDI-TPSA** **(13)**	*n*-Alkyl chains	C- and Si-based acetylenes	Hydrazine	Amperometric	200 s/[a] 200 s/[a] 200 s/[a] 200 s/[a]	10	I/I_o	2×10^4 2.1×10^4 1.7×10^3 3.9×10^4 at 10 ppm	[35]
PDI-C_{12} **PDI-BR_2C_{12}** **PDI-CN_2C_{12}** **(14)**	*n*-Alkyl chains	H, Br, and CN	Ammonia	Resistive	50 s/[a] 100 s/[a] 600 s/[a]	1%	$\Delta R/R_o \times 100$	40% 0.075% 0.75 at 1% NH_3,	[21]
PDI-SOT **PDI-STB** **PDI-SPP** **(15)**	*n*-Alkyl chains	Alkyl- and phenyl-based acetylenes	Ammonia	Amperometric	24 s/12 s 24 s/10 s 17 s/6 s	1.51 0.95 0.48	$\Delta I \times 100$	0.9% 1.8% 4.5% at 100 ppm	[23]

Table 1. *Cont.*

Material Description (Structure #)	N-Imide Structural Variations	Core Structural Variations	Analyte	Electrical Mode of Sensor	Response/Recovery Time	LOD (ppm)	Response Calculation	Response	Ref
PDI PDI-N PDI-2N (16)	*n*-Alkyl substituents with chiral center	H and NO_2	Ammonia	Amperometric	25 s/2 s 26 s/2 s 36 s/2 s	5 3 2	$\Delta I \times 100$	1% 6.6% 5.6% at 5 ppm	[86]
PDI PDI-S PDI-Se (17)	*n*-Alkyl substituents with chiral center	Bridged S and Se	Ammonia	Amperometric	80 s/[a] 80 s/[a] 80 s/[a]	100	I/I_o	250 320 100 at 100 ppm	[87]
PDI-HIS (18)	*n*-Alkyl and histidine substituents	NA	Ammonia	Amperometric	5 s/20 s	0.56	$\Delta I/I_o \times 100$	3500 at 100 ppm	[24]
PDI-C_4F_7/LuPC_2 (19)	Fluorinated alkyl substituents	NA	Ammonia	Amperometric	[a]	10	$\Delta I/I_o$	33% at 100 ppm	[88]
PDI-MA (20)	Phenyl substituents containing amine	NA	Nitromethane	Amperometric	10 s/75 s	36,000 ppm	$\Delta I/I_o$ (%)	65% at 36,000 ppm	[89]
PDI-MP (21)	*N*-atom containing cycloalkyls	NA	Hydrogen peroxide	Resistive	19.5 s/[a]	0.6	$\Delta I/I_o$ (%)	50% at 20 ppm	[90]
PDI-PR PDI-PS (22)	Sulphonic acid complex of tertiary amine	NA	Ethylenediamine	Amperometric	9 s/7 s 11 s/14 s	1.07 0.86	$\Delta I \times 100$	1.5% 17% at 100 ppm	[91]
PDI-TC (23)	*n*-Alkyl and -OH substituted alkyl chains	Cl	Trimethyl-amine	Amperometric	325 s/510 s	0.2	$(I - I_o)/I_o \times 100$	18% at 10 ppm	[92]
PDI-PB (24)	n-Ethylhexyl	Br and Aza-cyclobutane	Pyridine	Resistive	1 s/[a]	375 pbb	$\Delta R/R_o \times 100$	2200% at 375 ppm	[93]

[a] Response or recovery times are not reported in the reference. Response calculation techniques are given with slight modification as reported, where R_o = initial resistance; R = resistance after analyte exposure; ΔR = change in resistance; I_o = initial current; I = current after analyte exposure; and ΔI = change in current.

Sensitivity is the minimum fractional variation in the output signal ($\Delta I/I_0$ and $\Delta R/R_0$) of a sensor upon exposure to an analyte, where I_0 and R_0 are the current and resistance, before exposure to the target gas respectively, and ΔI and ΔR are the change in the current and resistance, respectively, when the sensor is exposed to a certain quantity of target analyte. Selectivity is one of the most significant sensing parameters because there are a variety of interfering gases that are detrimental to accurate target analyte detection. Selectivity is the capability of a sensor to recognize a specific target gas among a mixture of gases. Selectivity is generally examined by comparing the cross-sensitivity towards several analytes at a particular concentration. Stability is another critical factor that is defined as the capability of a sensor to repeat the same results for a target gas for a fixed period of time. Hysteresis is a variance of the sensor output at a specified point of the input signal when it is approached from the opposite direction. Low levels of hysteresis are desirable. Response and recovery times are measured to quantify the sensing speed of a sensor. Generally, the response time is defined as the time taken for the current/resistance to change from the baseline current/resistance (I_0/R_0) to 90% of the maximum change in current/resistance at a specific level of the target analyte. On the other hand, the recovery time is defined as the time taken by a sensor to reach the baseline state after completely removing the target analyte. The LOD is the lowest analyte concentration that a sensor can sense or detect reliably. The LOD is generally defined based on signal to noise ratio (S/N) and is typically determined by an S/N of 3:1. The working temperature is the temperature at which a sensor can work effectively with high sensitivity and stability. Table 1 aims to summarize many of the reported PDI based gas sensors of merit and their overall performance.

3. Hydrazine Sensing

Hydrazine is a hygroscopic gas that is widely used in pharmaceutical industries, rocket propulsion systems, and photographic chemicals [94]. Though hydrazine is an important industrial material that is used as a hydrogen source, it is also generally known as a hepatotoxic neurotoxin, a carcinogen, and very harmful to human health [95]. As a result and through oral or dermal contact, consumption, or inhalation, hydrazine primarily affects the central nervous system, liver, lungs, and kidneys of living organisms. Additionally, it can seriously affect the reproductive system of living organisms. The minimum permissible limit of hydrazine in the environment is ten parts per billion (ppb), recommended by the United States Environmental Protection Agency (USEPA) [96]. Therefore, the realization of reliable gas sensors for hydrazine is essential for those industries using commercial quantities. Most PDI-based hydrazine sensors have been developed in resistive and amperometric configurations, and their sensing parameters are summarized in Table 1.

In 2009, Huang et al. [48] prepared self-assembled one-dimensional (1D) nanotubes and nanorods containing PDI-I (structure 1; Figure 10) by evaporating an aqueous solution of PDI-I and employed them in resistive sensors. Different morphologies, including nanotubes with a diameter of 100–300 nm and nanorods with a diameter of 200–300 nm (Figure 3a,b), were achieved using a variety of solvents and preparations. The self-assembled nanotubes and nanorods showed high sensitivities (100 and 1000 (R/R_0), respectively) towards 10 ppm of hydrazine and phenylhydrazine vapor with moderate responses (50 and 100, respectively) to 10 ppm of triethylamine vapors. The resistance of PDI-I nanotubes and nanorods was recorded using a four-probe technique. A decrease in the resistance of the nanotubes and nanorods was noted upon exposure to hydrazine vapors (10 ppm), which was due to the presence of long-range π-electron delocalization via the surface doping of hydrazine. The change in the resistance of the nanorods was modest compared to the nanotubes (Figure 3c), though both demonstrated good response times (180 and 102 s, respectively) towards hydrazine.

The same research group also synthesized symmetrical sugar-based PDI derivative PDI-BAG (structure 2; Figure 10) and studied the effect of various solvents on the morphology of PDI-BAG [56]. Their study revealed that PDI-BAG has a good solubility in *N,N*-dimethylformamide (DMF) and that self-assembled nanoribbons and nanofibers of

PDI-BAG were formed in different ratios of water and DMF (Figure 4a,b). Sensors based on the nanoribbon structure displayed a better sensing performance towards hydrazine vapor when compared with nanofiber bundles. The nanoribbons showed a higher sensitivity (10^4 R_o/R_s) than the nanofiber bundles (10^2 R_o/R_s), as shown in Figure 4c, due to better organization of the PDI-BAG molecules. The better sensitivity of the nanoribbons towards hydrazine appeared to be at the expense of a slower response (400 s). Furthermore, the prepared chiral nanoribbons showed higher sensitivity than the achiral nanostructures, which was due to high crystallinity and single handed chirality [48]. Further studies on a series of sugar-based PDIs—OBAG, DBAG, and HBAG (structure 3; Figure 10)—bearing various achiral alkyl chains to modulate hydrophilic and hydrophobic interactions to adjust the supramolecular helicity by varying alkyl chain length as a way to explore the design principles to optimal results was also reported [80]. The helical nanostructures of all the three molecules were obtained from tetrahydrofuran (THF)/H_2O. The responses of OBAG-, DBAG-, and HBAG-based sensors were recorded as a function of change in current towards hydrazine vapor. The DBAG-based sensor displayed the highest sensitivity (5.2×10^4 I/I_o) when compared with OBAG and HBAG (2.5×10^4 and 4×10^4 I/I_o, respectively) (Figure 4d), mainly due to the smaller rotation angle (28° or −30°) between adjacent PDI molecules in comparison to those of OBAG (32° or −32°) and HBAG (30° or −31°). The smaller rotation angle for DBAG indicates that a larger π–π overlap of the adjacent perylene core is expected to increase mobility and lead to maximum current (five orders of magnitude different). All the sensors displayed a comparatively high sensitivity (10^4 I/I_o) and quick response (350 s) than the previously reported sugar-based PDI derivatives [56], though there was no investigation regarding selectivity, recovery, and reproducibility to allow for a further comparison.

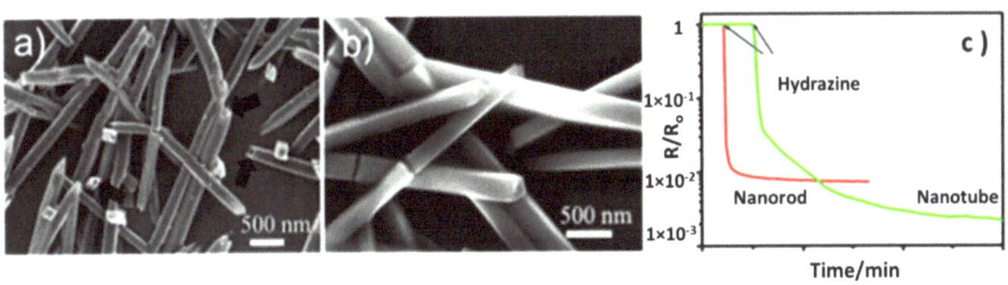

Figure 3. SEM images of (**a**) nanotubes and (**b**) nanorods of PDI-I. (**c**) Resistance modulation (R/R_o) vs. time (t) curve measured on a single nanotube and nanorod hydrazine vapor. Adapted with permission [48].

The effect of different electron-donor and electron-withdrawing functional groups on the sensing performance of PDIs was also ben investigated [22]. Two core-substituted, tetrachlorinated PDI derivatives, PDI-ClC_{12} and PDI-ClC_4F_7, bearing electron-donor dodecyl and electron-withdrawing perfluorobutyl groups at the imide positions, as well as one core-unsubstituted PDI-C_{12} (structure 4; Figure 10), were synthesized for hydrazine vapor sensing. Interestingly for the design and development of suitable sensors using PDIs, the resistance of both the core-substituted PDIs sensors decreased sharply (10^4 R/R_o) upon exposure to hydrazine vapor (10 ppm) when compared to the core-unsubstituted PDI-C_{12} (100 R/R_o). The sensing performance of these PDIs was shown to be independent of their morphologies and surface areas but very much aligned to the difference in reduction potentials, caused by different core-substituted groups, indicating the electronics dominate structure. For example, PDI-ClC_{12} and PDI-ClC_4F_7 presented similar sensing results (10^4 at 10 ppm) due to similar LUMO energy levels (−4.22 and −4.28 eV, respectively). It was also reported that different imide substituents have a secondary effect on the sensing performance. The results showed that PDI derivatives with lower reduction potentials enable a highly sufficient charge exchange and eventually lead to high gas sensitivity. PDI-ClC_{12}- and PDI-ClC_4F_7-based sensors showed slow response (600 s), and no investigation on recovery time and repeatability was reported. The good selectivity of these PDI sensors towards hydrazine vapor was quantified when compared with carbon monoxide, hydrogen sulphide, and hydrogen vapor.

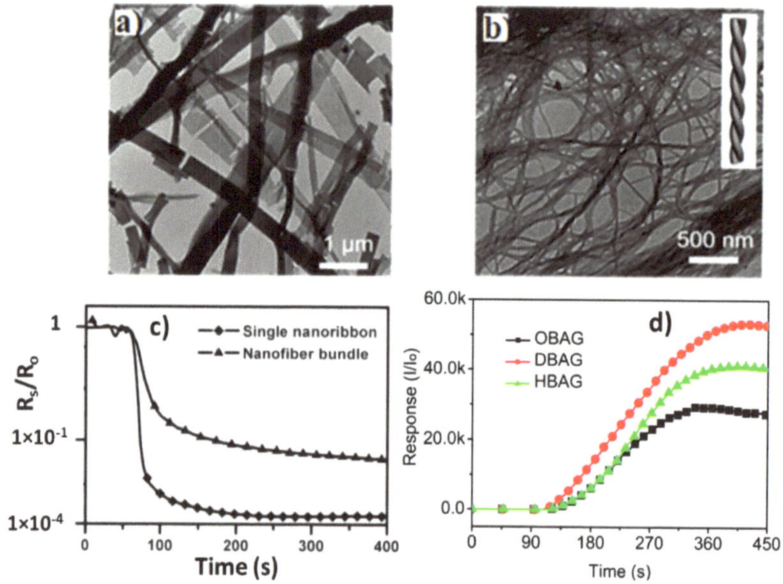

Figure 4. TEM images of PDI-BAG (**a**) nanoribbons and (**b**) nanofiber obtained from H_2O/DMF (*N,N*-dimethylformamide) mixed solvents with different volume ratios of 20/80 and 60/40, respectively. (**c**) Sensing responses of single nanoribbon and nanofiber bundles after the injection of hydrazine vapor. (**d**) Sensing responses of OBAG, DBAG, and HBAG devices obtained from THF/H_2O (20/80 *v/v*) in hydrazine vapor. Adapted with permission [56,80].

Wang et al. [81] also investigated the relationship between the reduction potential of PDI derivatives and its sensing properties towards hydrazine vapor and also studied the effect of different humidity levels, operating temperatures, and film thickness on the sensing performance of the sensor. They synthesized a twisted PDI derivative (PDI-d) (structure 5; Figure 10) exhibiting a low reduction potential. The PDI-d device was continuously exposed to hydrazine vapor of different concentrations while recording the real-time resistance. Prior to the exposure to hydrazine vapor, the resistance in air was noted as 0.75 GΩ, which was significantly decreased to 0.42 GΩ after the exposure to 5 ppm of hydrazine vapor. When the hydrazine vapor was removed, the resistance returned to its initial value within 3 s, which indicated that the PDI-d-based sensor had a fast response and continued cycles showed a high reproducibility. Difficulties arose at a high concentration (800 ppm) of hydrazine, which led to instability in the device, which was attributed to the variation in the film morphology demonstrated by SEM images before and after exposure to 800 ppm of hydrazine vapor (Figure 5a,b). The effect of relative humidity levels, temperatures, and film thickness on the sensing performance of the PDI-d film device was investigated in these examples, in detail. The response of the device towards hydrazine vapor slightly improved (within 10%) with the increase of the relative humidity (Figure 5c). It is believed that a higher humidity is conducive for the hydrolyzation of hydrazine or, in fact, the reduction in the nucleophilicity of hydrazine. The response of the sensor was almost the same at the temperature range of 0–60 °C (Figure 5d). To study the effect of film thickness on the sensing performance, volumes of 20 and 50 μL of PDI-d in dichloromethane were drop-casted on IDEs. The thick film showed an enhanced sensitivity and a slow response/recovery, which can be attributed to the greater amount of PDI-d radical anions produced on the thicker film, therefore requiring a longer time for the transition between PDI-d and PDI-d$^-$. Additionally, a comparison between the sensing performance of unsubstituted-PDI with higher reduction potential and PDI-d confirmed that reduction potential plays a critical role in the sensing performance of PDI-based sensors.

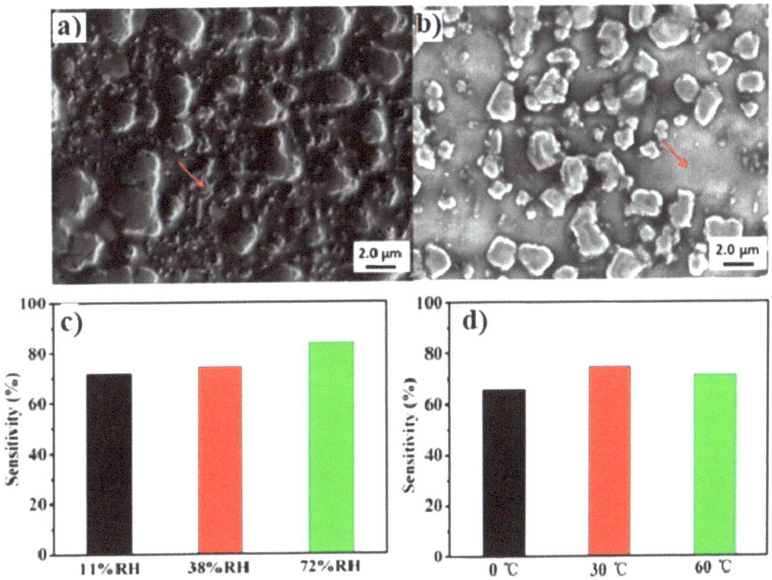

Figure 5. SEM images of a PDI-d film (**a**) before and (**b**) after treatment with 800 ppm of hydrazine vapor for 20 h. Red arrows indicates micro/nano particles that disappear after exposure to the target gas, indicating that the surface of the film is affected by the target gas. The effect of (**c**) relative humidity and (**d**) temperature on the sensing properties of a PDI-d film device. Adapted with permission [81].

The effect of core-substituted groups on the sensing properties of PDI is critical to our understanding of the tolerance and enhanced capability based on the ease of functionalization of organic compounds [82]. By way of example, three different derivatives of PDI—PDI-C_{10}, PDI-Br_2C_{10}, and PDI-BP_2C_{10} (structure 6; Figure 10)—were synthesized in order to study the effect of bay-substituted groups on their gas-sensing performance. It was reported that the sensitivity towards 10 ppm of hydrazine vapor was higher for the core-phenoxy-substituted PDI-BP_2C_{10} (10^4 I/I_o) and lower for the core-brominated PDI-Br_2C_{10} (10^2 I/I_o) when compared with the sensing performance of the core-unsubstituted PDI-C_{10} (10^3 I/I_o). The different responses to hydrazine for PDI-Br_2C_{10} and PDI-C_{10} gas sensors were again proposed to be less reliant on their morphologies and surface areas, and they were attributed to the twisted skeleton of PDI-Br_2C_{10} that derived from non-bonded repulsions between the bromine atoms that resided in the core-area. However, a lower activation energy level and a higher delocalized system efficiently made up for deficiencies in carrier transport that originated from the distorted packing of perylene units and led to only one order of magnitude in current increase for PDI-BP_2C_{10} when compared with PDI-C_{10}. The sensor using the PDI-BP_2C_{10} active layer showed a low LOD (0.75 ppm), good sensitivity (10^4), and better response and recovery times (450 s/150 s) compared with previously reported PDI-ClC_{12}- and PDI-ClC_4F_7-based sensors [22]. From these results, it can be postulated that sensing performance is less dependent on the morphologies and surface area of the sensors. Furthermore, these sensors were extremely selective towards hydrazine vapor when compared with carbon monoxide, hydrogen, nitrogen dioxide, and ozone vapor. These results demonstrated that bay-substituted groups have a significant influence on the sensitivity and stability of PDI-sensing devices.

The Huang group further investigated the effect of core-substituted aromatic groups on the sensing performance of PDIs [58]. For that purpose, they synthesized four different core-substituted aromatic PDI derivatives—DEY, DSPY, DTPY, and DFPY (structure 7; Figure 10)—to explore improved hydrazine sensing. The experimental results revealed that current increased five times towards 10 ppm of hydrazine for PDI-DEY compared to the PDI-DSPY -, PDI-DFPY-, and PDI-DPTY-based sensors. The results reconfirmed that sensor performance was not dependent on the morphologies and surface areas of the materials. The high sensitivity (10^5 I/I_o) of the PDI-DEY-based sensor was attributed to a lower LUMO energy level (−3.68 eV) and activation energy with a smaller interplanar spacing. The aromatic groups at the bay positions appeared to have a prominent effect on the PDI-sensing performance by modulating the energy level, stacking modes, and interplanar spacing. Though the sensors showed an enhanced sensitivity (10^4–10^5 I/I_o), they did present a slower response rate (800 s) and a higher LOD (10 ppm) than the PDI-ClC$_{12}$- and PDI-BP$_2$C$_{10}$-based sensors, and there was no investigation into the recovery, repeatability, and selectivity of the prepared sensors. Taking this further, the team investigated the effect of electron-withdrawing groups on the hydrazine-sensing performance of PDI-based sensors to complement the earlier work on electron-donating groups within PDI-BP$_2$C$_{10}$. Two PDI derivatives, namely PDI-CNC$_{10}$ and PDI-ClC$_{10}$ (structure 8; Figure 10), differentiated by electron-withdrawing groups, were synthesized and formed regular and smooth 1D micro/nanorods (Figure 6a,b) from a chloroform/methanol solution [34]. The sensing studies undertaken in comparison revealed that PDI-ClC$_{10}$ was more sensitive (10^4 I/I_o) than PDI-CNC$_{10}$ (10^3 I/I_o), but the PDI-CNC$_{10}$ derivative exhibited a better stability in an open environment due to its low lying LUMO level (−4.36 eV) and highly crystalline structure. Both the PDI-ClC$_{10}$ and PDI-CNC$_{10}$ devices showed low LOD (at 1 and 2 ppm, respectively) and a better response time (350 s) than the PDI-ClC$_4$F$_7$-based sensor. The repeatability and reproducibility (Figure 6c,d) of both sensors were found to be excellent, and the sensor showed the highest selectivity towards hydrazine in presence of carbon monoxide and ammonia. The results revealed that core-substituted electron-withdrawing groups have a substantial and positive effect on the sensitivity and stability of PDI-based gas-sensing devices for future design consideration.

In a complementary work, Liu et al. [50] also investigated the effect of core-substituted groups on the sensing performance of PDI-based sensors against hydrazine through PDI-pyrimidine and PDI-ph (structure 9; Figure 10). The PDI-pyrimidine-sensor displayed a high sensitivity (56%), significant reproducibility, excellent selectivity, and better long term stability towards hydrazine vapor. The response and recovery times of the sensor towards 10 ppm of hydrazine vapor were calculated to be as low as 0.7 and 0.8 s, respectively. From cyclic voltammetry and absorption studies, it was confirmed that the redox potential of PDI played an important role in the sensing performance. The effect of humidity and temperature was also investigated. The sensors showed a high sensitivity at high humidity levels from 70% to 96% but a slower response/recovery (Figure 7a), which was most likely due to the solvation of hydrazine by the water. While the sensor showed stability through a consistent response for an operating temperature range of 20–60 °C, elevated temperatures (>70 °C) caused a significant reduction in the sensitivity (Figure 7b). It is believed that at a high temperature (70 °C), the interaction between hydrazine vapor and pyrimidine-PDIs is reduced because the adsorption is exothermic. As a result, the sensitivity of the sensor is decreased. The effect of film thickness was also investigated in this study. For that purpose, three sensors with different thicknesses (8, 14, and 25 μm) were prepared, among which the thicker sensors showed higher sensitivities (~88.5%) but concomitant slower response and recovery times (5.3 s and 8.2 s, respectively) in comparison (Figure 7c). Sensitivity studies indicated this sensor showed the highest selectivity for hydrazine when compared to other reductive vapors such as ammonia, *t*-butylamine, *n*-butylamine, diethylamine, triethylamine, and aniline (Figure 7d). The enhanced performance of pyrimidine-PDI-based sensors yielding better sensitivity, selectivity, long-term stability, and short response/recovery times than the previously reported PDI-based sensors was an important advance in the design of organic sensors.

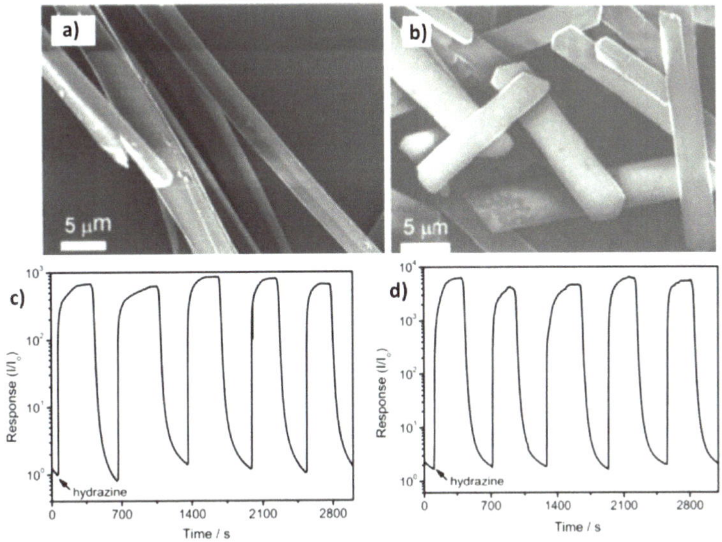

Figure 6. SEM images of (**a**) PDI-CNC$_{10}$ and (**b**) PDI-ClC$_{10}$ illustrating the similarity in a nanorod structure. Dynamic responses of (**c**) PDI-CNC$_{10}$ and (**d**) PDI-ClC$_{10}$ gas sensors towards 8 ppm of hydrazine vapor show the sensitivity increase of the latter based on the response value. Adapted with permission [34].

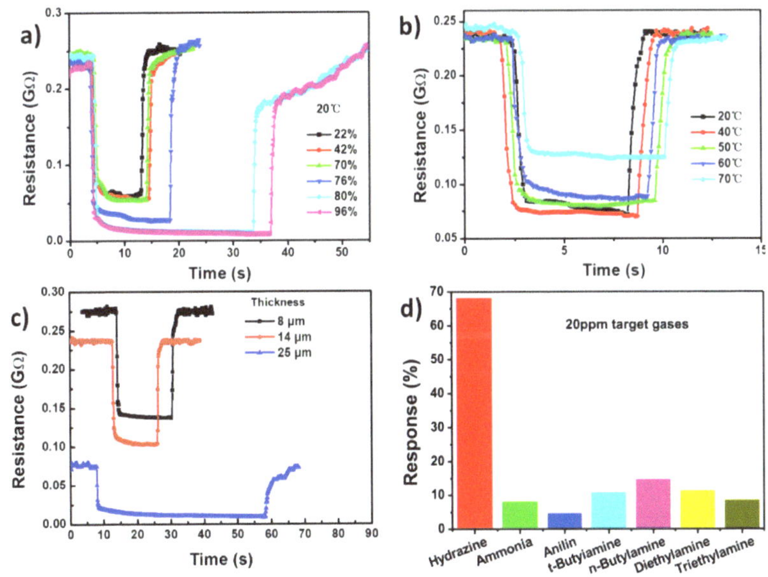

Figure 7. The effect of (**a**) relative humidity, (**b**) operating temperature, and (**c**) film thickness on the sensing properties of a PDI-pyrimidine sensor. (**d**) The response of a PDI-pyrimidine sensor towards different analytes (20 ppm). Adapted with permission [50].

With the advent of positive effects of core-substitution on the performance of PDI sensors through torsion strain, added steric effects, which also change the molecular ability to stack effectively, were evaluated in a simple series PH, TPH, and OPH (structure 10; Figure 10), in which the 4-alkylaryl groups differed from H to *t*-butyl to *t*-octyl, respectively [83]. With reference to hydrazine vapor sensing, among these PDI derivatives, OPH displayed the highest response (1800 I/I_0 at 10 ppm) (Figure 8d) due to its smallest torsion angle (25.80°), better crystalline structure Figure 8c), and smallest activation energy (1.03 eV) within a face-to-face stacking mode. An ordered crystalline structure (Figure 8b) with a smaller interplanar spacing and modest activation energy (1.04 eV) contributed to the average performance (520 I/I_0) of TPH. In contrast, the largest activation energy (1.05 eV) and torsion angle (26.14°) with a meager crystalline structure (Figure 8a) caused PH to be the least-efficient sensor (130 I/I_0). All three sensors, PH, TPH and OPH, displayed low LOD values (0.6, 0.5, and 0.5 ppm, respectively) and quick response and recovery times (116/19, 163/52, and 170 s/58 s, respectively) in comparison with previously reported PDI-based sensors. This study provides useful information on new factors such as torsion angle, morphology, and activation energy that could lead to the enhanced sensing performance of PDI-based sensors that are remote from the electronic function of the active material. To investigate the improved sensing performance of PDIs by further modulating the intermolecular interactions of PDI molecules, a series of stereochemically defined PDI derivatives—OT, ROT, and SOT (structure 11; Figure 10)—bearing pairs of racemic and *R*- and *S*-chiral hydroxyalkyl chains were synthesized [84]. While sensing gases should not rely on a chiral auxiliary, either as enantiomers or diastereomers, because most gases are achiral species, the hydrazine-sensing responses of PDI-OT, PDI-ROT, and PDI-SOT devices were recorded using a two-probe technique. Upon exposure to 10 ppm of hydrazine vapor, it was found that, surprisingly, the increase in the current for PDI-ROT was approximately 1.5 times that for its enantiomer PDI-SOT and twice that of the racemic mix of diastereomers PDI-OT (Figure 9d). All three sensors showed a response time of approximately 350 s, and there was no investigation about the recovery, repeatability, and selectivity. The difference in enantiomeric activity for PDI-ROT was rationalized by two factors: (a) a smaller torsional angle of 15.55 and, importantly, (b) the generation of a continuous and uniform film surface (Figure 9b). Furthermore, the PDI-ROT sensor exhibited an improved sensitivity than the PH, TPH, and OPH sensors, and this indeed may be attributable to a small torsional angle or additional conjugation. The slower response and higher LOD is more likely to be attributed to a poorer surface morphology than the PH, TPH, and OPH sensors. The results suggest that the sensitivity of these devices is independent on film thickness, as shown in Figure 9e.

The sensitivity of PDI-based sensors can also be improved by incorporating electron-donating substituents into the core locations of PDI [85]. Three PDI derivatives—PDI-PY, PDI-PI, and PDI-HE (structure 12; Figure 10)—ere synthesized with piperidinyl, pyrrolidinyl, and *n*-hexylamino groups at the bay 1,7-positions of the perylene skeleton, respectively. Experimental results revealed that the hexylamino-based sensor performed best, generating quick response and recovery times (both 7 s, respectively), a low LOD (0.17 ppm), and the highest response (1.5×10^5 I/I_0). This performance was again attributed to its higher electron density as a result of the EDG, the smallest torsional angle (16.83°), and a crystalline structure affording a tighter π–π distance of 3.55 Å. PDI-PY sensors bearing a piperidinyl group afforded inferior sensing results (1.5×10^4 I/I_0) because of its poor crystalline structure, weaker π–π orbital overlap, and resulting higher interplanar spacing of 3.62 Å as a result of packing the piperidyl groups. Selectivity towards hydrazine over ammonia and other volatile gases was displayed.

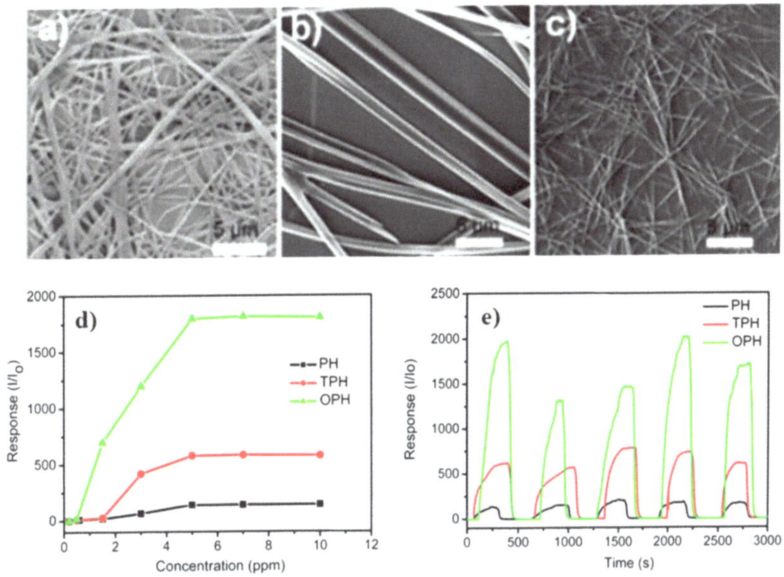

Figure 8. SEM images of microstructures of the (**a**) PH, (**b**) TPH, and (**c**) OPH PDIs. (**d**) The response modulation–concentration curves show an enhancement in OPH and (**e**) the dynamic responses of the PH, TPH, and OPH devices in hydrazine vapor (5 ppm). Adapted with permission [83].

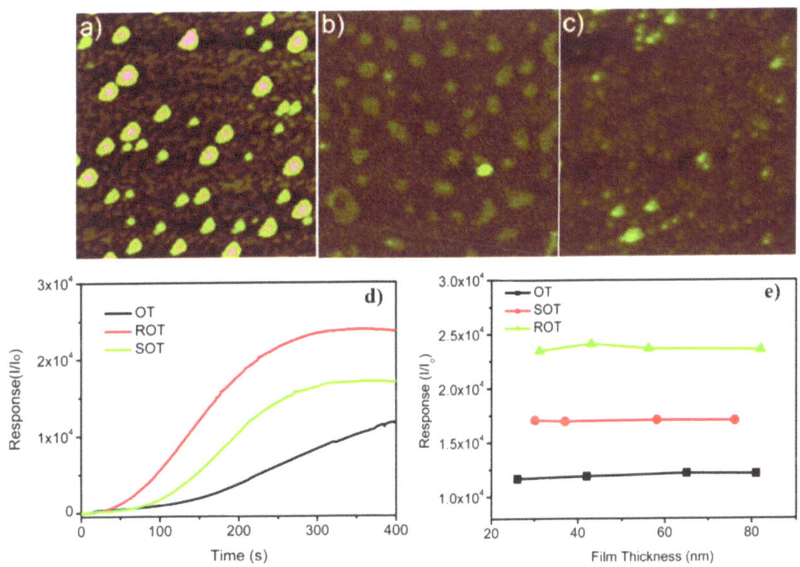

Figure 9. AFM images of thin films (2 × 2 µm) of (**a**) OT, (**b**) ROT, and (**c**) SOT devices. (**d**) Current modulation (I/I_o)–time (t) and (**e**) response modulation–film thickness curves of OT, ROT, and SOT devices in saturated hydrazine vapor (10 ppm). Adapted with permission [84].

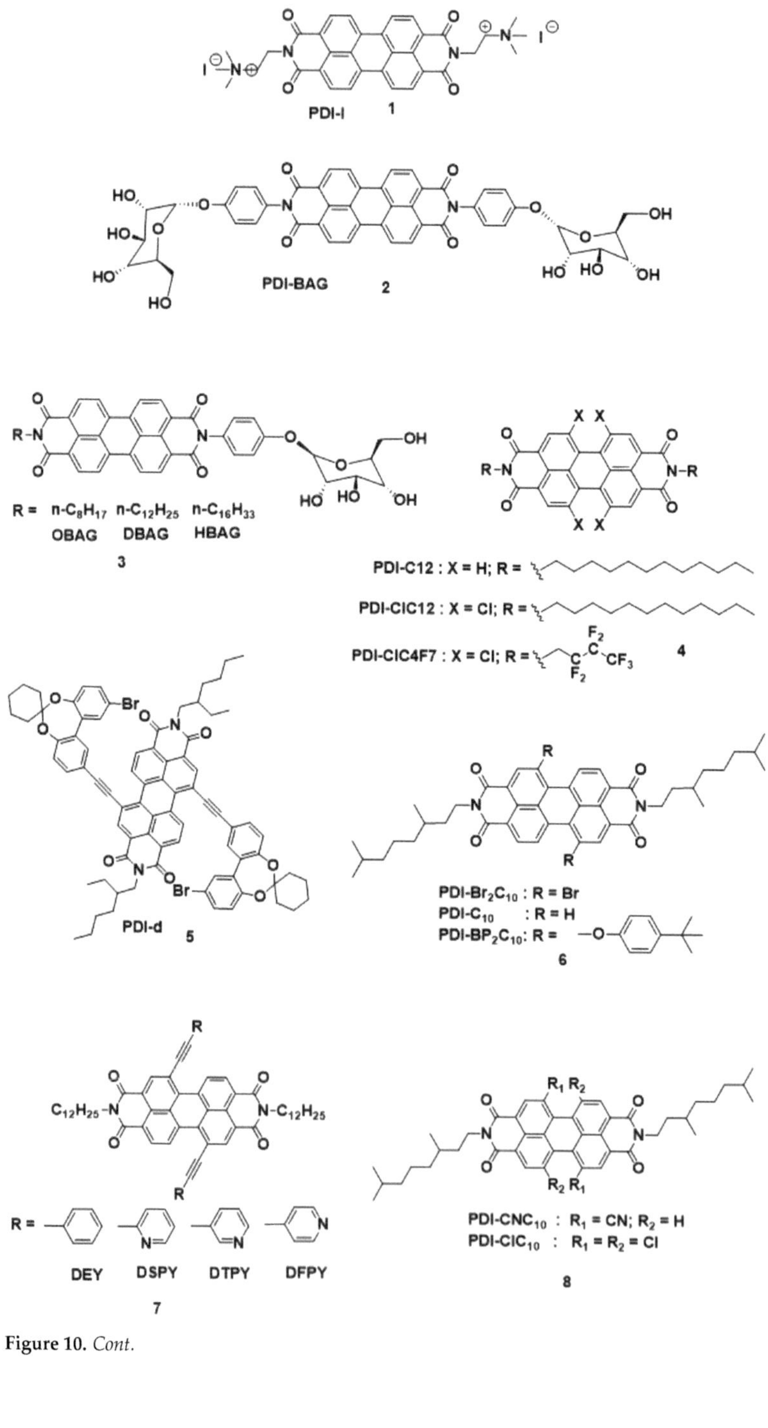

Figure 10. *Cont.*

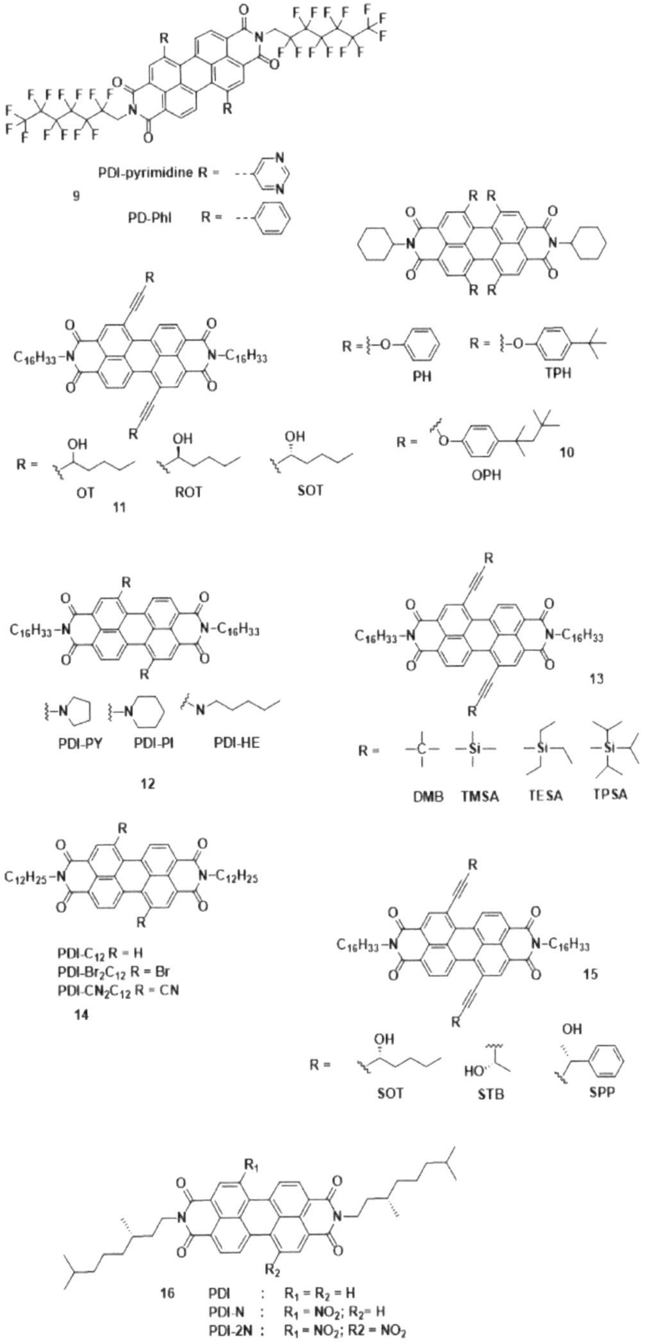

Figure 10. *Cont.*

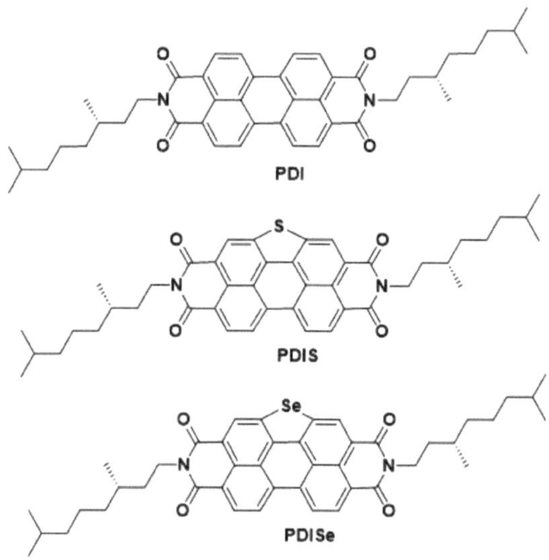

17

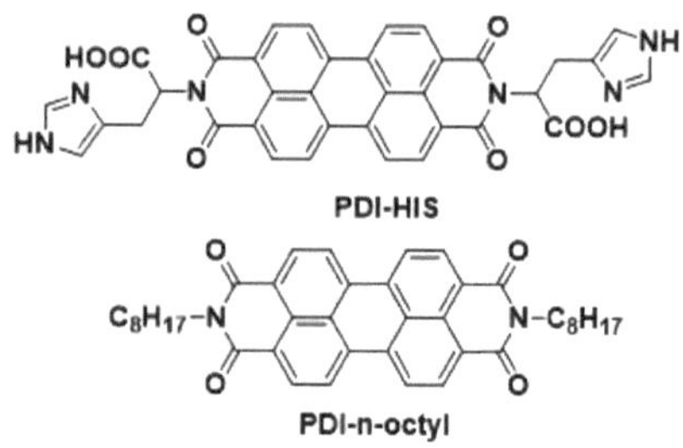

18

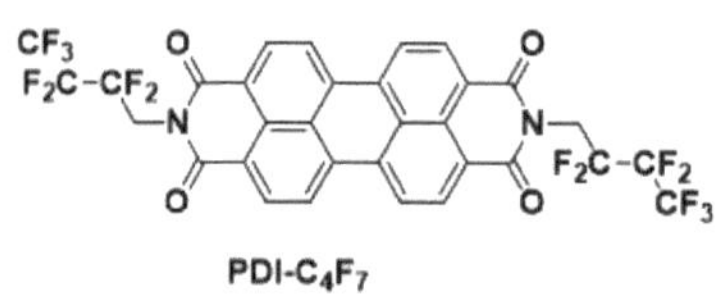

19

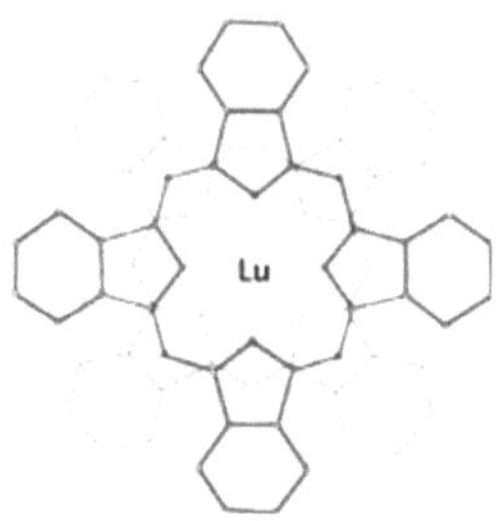

LuPc2

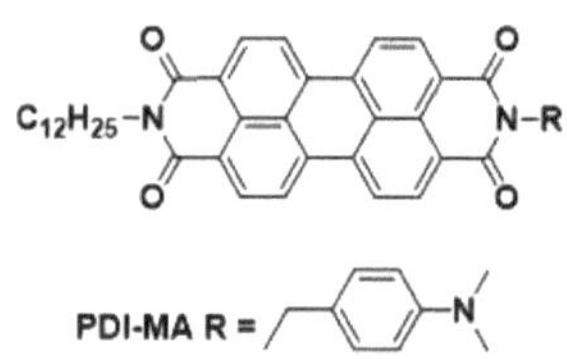

20

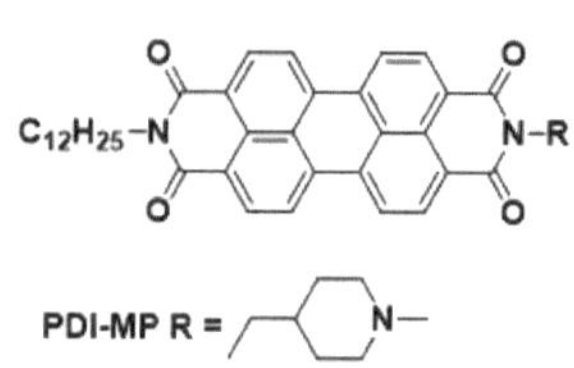

21

Figure 10. *Cont.*

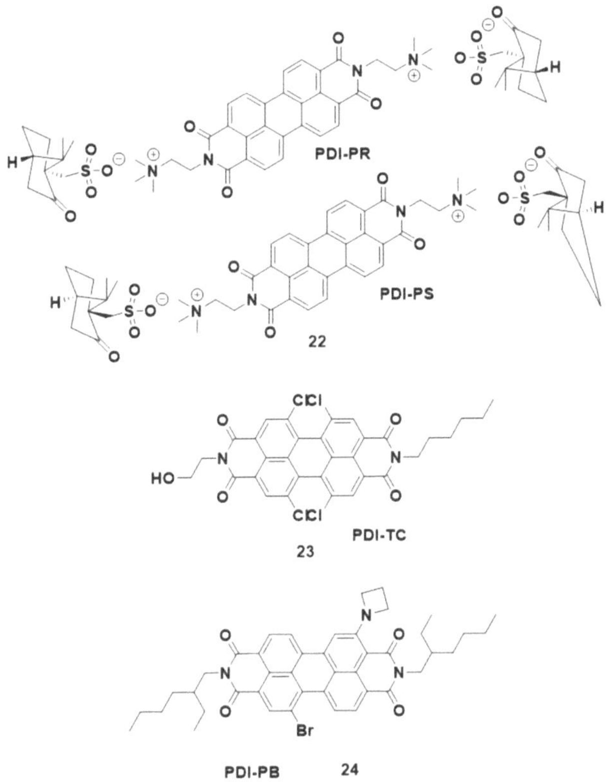

Figure 10. Chemical structures of the reported PDIs for hydrazine, ammonia, and other analyte sensing.

The final comparison group in this section looks at the alkylsilane-substituted PDIs, TMSA, TESA, and TPSA (structure 13; Figure 10) [35], in which the 1,7-acetylene PDI core is substituted at the terminal acetylenic positions with the alkylsilanes. To understand the effect of silyl chains on the sensing performance of devices, another PDI derivative, DMB bearing simple isostructural carbon frameworks, was also synthesized. An investigation into the sensing performance of PDI derivatives towards hydrazine vapor exhibited an increased current of around four orders of magnitude for the better performing TPSA derivative and 20 times higher than the TESA derivative. All the sensors showed a similar response time of approximately 200 s, which was better in comparison with the PDI-DEY-0 and PDI-DAG-based sensors. The authors intimated that from the sensing results, the sensing performance of these PDI-based sensors are independent on the surface morphologies and surface areas, and it can be attributed to the difference in stacking modes, torsional angle, and interplanar spacing, thus providing a consistent message to other studies. Again, the ordered layer structure appears to contribute to TPSA having an effective sensing performance. A similar face-to-face packing mode and torsional angle of DMB and TMSA led both the derivatives to show nearly the same sensing results, regardless of the differences in their interplanar spacing. However, a larger torsional angle and head-to-tail aggregated modes with larger interplanar spacing induced less-efficient sensing results. This study suggests that the lengths and branched points of the alkylsilane on core areas have a consequential effect on the sensing behavior of PTDCI-based gas sensors.

From the above discussion, we can summarize that the specific design features that influence the sensing performance of the core-substituted PDI derivatives are mainly derived from the LUMO energy levels, activation energy, and torsional angles. Furthermore, surface morphology generally appears to have a weak influence on the sensing performance of core-substituted PDI-based sensors for alike systems when compared to the effect of LUMO energy levels, activation energy, and torsional angles. Of course, systems with a significant disruption of stacking perform worse. The stability of PDI-based sensors also appears to be improved by core-substitution with electron-withdrawing functional groups, thus enhancing the *n*-type properties. Moreover, these functionalities, i.e., functionalities that promote hydrogen binding and modulate torsion angles, interplanar spacing, band gap, and stacking modes, can also significantly impact sensing performance through cooperative effects.

4. Ammonia Sensing

Ammonia is a critical commodity chemical with extensive applications in fertilizers, synthetic fibers, drugs, generating ammonium salts, and the manufacturing of plastics, but it is a highly corrosive and poisonous gas that can be easily spread into the environment [97]. Moreover, ammonium salts such as ammonium nitrate (NH_4NO_3) are found in various explosives and slowly dissociate to release trace amounts of ammonia that are critical to be detected and monitored to avoid accidents [98]. Though ammonia has applications in numerous research areas, its exposure in high concentrations is deadly to human health in its extreme, leading to prolonged effects based on edema and airway destruction through burning. The lower limit of human perception by smell is around 50 ppm for ammonia [24]. However, even underneath this limit, it is irritating to skin, eyes, and the respiratory system, thus leading to cough and nose/throat irritation [99]. The permissible limits of 25 ppm over an 8 h period and 35 ppm over a 10 min period have been recommended and legislated by Health and Safety Executive (HSE), London [100]. To date, PDI-based sensors have been developed as both resistive and amperometric devices for ammonia-sensing applications, as summarized in Table 1. This is not surprising given the strong similarity between ammonia and hydrazine as a reducing gas, with a similar magnitude of nucleophilicity regardless of any quantification of the alpha-effect [101].

To illustrate the point, the three electron-deficient PDI derivatives—PDI-C_{12}, PDI-Br_2C_{12}, and PDI-CN_2C_{12} (structure 14; Figure 10)—that differ by the nature of the bay substitution with H, Br, or CN were identified and trialed for sensing ammonia gas [21] as single crystalline micro/nanostructures. The micro/nanostructures of PDI-C_{12}, PDI-Br_2C_{12}, and PDI-CN_2C_{12} were prepared by a precipitation method in a chloroform/methanol binary solution (0.1 mg/mL). SEM images showed that all PDI compounds in this series could self-assemble into one-dimensional micro/nanostructures despite the influence of the bay groups. PDI-C_{12} molecules showed ribbon-like structures (Figure 11a) with an average width of 1–2 μm and lengths up to tens of micrometers, PDI-CN_2C_{12} also self-assembled into a belt like morphology (Figure 11c) with an average diameter of 200–400 nm, and PDI-Br_2C_{12} molecules showed self-assembled nanorods (Figure 11b) with a diameter of approximately 200–300 nm and a length up to tens of micrometers. The devices were fabricated with a simple resistive-type structure. Upon exposing the three sensors to 1% ammonia, the resistance of PDI-CN_2C_{12} nanobelts decreased quickly (50 s) to reach a stable value and demonstrated a good sensitivity (40%) (Figure 11d). However, no notable variations were observed in the resistance of PDI-C_{12} micro belts and PDI-Br_2C_{12} nanorods. The applicability of the PDI-CN_2C_{12} sensor under an ambient atmosphere was further studied. The results indicated that while the sensitivity reduced to 37% in the first 24 h, a constant value of 37% was maintained after its exposure to air for seven days. The superior response to ammonia gas by PDI-CN_2C_{12} was attributed to the difference in the reduction potentials as a result of the insertion of the strong electron-withdrawing nitrile group, which caused a significant increase in current upon exposure to ammonia. The low LUMO energy level (−4.36 eV) and the higher-order arrangement in nanobelts made PDI-CN_2C_{12} more sensitive and stable. Though the PDI-CN_2C_{12}-based sensor showed good sensitivity and quick response, but there was no investigation into the selectivity of the sensor.

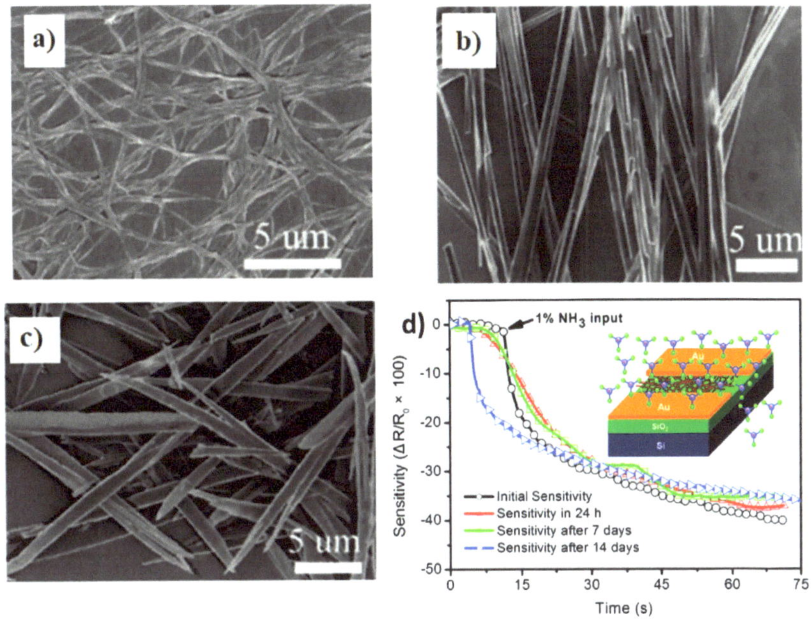

Figure 11. SEM images of (**a**) PDI-C_{12}, (**b**) PDI-Br_2C_{12}, and (**c**) PDI-CN_2C_{12} illustrating subtle changes in the nanostructure as a result of bay substitutions. (**d**) Sensitivity curves of PDI-CN_2C_{12} showing the stability of the system over time. Adapted with permission [21].

The effect of chirality-induced morphology through core-substitution on the sensing performance of PDIs [23] has been investigated through the three PDI derivatives—SOT, STB, and SPP (structure 15; Figure 10)—that differ by the position of hydroxy groups on the alkyl/aryl chains. All three PDI-based sensors showed quicker responses (>24 s), higher selectivities, and lower detection limits (>1.51 ppm) towards ammonia than the PDI-CN_2C_{12}-based sensor [21]. Among these derivatives, SPP exhibited the highest sensitivity (4.5%), the lowest LOD (0.48 ppm), and the quickest response and recovery times (17 and 6 s, respectively) (Figure 12d), mainly attributed to its lower LUMO energy level (−4.13 eV) and uniform crystalline structure (Figure 12c). SOT showed a poor sensing performance with a response of 0.9% due to its highest LUMO level (−4.06 eV) and poor crystallinity (Figure 12a). To explore the selectivity of the prepared sensors, all the sensors were exposed to ammonia (100 ppm), ethylamine, butylamine, diethylamine, triethylamine, dibutylamine, aniline, 2,4-dimethylaniline, and cyclohexylamine at a concentration of 1000 ppm. Interestingly, compared with other amines, all three sensors showed the highest selectivity towards ammonia, as shown in Figure 12e.

Two PDI derivatives bearing NO_2 groups—PDI-N and PDI-2N (structure 16; Figure 10)—were prepared to further investigate the effect of reduction potential and the LUMO energy level on the sensitivity and stability of PDI-based sensors [86] for ammonia against a model PDI system. Among these derivatives, the PDI-2N-based sensor presented the smallest detection limit of 2 ppm, which was attributed to its reduction potential of −0.36 eV, and the greatest ambient stability as a result of its low LUMO level of −4.35 eV. However, the large torsional angle (17.30°) and interplanar spacing (3.64 Å) reduced the π–π orbital overlap between the adjacent PDI-2N molecules, thus suppressing mobility and consequently moderating sensitivity—leading to an important design principle. A more positive reduction potential and ability for modest π–π orbital overlap made the PDI-N device more sensitive (6.6%) than PDI (1%) and PDI-2N (5.6%) devices. It was postulated that a negative reduction potential of −0.73 eV caused a lack of charge carrier density and induced an inferior sensing performance of the PDI sensor, although the structure of this device showed the highest π–π orbital overlap. The experimental results suggested that the detection limit and ambient stability are closely related to the reduction potential and LUMO energy level.

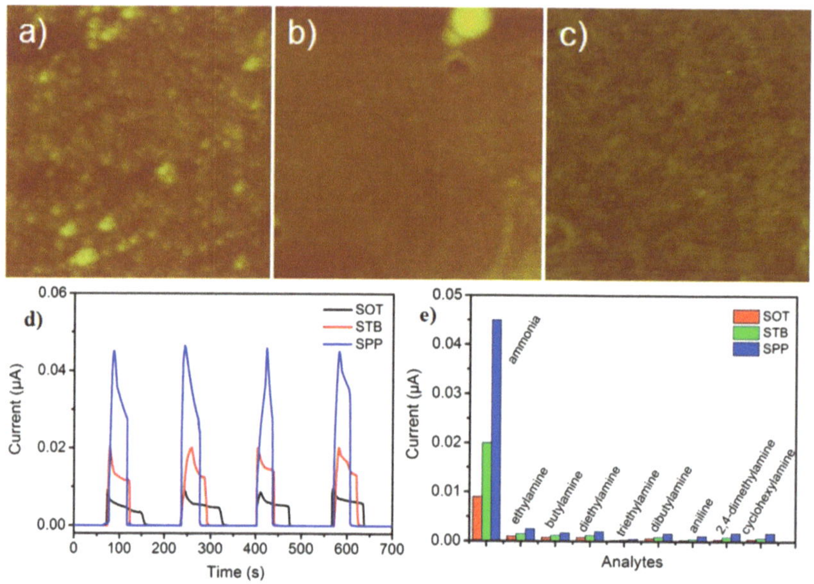

Figure 12. AFM images of (**a**) SOT, (**b**) STB, and (**c**) SPP thin films (2 × 2 μm) illustrating a similar morphology. (**d**) Dynamic current change of SOT, STB, and SPP devices in ammonia (100 ppm). (**e**) Current change for SOT, STB, and SPP devices exposed to numerous amine analytes. Adapted with permission [23].

The effect of chalcogenide inclusion on the sensing properties of PDIs has also been explored through the small series of PDI, PDIS, and PDISe (structure 17; Figure 10), where PDIS and PDISe derivatives represent the inclusion of sulfur (S) and selenium (Se) heterocycles, respectively, to the bay positions [87]. The sensing response of PDI, PDIS, and PDISe to 100 ppm of ammonia gas was studied by using the two-probe method. Interestingly, when exposed to 100 ppm of ammonia gas, the sulfur derivative showed an increase in current magnitude compared to the carbocyclic PDI, while the selenide showed a decrease compared to the same PDI. The PDIS ultimately showed the best sensing performance, with response magnitudes of 1.5 and 3 times higher than PDI and PDISe, respectively. The high response was mainly attributed to the electron donation from ammonia through donor–acceptor complexation towards perylene, followed by effective long-range electron transport via π–π delocalization that led to current variations. The small rotational angle and the large π–π overlap of vicinal PDIS cores were also attributed to improved electron mobility and the improvement in the sensing performance.

While much variation has occurred on the core of PDI, there are some examples to sense ammonia using functionalization at the imide nitrogens [24]. In a particular example, histidine moieties were attached to produce PDI-HIS, the sensing ability of which was compared to a bis(*n*-octyl)PDI (structure 18; Figure 10). The fabricated amperometric-type sensors were kept in a chamber, and various amounts of ammonia (2–100 ppm) were injected into the test chamber. The PDI-HIS-based sensor showed enhanced performance when compared with PDI-*n*-octyl-based sensor. A significant response (3500%) to 100 ppm of ammonia was recorded for PDI-HIS with a concomitant LOD of 0.56 ppm. To evaluate the selectivity of the PDI-HIS-based sensor, similar sensing experiments were also carried out using various analytes and different volatile organic solvents, including hexane, acetone, ethyl acetate, isopropanol, ethanol, methanol, and chloroform, together with ammonia. Interestingly, the response to 1000 ppm of common organic analytes was less in comparison to 200 ppm of ammonia (Figure 14a,b), cementing this as a good example of an organic-based sensor. By increasing the humidity, the sensitivity was shown to decrease, which could have been due to the weak adsorption/diffusion of ammonia gas on the sensor surface (Figure 13c). The effect of film thickness on the sensitivity of PDI-HIS was also investigated. Three sensors with different film thicknesses (30, 60, and 100 nm) were prepared and studied for their sensing response towards 100 ppm of ammonia (Figure 13e,d). It was confirmed that the sensitivity rose by three times with slight increase in the response/recovery time (from 1 to 4 s) when the film thickness was increased from 30 to 60 nm; however, there did not appear to be a linear relationship between humidity and sensitivity. Upon a further increase in film thickness from 60 to 100 nm, the sensitivity marginally increased and was greatly offset by the larger increase to the response/recovery times (26 s/16 s, respectively). The slow response in a thicker film is not uncommon and may be ascribed to the need for the diffusion of a large number of ammonia molecules into the film surface and their entrapment upon release. The lower redox potential of PDI-HIS, together with the presence of ionizable groups at the imide positions made the PDI-HIS-based sensors more efficient when compared with the sensors based on PDI-*n*-octyl.

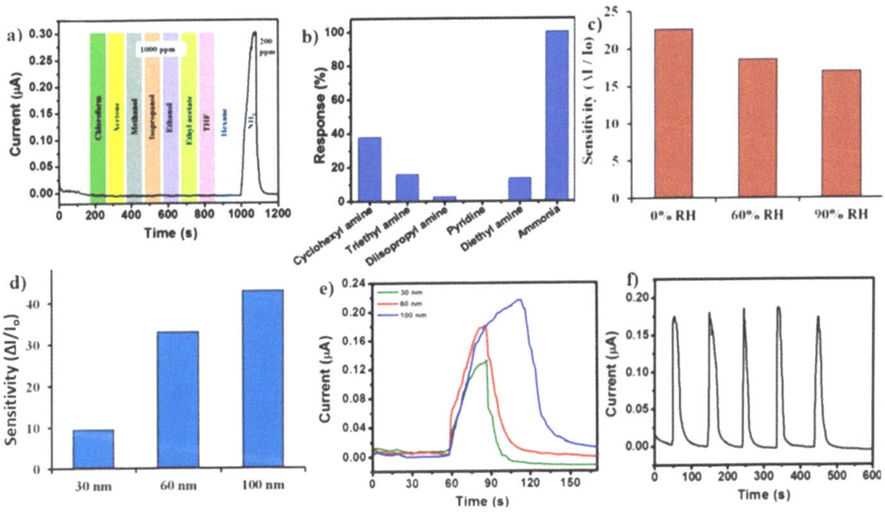

Figure 13. Response of a PDI-HIS sensor towards vapor of various common types of (**a**) organic solvents and (**b**) amines. Concentrations of ammonia and other analytes used were 200 and 1000 ppm, respectively. (**c**) Sensitivity of PDI-HIS at 0%, 60%, and 90% relative humidity levels. (**d**) Effect of film thickness on sensitivity of the sensor. (**e**) Current–time response observed in three devices with different film thicknesses upon exposure to 100 ppm of ammonia. (**f**) Recyclability study with 100 ppm of ammonia. Adapted with permission [24].

While much of the discussion has investigated changes to active materials as opposed to the nature of the sensor device, there is a literature example of a prepared molecular semiconductor-doped insulator (MSDI) heterojunctions for ammonia sensing [88]. A fluorinated PDI derivative, C_4F_7-PDI, as an *n*-type material (bottom layer of the device), and lutetium bisphthalocyanine, $LuPc_2$, as a *p*-type semiconductor (top layer), were used in the preparation of the MSDI heterojunction (structure 19, Figure 10). The heterojunction response to ammonia was compared to that from a sensor based solely on $LuPc_2$. In stark contrast to each other, exposure to ammonia increased the current for the MSDI-based sensor, whereas the same exposure caused a decrease in the current of $LuPc_2$-based sensor, thus illustrating a simple logic based 'ON' vs. 'OFF,' or 'bright' vs. 'dim,' response. The C_4F_7-PDI/$LuPc_2$ (50 nm/50 nm) MSDI showed a modest response to ammonia with a sensitivity of up 33% to 100 ppm of ammonia at 50% of relative humidity compared to the control.

From the above discussion, there appears to be unique design features that positively influence the sensing performance of PDI-based sensors for ammonia by optimizing the reduction potential, optimizing the thickness of the sensing layer, and attaching a particular functional group that can enhance hydrogen bonding. Furthermore, the reduction potential, LUMO energy level, crystal structure, and π–π stacking can drastically affect the LOD, sensitivity, and stability of PDI-based sensors. Additionally, the judicial choice of functionality to either the core or imide positions needs further exploration because these changes have been shown to dramatically improve the sensitivity and selectivity of PDI-based sensors.

5. Some Other Analyte-Sensing Applications

PDI derivatives have also been developed and tested towards other analytes, including nitromethane, hydrogen peroxide, aniline, ethylenediamine, and trimethylamine [89–93]. The sensing performance of these sensors are also summarized in Table 1 in comparison to hydrazine and ammonia sensor performance.

Ultrathin nanoribbons formed from the amphiphilic electron donor–acceptor (D–A) PDI-MA (structure 20; Figure 10), comprising PDI as the backbone scaffold bearing dimethylaniline groups, were used for sensing various nitro-compounds such as nitromethane, nitrobenzene, 4-nitrotoluene, and 1-chloro-4-nitrobenzene (Figure 14a) [89]. The current of the sensor was decreased by about 65% upon exposure to a high concentration of nitromethane (36,000 ppm). The nanoribbons also exhibited a fast response (10 s) with a quick recovery (75 s) (Figure 14c) due to the intrinsic volatility of the nitromethane. Various organic solvents and some reductive reagents, such as aniline, were also sensed. Some significant current variations were observed for organic solvents, such as hexane, toluene, and chloroform, thus suggesting a lower selectivity. It is postulated that this response is likely due to the solubility of PDI molecules in these solvents. PDI-MA nanofibers afford a mesh-like, highly porous surface that can offer a high surface area to the target analytes and the beneficial diffusion of the analytes through the 3D network.

Wu et al. [90] also synthesized nanoribbons (Figure 14b) of 1-methylpiperidine-substituted PDIs (PDI-MP; structure 21; Figure 10). The nanoribbons were drop-cast over interdigitated electrodes and tested towards oxidizing hydrogen peroxide (H_2O_2) vapor. When the PDI-MP sensor was exposed to 18.5 ppm of H_2O_2 vapor, a quick decrease in the current (50% reduction) was recorded, as shown in Figure 14d. The sensor also displayed a quick response of 19.5 s, which was attributed to the high surface area of nanoribbons, and the sensor was able to sense a concentration of 0.6 ppm of H_2O_2. The irreversible response towards H_2O_2 vapor was ascribed to the strong surface binding of H_2O_2 and the permanent oxidation of the 1-methylpiperidine group by H_2O_2.

The effect of the chiral auxiliary camphorsulfonic acid (CSA) on the sensing performance of PDI was also studied [91]. Two PDI derivatives, PR and PS (structure 22; Figure 10), were synthesized as active materials for the fabrication of amperometric-type sensors for sensing ammonia, ethylamine, butylamine, diethylamine, triethylamine, ethylenediamine, and hydrazine. The PDI-PR- and PDI-PS-based sensors displayed sensitivity (1.5% and 17%, respectively), quick response and recovery times (9/7 and 11/14 s, respectively), and excellent selectivity towards ethylenediamine vapor with detection limits as low as 1.07 and 0.86 ppm, respectively. The PDI-PR-based sensor showed less sensitivity than the PDI-PR-based sensor due to the loss of π–π interactions. The authors suggested that strong hydrogen bonding interactions between camphorsulfonic acid and ethylenediamine led to an enhanced sensing performance.

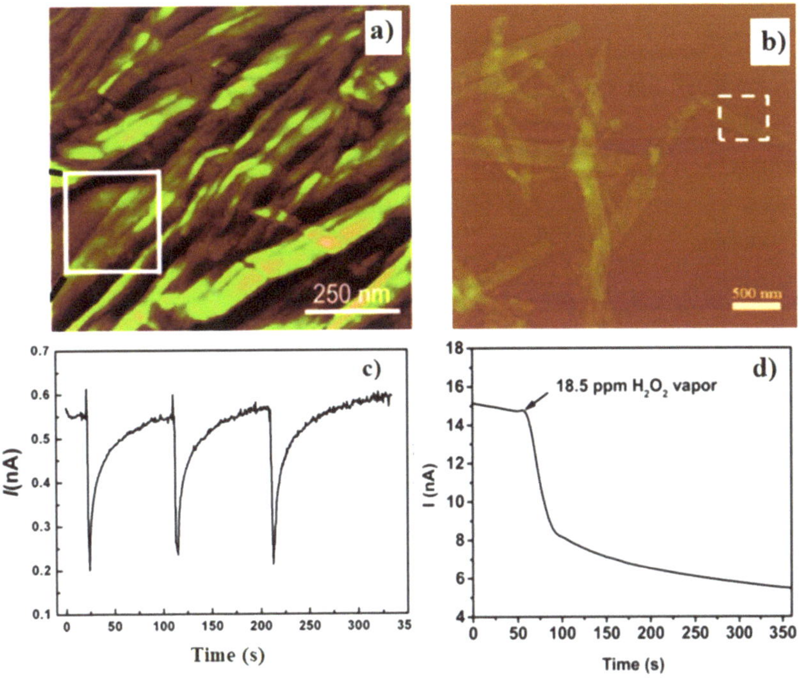

Figure 14. AFM images of (**a**) PDI-MA and (**b**) PDI-MP illustrating the ribbon-like morphology. Current change in PDI-MA nanoribbons in response to (**c**) nitromethane (36000 ppm) and (**d**) H_2O_2 (18.5 ppm). Adapted with permission [89,90].

Recently, Zhu et al. prepared a heterojunction of PDI and cadmium sulphide (CdS) (Figure 15a) to sense trimethylamine (TMA) vapor [92]. An amphiphilic PDI derivative, PDI-TC (structure 23; Figure 10), was synthesized and used to fabricate an amperometric gas sensor. The PDI-TC/CDS sensor showed a higher sensitivity (18% at 10 ppm), a lower LOD (0.2 ppm), better stability and reproducibility (Figure 15b), and better selectivity towards TMA when compared with the sensor based on TC-PDI (no notable response at 10 ppm). The results clarified the synergistic effect between organic and inorganic materials and confirmed that *n*–*n* heterojunctions are a viable approach to improve the sensing properties of TMA gas sensors.

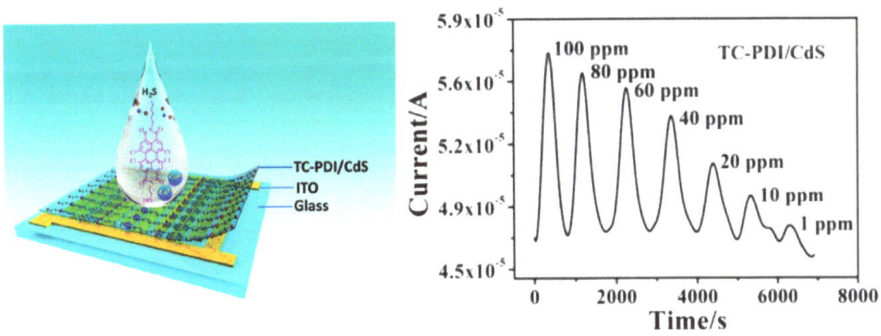

Figure 15. (**a**) Schematic diagram of TC-PDI/CdS (cadmium sulphide) heterojunction. (**b**) Response–recovery curves of TC-PDI/CdS sensor to different trimethylamine (TMA) concentrations. Adapted with permission [92].

The last example we discuss here is a novel, core PDI-PB derivative containing a core-substituted azacyclobutane (PDI-PB; structure 24; Figure 10) [93]. The self-assembled nano-network structure was prepared in chloroform and methanol, and it was used in the fabrication of resistive sensor for the detection of pyridine and its derivatives. The PDI-PB sensor showed a low LOD of 375 ppb, a quick response time (1 s), an excellent recyclability, and a high sensitivity (2200%) towards 375 ppm of pyridine. The results obtained for the PDI-PB-based sensor suggests that a self-assembled nano-networked structure can improve the interaction between the target analyte and the sensing, leading to enhanced sensing performance, if designed appropriately.

6. Conclusions and Future Outlook

Through this review, we have demonstrated that organic-derived sensors have the capability of ushering new technology, and this is particularly true for the class of rylene aromatics of which PDI-based gas sensors are a part. Their strong evolution as active materials for electronic sensors or as part of heterojunction materials has been demonstrated to be effective in the sensing of hazardous and reducing gases such as hydrazine and ammonia. Compared to gas sensors based on inorganic materials, PDI-based gas sensors possess several advantages including solution-processability, mechanical flexibility, room-temperature operability, low power consumption, and portability. Through this series of highlighted compounds, it is worthwhile to mention that some general rules can be articulated to correlate sensing performance with the structural aspects or photophysical parameters of a given sensor family—in this case, prominent n-type semiconductor materials.

Overall, it is the redox properties of the active material, film morphology, and self-assembly of the PDI derivatives within the active layer that appear to play a significant role in gas-sensing performance. Most of the PDI derivatives with a uniform crystalline structure have demonstrated better sensitivity compared to derivatives with a poor crystalline structure. Therefore, single micro/nanostructures and ultra-thin nanotubes/nanorods have been continually employed to fabricate gas sensors with higher sensitivities, swifter responses, lower LOD values, and quick recovery times (a few seconds). By incorporating a variety of functional groups at the bay and imide positions of PDI, a number of novel materials have been developed to fabricate gas sensors with enhanced sensitivity, selectivity, and stability. The core substitutions of PDIs with different electron-donating and electron-withdrawing groups have also been shown to provide a significant impact on the sensitivity and stability of a given sensor as the electron-donating/withdrawing groups decrease the LUMO energy level and the reduction potential, which allows for the efficient transport of electrons by reducing the charge injection barrier to enhance the performance of a sensor. This is one of the main reasons why the core-substituted PDIs show a high sensitivity and better stability.

Though the development of PDI-based sensors is noteworthy, there are a number of design issues that need to be resolved for the mass-uptake of PDI-based sensors. Firstly, there are conflicts among some electrical properties of the chemical sensors such that while stronger interactions between the target analytes and sensors lead to superior sensitivity, they also confine the recovery to the initial state. The selectivity of a specific gas in a mixture of gases is still not resolved, but it is likely to be enhanced through other technologies including membranes and size-exclusion phenomena. The sensing mechanism of most of the PDI-based sensors is based on either the chemical or physical adsorption of target analytes. Several analytes can be adsorbed on the PDI's active layers to give similar resistance/conductance changes. For example, the adsorption of a reducing gas, including ammonia and hydrazine, increases the conductivity of PDI sensors. Consequently, different analytes may cause the same sensing response. The functionalization of PDI with different functional groups, e.g., -COOH, -OH, -F, -Br, and -NH_2, can be considered as a suitable method to resolve this issue. Functionalization will enhance the sensing performance by improving the hydrogen bonding or dipole–dipole interaction between the PDI's active layer and the target gas. However, functionalization may affect the sensor sensitivity and response/recovery time; hence, a balance must be considered between selectivity and sensitivity. Stability is another issue for PDI-based sensors when compared with metal oxide-based sensors. Here, we have shown that stability can also be improved by core substitution with different functional groups.

Several other factors, such as film thickness, humidity, and temperature, may also influence the performance of PDI-based gas sensors in terms of sensitivity, stability, response and recovery time, and lifetime—though these are controlled. It can be argued that reducing the thickness of a sensing layer is helpful for an analyte to adsorb more quickly and to provide a swifter response. Similarly, thicker films can provide more adsorption/diffusion sites for analytes, which can increase sensitivity. Thinner films can help to induce crystallinity, which in fact is advantageous for high charge carriers' mobility and, hence, for high-performing electrical gas sensors. Thermal evaporation techniques can simply tune the film thickness by controlling the deposition rate, and the prepared ultrathin film has a superior quality with low-density defects. However, the requirement of a particular vacuum apparatus increases manufacturing cost and confines commercial processing. A change in the humidity level may also affect the sensitivity of a PDI-based sensor towards various analytes, as would more saturated mixed gases containing compounds that might act as good solvents for the active layer compound. For example, there have been reports that increasing the humidity level decreases the sensitivity towards ammonia vapor [24]. This is mainly due to the fact that there is a strong competition between the ammonia and water molecules to diffuse/adsorb through the sensing layer. Likewise, other reports have indicated that increasing the humidity level increases the sensitivity towards hydrazine vapor [50,81]. These reports reason that high humidity can be conductive for the hydration of hydrazine, and, thus, the active layer of PDIs will adsorb more hydrazine.

It is apparent that the influence of these factors on PDI-based electrical gas sensors has been rarely investigated. Thus, the effect of these factors on sensing performance needs a systematic study. The challenges moving forward might be overcome using other members of the rylene family—either smaller, more functionalizable and processable members or larger more aggregatable derivatives. We hope this review assists the readers in a judicial choice of strategy with which to affect significant advances in this novel field.

Author Contributions: Original draft preparation: S.A.; review, suggestions, and editions: A.G., M.S., and S.J.L. All authors have read and agreed to the published version of the manuscript.

Funding: We would like to gratefully acknowledge support from the Australian Research Council's Discovery Grant Scheme through DP170104477 and DP170102145. The authors acknowledge the funding support from the Faculty of Science, Engineering and Technology, Swinburne University of Technology. S Ali acknowledges PhD scholarship under the Swinburne Postgraduate Research Award (SUPRA) program.

Institutional Review Board Statement: Not applicable.

Informed Consent Statement: Not applicable.

Data Availability Statement: Data sharing not applicable.

Acknowledgments: We would like to gratefully Mohammed A. Jameel for his help in formatting the graphs and figures.

Conflicts of Interest: The authors declare no conflict of interests.

References

1. Zandi, A.; Gilani, A.; Koohsorkhi, J. An optimized resistive CNT-based gas sensor with a novel configuration by top electrical contact. *Diam. Relat. Mater.* **2019**, *93*, 224–232. [CrossRef]
2. Sun, J.; Chang, J.; Zhang, Q.; Xie, Y.; Zhang, Z.; Feng, Y.; Sun, H. Recent progress in research on TDLAS fiber optical gas sensor performance improvement. In *Second Symposium on Novel Technology of X-Ray Imaging*; International Society for Optics and Photonics: Bellingham, WA, USA, 2019; Volume 11068, p. 110680L.
3. Suma, M.; Prasad, M.; Gaddam, V.; Rajanna, K.; Nayak, M. Development of a Novel Acoustic Sensor using Sputtered ZnO Thin Film. *J. Pure Appl. Ind. Phys.* **2019**, *9*, 1–7.

4. Gao, L.; Zhang, J.; Zhai, L.; Wang, X.; Fan, L.; Hu, T. Gas adsorption and fluorescent sensing properties of two porous lanthanide metal–organic frameworks based on 3, 5-bis (2-carboxy-phenoxy)-benzoic acid. *Polyhedron* **2019**, *165*, 171–176. [CrossRef]
5. Shafiei, M.; Spizzirri, P.G.; Arsat, R.; Yu, J.; du Plessis, J.; Dubin, S.; Kaner, R.B.; Kalantar-Zadeh, K.; Wlodarski, W. Platinum/graphene nanosheet/SiC contacts and their application for hydrogen gas sensing. *J. Phys. Chem. C* **2010**, *114*, 13796–13801. [CrossRef]
6. Shafiei, M.; Sadek, A.Z.; Yu, J.; Latham, K.; Breedon, M.; McCulloch, D.; Kalantar-zadeh, K.; Wlodarski, W. A hydrogen gas sensor based on Pt/nanostructured WO3/SiC Schottky diode. *Sens. Lett.* **2011**, *9*, 11–15. [CrossRef]
7. Shafiei, M.; Bradford, J.; Khan, H.; Piloto, C.; Wlodarski, W.; Li, Y.; Motta, N. Low-operating temperature NO2 gas sensors based on hybrid two-dimensional SnS2-reduced graphene oxide. *Appl. Surf. Sci.* **2018**, *462*, 330–336. [CrossRef]
8. Jian, Y.; Hu, W.; Zhao, Z.; Cheng, P.; Haick, H.; Yao, M.; Wu, W. Gas sensors based on chemi-resistive hybrid functional nanomaterials. *Nano-Micro Lett.* **2020**, *12*, 1–43. [CrossRef]
9. Chen, S.; Gao, N.; Bunes, B.R.; Zang, L. Tunable nanofibril heterojunctions for controlling interfacial charge transfer in chemiresistive gas sensors. *J. Mater. Chem. C* **2019**, *7*, 13709–13735. [CrossRef]
10. Adgate, J.L.; Goldstein, B.D.; McKenzie, L.M. Potential public health hazards, exposures and health effects from unconventional natural gas development. *Environ. Sci. Technol.* **2014**, *48*, 8307–8320. [CrossRef]
11. Wang, C.; Bunes, B.R.; Xu, M.; Wu, N.; Yang, X.; Gross, D.E.; Zang, L. Interfacial donor–acceptor nanofibril composites for selective alkane vapor detection. *ACS Sens.* **2016**, *1*, 552–559. [CrossRef]
12. Andre, R.S.; Sanfelice, R.C.; Pavinatto, A.; Mattoso, L.H.; Correa, D.S. Hybrid nanomaterials designed for volatile organic compounds sensors: A review. *Mater. Des.* **2018**, *156*, 154–166. [CrossRef]
13. Van Hieu, N.; Thuy, L.T.B.; Chien, N.D. Highly sensitive thin film NH3 gas sensor operating at room temperature based on SnO2/MWCNTs composite. *Sens. Actuators B* **2008**, *129*, 888–895. [CrossRef]
14. Koo, W.-T.; Jang, J.-S.; Kim, I.-D. Metal-Organic Frameworks for Chemiresistive Sensors. *Chem* **2019**, *5*, 1938–1963. [CrossRef]
15. Lee, J.-H. Gas sensors using hierarchical and hollow oxide nanostructures: Overview. *Sens. Actuators B* **2009**, *140*, 319–336. [CrossRef]
16. Wang, Y.; Gong, Q.; Miao, Q. Structured and Functionalized Organic Semiconductors for Chemical and Biological Sensors Based on Organic Field Effect Transistors. *Mater. Chem. Front.* **2020**, *4*, 3505–3520. [CrossRef]
17. Jones, T.; Bott, B.; Thorpe, S. Fast response metal phthalocynanine-based gas sensors. *Sens. Actuators* **1989**, *17*, 467–474. [CrossRef]
18. Bohrer, F.I.; Colesniuc, C.N.; Park, J.; Ruidiaz, M.E.; Schuller, I.K.; Kummel, A.C.; Trogler, W.C. Comparative gas sensing in cobalt, nickel, copper, zinc, and metal-free phthalocyanine chemiresistors. *J. Am. Chem. Soc.* **2008**, *131*, 478–485. [CrossRef]
19. Jiang, Y.; Huang, W.; Zhuang, X.; Tang, Y.; Yu, J. Thickness modulation on semiconductor towards high performance gas sensors based on organic thin film transistors. *MSEB* **2017**, *226*, 107–113. [CrossRef]
20. Dong, Z.; Kong, X.; Wu, Y.; Zhang, J.; Chen, Y. High-sensitive room-temperature NO_2 sensor based on a soluble n-type phthalocyanine semiconductor. *Inorg. Chem. Commun.* **2017**, *77*, 18–22. [CrossRef]
21. Huang, Y.; Fu, L.; Zou, W.; Zhang, F.; Wei, Z. Ammonia sensory properties based on single-crystalline micro/nanostructures of perylenediimide derivatives: Core-substituted effect. *J. Phys. Chem. C* **2011**, *115*, 10399–10404. [CrossRef]
22. Huang, Y.; Fu, L.; Zou, W.; Zhang, F. Probing the effect of substituted groups on sensory properties based on single-crystalline micro/nanostructures of perylenediimide dyes. *New J. Chem.* **2012**, *36*, 1080–1084. [CrossRef]
23. Zhang, S.; Wang, X.; Huang, Y.; Zhai, H.; Liu, Z. Ammonia sensing properties of perylene diimides: Effects of core-substituted chiral groups. *Sens. Actuators B* **2018**, *254*, 805–810. [CrossRef]
24. Kalita, A.; Hussain, S.; Malik, A.H.; Subbarao, N.V.; Iyer, P.K. Vapor phase sensing of ammonia at the sub-ppm level using a perylene diimide thin film device. *J. Mater. Chem. C* **2015**, *3*, 10767–10774. [CrossRef]
25. Nketia-Yawson, B.; Noh, Y.-Y. Organic thin film transistor with conjugated polymers for highly sensitive gas sensors. *Macromol. Res.* **2017**, *25*, 489–495. [CrossRef]
26. Li, B.; Lambeth, D.N. Chemical sensing using nanostructured polythiophene transistors. *Nano Lett.* **2008**, *8*, 3563–3567. [CrossRef]
27. Tiwari, S.; Singh, A.K.; Joshi, L.; Chakrabarti, P.; Takashima, W.; Kaneto, K.; Prakash, R. Poly-3-hexylthiophene based organic field-effect transistor: Detection of low concentration of ammonia. *Sens. Actuators B* **2012**, *171*, 962–968. [CrossRef]
28. Dai, M.-Z.; Lin, Y.-L.; Lin, H.-C.; Zan, H.-W.; Chang, K.-T.; Meng, H.-F.; Liao, J.-W.; Tsai, M.-J.; Cheng, H. Highly sensitive ammonia sensor with organic vertical nanojunctions for noninvasive detection of hepatic injury. *Anal. Chem.* **2013**, *85*, 3110–3117. [CrossRef]
29. Yang, Z.; Zhuang, X.; Han, S.; Yu, J. Hole-transporting polymer dilution driven high performance organic transistor-based NO_2 gas sensor. *Mater. Lett.* **2019**, *236*, 285–288. [CrossRef]
30. Bertoni, C.; Naclerio, P.; Viviani, E.; Dal Zilio, S.; Carrato, S.; Fraleoni-Morgera, A. Nanostructured P3HT as a Promising Sensing Element for Real-Time, Dynamic Detection of Gaseous Acetone. *Sensors* **2019**, *19*, 1296. [CrossRef]
31. Dutta, G.K.; Kasthuri, S.; Marappan, G.; Jayaraman, S.V.; Sivalingam, Y.; Di Natale, C.; Nutalapati, V. Aggregation behavior in naphthalene-appended diketopyrrolopyrrole derivatives and its gas adsorption impact on surface potential. *J. Mater. Chem. C* **2019**, *7*, 9954–9965. [CrossRef]
32. Song, R.; Wang, Z.; Zhou, X.; Huang, L.; Chi, L. Gas-Sensing Performance and Operation Mechanism of Organic π-Conjugated Materials. *Chempluschem* **2019**, *84*, 1222–1234. [CrossRef] [PubMed]

33. Zhao, Y.; Guo, Y.; Liu, Y. 25th anniversary article: Recent advances in n-type and ambipolar organic field-effect transistors. *Adv. Mater.* **2013**, *25*, 5372–5391. [CrossRef] [PubMed]
34. Huang, Y.; Zhang, W.; Wang, J.; Fu, L.; Shi, J. Electron-withdrawing groups induced remarkable changes in sensory property based on single micro/nanostructure of perylenediimide derivatives. *Mater. Chem. Phys.* **2015**, *151*, 93–98. [CrossRef]
35. Huang, Y.; Zhang, W.; Zhai, H.; Li, C. Alkylsilane-functionalized perylenediimide derivatives with differential gas sensing properties. *J. Mater. Chem. C* **2015**, *3*, 466–472. [CrossRef]
36. Kozma, E.; Catellani, M.J.D. Perylene diimides based materials for organic solar cells. *Dyes Pigment.* **2013**, *98*, 160–179. [CrossRef]
37. Smith, H.M. *High Performance Pigments*; Wiley Online Library: Hoboken, NJ, USA, 2002.
38. Li, C.; Wonneberger, H. Perylene imides for organic photovoltaics: Yesterday, today, and tomorrow. *Adv. Mater.* **2012**, *24*, 613–636. [CrossRef] [PubMed]
39. Zhan, X.; Facchetti, A.; Barlow, S.; Marks, T.J.; Ratner, M.A.; Wasielewski, M.R.; Marder, S.R. Rylene and related diimides for organic electronics. *Adv. Mater.* **2011**, *23*, 268–284. [CrossRef]
40. Zang, L.; Che, Y.; Moore, J.S. One-dimensional self-assembly of planar π-conjugated molecules: Adaptable building blocks for organic nanodevices. *Acc. Chem. Res.* **2008**, *41*, 1596–1608. [CrossRef]
41. Würthner, F. Perylene bisimide dyes as versatile building blocks for functional supramolecular architectures. *Chem. Commun.* **2004**, 1564–1579. [CrossRef]
42. Huang, C.; Barlow, S.; Marder, S.R. Perylene-3, 4, 9, 10-tetracarboxylic acid diimides: Synthesis, physical properties, and use in organic electronics. *J. Org. Chem.* **2011**, *76*, 2386–2407. [CrossRef]
43. Schmidt, R.; Ling, M.M.; Oh, J.H.; Winkler, M.; Könemann, M.; Bao, Z.; Würthner, F. Core-Fluorinated Perylene Bisimide Dyes: Air Stable n-Channel Organic Semiconductors for Thin Film Transistors with Exceptionally High On-to-Off Current Ratios. *Adv. Mater.* **2007**, *19*, 3692–3695. [CrossRef]
44. Chen, S.; Slattum, P.; Wang, C.; Zang, L. Self-assembly of perylene imide molecules into 1D nanostructures: Methods, morphologies, and applications. *Chem. Rev.* **2015**, *115*, 11967–11998. [CrossRef] [PubMed]
45. Zang, L. Interfacial donor–acceptor engineering of nanofiber materials to achieve photoconductivity and applications. *Acc. Chem. Res.* **2015**, *48*, 2705–2714. [CrossRef] [PubMed]
46. Ego, C.; Marsitzky, D.; Becker, S.; Zhang, J.; Grimsdale, A.C.; Müllen, K.; MacKenzie, J.D.; Silva, C.; Friend, R.H. Attaching perylene dyes to polyfluorene: Three simple, efficient methods for facile color tuning of light-emitting polymers. *J. Am. Chem. Soc.* **2003**, *125*, 437–443. [CrossRef]
47. Chen, Z.; Zheng, Y.; Yan, H.; Facchetti, A. Naphthalenedicarboximide-vs perylenedicarboximide-based copolymers. Synthesis and semiconducting properties in bottom-gate n-channel organic transistors. *J. Am. Chem. Soc.* **2009**, *131*, 8–9. [CrossRef]
48. Huang, Y.; Quan, B.; Wei, Z.; Liu, G.; Sun, L. Self-assembled organic functional nanotubes and nanorods and their sensory properties. *J. Phys. Chem. C* **2009**, *113*, 3929–3933. [CrossRef]
49. Huang, Y.; Yuan, R.; Zhou, S. Gas phase-based growth of highly sensitive single-crystal rectangular micro-and nanotubes. *J. Mater. Chem.* **2012**, *22*, 883–888. [CrossRef]
50. Liu, H.; Wang, H.; Liu, G.; Pu, S.; Zhang, H. Ultrasensitive sensing of hydrazine vapor at sub-ppm level with pyrimidine-substituted perylene diimide film device. *Tetrahedron* **2019**, *75*, 1988–1996. [CrossRef]
51. Huang, Y.; Hu, J.; Kuang, W.; Wei, Z.; Faul, C.F. Modulating helicity through amphiphilicity—tuning supramolecular interactions for the controlled assembly of perylenes. *Chem. Commun.* **2011**, *47*, 5554–5556. [CrossRef]
52. Huang, Y.; Yan, Y.; Smarsly, B.; Wei, Z.; Faul, C. Helical supramolecular aggregates, liquid crystallinity and nanofibers of a perylenebisimide-chiral surfactant complex via ionic self-assembly. *J. Mater. Chem.* **2009**, *19*, 2356–2362. [CrossRef]
53. Schmidt, R.D.; Oh, J.H.; Sun, Y.-S.; Deppisch, M.; Krause, A.-M.; Radacki, K.; Braunschweig, H.; Könemann, M.; Erk, P.; Bao, Z. High-performance air-stable n-channel organic thin film transistors based on halogenated perylene bisimide semiconductors. *J. Am. Chem. Soc.* **2009**, *131*, 6215–6228. [CrossRef] [PubMed]
54. Delgado, M.C.R.; Kim, E.-G.; Filho, D.t.A.d.S.; Bredas, J.-L. Tuning the charge-transport parameters of perylene diimide single crystals via end and/or core functionalization: A density functional theory investigation. *J. Am. Chem. Soc.* **2010**, *132*, 3375–3387. [CrossRef] [PubMed]
55. Molinari, A.S.; Alves, H.; Chen, Z.; Facchetti, A.; Morpurgo, A.F. High electron mobility in vacuum and ambient for PDIF-CN2 single-crystal transistors. *J. Am. Chem. Soc.* **2009**, *131*, 2462–2463. [CrossRef] [PubMed]
56. Hu, J.; Kuang, W.; Deng, K.; Zou, W.; Huang, Y.; Wei, Z.; Faul, C.F. Self-Assembled Sugar-Substituted Perylene Diimide Nanostructures with Homochirality and High Gas Sensitivity. *Adv. Energy Mater.* **2012**, *22*, 4149–4158. [CrossRef]
57. Huang, L.; Wang, Z.; Zhu, X.; Chi, L. Electrical gas sensors based on structured organic ultra-thin films and nanocrystals on solid state substrates. *Nanoscale Horiz.* **2016**, *1*, 383–393. [CrossRef]
58. Huang, Y.; Zhang, W.; Wang, J.; Wei, Z. Probing the sensory property of perylenediimide derivatives in hydrazine gas: Core-substituted aromatic group effect. *ACS Appl. Mater. Interfaces* **2014**, *6*, 9307–9313. [CrossRef]
59. Muzikante, I.; Parra, V.; Dobulans, R.; Fonavs, E.; Latvels, J.; Bouvet, M. A novel gas sensor transducer based on phthalocyanine heterojunction devices. *Sensors* **2007**, *7*, 2984–2996. [CrossRef]
60. Bohrer, F.I.; Sharoni, A.; Colesniuc, C.; Park, J.; Schuller, I.K.; Kummel, A.C.; Trogler, W.C. Gas sensing mechanism in chemiresistive cobalt and metal-free phthalocyanine thin films. *J. Am. Chem. Soc.* **2007**, *129*, 5640–5646. [CrossRef]
61. Bakker, E.; Telting-Diaz, M. Electrochemical sensors. *Anal. Chem.* **2002**, *74*, 2781–2800. [CrossRef]

62. Deshpande, R. *Capacitors: Technology and Trends*; Tata McGraw-Hill Education: New York, NY, USA, 2012.
63. Sinclair, I. *Passive Components for Circuit Design*; Elsevier: Amsterdam, The Netherlands, 2000.
64. Zhang, C.; Chen, P.; Hu, W. Organic field-effect transistor-based gas sensors. *Chem. Soc. Rev.* **2015**, *44*, 2087–2107. [CrossRef]
65. Teradal, N.L.; Marx, S.; Morag, A.; Jelinek, R. Porous graphene oxide chemi-capacitor vapor sensor array. *J. Mater. Chem. C* **2017**, *5*, 1128–1135. [CrossRef]
66. Yao, M.S.; Tang, W.X.; Wang, G.E.; Nath, B.; Xu, G. MOF thin film-coated metal oxide nanowire array: Significantly improved chemiresistor sensor performance. *Adv. Mater.* **2016**, *28*, 5229–5234. [CrossRef] [PubMed]
67. Dai, J.; Ogbeide, O.; Macadam, N.; Sun, Q.; Yu, W.; Li, Y.; Su, B.-L.; Hasan, T.; Huang, X.; Huang, W. Printed gas sensors. *Chem. Soc. Rev.* **2020**, *49*, 1756–1789. [CrossRef]
68. Lundström, K.; Shivaraman, M.; Svensson, C. A hydrogen-sensitive Pd-gate MOS transistor. *J. Appl. Phys.* **1975**, *46*, 3876–3881. [CrossRef]
69. Torsi, L.; Dodabalapur, A.; Sabbatini, L.; Zambonin, P. Multi-parameter gas sensors based on organic thin-film-transistors. *Sens. Actuators B* **2000**, *67*, 312–316. [CrossRef]
70. Feliciano, G.T.; Sanz-Navarro, C.; Coutinho-Neto, M.D.; Ordejón, P.; Scheicher, R.H.; Rocha, A.R. Capacitive DNA detection driven by electronic charge fluctuations in a graphene nanopore. *Phys. Rev. Appl.* **2015**, *3*, 034003. [CrossRef]
71. Cui, H.; Cheng, C.; Lin, X.; Wu, J.; Chen, J.; Eda, S.; Yuan, Q. Rapid and sensitive detection of small biomolecule by capacitive sensing and low field AC electrothermal effect. *Sens. Actuators B* **2016**, *226*, 245–253. [CrossRef]
72. Tahir, M.; Sayyad, M.H.; Clark, J.; Wahab, F.; Aziz, F.; Shahid, M.; Munawar, M.A.; Chaudry, J.A.J.S.; Chemical, A.B. Humidity, light and temperature dependent characteristics of Au/N-BuHHPDI/Au surface type multifunctional sensor. *Sens. Actuators B* **2014**, *192*, 565–571. [CrossRef]
73. Yassine, O.; Shekhah, O.; Assen, A.H.; Belmabkhout, Y.; Salama, K.N.; Eddaoudi, M. H2S Sensors: Fumarate-Based fcu-MOF Thin Film Grown on a Capacitive Interdigitated Electrode. *Angew. Chem.* **2016**, *128*, 16111–16115. [CrossRef]
74. Lou, Z.; Li, L.; Wang, L.; Shen, G. Recent progress of self-powered sensing systems for wearable electronics. *Small* **2017**, *13*, 1701791. [CrossRef]
75. Altenberend, U.; Oprea, A.; Barsan, N.; Weimar, U. Contribution of polymeric swelling to the overall response of capacitive gas sensors. *Anal. Bioanal. Chem.* **2013**, *405*, 6445–6452. [CrossRef] [PubMed]
76. Pokhrel, S.; Simion, C.; Quemener, V.; Barsan, N.; Weimar, U. Investigations of conduction mechanism in Cr_2O_3 gas sensing thick films by ac impedance spectroscopy and work function changes measurements. *Sens. Actuators B* **2008**, *133*, 78–83. [CrossRef]
77. Neri, G. First fifty years of chemoresistive gas sensors. *Chemosensors* **2015**, *3*, 1–20. [CrossRef]
78. Bochenkov, V.; Sergeev, G. Sensitivity, selectivity, and stability of gas-sensitive metal-oxide nanostructures. *Metal Oxide Nanostruct. Their Appl.* **2010**, *3*, 31–52.
79. Torsi, L.; Magliulo, M.; Manoli, K.; Palazzo, G. Organic field-effect transistor sensors: A tutorial review. *Chem. Soc. Rev.* **2013**, *42*, 8612–8628. [CrossRef] [PubMed]
80. Huang, Y.; Wang, J.; Wei, Z. Modulating supramolecular helicity and electrical conductivity of perylene dyes through an achiral alkyl chain. *Chem. Commun.* **2014**, *50*, 8343–8345. [CrossRef]
81. Wang, J.; He, E.; Liu, X.; Yu, L.; Wang, H.; Zhang, R.; Zhang, H. High performance hydrazine vapor sensor based on redox mechanism of twisted perylene diimide derivative with lower reduction potential. *Sens. Actuators B* **2017**, *239*, 898–905. [CrossRef]
82. Huang, Y.; Wang, J.; Fu, L.; Kuang, W.; Shi, J. Effect of core-substituted groups on sensing properties based on single micro/nanorod of perylenediimide derivatives. *Sens. Actuators B* **2013**, *188*, 411–416. [CrossRef]
83. Liu, X.; Zhai, H.; Zhang, S.; Fu, J.; Huang, Y. Impact of alkyl chain and bay-phenoxy unit on the sensing performance of perylenediimide derivatives. *Sens. Actuators B* **2017**, *243*, 500–506. [CrossRef]
84. Li, C.; Zhai, H.; Liu, X.; Zhang, W.; Huang, Y. Effect of stereoisomerism of the alkyl chain on the gas sensing properties based on perylene dyes. *J. Mater. Chem. C* **2015**, *3*, 2778–2782. [CrossRef]
85. Huang, Y.; Zhang, S.; Zhong, G.; Li, C.; Liu, Z.; Jin, D. Highly responsive hydrazine sensors based on donor–acceptor perylene diimides: Impact of electron-donating groups. *Phys. Chem. Chem. Phys.* **2018**, *20*, 19037–19044. [CrossRef] [PubMed]
86. Wang, X.; Li, C.; Huang, Y.; Zhai, H.; Liu, Z.; Jin, D. Highly sensitive and stable perylene sensor for ammonia detection: A case study of structure-property relationships. *Sens. Actuators B* **2018**, *275*, 451–458. [CrossRef]
87. Deng, Q.; Zhou, E.; Huang, Y.; Qing, W.; Zhai, H.; Liu, Z.; Wei, Z. Chalcogen-substitution modulated supramolecular chirality and gas sensing properties in perylenediimides. *Chem. Commun.* **2019**, *55*, 4379–4382. [CrossRef] [PubMed]
88. Gaudillat, P.; Wannebroucq, A.; Suisse, J.-M.; Bouvet, M. Bias and humidity effects on the ammonia sensing of perylene derivative/lutetium bisphthalocyanine MSDI heterojunctions. *Sens. Actuators B* **2016**, *222*, 910–917. [CrossRef]
89. Che, Y.; Yang, X.; Liu, G.; Yu, C.; Ji, H.; Zuo, J.; Zhao, J.; Zang, L. Ultrathin n-type organic nanoribbons with high photoconductivity and application in optoelectronic vapor sensing of explosives. *J. Am. Chem. Soc.* **2010**, *132*, 5743–5750. [CrossRef]
90. Wu, N.; Wang, C.; Bunes, B.R.; Zhang, Y.; Slattum, P.M.; Yang, X.; Zang, L. Chemical self-doping of organic nanoribbons for high conductivity and potential application as chemiresistive sensor. *ACS Appl. Mater. Interfaces* **2016**, *8*, 12360–12368. [CrossRef]
91. Huang, Y.; Liu, X.; Wang, Q.; Fu, J.; Zhao, L.; Liu, Z.; Jin, D. Highly responsive ethylenediamine vapor sensor based on a perylenediimide–camphorsulfonic acid complex via ionic self-assembly. *J. Mater. Chem. C* **2017**, *5*, 7644–7651. [CrossRef]
92. Zhu, P.; Wang, Y.; Ma, P.; Li, S.; Fan, F.; Cui, K.; Ge, S.; Zhang, Y.; Yu, J. Low-Power and High-Performance Trimethylamine Gas Sensor Based on nn Heterojunction Microbelts of Perylene Diimide/CdS. *Anal. Chem.* **2019**, *91*, 5591–5598. [CrossRef]

93. Liu, Q.; Zhao, C.; Shao, X.; Wang, W.; Ji, X. High sensitive pyridine chemiresistive sensors based on azacyclobutane modified perylene bisimide derivatives. *Dyes Pigment.* **2021**, *185*, 108902. [CrossRef]
94. Ye, H.; Chen, L.; Wang, X.; Lu, D. A highly sensitive fluorescent probe for hydrazine detection: Synthesis, characterisation and application in living cells. *Int. J. Environ. Anal. Chem.* **2019**, 1–13. [CrossRef]
95. Wang, G.; Zhang, C.; He, X.; Li, Z.; Zhang, X.; Wang, L.; Fang, B. Detection of hydrazine based on Nano-Au deposited on Porous-TiO_2 film. *Electrochim. Acta* **2010**, *55*, 7204–7210. [CrossRef]
96. U.E.P. Agency. Integrated Risk Information System (IRIS) on Benzene. 2002. Available online: http://yosemite.epa.gov/sab/sabproduct.nsf/d21b76bff879fa0a8525735a00766807/d3bb75d4297ca46985257943005-22ace!OpenDocument (accessed on 8 October 2020).
97. Appl, M. Ammonia: Principles and industrial practice. *Future* **1999**, *245*, 12.
98. Pandey, S.; Goswami, G.K.; Nanda, K.K. Nanocomposite based flexible ultrasensitive resistive gas sensor for chemical reactions studies. *Sci. Rep.* **2013**, *3*, 2082. [CrossRef] [PubMed]
99. Wang, J.; Yang, P.; Wei, X. High-performance, room-temperature, and no-humidity-impact ammonia sensor based on heterogeneous nickel oxide and zinc oxide nanocrystals. *ACS Appl. Mater. Interfaces* **2015**, *7*, 3816–3824. [CrossRef] [PubMed]
100. Malins, C.; Doyle, A.; MacCraith, B.; Kvasnik, F.; Landl, M.; Šimon, P.; Kalvoda, L.; Lukaš, R.; Pufler, K.D.; Babusik, I. Personal ammonia sensor for industrial environments. *J. Environ. Monit.* **1999**, *1*, 417–422. [CrossRef] [PubMed]
101. Nigst, T.A.; Antipova, A.; Mayr, H. Nucleophilic reactivities of hydrazines and amines: The futile search for the α-effect in hydrazine reactivities. *J. Org. Chem.* **2012**, *77*, 8142–8155. [CrossRef] [PubMed]

chemosensors

MDPI

Communication

A Syringe-Based and Centrifugation-Free DNA Extraction Procedure for the Rapid Detection of Bacteria

Taehwi Yoon, Seokjoon Kim, Jung Ho Kim and Ki Soo Park *

Department of Biological Engineering, College of Engineering, Konkuk University, Seoul 05029, Korea; yth4784@konkuk.ac.kr (T.Y.); ghjghy@konkuk.ac.kr (S.K.); jung9511@konkuk.ac.kr (J.H.K.)
* Correspondence: akdong486@konkuk.ac.kr

Abstract: Several bacteria are known to cause food poisoning; therefore, diagnostic systems that detect bacteria are essential. Nucleic acid-based testing methods that involve polymerase chain reaction (PCR) amplification are of great interest due to their high sensitivity and specificity. Herein, we developed a syringe-based one-step DNA extraction device that streamlines the extraction of genomic DNA (gDNA) from bacteria within 2 min, enabling versatile application of nucleic acid-based testing in the field. Notably, the bolt-nut structured case coupled with the syringe enables control of the volume of solution dispensed for enabling DNA extraction without the need for bulky centrifuge equipment. Using the proposed system, the gDNA of a model bacterium, *Escherichia coli,* was extracted at a good quantity and quality and amplified via PCR. The DNA extracted was comparable to that extracted via a centrifugation-based procedure. In addition, bacteria that were artificially spiked in common samples, including a work cloth, a work bench, and meat, were successfully detected with high accuracy.

Keywords: DNA extraction; syringe-based device; centrifugation-free extraction; food contamination; quantitative polymerase chain reaction

Citation: Yoon, T.; Kim, S.; Kim, J.H.; Park, K.S. A Syringe-Based and Centrifugation-Free DNA Extraction Procedure for the Rapid Detection of Bacteria. *Chemosensors* **2021**, *9*, 167. https://doi.org/10.3390/chemosensors9070167

Academic Editor: Pietro Salvo

Received: 1 June 2021
Accepted: 30 June 2021
Published: 3 July 2021

Publisher's Note: MDPI stays neutral with regard to jurisdictional claims in published maps and institutional affiliations.

1. Introduction

Food poisoning caused by bacteria or viruses is a major public health problem that additionally incurs various social problems [1]. To prevent the risk of food poisoning, the development of diagnostic systems that detect specific target bacteria is essential. In particular, coliforms that produce the enzyme β-galactosidase are widely used as an indicator of the microbial safety of food, processed products, and drinking water [2,3]. *Escherichia coli* (*E. coli*) is one of the most studied coliforms and several studies have attempted to develop diagnostic systems for *E. coli*. The standard method used to discriminate *E. coli* involves microbial culturing using selective media, such as MacConkey agar, that contains different chromogenic or fluorogenic substrates [4–6]. However, culture-based methods require 1–2 days or more to generate results [7], and may produce false-positive results. For example, lactose-fermenting bacteria, such as *Lactobacillus* spp., which are non-coliform bacteria, can generate the same results as those obtained using *E. coli*, resulting in false-positive results because MacConkey agar can only differentiate bacteria based on their ability to ferment lactose. Therefore, polymerase chain reaction (PCR)-based nucleic acid testing, especially quantitative PCR (qPCR), owing to its high accuracy, is widely used as an alternative method to detect bacteria, such as *E. coli* and other coliforms, that cause food poisoning [8–11]. The first step in qPCR involves the extraction of genomic DNA (gDNA) from target bacteria [12,13]. Two representative methods are widely used for gDNA extraction: organic extraction and solid-phase extraction [14]. The organic extraction method is inexpensive and can extract a large amount of gDNA. However, it requires toxic chemicals and is time-consuming, owing to several steps, such as phenol–chloroform extraction, ethanol precipitation, and washing. In contrast, the solid-phase extraction method, which uses a silica membrane for interacting with DNA in the presence of chaotropic salts, presents a better option for the rapid extraction of gDNA because it is performed using a single tube and is a relatively simple procedure [15,16]. Several commercial kits manufactured and marketed by different companies are available. In principle, gDNA that is bound to silica membranes in the presence of chaotropic salts via the formation of salt bridges is eluted by changing the salt concentration after multiple washing steps [17]. While it is a simple and quick operation, centrifugation is required in each step, which is not desirable in field applications, especially in a resource-limited setting [18]. In addition, there is an attempt to skip DNA extraction, which is termed Direct-PCR. As the name indicates, this method can directly amplify target nucleic acids without the need for DNA extraction from the sample [19,20]. Because it simplifies the whole assay procedure and shortens the total analysis time, it has attracted special attention as an excellent alternative to the traditional DNA detection methods. However, Direct-PCR buffer that effectively lyses the sample and neutralizes PCR inhibitors is not compatible with other DNA amplification methods, such as recombinase polymerase amplification and helicase dependent amplification. Furthermore, Direct-PCR reagents are expensive and cannot completely minimize some PCR inhibitors, leading to inefficient PCR amplification and reduced detection sensitivity.

In this study, we developed a centrifugation-free and syringe-based one-step DNA extraction device for in-field detection of bacteria that cause food poisoning, which maximizes user convenience and reduces the duration of the overall assay. A bolt and nut-structured case was designed to fit a syringe, allowing the user to control the volume of the solution dispensed while avoiding the need for bulky centrifugation equipment. As a model target bacterium, *E. coli* was selected and its gDNA was extracted using both the proposed and standard, centrifugation-based procedures. We confirmed that gDNA extracted using the syringe-based system was of a suitable quantity and quality required for subsequent qPCR analysis, and was comparable to gDNA extracted via centrifugation-based assays. In addition, bacteria spiked in various samples were successfully detected with high accuracy.

2. Materials and Methods

2.1. Primer Design

DNA primers (forward primer: 5′-GCCATTGCACCGACAAAACT-3′; reverse primer: 5′-ACCAAGCATTCCGCCGATAA-3′) were designed using the Basic Logical Alignment Search Tool of the National Center for Biotechnology Information (National Institutes of Health, Bethesda, MD, USA) and were synthesized by Bionics (Seoul, Korea). The DNA primers were specific to the *ybbW* gene of *E. coli* that encodes a putative allantoin transporter and is recently found to be present in most strains of *E. coli* [21,22].

2.2. Bacterial Cultivation and Centrifugation-Based DNA Extraction

All bacterial strains used in this study (*E. coli* (KCTC 2441), *Enterobacter cloacae* (KCTC 2519), *Klebsiella pneumoniae* (ATCC 70063), *Pseudomonas aeruginosa* (ATCC 27853), and *Staphylococcus aureus* (ATCC 29213)) were purchased from the Korean Collection for Type Cultures (Daejeon, Korea). After the bacteria were grown in Luria-Bertani (LB) medium (BD, Franklin Lakes, NJ, USA) at 37 °C for 24 h, the culture solution was centrifuged at 5000× *g* for 5 min to obtain the cell pellet. The gDNA was then isolated using the Total DNA Extraction S&V kit (Bionics) according to the manufacturer's instructions. Briefly, a bacterial cell lysate was prepared by adding 200 μL of lysis buffer and 20 μL of Proteinase K to the bacteria cells, which were incubated at 56 °C for 30 min and subsequently mixed with 500 μL of GDX buffer. Subsequently, binding, washing, and elution were performed to obtain the purified gDNA. Between each step, centrifugation was performed at 13,000× *g* for 2 min. The amount of extracted DNA was evaluated using a Nanodrop instrument (Spectramax iD5 multi-mode microplate reader; Molecular Devices, San José, CA, USA).

2.3. Fabrication of Syringe-Based DNA Extraction Device

The syringe-based DNA extraction device contained three major components. The first component is a silica membrane filter that binds to gDNA at a high salt concentration.

The second component is a syringe (Korea Vaccine, Seoul, Korea) whose inner space is compartmentalized to sequentially flush the bacterial cell lysate, wash buffer, and elution buffer. A punctuated rubber packing at the center with an inverted bowl shape was placed inside the syringe (Figure 1A). The upper compartment was filled with 200 μL of elution buffer, and the lower compartment was filled with 750 μL of wash buffer. The final component involved a bolt and nut-structured case constructed using a Creality Ender 5 3D printer (Shenzhen, China), and was coupled with the syringe for enabling consistent control of the solution dispensed. The silica membrane filter, wash buffer, and elution buffer were all obtained from the Total DNA extraction S&V Kit.

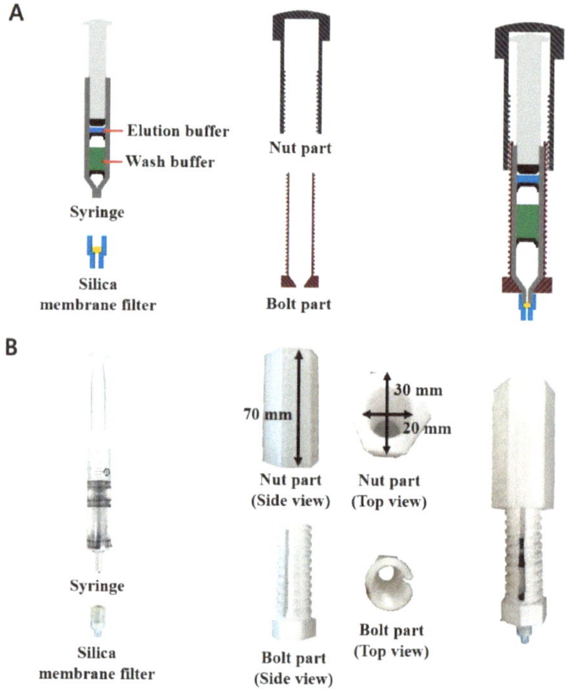

Figure 1. Syringe-based one-step DNA extraction device. (**A**) Schematic illustration of each component of the syringe-based DNA extraction device. The elution buffer and wash buffer fill the syringe, compartmentalized with the inverted bowl-shaped rubber packing. (**B**) Photographs of each component of the syringe-based DNA extraction device.

2.4. *Operation of Syringe-Based DNA Extraction Device*

Bacterial cell lysates were prepared as described in Section 2.2. The prepared bacterial cell lysates were then drawn using the syringe, which was then assembled together with the bolt and nut-structured case and silica membrane filter. By turning the bolt and nut-structured case, the bacterial cell lysate and 750 μL of wash buffer present in the syringe were sequentially ejected through the silica membrane filter. Finally, 200 μL of elution buffer present in the syringe was ejected through the silica membrane filter to collect the purified gDNA, which was then analyzed via qPCR.

2.5. *qPCR Assay*

qPCR assays were performed using a CFX Connect Real-Time PCR Detection System (Bio-Rad, Hercules, CA, USA); the reaction mix (total volume, 10 μL) comprised TOPreal qPCR 1X PreMIX (SYBR Green with low ROX) (Enzynomics, Korea), forward/reverse DNA primers (100 nM), and the extracted gDNA at different concentrations. Amplification conditions were as follows: 10 min at 95 °C, followed by 40 cycles of 15 s at 95 °C, 15 s at 58 °C, and 30 s at 72 °C.

2.6. Spike-and-Recovery Analysis

Three model samples (size, 10 cm × 10 cm) found in the food industry, including work clothes (sample 1), work bench (sample 2), and meat (sample 3), were spiked with cultures of *E. coli* at different concentrations. Specifically, the solution containing *E. coli* was poured onto our model samples, which were then distributed with a pipette and dried. Subsequently, the spiked *E. coli* were collected using pipette swabs, which were immediately submerged in 10 mL of saline containing 0.85% sodium chloride (3M Pipette Swab Plus; 3M, Saint Paul, MN, USA). Finally, 1 mL of the solution was spread onto LB agar plates, and the number of cells was determined as colony-forming units (CFU)/mL by plating 10-fold serial dilutions. Simultaneously, 1 mL of solution was used to extract gDNA using the syringe-based DNA extraction device, which was analyzed via qPCR as described in Section 2.5.

3. Results and Discussion

3.1. Design of the Syringe-Based DNA Extraction Device

Figure 1A shows the schematic illustration of the syringe-based one-step DNA extraction device comprising three major components: (i) a silica membrane filter that is capable of binding to gDNA at a high salt concentration; (ii) a syringe where the inner space is filled with wash and elution buffers in each compartment by the inverted bowl-shaped rubber packing with a small hole at the center; (iii) a bolt and nut-structured case that enables the users to eject the solution through a silica membrane filter in a controlled and consistent manner. Figure 1B shows photographs of the syringe-based DNA extraction device. It is a portable, one-step device for the in-field detection of target bacteria.

The operational procedure for extracting gDNA from bacterial cell lysates is illustrated in Figure 2A. In the first step (Step 1), the bacterial cell lysate is sucked into a syringe. Next, a silica membrane filter is assembled to the syringe inlet, and the cell lysate and wash buffer present in the syringe are sequentially ejected through a silica membrane filter by turning the bolt and nut-structured case (Steps 2–4). The rubber packing that contains a small hole at the center was designed to contain an inverted bowl (arcuate) shape to prevent the solution from leaking in the absence of pressure. The arch-shaped rubber packing moves to the bottom of the syringe and changes into the original bowl shape only when pressure is applied, thereby discharging the solution through the silica membrane filter (Figure S1). In addition, we created an empty space for air in the compartment containing the wash buffer because it is advisable to dry the silica membrane filter after the washing step. As shown in Figure 2B, the proposed syringe-based DNA extraction system assembled with the bolt and nut-structured case allows for simple and consistent control of solution dispensed during the DNA extraction procedure. Overall, all the steps are performed using a single tube within 2 min, and the assay does not require bulky centrifuge devices. Therefore, the syringe-based one-step DNA extraction device is suitable for in-field diagnosis of target bacteria.

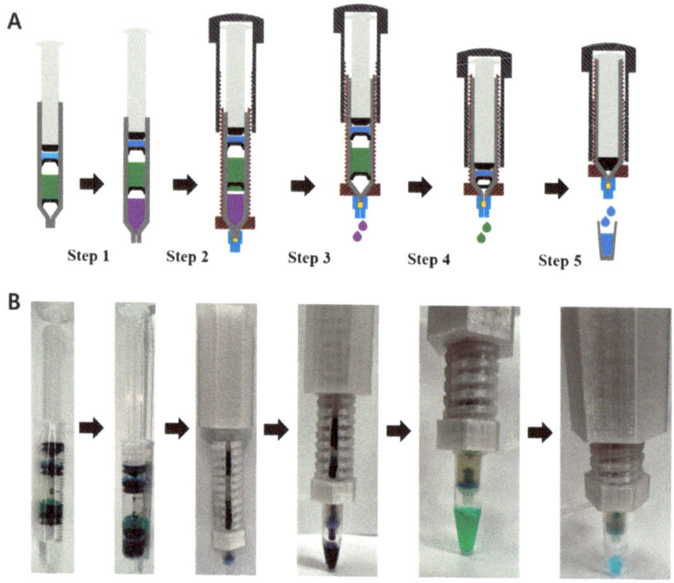

Figure 2. Operation procedure of the syringe-based one-step DNA extraction device. (**A**) Schematic illustration of the DNA extraction procedure using the syringe-based DNA extraction device. (**B**) Photographs of the syringe-based DNA extraction device at each step of the extraction procedure.

3.2. Performance of Syringe-Based DNA Extraction Device

We used the proposed system to extract gDNA from a model bacterium, *E. coli*, and estimated the gDNA amount (45.23 ng/µL) and the ratio of absorbance at 260 and 280 nm (A260/280; 1.64) (Table 1). For comparison of efficiency, we extracted gDNA from *E. coli* at the same number (2.2×10^8 CFU) using a commercial centrifugation-based DNA extraction kit (see the details in Section 2). As shown in Table 1, there was no significant difference in gDNA extracted using the syringe-based and centrifugation-based methods. Although the accuracy of the syringe-based method was relatively lower than the centrifugation-based one, there were no issues with the subsequent PCR amplification.

Table 1. DNA extracted from *Escherichia coli* using syringe-based and centrifugation-based extraction methods.

DNA Extraction Method	*E. coli* (Cell Number, CFU)	A260/A280	Concentration (ng/µL)	Time
Syringe-based method	2.2×10^8	1.64 ± 0.23	45.23 ± 11.7	~2 min
Centrifuge-based method	2.2×10^8	1.72 ± 0.09	30.83 ± 3.6	~30 min

Data are expressed as mean ± standard deviation; $n = 3$ technical replicates. A260/A280, ratio of absorbance at 260 nm to absorbance at 280 nm; CFU, colony-forming unit.

3.3. qPCR Analysis of Extracted gDNA

The extracted *E. coli* gDNA at different amounts was analyzed via qPCR. As shown in Figure 3A, a linear relationship was observed for the logarithmic values of *E. coli* gDNA concentrations ranging from 20 fg/µL to 2 ng/µL over 6 orders of magnitude ($R^2 = 0.9906$). Similarly, qPCR was performed using different amounts of the extracted *E. coli* gDNA, which were compared using the syringe-based DNA extraction device. Both systems were found to have a high correlation ($R^2 = 0.9875$) and could detect a very low concentration of DNA (20 fg/µL = 4 copies/µL) based on qPCR analysis (Figure 3B). Therefore, gDNA extracted using the syringe-based device was of excellent quality, and there was a minor loss of gDNA during the extraction process.

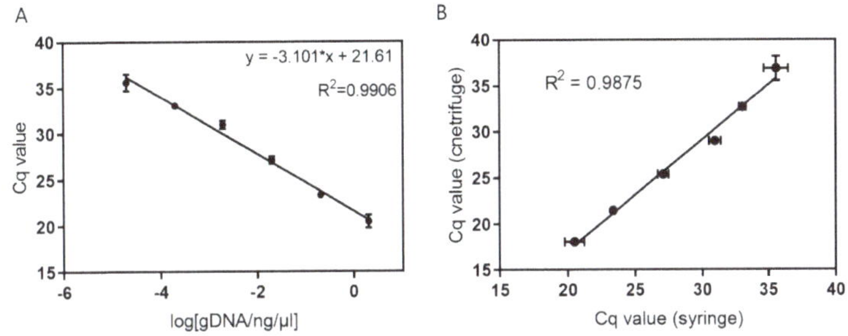

Figure 3. Detection sensitivity of *Escherichia coli* gDNA. (**A**) Linear correlation of the logarithmic value of *Escherichia coli* gDNA concentration with Cq values. (**B**) Comparison of Cq values obtained using the syringe-based (x-axis) and centrifugation-based (y-axis) extraction procedures. Data are expressed as mean $\pm$ standard deviation; n = 3 technical replicates. gDNA, genomic DNA; Cq, cycle quantification value.

To evaluate the selectivity of the procedure, we performed qPCR using gDNA extracted from different bacteria (*E. cloacae*, *K. pneumoniae*, *P. aeruginosa*, *S. aureus*, and *E. coli*). Since the DNA primer was specifically designed for the *ybbW* gene of *E. coli*, it was expected that PCR amplification would occur only in the presence of *E. coli* gDNA. As shown in Figure 4, PCR amplification products were only obtained for *E. coli*, and were clearly distinguished from the PCR products of other bacteria (*E. cloacae*, *K. pneumoniae*, *P. aeruginosa*, and *S. aureus*). In addition, the change in the quantification cycle (ΔCq) value of *E. coli*, calculated by subtracting the Cq value in the presence of *E. coli* gDNA from 40 in a non-template control, was found to be 13.07, which is significantly higher than the ΔCq values of other bacteria. These results indicate that gDNA extracted using the syringe-based device is suitable for performing subsequent amplification using specific DNA primers, demonstrating that the extracted gDNA can be used in downstream applications.

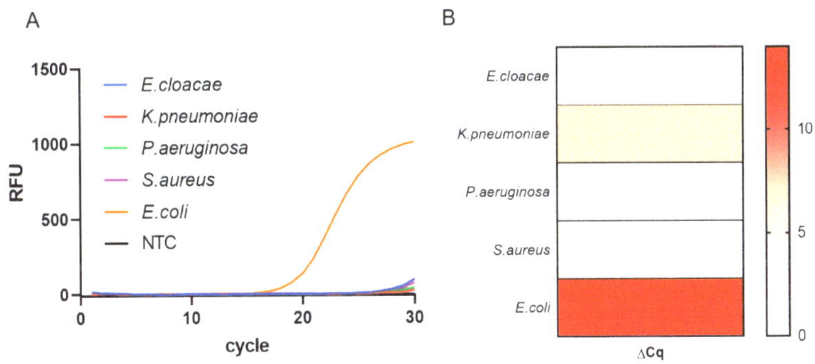

Figure 4. Detection specificity of *Escherichia coli* gDNA. (**A**) Real-time polymerase chain reaction amplification curves of gDNA (2 ng/µL) extracted from different bacteria. (**B**) Heatmap of delta Cq values of gDNA (2 ng/µL) extracted from different bacteria. The delta Cq was calculated by subtracting the Cq value of the control from the Cq value of each sample. Data are expressed as mean $\pm$ standard deviation; n = 3 technical replicates. gDNA, genomic DNA; Cq, cycle quantification; RFU, relative fluorescence unit; NTC, non-template control.

3.4. *Spike-and-Recovery Analysis*

Finally, a spike-and-recovery test was performed to determine the utility of the syringe-based one-step DNA extraction device in practical applications. Three samples (work cloth, work bench, and meat) that are susceptible to contamination in food factories were selected and spiked with *E. coli* at different concentrations. After performing a lysis of the collected samples, the gDNA was extracted using our syringe-based device and analyzed via both qPCR and cell culturing (see the details in Section 2). The number of cells obtained using the qPCR and culturing methods was similar, with excellent recovery rates of 98–103% and a relative standard deviation (RSD) of <15% (Table 2). These results confirm that the proposed syringe-based device can be used to extract DNA and detect bacteria present in common samples.

Table 2. Results of spike-and-recovery assay using syringe-based DNA extraction device.

Samples	Cell Count in Spiked Sample (CFU/mL)	Cell Count Detected Via Syringe Method (CFU/mL)	Recovery (%)	RSD (%)
Sample 1 (work cloth)	2.6×10^5	2.68×10^5	103.00	0.37
Sample 2 (work bench)	3.3×10^7	3.34×10^7	101.23	14.93
Sample 3 (meat)	4.5×10^6	4.41×10^6	98.08	5.96

CFU, colony-forming unit; RSD, relative standard deviation.

4. Conclusions

In summary, we developed a simple, syringe-based one-step DNA extraction device comprising a silica membrane filter, a commercially available syringe, and a specially designed bolt and nut-structured case. The proposed system was used to extract gDNA of good quantity and quality, which was successfully used in a downstream application, qPCR. In addition, *E. coli* present in spiked samples were successfully detected with excellent recovery rates and RSD. The performance of our system is comparable to that of commercial, centrifugation-based procedures; however, the syringe-based method completes the extraction process within 2 min without the need for bulky centrifuge equipment. We believe that this portable, one-step device in combination with isothermal nucleic acid amplification methods has practical potential for use in the on-field detection of various bacteria, especially in resource-limited settings.

Supplementary Materials: The following are available online at https://www.mdpi.com/article/10.3390/chemosensors9070167/s1, Figure S1: The operational procedure showing how different solutions are sequentially dispensed. Green, red, and blue arrows indicate the pressure uniformly distributed over the rubber packing, the pressure concentrated on the center of the rugger packing, and the pressure that does not applied on the rubber packing, respectively.

Author Contributions: Conceptualization, T.Y. and S.K.; methodology, T.Y., S.K. and J.H.K.; validation, T.Y., S.K. and J.H.K.; formal analysis, T.Y. and S.K.; investigation, T.Y., S.K. and J.H.K.; writing-original draft preparation, T.Y.; writing-review and editing, K.S.P.; supervision, K.S.P.; funding acquisition, K.S.P. All authors have read and agreed to be published version of the manuscript.

Funding: This research was supported by the National Research Foundation of Korea (NRF) grant funded by the Korean government (MSIT) (No. NRF-2020R1C1C1012275), and Korea Institute of Energy Technology Evaluation and Planning (KETEP) and the Ministry of Trade, Industry and Energy (MOTIE, 20194010201900).

Institutional Review Board Statement: Not applicable.

Informed Consent Statement: Not applicable.

Data Availability Statement: Not applicable.

Conflicts of Interest: The authors declare no conflict of interest.

References

1. World Health Organization. *WHO Estimates of the Global Burden of Foodborne Diseases: Foodborne Disease Burden Epidemiology Reference Group*; World Health Organization: Geneva, Switzerland, 2015; p. 225.
2. Christen, L.; Davidson, J.; McAllister, S.; Roth, L. *Coliform and Other Indicator Bacteria: In Standard Methods for the Examination of Dairy Products*; RT Marshal; American Public Health Association: Washington, DC, USA, 1992.
3. Leclerc, H.; Mossel, D.; Edberg, S.; Struijk, C. Advances in the bacteriology of the coliform group: Their suitability as markers of microbial water safety. *Annu. Rev. Microbiol.* **2001**, *55*, 201–234. [CrossRef] [PubMed]
4. Chapman, P.; Siddons, C.A.; Zadik, P.; Jewes, L. An improved selective medium for the isolation of Escherichia coli O 157. *J. Med. Microbiol.* **1991**, *35*, 107–110. [CrossRef] [PubMed]
5. Klein, H.; Fung, D.Y. Identification and quantification of fecal coliforms using violet red bile agar at elevated temperature. *J. Milk Food Technol.* **1976**, *39*, 768–770. [CrossRef]
6. Edberg, S.C.; Allen, M.J.; Smith, D.B. National field evaluation of a defined substrate method for the simultaneous enumeration of total coliforms and Escherichia coli from drinking water: Comparison with the standard multiple tube fermentation method. *Appl. Environ. Microbiol.* **1988**, *54*, 1595–1601. [CrossRef] [PubMed]
7. Hartman, P.A. Further studies on the selectivity of violet red bile agar. *J. Milk Food Technol.* **1960**, *23*, 45–48. [CrossRef]
8. Molina, F.; López-Acedo, E.; Tabla, R.; Roa, I.; Gómez, A.; Rebollo, J.E. Improved detection of Escherichia coli and coliform bacteria by multiplex PCR. *BMC Biotechnol.* **2015**, *15*, 1–9. [CrossRef] [PubMed]
9. Reza, Z.M.; Mohammad, A.; Salomeh, K.; Reza, A.G.; Hossein, S.; Maryam, S.; Azam, A.; Mana, S.; Negin, N.; Reza, K.A. Rapid detection of coliforms in drinking water of Arak city using multiplex PCR method in comparison with the standard method of culture (Most Probably Number). *Asian Pac. J. Trop. Biomed.* **2014**, *4*, 404–409.
10. Hwang, S.H.; Kwon, W.Y.; Eun, H.; Jeong, S.; Park, J.S.; Kim, K.J.; Kim, H.J.; Lee, S.H.; Park, K.; Yoon, J.-J. The use of a 2-aminopurine-containing split G-quadruplex for sequence-specific DNA detection. *Artif. Cells Nanomed. Biotechnol.* **2018**, *46*, S950–S955. [CrossRef] [PubMed]
11. Holst-Jensen, A.; Rønning, S.B.; Løvseth, A.; Berdal, K.G. PCR technology for screening and quantification of genetically modified organisms (GMOs). *Anal. Bioanal. Chem.* **2003**, *375*, 985–993. [CrossRef] [PubMed]
12. Cenis, J. Rapid extraction of fungal DNA for PCR amplification. *Nucleic Acids Res.* **1992**, *20*, 2380. [CrossRef] [PubMed]
13. Bustin, S.A. Why the need for qPCR publication guidelines? —*The case for MIQE. Methods* **2010**, *50*, 217–226. [PubMed]
14. Doyle, J.J.; Doyle, J.L. A rapid DNA isolation procedure for small quantities of fresh leaf tissue. *Phytochemistry* **1987**, *19*, 11–15.
15. Boom, R.; Sol, C.; Salimans, M.; Jansen, C.; Wertheim-van Dillen, P.; Van der Noordaa, J. Rapid and simple method for purification of nucleic acids. *J. Clin. Microbiol.* **1990**, *28*, 495–503. [CrossRef] [PubMed]
16. Hill-Ambroz, K.L.; Brown-Guedira, G.L.; Fellers, J.P. Modified rapid DNA extraction protocol for high throughput microsatellite analysis in wheat. *Crop Sci.* **2002**, *42*, 2088–2091. [CrossRef]
17. Aljanabi, S.M.; Martinez, I. Universal and rapid salt-extraction of high quality genomic DNA for PCR-based techniques. *Nucleic Acids Res.* **1997**, *25*, 4692–4693. [CrossRef] [PubMed]
18. Mogg, R.; Bond, J. A cheap, reliable and rapid method of extracting high-quality DNA from plants. *Mol. Ecol. Notes* **2003**, *3*, 666–668. [CrossRef]
19. Ben-Amar, A.; Oueslati, S.; Mliki, A. Universal direct PCR amplification system: A time- and cost-effective tool for high-throughput applications. *3 Biotech.* **2017**, *7*, 246. [CrossRef] [PubMed]
20. Cavanaugh, S.E.; Bathrick, A. Direct PCR amplification of forensic touch and other challenging DNA samples: A review. *Forensic Sci. Int. Genet.* **2018**, *32*, 40–49. [CrossRef] [PubMed]
21. Walker, D.I.; McQuillan, J.; Taiwo, M.; Parks, R.; Stenton, C.A.; Morgan, H.; Mowlem, M.C.; Lees, D.N. A highly specific Escherichia coli qPCR and its comparison with existing methods for environmental waters. *Water Res.* **2017**, *126*, 101–110. [CrossRef] [PubMed]
22. Kim, J.M.; Park, J.S.; Yoon, T.H.; Park, J.; Park, K.S. Nucleic acid lateral flow assay for simultaneous detection of hygiene indicator bacteria. *Anal. Bioanal. Chem.* **2021**, *413*, 1–9.

chemosensors

MDPI

Review

Algal Biosensors for Detection of Potentially Toxic Pollutants and Validation by Advanced Methods: A Brief Review

Diego Serrasol do Amaral [1], Luana Vaz Tholozan [2], Daisa Hakbart Bonemann [1], Cristina Jansen-Alves [1], Wiliam Boschetti [3], Diogo La Rosa Novo [3], Neftali Lenin Villarreal Carreno [2] and Claudio Martin Pereira de Pereira [1,*]

1 Laboratory of Innovation and Solutions in Chemistry, Bioforensics Research Group, Federal University of Pelotas, Pelotas 96160-000, Brazil; diego.serrasol@ufpel.edu.br (D.S.d.A.); daisa.bonemann@ufpel.edu.br (D.H.B.); crisjansen4@gmail.com (C.J.-A.)

2 Graduate Program in Materials Science and Engineering, Technology Development Center, Federal University of Pelotas, Pelotas 96010-000, Brazil; luana.tholozan@ufpel.edu.br (L.V.T.); neftali@ufpel.edu.br (N.L.V.C.)

3 Graduate Program in Chemistry, Center for Chemical, Pharmaceutical, and Food Sciences, Federal University of Pelotas, Pelotas 96160-000, Brazil; wiliam.boschetti@edu.ufpel.br (W.B.); diogo.novo@ufpel.edu.br (D.L.R.N.)

* Correspondence: claudiochemistry@gmail.com

Abstract: The presence of potentially toxic pollutants, such as pesticides and metal ions, even at low concentrations, can significantly impact aquatic environmental health. This pollution is a globally widespread problem and requires fast and reliable analysis, especially for in-situ identification/quantification. Atomic absorption spectrometry and plasma-based spectrometry techniques have been considered the most analytical tools used to monitor potentially toxic metal ions in aquatic media and other related matrices. The dynamics of global climate change and its correlation with pollution, especially from anthropogenic sources, have encouraged the development of other faster analytical tools for monitoring these pollutants. A noteworthy alternative for determining potentially toxic pollutants is using algae-based biosensors, resulting in a cost reduction and simplification of environmental analysis, enabling a more reliable comprehension of the role of humans in climate change. These biosensors, which may not have the highest sensitivity in quantification, have demonstrated remarkable potential in the identification of potentially toxic pollutants and several field applications. Biosensors can be an excellent biotechnology solution for monitoring global environmental changes. Thus, this review highlights the main advances in developing and comparing algae-based biosensors and other analytical possibilities for the identification of potentially toxic pollutants and their possible applications in environmental analysis.

Keywords: algae-based biosensors; metal ions determination; environmental analysis; aquatic environmental health; herbicide determination

Citation: Amaral, D.S.d.; Tholozan, L.V.; Bonemann, D.H.; Jansen-Alves, C.; Boschetti, W.; Novo, D.L.R.; Carreno, N.L.V.; Pereira, C.M.P.d. Algal Biosensors for Detection of Potentially Toxic Pollutants and Validation by Advanced Methods: A Brief Review. *Chemosensors* **2024**, *12*, 235. https://doi.org/10.3390/chemosensors12110235

Received: 30 September 2024
Revised: 30 October 2024
Accepted: 11 November 2024
Published: 13 November 2024

1. Introduction

Potentially toxic pollutants can bioaccumulate in the environment from both natural and human sources. Human activities, such as mining, smelting, mineral refining, and industrial production, are mainly responsible for most pollution in the natural environment. Industrial effluents may result in several risks to aquatic and human health contaminating water resources through runoff or intentional release of wastewater [1]. Mercury, lead, cadmium, chromium, and arsenic have no biological function and can cause mental, genetic, and morphological abnormalities in humans [2]. Furthermore, these toxic elements are persistent in the environment, and harmful to humans, animals, and plants even at low concentrations (Hg: 0.003 mg L^{-1}; Pb: 6 mg L^{-1}; Cd: 10 mg L^{-1}; As and Cr: 0.05 mg L^{-1}) [2]. Many herbicides, such as simazine, exhibit similar environmental persistence and toxicity, posing a significant risk to aquatic ecosystems by inhibiting photosynthetic activity [3,4].

Furthermore, the widespread use of these chemicals can lead to groundwater contamination, further complicating the challenge of maintaining safe water quality standards [5,6]. The bioaccumulation of potentially toxic elements can cause health problems, such as intoxication, cancer, and mental disorders [4].

Chemical analyses of water and sediments have been conducted to monitor toxic elements in aquatic environments [7,8]. The analytical techniques most frequently used to identify and quantify the presence of elements in aquatic environments are Flame Atomic Absorption Spectrometry (F AAS), Graphite Furnace Atomic Absorption Spectrometry (GF AAS), Flame Atomic Emission Spectrometry (F AES), Inductively Coupled Plasma Optical Emission Spectrometry (ICP OES), Microwave Induced Plasma Optical Emission Spectrometry (MIP OES), and Inductively Coupled Plasma Mass Spectrometry (ICP–MS) [8,9]. Although those analytical techniques have been reported for metal ion determination in water and sediment samples, with high accuracy and precision, they can be considered expensive, requiring trained personnel, and laboratory settings [5,10,11]. In this sense, research into innovative and economical alternatives for the specific and rapid detection of these pollutants is crucial.

In this context, biosensors have emerged as a modern analytical strategy for identifying and quantifying pollutants in aquatic environments. Unlike physicochemical analyzers, which quantify toxic substances, biosensors can measure toxicity. They offer several advantages, including lower analysis costs and faster response than spectrometric techniques [12,13]. Biosensors work as devices connected to biological recognition elements, which are associated with a transducer that converts a stimulus into a measurable unit of energy. Whole algae, cyanobacteria—or algae components—can be utilized as biological recognition elements in biosensors [12].

Algae is a heterogeneous group inhabiting various planet regions: oceans, rivers, lakes, soils, and organism surfaces [14,15]. There are two main types of algae: macroalgae, also known as seaweeds, inhabiting coastal zones and encompass varieties (green, brown, and red algae); and microalgae, found in both benthic and coastal habitats, distributed throughout the oceans as phytoplankton [16]. Furthermore, algae are a promising alternative for carbon sequestration due to photosynthesis [15].

Microalgae possess physiological and biochemical characteristics that make them promising candidates for application in biosensors aimed at monitoring heavy metals and pesticides in contaminated environments [17–19]. These unicellular organisms are able to detect and respond to toxic contaminants through well-established detoxification mechanisms, which include initial adsorption at the cell wall, followed by intracellular accumulation and activation of specific antioxidant systems. In response to heavy metals, species such as *Chlorella vulgaris* and *Scenedesmus obliquus* perform passive adsorption on the cell surface, utilizing functional groups such as carboxyl and hydroxyl to retain metal ions. Metals may also be chelated by metallothioneins and phytochelatins, which protect the cell by sequestering these ions in cellular compartments such as vacuoles [19]. Additionally, microalgae activate antioxidant enzymes such as superoxide dismutase (SOD) and catalase (CAT), which neutralize reactive oxygen species (ROS) generated by oxidative stress, thereby preserving cellular viability [17].

For pesticides, microalgae like *Scenedesmus obliquus* promote compound decomposition into less toxic by-products, such as phthalic acid, via the action of enzymes like glutathione S-transferase (GST). This process is further supported by antioxidants, including glutathione (GSH) and the ascorbate cycle, which mitigate the oxidative stress induced by the pesticide, demonstrating the ability of these microalgae to act as effective biosensors for environmental monitoring of both organic and inorganic pollutants [18]. Algae can be used as biosensors for chemical elements, providing an accurate analysis of the potential risk to human and environmental well-being [20–22]. Thus, algae-based biosensors for the detection of potentially toxic metal ions are an interesting topic aiming at environmental and agroecological contamination problems. This review article is focused on using algae or their components for application in biosensors for the detection of potentially toxic elements (Figure 1). This review highlights the latest advances in biosensors derived from algae and their components for potentially toxic pollutant detection *in situ* or *ex situ*.

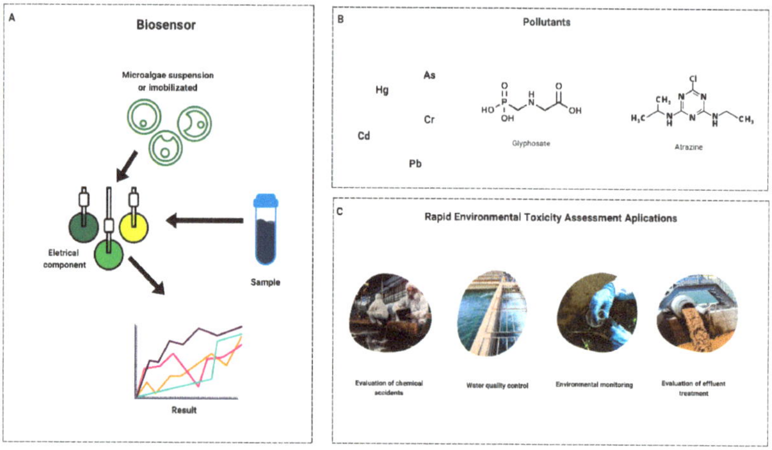

Figure 1. Biosensors based on algae and their applications. (**A**) Biosensor components. (**B**) Toxic elements. (**C**) Locals for in situ use biosensors based on algae.

2. Possibilities for Metal Ions and Pesticides Determination: Advantages and Limitation

Classical analytical methods for detecting potentially toxic metal ions, such as gravimetric and volumetric assays, were widely used in the mid-20th century. Gravimetric analysis, which measures the mass of an isolated substance in an insoluble form, and volumetric analysis, which determines the concentration of the substance based on the volume of a reagent solution, were fundamental to analytical chemistry [20–25]. However, these methods have limitations, including low sensitivity, detection limits above 10 mg L^{-1}, and the need for significant quantities of reagents, making them less suitable for modern environmental analysis [26,27].

Recent advancements in analytical technologies have addressed the limitations of classical methods for the identification and quantification of potentially toxic metals, yielding reliable results characterized by precision, accuracy, repeatability, and reproducibility. Atomic spectrometry techniques, including atomic absorption spectrometry (AAS) and plasma-based methods, play a crucial role in metal determination [7]. The choice of analytical technique depends on factors such as the number of elements to be determined, sample form (solid or solution), sensitivity, linear range, interferences, and the cost and maintenance of the instrumentation [8,28]. AAS employs various atomizers, such as flame or graphite furnace, which differ in sensitivity and operational efficiency; the graphite furnace offers higher sensitivity with lower detection limits (in µg L^{-1}) compared to the flame atomizer (in mg L^{-1}), albeit at a higher operational cost [7].

Plasma-based analytical techniques, such as inductively coupled plasma optical emission spectrometry (ICP OES) and inductively coupled plasma mass spectrometry (ICP-MS), have gained prominence due to their high sensitivity and multielemental capabilities, enabling the determination of concentrations ranging from µg L^{-1} to mg L^{-1} [9,29]. Despite their advantages, these techniques are often hindered by high acquisition and operational costs due to substantial argon gas consumption [29]. Given the ongoing exploration of new analytical methods, biosensors have emerged as a promising alternative, offering significant benefits in application, cost, versatility, and robustness by detecting signals produced from the interaction between analytes and biological components, which correlates with metal concentrations [5,30].

The detection of pesticides in environmental ecosystems can be achieved through various analytical techniques, including gas chromatography (GC) and liquid chromatography (LC). GC is predominantly employed for volatile and non-polar compounds, utilizing detectors such as electron capture detectors (ECD), flame photometric detectors (FPD), and mass spectrometry (MS) for enhanced sensitivity [31,32]. High-performance liquid chromatography (HPLC), on the other hand, is particularly suitable for high-polarity pesticides and can be paired with classical detectors such as ultraviolet (UV) and fluorescence, as well as MS, which offers exceptional sensitivity without the need for derivatization [33].

In addition, capillary electrophoresis (CE) is an effective analytical technique that requires minimal reagents and sample volumes, providing high separation efficiency. Despite its limitations in sensitivity due to the small capillary diameter, CE is often coupled with sensitive detection methods, including MS (Sánchez-Hernández et al., 2014) [34], and integrated with pre-concentration techniques. Recent studies have demonstrated the application of innovative approaches, such as electrokinetic supercharging (EKS) and biomimetic immunoassays (BI-CE), for efficient pesticide detection, offering more accessible alternatives to conventional methods [32,35,36].

Recent developments in plasma-based techniques, such as Inductively Coupled Plasma Optical Emission Spectrometry (ICP-OES) and Inductively Coupled Plasma Mass Spectrometry (ICP-MS), have significantly improved detection capabilities. These methods allow multielement analysis with detection limits down to $\mu g\ L^{-1}$ or less but also have high operational costs [37–42]. Challenges of these advanced techniques include their susceptibility to spectral and non-spectral interferences and the need for extensive sample preparation [42–45]. In response to these challenges, there is increasing interest in the development of alternative methods, such as algal biosensors. These biosensors provide a sustainable and cost-effective solution for in situ metal ion and pesticide detection by leveraging biological interactions for sensitive and robust environmental monitoring [3,6,46].

3. Algae-Based Biosensors

3.1. Pollutants Capture Properties of Algae

Algae, as photosynthetic organisms found in various aquatic habitats, are essential for the health and biodiversity of aquatic ecosystems [47]. They act as primary producers in food chains, provide nourishment for invertebrates and fish, and contribute to nutrient recycling and oxygen production [48]. In addition, algae help maintain the stability of coastal environments by filtering pollutants and preventing coastal erosion [49]. Due to their rich composition of vitamins, proteins, lipids, pigments, polysaccharides, and bioactive compounds, algae are utilized in various industries such as food, cosmetics, and pharmaceuticals [50,51].

Algae have only been used in developing biosensors in the last two decades [14]. In this sense, the use of algae as biosensors has been reported as a low-cost and eco-friendly option for environmental monitoring [52–55]. Their cell walls rich in polysaccharides, proteins, and lipids (Figure 2) give them a high capacity for adsorbing contaminant molecules, such as metal ions and pesticides, through functional groups that offer negatively charged sites: carboxyl, hydroxyl, amino, sulfate, and carboxylate; while physiological processes, such as active transport, contribute to the internalization of these substances [56–58]. The carboxyl group is recognized as a key factor in adsorption, with carboxylic moieties able to bind metal ions, such as Cd^{2+}, through two distinct mechanisms: either by replacing the H^{+} in the carboxyl group (type-I) or, after prolonged exposure, forming carboxylates (type-II) [59]. This adsorption process follows a biphasic mechanism. In the initial, rapid phase, metal ions are adsorbed by various components of the microalgal cell wall and its functional groups [60]. This non-metabolic phase is influenced by several factors, including pH, temperature, metal concentration, biosorbent dosage, and contact time, with the microalgal strain, contact time, and pH being the most critical parameters [61], while temperature primarily affects the kinetics rather than the capacity [62].

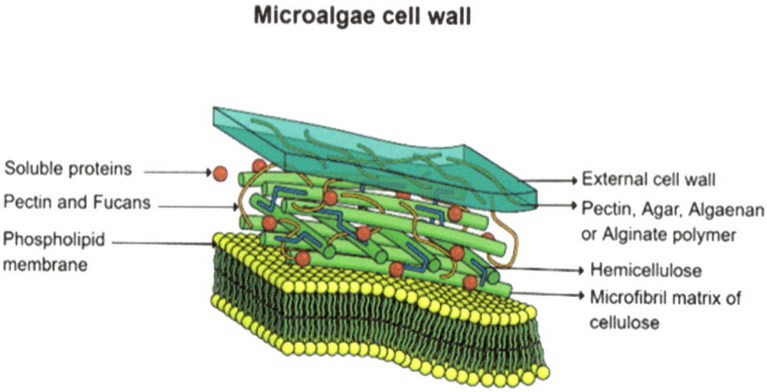

Figure 2. Microalgae cell wall and its components (adapted [63]).

Additionally, the nature of the metal ion also affects the process; cationic metal uptake is enhanced at higher pH, while anionic ion removal is more efficient at lower pH [64,65]. In the second, metabolic phase, metal ions are absorbed and accumulated within the cells through active transport and binding to proteins like phytochelatins or metallothioneins, making this phase slower [17,18,65]. Unlike the first phase, the second phase is dependent on factors such as temperature and the metabolic state, as active metal transport requires energy [19]. This affinity for metal ions has led to the development of algae-based biosensors, such as whole-cell biosensors constructed utilizing the microalga *Chlorella vulgaris* as the recognition component, with its fluorescence response employed as the measurement parameter to detect the presence of titanium dioxide (TiO_2) and silver (Ag) nanoparticles in water. This biosensor demonstrated significant sensitivity, with a detection limit of 1×10^{-3} mg L^{-1} [66–71].

Similarly for pesticides, hydroxyl, carboxyl, and amine groups have been identified as the predominant active surface groups in the adsorption of 2,4-dichlorophenoxyacetic acid (2,4-D) by *Gracilaria verrucosa* [67]. The cell wall of microalgae is composed of a fibrillar matrix of carbohydrates, intercellular spaces, and sulfated polysaccharides, which facilitate the adsorption of organic contaminants from water [68,69]. Pesticide adsorption depends on two aspects: the growth of the microalgae, and the chemical structure of the pesticide and organism-related factors (microalgae) [70]. Therefore, several studies use microalgae for bioremediation and the development of sensors for pesticides. Moro et al. (2018) used the inhibition of microalgae fluorescence to identify pesticides in seawater. In this study, fluorescence parameters of several species of microalgae were analyzed in the presence of three common marine pesticides that act as photosynthesis inhibitors. The three pollutants were detected in a 10-min interval, at concentrations ranging from ng L^{-1} to μg L^{-1}. The different species of microalgae demonstrated slightly different sensitivities to pesticides, with *Chlorella mirabilis* being the most sensitive [71].

3.2. Features of Algae-Based Biosensor

Algae-based biosensors were initially employed in the 1980s and 1990s, involving the cultivation of algae that underwent alterations in their cellular growth and photosynthetic activity in the presence of pollutants [12,14]. Biosensors are currently applied in a wide range of areas such as industrial process control, food control, and environmental monitoring [5,6]. Biosensors are considered highly promising tools for the detection of potentially toxic pollutants due to their specificity, low cost, portability, real-time monitoring capability, fast response time, ease of handling, compactness, sensitivity, user-friendliness, and reliability [72,73]. Furthermore, biosensors have a fast response time, are portable, and can be used for in situ applications, as is their ability to assess the biotoxicity and bioavailable of pollutants in living beings. Biosensing technology represents a synergistic combination of nanotechnology, biotechnology, and microelectronics. These analytical devices rely primarily on a biological sensing component capable of recognizing the target analyte with the signal transduced by an appropriate transduction element [72–75]. The construction of an effective biosensor involves several steps: selection of the bioreceptor, characterization and adaptation to be used as a sensor, selection of a transducer, system assembly, signal amplification, and verification of the real efficiency of the biosensor. A bioreceptor or biological recognition element is a biological entity that specifically reacts with the molecule to be analyzed and produces a measurable signal. Meanwhile, the transducer is the device that converts the signal from the bioreceptor into electrical energy. Biosensors can utilize different analytes, biological receptors, and transducers. The careful selection of each biosensor component is crucial, as it directly affects the sensitivity and specificity of detection; therefore, optimization is necessary to adjust the biosensor to the specific pollutant and environmental conditions [12,14,74].

Conceptually, all biomolecules and molecular assemblies have the ability to recognize target analytes and then become a bio-receiver. The first elements to be used in biorecognition for biosensors were elements of a living system [75]. Biorecognition elements mainly include whole cells, enzymes, molecularly imprinted polymers, and others, such as aptamers and antibodies, with greater application in the healthcare field [74]. Bacteria [76], yeasts [77], cyanobacteria [78], and microalgae [66] have been utilized as bioreceptors in these devices' fabrication. The principle of using bio-recognition elements in algae-based biosensors for the detection of potentially toxic pollutants is based on the ability of these elements to selectively and specifically bind to the target compound [79]. When the algae binds to the toxic compound, it triggers a signal detected and quantified by the biosensor. This signal is typically a change in fluorescence, conductivity, or other physical or chemical property that can be measured to determine the concentration of the pesticides, chemicals, toxic compounds, and metal ions in the sample [80–83].

Among the various biosensors, those based on algae have shown a good perspective on potentially toxic pollutant detection. These biosensors are adaptable to adverse conditions, reproducible with low nutrient requirements, and more stable than enzyme-based biosensors. Microalgae biosensors utilize intact microalgal cells to detect pesticides by monitoring changes in photosynthesis, measured through variations in oxygen production or chlorophyll fluorescence. A paper-based biosensor with whole cells of the microalgae *Chlamydomonas reinhardtii* immobilized on a paper substrate for herbicide detection through chlorophyll fluorescence has been developed [48,54]. The results showed a decrease in variable fluorescence inversely proportional to the herbicide concentration (0.5–200 nmol L^{-1}), indicating a linear relationship in the measured dose–response curves and a limit of detection of 4 pmol L^{-1}. In addition, the biosensor demonstrated satisfactory storage stability for up to three weeks. Over the past two decades, microalgae-based biosensors have been extensively developed for pollutant detection due to their high sensitivity (in the μg L^{-1} range), sustainability, and ease of handling, since microalgae can be cultivated on a large scale at a low cost, and remain active in high-pH environments, reducing contamination by other microorganisms. By using bio-recognition elements specific to the target metal ion, algae-based biosensors can provide susceptible and selective detection of potentially toxic pollutants in environmental samples. This allows for rapid and reliable monitoring of water quality and the early detection of contamination events, enabling timely interventions to protect human health and the environment [54,75–78].

One of the main steps in biosensor development is the immobilization of the biological recognition element. In the case of an algae-based biosensor, the microalgae cells are immobilized, on a surface or solid matrix (as long their detection ability is maintained) or used as a cell suspension [14]. In addition, immobilization is crucial to secure the stability, reproducibility, practicability, and durability of the biosensor [78,79]. Different immobilization procedures have been adopted based on the suitability of the biological element and transducing system, involving physical methods like entrapment, encapsulation, and adsorption or chemical methods like covalent bonding and cross-linking [20,22,54,84,85].

Physical methods are primarily used to immobilize whole cells or cellular organelles. The thickness and porosity of membranes used for encapsulation and entrapment significantly affect the performance of biocomponents, often resulting in reduced sensitivity and slower response times [79,86]. In this sense, a study examining the fluorescence of various microalgae species (*C. vulgaris*, *P. subcapitata*, and *C. reinhardtii*) immobilized with a calcium alginate matrix further immobilized in an inorganic silica matrix as a biosensor demonstrated that the immobilized microalgae exhibited lower sensitivity in comparison with the algae in suspension (limits of detection 0.001 μmol L^{-1} and <0.001 μmol L^{-1}). These findings suggest that the immobilization of whole-cell biosensors does not compromise their functionality. Furthermore, photosynthetic activity measured by fluorescence decreased in a dose-dependent manner consistently across all three species [86], indicating that immobilization can protect cells (resulting in greater durability of the biosensor) and allow the study of highly toxic metal ions and in larger quantities without dilutions.

After preparation, the algae-based biosensor is exposed to the medium containing the analyte. At this stage, the immobilized algae are put directly in contact with the target substances (Figure 3). The high biological selectivity of the algal receptors enables the specific interaction of the analyte, such as metal ions, through biochemical interactions. Certain algae species demonstrate a remarkable affinity for metal ions such as Cd^{2+}, Cr^{6+}, and Zn^{2+} [46,52,55]. The interaction between the analyte and the biological components of the biosensor triggers biochemical changes in the algae, which are essential for subsequent signal transduction. These biochemical alterations, resulting from the algae–analyte interaction, are measured by detectable optical and electrochemical signals through physicochemical transducers. Electrochemical biosensors function by detecting chemical reactions through variations in the electrical properties of the solution. These systems require a three-electrode configuration: a working electrode, a counter electrode, and a reference electrode [80,82]. The working electrode serves as the transducer where the reaction takes place; the counter electrode completes the circuit and applies current to the working electrode; and the reference electrode maintains a stable potential. Electrochemical biosensors are further categorized into conductometric, potentiometric, and amperometric types [73]. Optical biosensors, on the other hand, comprise an optical transducer and bioreceptor molecules, converting biological events into electrical signals by inducing changes in the light's absorption, transmission, reflection, refraction, phase, amplitude, frequency, or polarization in response to the physicochemical alterations caused by biorecognition events [83].

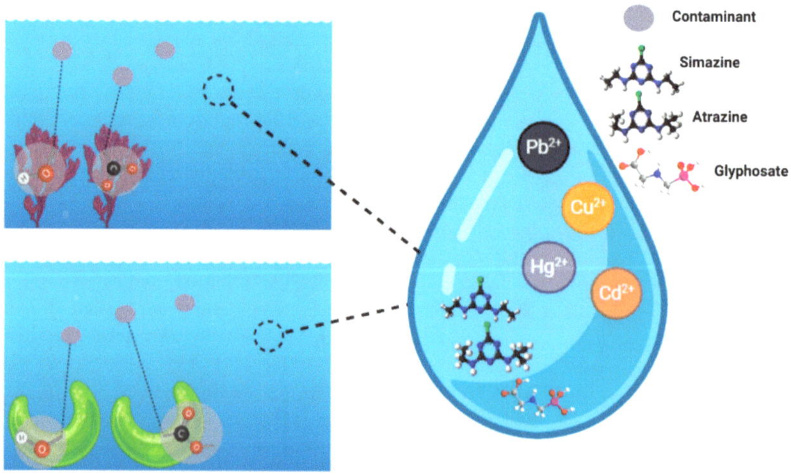

Figure 3. Chemical interaction between the algae and different contaminants.

Algae-based biosensors can be utilized for the specific detection of potentially polluting elements through specific response patterns. The identification of a specific element by means of a biosensor occurs through the comparative analysis of the response patterns generated by different strains of microalgae. Each element induces a distinct alteration in the fluorescence of the microalgae, which, when recorded and compared with a database of known patterns, allows for the identification of the contaminant. Podola and Melkonian

The use of the microalgae *Chlamydomonas reinhardtii* as a bioindicator for the presence of pesticides has been investigated [82]. Electrochemical biosensors to monitor the metabolic activity of the algae through the detection of O_2, H_2O_2, and H_3O^+/OH^- (pH-related ions) were carried out. The platinum black biosensor showed the highest sensitivity for O_2 and H_2O_2, whilst Pt/IrO_2 showed the greatest sensitivity for pH. In optical biosensors, this transduction typically involves measuring changes in the natural fluorescence of the algae. The fluorescence intensity varies according to the analyte concentration, allowing the indirect quantification of the target substance. Alternatively, electrochemical biosensors can monitor changes in current or potential generated by the biological interaction. The choice of transducer type depends on the specific properties of the analyte and the characteristics of the developed biosensor [22].

Algae-based biosensors can be utilized for the specific detection of potentially polluting elements through specific response patterns. The identification of a specific element by means of a biosensor occurs through the comparative analysis of the response patterns generated by different strains of microalgae. Each element induces a distinct alteration in the fluorescence of the microalgae, which, when recorded and compared with a database of known patterns, allows for the identification of the contaminant. Podola and Melkonian (2005) studied various strains of microalgae as biosensors, among which the strains *Tetraselmis cordiformis* and *Scherffelia dubia* demonstrated high sensitivity to herbicides such as diuron and isoproturon, creating a response pattern that enabled the precise identification of these compounds in water based on prior standardization with reference substances [87]. Another alternative for specific detection of potentially polluting elements is the use of genetically modified microalgae that are resistant to certain compounds [88,89]. Haigh-Flórez et al. (2014) developed a dual-head biosensor employing strains of *Dictyosphaerium chlorelloides*, which are sensitive and resistant to simazine, immobilized in porous silicone films. The sensitive strain shows a significant decrease in O_2 production in the presence of the herbicide, while the resistant strain remains unaffected, enabling selective detection of the compound. The photosynthetic activity of the strains was monitored using an integrated luminescent O_2 sensor. The device provides in situ herbicide concentration measurements every 180 min, with a detection limit of 12 $\mu g\ L^{-1}$ and a working range of 50–800 $\mu g\ L^{-1}$ [90].

Biosensors for specific detection and biosensors for general toxicity detection differ in terms of prior preparation and analysis methods. Biosensors using specific response patterns are designed to identify and quantify individual compounds, such as herbicides, through the analysis of differentiated responses from various strains of microalgae, previously evaluated against these contaminants and thus defining a standard response. This type of biosensor utilizes more than one microalgal strain and compares the variations in chlorophyll fluorescence of these microalgae exposed to standards and environmental samples [87]. In contrast, biosensors aimed at detecting general toxicity may use only one strain and assess the overall impact of all contaminants present in a sample, without comparison to previously established response patterns. These biosensors measure total toxicity, reflecting the aggregated response of the microalgae to the environmental sample, without determining which element is responsible for the generated toxicity [14].

3.3. Optical Biosensor Using Algal Fluorescence

A bioluminescent biosensor using microalgae *Chlorella vulgaris* as a biological component for detecting several metal ions in water has been developed [20]. The tests revealed a fast detection time of 15 min, and a broad pH tolerance of 6–8, and contributed to cell growth monitoring through the analysis of cell density. Optical biosensors were also developed, exploring the inhibition of electron transfer in terms of variations in chlorophyll fluorescence emission in the presence of pollutants [91]. In this context, algal biomass can be immobilized on surfaces to measure its fluorescence when in contact with metal ions.

A biosensor comprising of microalgae *Mesotaenium* sp. and cyanobacteria *Synechococcus* sp., immobilized in a 96-well microplate using silica, exhibited satisfactory performance in detecting Cd^{2+}, Cr^{6+}, and Zn^{2+} in solution, increasing the fluorescence values in 10 min of exposure. Also, the biosensors developed presented excellent storage stability of 4 weeks for microalgae and 8 weeks for cyanobacteria [46]. A whole-cell biosensor with the microalga *Scenedesmus subspicatus* was used to detect Cd^{2+}, Cu^{2+}, and Zn^{2+} in water has been reported [55]. The results revealed a satisfactory response, with limits of detection of 0.90, 0.91, and 0.88 mg L^{-1} for Cd, Cu, and Zn, respectively. Metal can influence fluorescence, making it a valuable indicator for biomonitoring environmental pollution, particularly in aquatic systems [92]. Consequently, algal biomass emerges as an economical and efficient material for the selective removal and recovery of metal ions from industrial wastewater or other sources [93,94]. The algae fluorescence can also be analyzed together with other parameters, such as cell growth and photosynthetic activity in oxygen production (Figure 4). These parameters are affected by the presence of contaminants in the cultivation medium [12].

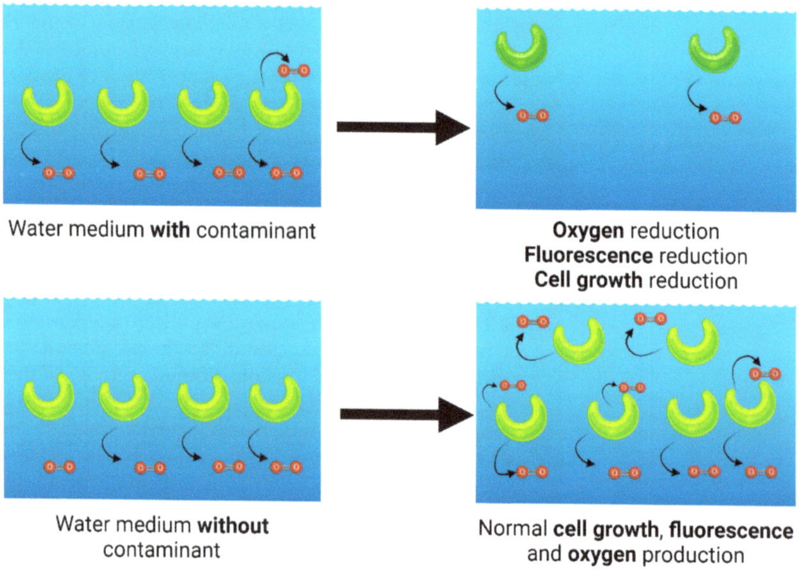

Figure 4. Example of parameters analyzed by an algae-based biosensor.

In addition to the fluorescence generated by photosynthesis, the fluorescence of other components that bind to metal ions can be quantified. A recent study developed an optical biosensor based on the intracellular fluorescence of microalgae genetically modified to facilitate the entry of metal ions into their cells [95]. The absorption and intracellular storage of metal ions in microalgae, focusing on the species *Chlamydomonas reinhardtii*, were investigated. The production of the biosensor through the insertion of a plasmid with fluorescent proteins fused to a metallothionein and in vitro metal ion binding studies was evaluated. The results highlighted the biosensor's ability to quantify free toxic metal ions in microalgae. Among the metal ions evaluated, the biosensor presented a higher sensitivity for Hg^{2+} followed by Cd^{2+}, Pb^{2+}, Zn^{2+}, and Cu^{2+}. This study demonstrates the potential of microalgae to be used in diverse and efficient ways as biosensors [95]. Along with fluorescence, dissolved oxygen is also an important parameter for evaluating the presence of pollutants, as it is directly related to the photosynthetic activity of the algae.

Another advantage of biosensors lies in their versatility, as they can be effectively utilized not only for environmental analysis but also for the detection of pesticide residues in food products. Liu et al. (2023) reported the successful creation of a microalgae-based, confinement-enhanced optical biosensor that achieved accurate detection of the pesticide atrazine at remarkably low concentrations. In this study, *Chlamydomonas reinhardtii* cells were confined in situ in microgel traps, eliminating common issues related to optical signal instability caused by cell movement and light scattering at high concentrations. This methodology facilitated spatial phase stabilization and reduced multiple scattering, enabling linear detection of atrazine over a range of 0.04–100 μg L^{-1}. Double-blind tests on commercial samples of corn and sugarcane juice demonstrated the system's high accuracy, with an average bias of 1.661 μg L^{-1} for corn and 3.144 μg L^{-1} for sugarcane samples, underscoring the biosensor's robustness for pesticide analysis in contaminated food products [96].

3.4. Electrochemical Biosensor by Algae Oxygen Production

Amperometric measurement and evaluation of photosynthetic activity based on oxygen production have proven to be a promising technique for developing algae-based environmental biosensors [97]. These biosensors measure the reduction of a redox mediator or monitor oxygen production through an electrode. Oxygen electrode-based biosensors have demonstrated good sensitivity, long operational life, and suitability for aquatic environmental monitoring [98]. Photosynthetic response as a function of light intensity, along with the amperometric activity, while the microalgae and cyanobacteria were in contact with the pollutant, has been investigated [73,78]. In this system, a sensor using *Pseudokirchneriella subcapitata, Desmodesmus quadricauda, Microcystis aeruginosa,* and *Synechococcus elongatus* in an electrochemical sensor for toxicological detection was developed, detecting the contaminant through changes in the electrical properties of the solution [80,82]. The material presented satisfactory results with a quick response time, affordability, ease of use, and the potential for development into a fully automated system [78].

Conductometric biosensors operate based on the production or consumption of ionic species during metabolic processes, causing changes in the electrical conductivity of the electrolytic solution, which can be measured to determine the sensor's response [52,80]. Potentiometric biosensors convert biological interactions between the sensor and the contaminant into measurable electrical signals [99]. On the other hand, amperometric biosensors measure changes in current resulting from the reduction or oxidation of electroactive species on the electrode surface, with the potential between the working and reference electrodes kept constant during current measurement [100].

A biosensor based on *Pseudokirchneriella subcapitata* to determine the presence of diuron, simetryn, simazine, and atrazine in water, was investigated, focused on oxygen production, along with fluorescence intensity [101]. In 10 min, the biosensor showed a limit of detection of 1.0 μg L^{-1} for diuron, and 10 μg L^{-1} for simetryn, simazine, and atrazine using a sample volume of only 200 μL, demonstrating that the material is fast, low-cost, and eco-friendly. The monitoring of atrazine in water was also investigated using a portable microalgae *Scenedesmus acutus* and *Monoraphidium contortum* sensor [86]. The microalgae were immobilized in polyelectrolyte-surfactant-carbon nanotube self-assembled material cast on a screen-printed graphite electrode, which was able to monitor the oxygen reduction through photosynthesis inhibition with a detection limit of 0.11 μmol L^{-1}, indicating a promising response for river samples. Also, the biosensor presented satisfactory stability, maintaining its integrity over 5 months of immersion in a freshwater algae medium at room temperature. The use of microbial fuel cells (MFC) to develop a photosynthetic MFC sensor for detecting the presence of formaldehyde in water has been studied [21]. To inoculate the MFCs, two species of microalgae from a pilot high-rate algal pond (HRAP) were used: *Scenedesmus obliquus* and *Chlorella luteoviridis*. The photo MFC sensor was able to demonstrate the interference of the contaminant in the photosynthetic activity of the algae, detecting the presence of formaldehyde in concentrations ranging from 0.001 to 0.02%. In addition, the electrogenic activity of the microalgae in the sensor demonstrated a cost-effective and rapid analytical method for detection compared to traditional biological analysis.

An electrochemical biosensor based on the green microalga *Scenedesmus* sp. MN738556 aimed to evaluate the biotoxicity of Cd^{2+} has been developed [52]. The biosensor was constructed by immobilizing *Scenedesmus* sp. microalgae on a glassy carbon electrode surface using bovine serum albumin. Its responses were based on the chronoamperometric currents generated by the activity of alkaline phosphatase. The sensitivity was verified with a half inhibition concentration of 0.26 μg L^{-1} for Cd^{2+} at pH 8.5 and 10^7 cells mL^{-1}. Meanwhile, an electrochemical biosensor using algae (*Desmodesmus quadricauda* and *Pseudokirchneriella subcapitata*) and cyanobacteria (*Microcystis aeruginosa* and *Synechococcus elongatus*) has been studied [78]. The strategy involved immobilizing the microalgae and cyanobacteria along with a mediator in a Sephadex gel structure, allowing the detection of microbial reduction of the mediator. The results showed a good correlation with the reference analytical methods, such as the Algal Growth Inhibition Test and the Microtox Test, highlighting the ease of handling and reduced measurement time as advantages of this new biosensor.

An alternative approach to the development of algae-based biosensors involves utilizing algal biomass as a sustainable substrate for the chemical coating of silver nanoparticles, which can be employed for the electrochemical detection of pesticides. According to Ameen et al. (2023), this study used self-grown biomass of *Spirulina platensis*, modified with chemical silver, as a powder amplifier in the construction of a carbon paste electrode, with 1-hexyl-3-methylimidazolium hexafluorophosphate as a conductive ligand. The determination of atropine was conducted using voltammetric methods, demonstrating that the electrochemical behavior of atropine is pH dependent, with pH 10.0 identified as the optimal condition for analysis. The fabricated sensor exhibited a linear response within the concentration range of 0.01–800 μM, achieving an impressive detection limit of 5 nM for atropine. The results also confirmed the stability, reproducibility, and selectivity of the sensor, as evidenced by recovery percentages of 94.48–101.58% for atropine sulfate vials and 98.01–101.3% for water samples [101].

Table 1 presents the main articles focused on using algae or their components for application in biosensors for the detection of potentially toxic metal ions, covering the period from 2014 to the present. As can be seen in Table 1, the genus of algae commonly used for biosensing is *Scenedesmus*, specifically from the species *Scenedesmus obliquus*, *Scenedesmus acutus*, *Scenedesmus* sp., and *Scenedesmus subspicatus*. However, research should continue comparing the performance of the strains of algae as cell biosensors. The limits of detection using algae-based biosensors range from 0.2 mmol L^{-1} to 0.9 mg L^{-1} for different contaminants, such as diuron, Cd, Cu, and Zn. Thus, in general, algae-based biosensors offer several advantages for application in biosensors for the detection of potentially toxic metal ions, such as sensitivity and selectivity, fast response, low production cost (less than 2 USD per-sample) [53], satisfactory stability, and efficiency, indicating these devices as an effective analytical tool for environmental monitoring.

Table 1. Demonstration of different strategies of biosensor use based on reported works over the last decade (2014–2024).

Algae	Objective	Algae Preparation	Detection Method	Main Results	Reference
Chlamydomonas reinhardtii	To use biosensor as an indicator of the presence of pesticides.	Dense algal solutions	Detection of O_2, H_2O_2, and H_3O^+/OH ions, species taking part in metabolic activities of algae.	Diuron herbicide detection was achieved with a sufficiently low limit of detection—0.2 mmol L^{-1};	[82]

Table 1. *Cont.*

Algae	Objective	Algae Preparation	Detection Method	Main Results	Reference
Porphyridium cruentum	To develop biosensor based on carbon paste electrode modified with *Porphyridium cruentum* biomass for the determination of As^{3+} in contaminated water.	The microalgal biomass was well-dispersed in appropriate quantities of graphite powder and mineral oil using pestle and mortar to obtain a homogenized paste. Biomass quantity of *Porphyridium cruentum* varied from 0.5% to 7.5% optimizing maximum As biosorption.	Differential pulse anodic stripping voltammetric technique	Suitable result was obtained at pH 6.0 with 0.1 mol L^{-1} HNO_3 solution as a stripping medium, allo wing biosorption–accumulation time of 8 min using 5% *Porphyridium cruentum* biomass in graphite–mineral oil paste. Linear range for As^{3+} detection with the modified electrode–biosensor was observed between 2.5 mg L^{-1} and 20 mg L^{-1}; Efficiency of the biosensor in the presence of different interference metal ions (Na^+, K^+, Ca^{2+}, and Mg^{2+}) ions was also evaluated; The application of *Porphyridium cruentum* modified biosensor was successfully used for the detection of As^{3+} in the binary metal (Fe^{3+}, Mn^{2+}, Cd^{2+}, Cu^{2+}, Ni^{2+}, Hg^{2+}, and Pb^{2+}) contaminated system; The accuracy of application of biosorption-based biosensors for the detection of As^{3+} is as low as 2.5 mg L^{-1}.	[13]
Chlorella vulgaris, Pseudokirchner-iella *subcapitata*, and *Chlamydomonas reinhardtii*	Evaluation of fluorescence of immobilized microalgae on the detection of chemicals in urban rainwaters.	Immobilized microalgae by encapsulation in a hybrid alginate/silica translucid hydrogel.	Fluorescence emission detected by a fluorometer, of chlorophyll.	Immobilized algae exhibited lower sensitivity than free algae in suspension (limit of detection 0.001 µmol L^{-1} and <0.001 µmol L^{-1}) at atrazine (pesticide), but higher fluorescence.	[84]
Paulschulzia pseudovolvox and species of cyanobacteria in the order *Chroococcales*	Investigation of the photosynthetic metabolism to be used to integrate and sense environmental signals and the effect of toxic compounds detected by the disruption of the reliable light-dependent electrogenic effect.	The algae were grown in a chamber for four weeks prior to experiments. The chamber was then converted into a Photosynthetic Microbial fuel cell by the addition of an assembly containing the cathode.	The light-dependent electrogenic activity was monitored using a potentiostat.	Initial results suggest *Paulschulzia pseudovolvox* to be more resistant to effects of the toxicants tested in *Chroococcales*, demonstrating the importance of using multiple species as they will present a different level of sensitivity to different analytes (copper, thallium, zinc, and glyphosate).	[102]

Table 1. *Cont.*

Algae	Objective	Algae Preparation	Detection Method	Main Results	Reference
Chlorella vulgaris	To use biosensor to detect the presence of metal ions (Cu, Pb, Cd, Na, Al, and Li) in water.	Immobilization of microalgae in agarose solution	Bioluminescence of microalgae was used as an indication parameter (λ = 350–650 nm).	pH tolerance (pH 6–pH 8), and able to produce signals at different cell densities (1×10^6 cells mL^{-1} to 9×10^6 cells mL^{-1}) and culture ages (day 1 to day 5); The biosensor showed high sensitivity to metal ions (Cu, Pb, Cd, Na, Al, and Li); The presence of these metal ions with concentrations of 0.001 mg L^{-1} to 10.000 mg L^{-1} could be detected rapidly within 15 min of exposure.	[20]
Chlamydomonas reinhardtii	Develop an algae biosensor for the optical detection of nano-encapsulated-atrazine in agriculture.	Algae were immobilized on a paper substrate soaked with an agar thin film and placed in a glass optical measurement cell.	Detected by following the variable fluorescence	Nanoencapsulated atrazine was detected by fluorescence, with an inverse and proportional decrease to the herbicide concentrations ranging from 0.5 nmol L^{-1} to 200 nmol L^{-1} with a limit of detection of 16 h; There was slight interference in the presence of 2 mg L^{-1} for Cu and 10 μg L^{-1} for As at safe limits, a slight matrix effect, and a recovery value of 96 ± 5% for 75 nmol L^{-1} nanoencapsulated atrazine in tap water; The suitability of the proposed paper-based optical biosensor as valid support in agriculture.	[54]
A mixed culture of microalgae that predominantly contains the strains (*Scenedesmus obliquus* and *Chlorella luteoviridisI*)	Develop an innovative single-chamber air system microbial fuel cell (MFC) photosynthetic cathode and analyze electrochemical performance of the device using formaldehyde as a response meter—toxic model. Use this technology to monitor water quality.	To initiate biofilm growth, green microalgae was injected into photo MFC and allowed seating and attachment to the anode under static conditions. Two hours an open circuit potential was allowed to develop, and then the photo MFC was connected to a 1000 U resistor to trigger the development of an electroactive biofilm.	Electrochemical analysis using two-electrode mode, with the anode as the working electrode and the cathode as the counter electrode.	The photo MFC demonstrated promising proof-of-concept capability for detecting formaldehyde between 0.001% and 0.02%. Through the measurement of the electrogenic activity of micro-algae in the photo MFC, detection of these contaminants could be rapid and cost-effective compared to biological assays (given the relationship below low cost and simple treatment of materials used in manufacturing) and onsite (due to the device's small size and portability).	[21]

Table 1. *Cont.*

Algae	Objective	Algae Preparation	Detection Method	Main Results	Reference
Cystoseira algae	Prepare new types of electrode materials for electrochemical sensing of vardenafil as an active substance.	*Cystoseira* algae dried were dispersed in 400 mL of distilled water and transferred to autoclave. Carbon electrode material was prepared by two-step cycled hydrothermal carbonization of *Cystoseira* algae at temperatures of 180 °C and 250 °C for 8 h in the autoclave.	Cyclic voltammetry, differential pulse voltammograms, and square wave voltammetry techniques.	Material exhibited high sensitivity with the limits of detection of 96.3 pmol L^{-1} at pH 1.0 H_2SO_4 solution; The proposed sensor was successfully applied in tablet formulation, human serum, and human urine samples.	[103]
Scenedesmus acutus and *Monoraphidium contortum* strains	The construction of a portable system based on reversible photosynthesis inhibition was used as an indication parameter (λ = 350–650 nm). Produced by herbicides on microalgae, using atrazine as a model compound.	The immobilization of microalgae in a polyelectrolyte-surfactant-carbon nanotube self-assembled material cast on a screen-printed graphite electrode.	Electrochemical experiments were carried out with a purpose-built potentiostat. Oxygen production was followed by chronoamperometry. Cyclic voltammetry was carried out using the same three-electrode system.	*Monoraphidium contortum* can perform as an efficient recognition element for the construction of biosensors sensitive to atrazine; The system presents a limit of detection of 0.11 µmol L^{-1}, showing an excellent performance in river samples. The sensor maintains its integrity after five months immersed in a freshwater algae medium at room temperature.	[85]
Chlorella sp.	Developed a living sensor for metal ion detection with nanocavity-enhanced photoelectrochemistry	Mix of *Chlorella* sp. with copper nanoparticles solution.	Photoelectrochemical measurements in chronoamperometry mode.	Microalgae sensor was exploited to detect potentially toxic metal ions, Cd, Cr, Fe, and Mn with a breakthrough limit of detection of 50 nmol L^{-1}.	[86]

Table 1. *Cont.*

Algae	Objective	Algae Preparation	Detection Method	Main Results	Reference
Mesotaenium sp. and a strain of *Synechococcus* sp.	Developed an optical microalgal-cyanobacterial array biosensor using microalgae, to detect Cd^{2+}, Cr^{6+}, and Zn^{2+} in aquatic systems.	Sol-gel immobilization mixture prepared by sodium silicate and colloidal silica	Optimum operational conditions for the biosensor array such as exposure time, storage stability, pH, and multiple metal ions effects.	10 min exposure time yielded optimum fluorescence values; Metal ions toxicity increased with decreasing pH, resulting in low relative fluorescence (%), and decreased with increasing pH, resulting in higher relative fluorescence (%); The optimum storage time for biosensor strains was 4 weeks for microalgal cultures and 8 weeks for cyanobacterial culture, at 4 °C storage temperature; The metal ion mixtures showed less effect on the inhibition of relative fluorescence (%) of microalgal/cyanobacterial cultures, displaying an antagonistic behavior among the metal ions tested.	[46]
Microcystis aeruginosa, *Synechococcus elongatus*, and strains *Desmodesmus quadricauda*, and *Pseudokirchneriella subcapitata*	Develop an alternative to using this electrochemical biosensor equipped with algae and cyanobacteria for toxicological investigations based on selected test chemicals.	Algae/cyanobacteria solution was centrifuged, and the pellets were diluted with 30 mL culture medium.	Potentiostatic measurement	The evaluation of the sensor signal is based on the current–time curves of a potentiostatic measurement produced by the detection of microbially reduced mediator molecules immobilized in a gel structure; The mediator molecules are reduced during the measurement process and produce a current signal, which rapidly provides information about the vigor and vitality of living bacteria, yeasts, fungi, or cells.	[78]

Table 1. *Cont.*

Algae	Objective	Algae Preparation	Detection Method	Main Results	Reference
Scenedesmus acutus and *Pseudokirchneriella subcapitata*	The developed sensor was based on green microalgae immobilized in an alginate matrix.	Immobilization in encapsulation (alginate beads)	Na Visual observation, and absorbance and fluorescence measurements.	After incubation with different pollutants for five days, naked-eye analysis by several observers proved to be a successful method for surveying algae's growth and establishing the limits of detection; Suitable limits of detection were 10 mg L^{-1} for technical-grade acid glyphosate, 15 mg L^{-1} for glyphosate-based formulation, 50 µg L^{-1} for atrazine formulation, 7.5 mg L^{-1} for Cu, and 250 µg L^{-1} for Cr; The use of the biosensor on the local samples also proved to be successful: strong intensity of green color in those samples from clean water sources.	[53]
Scenedesmus subspicatus	Develop a novel whole-cell biosensor using chlorophyll a fluorescence from a single species of microalga, Scenedesmus subspicatus, immobilized in an inorganic silica matrix, for detecting bio-availability of multi-metal ions in freshwater.	Immobilization in inorganic silica hydrogels using the sol-gel technique	Effective pH range, cell density, exposure time, and storage stability.	The optimum response for the biosensor was dependent on the pH of the matrix, cell concentration, and exposure time derived; The biosensor was operational for four weeks; The limit of detection for the algal biosensor was determined as 0.9, 0.91, and 0.88 mg L^{-1} for Cd, Cu, and Zn, respectively.	[55]
Scenedesmus sp.	Develop an electrochemical biosensor with microalgae to evaluate the biotoxicity of Cd^{2+} ions in freshwater.	This electrochemical biosensor was constructed by immobilization of microalgae–bovine serum albumin and crosslinked with glutaraldehyde in film on a glassy carbon electrode surface.	The chronoamperometric currents generated by alkaline phosphatase activity. The feasibility was evaluated, and the application of the biosensor was optimized for parameters such as pH and cell density.	*Scenedesmus* sp. biosensor is highly sensitive with a good selectivity at a 1:1 ratio for measuring the concentration of Cd^{2+} cations except in the presence of Hg^{2+}; This biosensor could respond with only one drop of an analyte (50 µL of 1000 µg L^{-1} Cd^{2+}), resulting in suitability for simple and on-site water toxicity testing.	[52]

Table 1. *Cont.*

Algae	Objective	Algae Preparation	Detection Method	Main Results	Reference
Chlamydomonas reinhardtii	Use a biosensor with transgenic microalgae for metal ion detection and quantification.	Suspension of cells in the exponential growth phase	Detection by fluorescence measurements using a spectrofluorometer.	Detection of various metal ions at low limits of detection—0.93 nmol L^{-1}; Algae have substantial buffering capacity for free potential metal ions in their cytosol, even at high external metal ions concentrations.	[95]

4. Challenges and Perspectives in the Development and Implementation of Biosensors

The application of algae-based biosensors for detecting potentially toxic pollutants is quite promising, and a key challenge is the attainment of high sensitivity, specificity, stability, and selectivity of the electrode response. The production costs of biosensors in large quantities must also be considered. It should be noted that the commercialization of biosensor technology has lagged significantly behind research results, as reflected by publications with various methods and materials that can be used. The logic behind the slow and limited transfer of technology can be attributed to cost considerations and some obstacles related to the regulation and compliance with health legislation.

Meanwhile, analytical techniques such as spectrometric techniques guarantee excellent results in the detection of potentially toxic metal ions, with some of them allowing the quantification of toxic metal ions at trace concentrations. However, there is a growing interest in developing alternative analytical methods, offering a sustainable, efficient, and cost-effective solution for potential toxic element detection, as is the case with metal ions. Biosensors emerge as an excellent alternative considering their costs, versatility, and robustness for the quantification of metal ions in environmental samples. In particular, optical and electrochemical biosensors exhibit significant analytical performance characterized by sensitivity, stability, and detection limits, making them promising tools for environmental monitoring. Optical biosensors have demonstrated high sensitivity, with detection limits ranging from 0.001 to 0.11 μmol L^{-1} for various contaminants, while electrochemical biosensors display detection limits as low as 0.11 μmol L^{-1} for atrazine, with long-term stability of up to five months [21,84,85].

Algae-based biosensors exhibit notable versatility, allowing for application in diverse environmental contexts, while their potential for low-cost production enhances their accessibility. However, limitations in reproducibility and specificity restrict their effectiveness in detecting multiple pollutants, and the complexity of optimizing immobilization techniques and simplified transducers presents additional challenges for performance improvement. Advances in nanotechnology offer promising avenues to improve biosensor functionality and facilitate the creation of innovative sensing platforms.

In terms of future perspectives, the success of using whole algae in biosensor production is noteworthy, but other new materials to optimize the performance of metal ion sensors for specific applications are still necessary. Notably, advances in nanotechnology show great promise in enhancing biosensor performance and enabling the development of novel materials and detection platforms. To address this issue, there is a growing trend toward developing user-friendly biosensors that can be easily assembled and utilized by individuals without a scientific background [53]. This shift towards more accessible technology signals a promising direction for the future of biosensor development. In addition, bioinformatics tools and data analysis algorithms can improve the accuracy and reliability of biosensor measurements, and artificial intelligence facilitates the analysis of complex data sets generated by biosensors, leading to faster and more accurate results.

5. Conclusions

In summary, algae represent a promising component of life and biomass on our planet. In addition to being promising in chemical and pharmaceutical biotechnology, algae offer a unique range of bio-compounds that enable this biomass to work within a fully sustainable plan, producing oxygen alongside biosensor performance. This alternative could overcome some disadvantages of powerful and well-consolidated analytical techniques, such as spectrometric and electrochemical techniques, as well as the recent integration of artificial intelligence to support quantitative or qualitative analytical profiles. Improving environmental quality has become a primary objective of research worldwide, inherently tied to controlling and monitoring environmental parameters. These issues demand continuous, rapid, and sensitive monitoring systems capable of detecting toxic pollutants in water and sediment samples, given the extremely negative effects on the environment and living organisms. In this regard, with the advances and issues of sustainable biosensors, algae biomass holds significant potential for new technologies and innovative applications.

Funding: This research was funded by by FAPERGS (Research Support Foundation of the Rio Grande do Sul State 22/2551-0000840-2), Coordination for Improvement of Higher-Level Personnel (CAPES), and the Forensic National Institute of Science and Technology Grant number (465450/2014-8).

Acknowledgments: The authors would like to thank the Coordenação de Aperfeiçoamento de Pessoal de Nível Superior—Brasil (CAPES), the "Inova Clusters Tecnológicos Project", Innovat Hub B^3 from "Fundação de Amparo à Pesquisa do Estado do Rio Grande do Sul (FAPERGS)"—finance code 22/2551-0000840-2, and Fapergs/PqG 21/2551-0002237-0.

Conflicts of Interest: The authors declare no conflict of interest.

References

1. Ali, H.; Khan, E.; Ilahi, I. Environmental chemistry and ecotoxicology of hazardous heavy metals: Environmental persistence, toxicity, and bioaccumulation. *J. Chem.* **2019**, *2019*, 6730305. [CrossRef]
2. Briffa, J.; Sinagra, E.; Blundell, R. Heavy metal pollution in the environment and their toxicological effects on humans. *Heliyon* **2020**, *6*, 9. [CrossRef] [PubMed]
3. Tyohemba, R.L.; Pillay, L.; Humphries, M.S. Bioaccumulation of current-use herbicides in fish from a global biodiversity hotspot: Lake St Lucia, South Africa. *Chemosphere* **2021**, *284*, 131407. [CrossRef] [PubMed]
4. Curran, W.S. Persistence of herbicides in soil. *Crops Soils* **2016**, *49*, 16–21. [CrossRef]
5. Velusamy, K.; Periyasamy, S.; Kumar, P.S.; Rangasamy, G.; Pauline, J.M.N.; Ramaraju, P.; Mohanasundaram, S.; Vo, D.-V.N. Biosensor for heavy metals detection in wastewater: A review. *Food Chem. Toxicol.* **2022**, *168*, 113307. [CrossRef]
6. Verma, N.; Kaur, G. Trends on biosensing systems for heavy metal detection. In *Comprehensive Analytical Chemistry*; Elsevier: Amsterdam, The Netherlands, 2016; Volume 74, pp. 33–71. [CrossRef]
7. Malik, L.A.; Bashir, A.; Qureashi, A.; Pandith, A.H. Detection and removal of heavy metal ions: A review. *Environ. Chem. Lett.* **2019**, *17*, 1495–1521. [CrossRef]
8. Singh, A.; Shah, S.S.; Sharma, C.; Gupta, V.; Sundramoorthy, A.K.; Kumar, P.; Arya, S. An approach towards different techniques for detection of heavy metal ions and their removal from wastewater. *J. Environ. Chem. Eng.* **2024**, *12*, 113032. [CrossRef]
9. Wilschefski, S.C.; Baxter, M.R. Inductively coupled plasma mass spectrometry: Introduction to analytical aspects. *Clin. Biochem. Rev.* **2019**, *40*, 115. [CrossRef]
10. Maciel, E.V.S.; Soares, E.V.; de Toffoli, A.L.; da Silva, J.A.F.; Neto, E.S.; Nazario, C.E.D.; Lanças, F.M. New Materials in Sample Preparation: Recent Advances and Future Trends. *TrAC Trends Anal. Chem.* **2019**, *119*, 115633. [CrossRef]
11. Muller, A.; Pozebon, D.; Dressler, V.L. Advances of nitrogen microwave plasma for optical emission spectrometry and applications in elemental analysis: A review. *J. Anal. At. Spectrom.* **2020**, *35*, 2113–2131. [CrossRef]
12. Verma, N.; Singh, M. Biosensors for heavy metals. *Biometals* **2005**, *18*, 121–129. [CrossRef] [PubMed]
13. Zaib, M.; Saeed, A.; Hussain, I.; Athar, M.M.; Iqbal, M. Voltammetric detection of As(III) with Porphyridium cruentum based modified carbon paste electrode biosensor. *Biosens. Bioelectron.* **2014**, *62*, 242–248. [CrossRef] [PubMed]
14. Antonacci, A.; Scognamiglio, V. Biotechnological Advances in the Design of Algae-Based Biosensors. *Trends Biotechnol.* **2019**, *38*, 334–347. [CrossRef] [PubMed]
15. Singh, U.B.; Ahluwalia, A.S. Microalgae: A promising tool for carbon sequestration. *Mitig. Adapt. Strateg. Glob. Chang.* **2013**, *18*, 73–95. [CrossRef]
16. Galal El-Din Thabet Shams El-Din, N.; Rashedy, S.H. Biodiversity of Seaweeds in the Red Sea. In *Biodiversity of Seaweeds in the Egyptian Marine Waters: The Mediterranean Sea, Red Sea and Suez Canal*; Springer Nature: Cham, Switzerland, 2023; pp. 105–199. [CrossRef]

17. Hamed, S.M.; Okla, M.K.; Al-Saadi, L.S.; Hozzein, W.N.; Mohamed, H.S.; Selim, S.; AbdElgawad, H. Evaluation of the phycoremediation potential of microalgae for captan removal: Comprehensive analysis on toxicity, detoxification and antioxidants modulation. *J. Hazard. Mater.* **2022**, *427*, 128177. [CrossRef]
18. Xiao, X.; Li, W.; Jin, M.; Zhang, L.; Qin, L.; Geng, W. Responses and tolerance mechanisms of microalgae to heavy metal stress: A review. *Mar. Environ. Res.* **2023**, *183*, 105805. [CrossRef]
19. Tripathi, S.; Poluri, K.M. Heavy metal detoxification mechanisms by microalgae: Insights from transcriptomics analysis. *Environ. Pollut.* **2021**, *285*, 117443. [CrossRef]
20. Wong, L.S.; Judge, S.K.; Voon, B.W.N.; Tee, L.J.; Tan, K.Y.; Murti, M.; Chai, M.K. Bioluminescent Microalgae-Based Biosensor for Metal Detection in Water. *IEEE Sens. J.* **2018**, *18*, 2091–2096. [CrossRef]
21. Chouler, J.; Monti, M.D.; Morgan, W.J.; Cameron, P.J.; Di Lorenzo, M. A photosynthetic toxicity biosensor for water. *Electrochim. Acta* **2019**, *309*, 392–401. [CrossRef]
22. Huang, C.W.; Lin, C.; Nguyen, M.K.; Hussain, A.; Bui, X.T.; Ngo, H.H. A review of biosensor for environmental monitoring: Principle, application, and corresponding achievement of sustainable development goals. *Bioengineered* **2023**, *14*, 58–80. [CrossRef]
23. Taimni, I.K.; Srivastava, M.N. Separations involving sulphides: Separation of selenium, arsenic, antimony and mercury from barium. *Anal. Chim. Acta* **1956**, *15*, 517–520. [CrossRef]
24. Beamish, F.E. A critical review of titrimetric methods for determination of the noble metals—II. *Talanta* **1966**, *13*, 1053–1068. [CrossRef] [PubMed]
25. Ellis, C.M.; Vogel, A.I. Studies on hypovanadous salts as analytical reagents. *Analyst* **1956**, *81*, 693–703. [CrossRef]
26. Beamish, F.E.; Westland, A.D. Inorganic gravimetric and volumetric analysis. *Anal. Chem.* **1958**, *30*, 805–822. [CrossRef]
27. Beamish, F.E. A critical review of gravimetric methods for determination of the noble metals—II. *Talanta* **1966**, *13*, 773–801. [CrossRef]
28. Smichowski, P.; Londonio, A. The role of analytical techniques in the determination of metals and metalloids in dietary supplements: A review. *Microchem. J.* **2018**, *136*, 113–120. [CrossRef]
29. Sharma, I. ICP-OES: An advance tool in biological research. *Open J. Environ. Biol.* **2020**, *5*, 027–033. [CrossRef]
30. Ma, Z.; Meliana, C.; Munawaroh, H.S.H.; Karaman, C.; Karimi-Maleh, H.; Low, S.S.; Show, P.L. Recent advances in the analytical strategies of microbial biosensor for detection of pollutants. *Chemosphere* **2022**, *306*, 135515. [CrossRef]
31. Nodeh, H.R.; Ibrahim, W.A.W.; Kamboh, M.A.; Sanagi, M.M. New magnetic graphene-based inorganic–organic sol-gel hybrid nanocomposite for simultaneous analysis of polar and non-polar organophosphorus pesticides from water samples using solid-phase extraction. *Chemosphere* **2017**, *166*, 21–30. [CrossRef]
32. Tadeo, J.L.; Sánchez-Brunete, C.; Albero, B.; García-Valcárcel, A.I. Application of ultrasound-assisted extraction to the determination of contaminants in food and soil samples. *J. Chromatogr. A* **2010**, *1217*, 2415–2440. [CrossRef]
33. Sánchez-Hernández, L.; Hernández-Domínguez, D.; Bernal, J.; Neusüß, C.; Martín, M.T.; Bernal, J.L. Capillary electrophoresis–mass spectrometry as a new approach to analyze neonicotinoid insecticides. *J. Chromatogr. A* **2014**, *1359*, 317–324. [CrossRef] [PubMed]
34. Li, J.; Lu, J.; Qiao, X.; Xu, Z. A study on biomimetic immunoassay-capillary electrophoresis method based on molecularly imprinted polymer for determination of trace trichlorfon residue in vegetables. *Food Chem.* **2017**, *221*, 1285–1290. [CrossRef] [PubMed]
35. Chui, M.Q.; Thang, L.Y.; See, H.H. Integration of the free liquid membrane into electrokinetic supercharging–capillary electrophoresis for the determination of cationic herbicides in environmental water samples. *J. Chromatogr. A* **2017**, *1481*, 145–151. [CrossRef] [PubMed]
36. Moawed, E.A.; Radwan, A.M. Application of acid modified polyurethane foam surface for detection and removing of organochlorine pesticides from wastewater. *J. Chromatogr. B* **2017**, *1044*, 95–102. [CrossRef] [PubMed]
37. Welz, B.; Sperling, M. *Atomic Absorption Spectrometry*, 3rd ed.; John Wiley & Sons: Hoboken, NJ, USA, 2008. [CrossRef]
38. Douvris, C.; Trey, V.; Bussan, D.; Bartzas, G.; Thomas, R. How ICP-OES changed the face of trace element analysis: Review of the global application landscape. *Sci. Total Environ.* **2023**, *905*, 167242. [CrossRef]
39. Smirnova, S.V.; Ilin, D.V.; Pletnev, I.V. Extraction and ICP-OES determination of heavy metals using tetrabutylammonium bromide aqueous biphasic system and oleophilic collector. *Talanta* **2021**, *221*, 121485. [CrossRef]
40. Yu, Y.; Li, Y. New insight into the negative difference effect in aluminium corrosion using in-situ electrochemical ICP-OES. *Corros. Sci.* **2020**, *168*, 108568. [CrossRef]
41. Yüksel, B.; Ustaoğlu, F.; Arıca, E. Impacts of a garbage disposal facility on the water quality of çavuşlu stream in Giresun, Turkey: A health risk assessment study by a validated ICP-MS assay. *Aquat. Sci. Eng.* **2021**, *36*, 181–192. [CrossRef]
42. Soodan, R.K.; Pakade, Y.B.; Nagpal, A.; Katnoria, J.K. Analytical techniques for estimation of heavy metals in soil ecosystem: A tabulated review. *Talanta* **2014**, *125*, 405–410. [CrossRef]
43. Tobolkina, E.; Rudaz, S. Capillary Electrophoresis Instruments for Medical Applications and Falsified Drug Analysis/Quality Control in Developing Countries. *Anal. Chem.* **2021**, *93*, 8107–8115. [CrossRef]
44. Hyder, A.; Buledi, J.A.; Nawaz, M.; Rajpar, D.B.; Shah, Z.-U.; Orooji, Y.; Yola, M.L.; Karimi-Maleh, H.; Lin, H.; Solangi, A.R. Identification of heavy metal ions from aqueous environment through gold, silver, and copper nanoparticles: An excellent colorimetric approach. *Environ. Res.* **2022**, *205*, 112475. [CrossRef] [PubMed]

45. Jofre, F.C.; Savio, M. Infrared Radiation: A Sustainable and Promising Sample Preparation Alternative for Inorganic Analysis. *TrAC Trends Anal. Chem.* **2024**, *170*, 117469. [CrossRef]
46. Rathnayake, I.V.N.; Munagamage, T.; Pathirathne, A.; Megharaj, M. Whole cell microalgal-cyanobacterial array biosensor for monitoring Cd, Cr and Zn in aquatic systems. *Water Sci. Technol.* **2021**, *84*, 1579–1593. [CrossRef] [PubMed]
47. Rosa, M.D.H.; Alves, C.J.; dos Santos, F.N.; de Souza, A.O.; Zavareze, E.D.R.; Pinto, E.; de Pereira, C.M.P. Macroalgae and microalgae biomass as feedstock for products applied to bioenergy and food industry: A brief review. *Energies* **2023**, *16*, 1820. [CrossRef]
48. Bourgougnon, N.; Burlot, A.-S.; Jacquin, A.-G. Algae for global sustainability? In *Advances in Botanical Research*; Academic Press: Cambridge, MA, USA, 2021; Volume 100, pp. 145–212. [CrossRef]
49. Perricone, V.; Mutalipassi, M.; Mele, A.; Buono, M.; Vicinanza, D.; Contestabile, P. Nature-based and bioinspired solutions for coastal protection: An overview among key ecosystems and a promising pathway for new functional and sustainable designs. *ICES J. Mar. Sci.* **2023**, *80*, 1218–1239. [CrossRef]
50. Ariede, M.B.; Candido, T.M.; Jacome, A.L.M.; Velasco, M.V.R.; Carvalho, J.C.M.; Baby, A.R. Cosmetic attributes of algae—A review. *Algal Res.* **2017**, *25*, 483–487. [CrossRef]
51. Chojnacka, K.; Wieczorek, P.P.; Schroeder, G.; Michalak, I. *Algae Biomass: Characteristics and Applications: Towards Algae-Based Products*; Springer: Cham, Switzerland, 2018; Volume 8, p. 143. [CrossRef]
52. Iravani, S.; Ahmadzadeh, H.; Sajjadizadeh, H.S.; Rounaghi, G. *Scenedesmus* sp. MN738556 as a whole cell biosensor for sub-part per billion detections of Cd^{2+} cations in freshwater sources. *Microchem. J.* **2024**, *197*, 109836. [CrossRef]
53. Prudkin-Silva, C.; Lanzarotti, E.; Álvarez, L.; Vallerga, M.B.; Factorovich, M.; Morzan, U.N.; Urdampilleta, C.M. A cost-effective algae-based biosensor for water quality analysis: Development and testing in collaboration with peasant communities. *Environ. Technol. Innov.* **2021**, *22*, 101479. [CrossRef]
54. Scognamiglio, V.; Antonacci, A.; Arduini, F.; Moscone, D.; Campos, E.V.R.; Fraceto, L.F.; Palleschi, G. An eco-designed paper-based algal biosensor for nanoformulated herbicide optical detection. *J. Hazard. Mater.* **2019**, *373*, 483–492. [CrossRef]
55. Rathnayake, I.V.N.; Megharaj, M.; Naidu, R. Sol–gel immobilized optical microalgal biosensor for monitoring Cd, Cu and Zn bioavailability in freshwater. *Bull. Environ. Contam. Toxicol.* **2023**, *110*, 73. [CrossRef]
56. Bayramoglu, G.; Yalçın, E.; Arica, M.Y. Preparation of effective green sorbents using O. Princeps alga biomass with different compositions of amine groups: Comparison to adsorption performances for removal of a model acid dye. *J. Mol. Liq.* **2022**, *347*, 118375. [CrossRef]
57. Spain, O.; Plöhn, M.; Funk, C. The cell wall of green microalgae and its role in heavy metal removal. *Physiol. Plant.* **2021**, *173*, 526–535. [CrossRef] [PubMed]
58. Nie, J.; Sun, Y.; Zhou, Y.; Kumar, M.; Usman, M.; Li, J.; Tsang, D.C. Bioremediation of water containing pesticides by microalgae: Mechanisms, methods, and prospects for future research. *Sci. Total Environ.* **2020**, *707*, 136080. [CrossRef] [PubMed]
59. Doshi, H.; Ray, A.; Kothari, I.L. Biosorption of cadmium by live and dead Spirulina: IR spectroscopic, kinetics, and SEM studies. *Curr. Microbiol.* **2007**, *54*, 213–218. [CrossRef]
60. Kotrba, P.; Mackova, M.; Macek, T. Microbial biosorption of metals. In *Microbial Biosorption of Metals*; Springer: Berlin/Heidelberg, Germany, 2007. [CrossRef]
61. Pradhan, D.; Sukla, L.B.; Mishra, B.B.; Devi, N. Biosorption for removal of hexavalent chromium using microalgae *Scenedesmus* sp. *J. Clean. Prod.* **2019**, *209*, 617–629. [CrossRef]
62. Sedlakova-Kadukova, J.; Kopcakova, A.; Gresakova, L.; Godany, A.; Pristas, P. Bioaccumulation and biosorption of zinc by a novel Streptomyces K11 strain isolated from highly alkaline aluminium brown mud disposal site. *Ecotoxicol. Environ. Saf.* **2019**, *167*, 204–211. [CrossRef]
63. Zabed, H.M.; Akter, S.; Yun, J.; Zhang, G.; Awad, F.N.; Qi, X.; Sahu, J.N. Recent advances in biological pretreatment of microalgae and lignocellulosic biomass for biofuel production. *Renew. Sustain. Energy Rev.* **2019**, *105*, 105–128. [CrossRef]
64. Sibi, G. Factors influencing heavy metal removal by microalgae—A review. *J. Crit. Rev* **2019**, *6*, 29–32. [CrossRef]
65. Singh, D.V.; Singh, R.P. Algal consortia based metal detoxification of municipal wastewater: Implication on photosynthetic performance, lipid production, and defense responses. *Sci. Total Environ.* **2022**, *814*, 151928. [CrossRef]
66. Thenarasu, A.; Chai, M.K.; Tan, Y.H.; Wong, L.S.; Rajamani, R.; Djearamane, S. Study of Chlorella vulgaris from Different Growth Phases as Biosensor for Detection of Titanium and Silver Nanoparticles in Water. *Nat. Environ. Pollut. Technol.* **2024**, *23*, 959–969. [CrossRef]
67. Ata, A.; Nalcaci, O.O.; Ovez, B. Macro algae Gracilaria verrucosa as a biosorbent: A study of sorption mechanisms. *Algal Res.* **2012**, *1*, 194–204. [CrossRef]
68. Hammed, A.M.; Prajapati, S.K.; Simsek, S.; Simsek, H.; Hammed, A.M.; Prajapati, S.K.; Simsek, H. Growth regime and environmental remediation of microalgae. *Algae* **2016**, *31*, 189–204. [CrossRef]
69. Qiu, Y.W.; Zeng, E.Y.; Qiu, H.; Yu, K.; Cai, S. Bioconcentration of polybrominated diphenyl ethers and organochlorine pesticides in algae is an important contaminant route to higher trophic levels. *Sci. Total Environ.* **2017**, *579*, 1885–1893. [CrossRef] [PubMed]
70. Rath, B. Microalgal bioremediation: Current practices and perspectives. *J. Biochem. Technol.* **2012**, *3*, 299–304.
71. Moro, L.; Pezzotti, G.; Turemis, M.; Sanchís, J.; Farré, M.; Denaro, R.; Giardi, M.T. Fast pesticide pre-screening in marine environment using a green microalgae-based optical bioassay. *Mar. Pollut. Bull.* **2018**, *129*, 212–221. [CrossRef]

72. Malhotra, B.D.; Ali, M.A. Nanomaterials in biosensors: Fundamentals and applications. In *Nanomaterials for Biosensors*; Elsevier: Amsterdam, The Netherlands, 2018; Volume 1, pp. 1–74. [CrossRef]
73. Chadha, U.; Bhardwaj, P.; Agarwal, R.; Rawat, P.; Agarwal, R.; Gupta, I.; Panjwani, M.; Singh, S.; Ahuja, C.; Selvaraj, S.K.; et al. Recent progress and growth in biosensors technology: A critical review. *J. Ind. Eng. Chem.* **2022**, *109*, 21–51. [CrossRef]
74. Bazin, I.; Tria, S.A.; Hayat, A.; Marty, J.-L. New biorecognition molecules in biosensors for the detection of toxins. *Biosens. Bioelectron.* **2017**, *87*, 285–298. [CrossRef]
75. Purohit, B.; Vernekar, P.R.; Shetti, N.P.; Chandra, P. Biosensor nanoengineering: Design, operation, and implementation for biomolecular analysis. *Sens. Int.* **2020**, *1*, 100040. [CrossRef]
76. Xu, X.; Ying, Y. Microbial biosensors for environmental monitoring and food analysis. *Food Rev. Int.* **2011**, *27*, 300–329. [CrossRef]
77. Fan, C.; Zhang, D.; Mo, Q.; Yuan, J. Engineering Saccharomyces cerevisiae-based biosensors for copper detection. *Microb. Biotechnol.* **2022**, *15*, 2854–2860. [CrossRef]
78. Pescheck, M.; Schweizer, A.; Bláha, L. Innovative electrochemical biosensor for toxicological investigations on algae and cyanobacteria. *Bioelectrochemistry* **2021**, *143*, 107926. [CrossRef] [PubMed]
79. Liu, Q.; Wu, H.; Chen, J.; Guo, B.; Zhao, X.; Lin, H.; Huang, C. Adsorption mechanism of trace heavy metals on microplastics and simulating their effect on microalgae in river. *Environ. Res.* **2022**, *214*, 113777. [CrossRef] [PubMed]
80. Antonacci, A.; Frisulli, V.; Carvalho, L.B.; Fraceto, L.F.; Miranda, B.; De Stefano, L.; Scognamiglio, V. An all-green photo-electrochemical biosensor using microalgae immobilized on eco-designed lignin-based screen-printed electrodes to detect sustainable nanoherbicides. *Int. J. Mol. Sci.* **2023**, *24*, 10088. [CrossRef] [PubMed]
81. Gosset, A.; Oestreicher, V.; Perullini, M.; Bilmes, S.A.; Jobbágy, M.; Dulhoste, S.; Bayard, R.; Durrieu, C. Optimization of sensors based on encapsulated algae for pesticide detection in water. *Anal. Methods* **2019**, *11*, 6193–6203. [CrossRef]
82. Tsopela, A.; Lale, A.; Vanhove, E.; Reynes, O.; Séguy, I.; Temple-Boyer, P.; Launay, J. Integrated electrochemical biosensor based on algal metabolism for water toxicity analysis. *Biosens. Bioelectron.* **2014**, *61*, 290–297. [CrossRef] [PubMed]
83. Allouzi, M.M.A.; Allouzi, S.; Al-Salaheen, B.; Khoo, K.S.; Rajendran, S.; Sankaran, R.; Show, P.L. Current advances and future trend of nanotechnology as microalgae-based biosensor. *Biochem. Eng. J.* **2022**, *187*, 108653. [CrossRef]
84. Roxby, D.N.; Rivy, H.; Gong, C.; Gong, X.; Yuan, Z.; Chang, G.-E.; Chen, Y.-C. Microalgae living sensor for metal ion detection with nanocavity-enhanced photoelectrochemistry. *Biosens. Bioelectron.* **2020**, *165*, 112420. [CrossRef]
85. Boron, I.; Juarez, A.; Battaglini, F. Portable microalgal biosensor for herbicide monitoring. *ChemElectroChem* **2020**, *7*, 1623–1630. [CrossRef]
86. Durrieu, C.; Ferro, Y.; Perullini, M.; Gosset, A.; Jobbágy, M.; Bilmes, S.A. Feasibility of using a translucid inorganic hydrogel to build a biosensor using immobilized algal cells. *Environ. Sci. Pollut. Res.* **2016**, *23*, 9–13. [CrossRef]
87. Podola, B.; Melkonian, M. Selective real-time herbicide monitoring by an array chip biosensor employing diverse microalgae. *J. Appl. Phycol.* **2005**, *17*, 261–271. [CrossRef]
88. Altamirano, M.; Garcia-Villada, L.; Agrelo, M.; Sanchez-Martin, L.; Martin-Otero, L.; Flores-Moya, A.; Rico, M.; Lopez-Rodas, V.; Costas, E. A novel approach to improve specificity of algal biosensors using wild-type and resistant mutants: An application to detect TNT. *Biosens. Bioelectron.* **2004**, *19*, 1319–1323. [CrossRef] [PubMed]
89. Cheng, S.Y.; Show, P.L.; Lau, B.F.; Chang, J.-S.; Ling, T.C. New prospects for modified algae in heavy metal adsorption. *Trends Biotechnol.* **2019**, *37*, 1255–1268. [CrossRef] [PubMed]
90. Haigh-Flórez, D.; de la Hera, C.; Costas, E.; Orellana, G. Microalgae dual-head biosensors for selective detection of herbicides with fiber-optic luminescent O_2 transduction. *Biosens. Bioelectron.* **2014**, *54*, 484–491. [CrossRef] [PubMed]
91. Mandal, R.; Dutta, G. From photosynthesis to biosensing: Chlorophyll proves to be a versatile molecule. *Sens. Int.* **2020**, *1*, 100058. [CrossRef]
92. Buonasera, K.; Lambreva, M.; Rea, G.; Touloupakis, E.; Giardi, M.T. Technological applications of chlorophyll a fluorescence for the assessment of environmental pollutants. *Anal. Bioanal. Chem.* **2011**, *401*, 1139–1151. [CrossRef]
93. Bilal, M.; Rasheed, T.; Sosa-Hernández, J.E.; Raza, A.; Nabeel, F.; Iqbal, H.M. Biosorption: An interplay between marine algae and potentially toxic elements—A review. *Mar. Drugs* **2018**, *16*, 65. [CrossRef]
94. Danouche, M.; El Ghachtouli, N.; El Arroussi, H. Phycoremediation mechanisms of heavy metals using living green microalgae: Physicochemical and molecular approaches for enhancing selectivity and removal capacity. *Heliyon* **2021**, *7*, e07609. [CrossRef]
95. Rajamani, S.; Torres, M.; Falcao, V.; Ewalt Gray, J.; Coury, D.A.; Colepicolo, P.; Sayre, R. Noninvasive evaluation of heavy metal uptake and storage in micoralgae using a fluorescence resonance energy transfer-based heavy metal biosensor. *Plant Physiol.* **2014**, *164*, 1059–1067. [CrossRef]
96. Liu, Y.; Chen, L.; Yu, L.; Yang, C.; Zhu, J.; Wang, J.; Zheng, J.; Wang, F.; He, G.; Jiang, F.; et al. Confinement-enhanced microalgal individuals biosensing for digital atrazine assay. *Biosens. Bioelectron.* **2023**, *241*, 115647. [CrossRef]
97. Buyukharman, M.; Gover, T.; Gumus, A.; Gumus, S.; Yildiz, H.B. Design of a Novel Green Algae-Based Biological Photovoltaic Cell with High Photocurrent and a Photoelectrochemical Biosensing Approach Utilizing the BPV for Pesticide Analysis in Water. *ChemistrySelect* **2024**, *9*, e202304943. [CrossRef]
98. Zhou, Y.; Zheng, S.; Qin, W. Electrochemical biochemical oxygen demand biosensors and their applications in aquatic environmental monitoring. *Sens. Bio-Sens. Res.* **2024**, *44*, 100642. [CrossRef]

99. Hassani, S.; Akmal, M.R.; Maghsoudi, A.S.; Rahmani, S.; Vakhshiteh, F.; Norouzi, P.; Ganjali, M.R.; Abdollahi, M. High-performance voltammetric aptasensing platform for ultrasensitive detection of bisphenol A as an environmental pollutant. *Front. Bioeng. Biotechnol.* **2023**, *11*, 1134601. [CrossRef] [PubMed]
100. Modak, N.; Friebe, V.M. Amperometric biosensors: Harnessing photosynthetic reaction centers for herbicide detection. *Curr. Opin. Electrochem.* **2023**, *42*, 101414. [CrossRef]
101. Ameen, F.; Hamidian, Y.; Mostafazadeh, R.; Darabi, R.; Erk, N.; Islam, M.A.; Orfali, R. A novel atropine electrochemical sensor based on silver nano particle-coated Spirulina platensis multicellular blue-green microalga. *Chemosphere* **2023**, *324*, 138180. [CrossRef] [PubMed]
102. Labro, J.; Craig, T.; Wood, S.A.; Packer, M.A. Demonstration of the use of a photosynthetic microbial fuel cell as an environmental biosensor. *Int. J. Nanotechnol.* **2017**, *14*, 213. [CrossRef]
103. Bilge, S.; Bakirhan, N.K.; Donar, Y.O.; Sinag, A.; Ozkan, S.A.; Okudan, E.S. Green synthesis of carbon based biosensor materials from algal biomass for the sensitive detection of vardenafil. *J. Electroanal. Chem.* **2020**, *871*, 114286. [CrossRef]

MDPI AG
Grosspeteranlage 5
4052 Basel
Switzerland
Tel.: +41 61 683 77 34

Chemosensors Editorial Office
E-mail: chemosensors@mdpi.com
www.mdpi.com/journal/chemosensors

www.ingramcontent.com/pod-product-compliance
Lightning Source LLC
Chambersburg PA
CBHW041422210926
51850CB00013B/3334
* 9 7 8 3 7 2 5 8 3 7 6 7 0 *